SOME MILESTONES IN MOLECULAR BIOLOGY AND RECOMB DNA TECHNOLOGY

1869 Miescher discovers DNA in trout sperm.

1882 Flemming describes chromosomes in dividing cells.

1917 Term *biotechnology* first used.

1943 Penicillin produced industrially.

1944 Avery, MacLeod, and McCarty demonstrate DNA is genetic material in bacteria.

1952 Hershey and Chase determine that T2 bacteriophage DNA is the material that enters host cells.

1953 Watson and Crick elucidate structure of DNA from Franklin and Wilkins' x-ray diffraction patterns.

1957 Kornberg isolates DNA polymerase I.

1958 The mode of DNA replication demonstrated experimentally.

1959 RNA polymerase discovered.

1960 mRNA and role in encoding information for amino acids discovered.

1961 Synthetic mRNA (UUU) used to begin to decipher the genetic code.

DNA renaturation discovered.

1962 Existence of restriction endonucleases in bacteria demonstrated.

Watson, Crick, and Wilkins awarded Nobel Prize for discovering structure of DNA.

1965 Discovery that plasmids encode antibiotic resistance.

1966 Genetic code completely elucidated.

1967 DNA ligase isolated.

1968 Nirenberg, Khorana, and Holley awarded Nobel Prize for elucidating genetic code.

1970 First restriction endonuclease isolated.

1972 DNA ligase joins two DNA fragments, creating first recombinant DNA molecules.

1973 DNA inserted into plasmid vector and transferred to host *E. coli* cell for propagation; cloning methods established in bacteria.

Potential hazards of recombinant DNA technology raise concerns.

1974 World moratorium on some recombinant DNA experiments.

1975 Development of experimental guidelines for recombinant DNA technology discussed at the Asilomar Conference on Recombinant DNA Molecules.

Southern blotting and hybridization developed for detecting specific DNA sequences.

Monoclonal antibodies produced.

1976 National Institutes of Health prepares first guidelines for physical and biological containment.

DNA sequencing methods developed.

1977 Genentech, the first biotechnology firm, established.

Introns discovered.

1978 Somatostatin, first human hormone produced by recombinant DNA technology.

Human insulin produced by recombinant DNA technology.

Arber, Smith, and Nathans awarded Nobel Prize for discovering restriction endonucleases.

1979 NIH guidelines relaxed.

1980 Landmark *Diamond v. Chakrabarty* case leads to first patent of genetically engineered microorganism.

Nobel Prize to Sanger and Gilbert for developing DNA sequencing methods; to Berg for first DNA cloning methods.

1981 First diagnostic kit based on monoclonal antibodies approved in U.S.

Transgenic mice and *Drosophila* fruit flies produced.

1982 First animal vaccine produced by recombinant DNA technology in Europe.

1983 Plants transformed by genetically modified Ti plasmids.

1985 Polymerase chain reaction developed for *in vitro* amplification of DNA.

DNA fingerprinting first developed by Alec Jeffreys.

1986 Research on the *in vitro* formation of tissues using degradable polymer scaffolds.

1988 Genetically engineered mouse that developed cancer patented.

1990s Automated DNA sequencing technology developed and refined.

1990 First gene therapy trials approved in U.S.

Human Genome Project begins in U.S.

1992 Development of automated DNA sequencing technologies.

1993 Nobel Prize to Sharp and Roberts for discovering introns and RNA splicing; to Mullis and Smith for developing polymerase chain reaction.

1995 Completion of *Haemophilus influenzae* DNA sequence, the first free-living organism to be sequenced; the complete DNA sequence of *Mycoplasma genitalium*; after 1995 the complete genomic sequences have since been published or made available for *Methanococcus jannaschii*, *Synechocystis* sp., *Mycoplasma pneumoniae*, *Helicobacter pylori*, *Escherichia coli*, *Archaeoglobus fulgidis*, *Methanobacterium thermoautotrophicum*, and the yeast *Saccharomyces cerevisiae*, the first eukaryotic organism to be completely sequenced.

1996 Focus on the large-scale sequencing of the human genome.

1997 Nuclear transplantation experiment using nucleus from a differentiated cell produces a lamb called Dolly; expression of human proteins from sheep clones.

Experiment led to the cloning of many different animals.

DNA microarray technology and DNA chips developed as a tool to identify disease-causing genes.

First genetically engineered weed- and insect-resistant crops commercialized as Roundup Ready soybeans and Bollgard Insect-Protected Cotton.

Genetically engineered crops grown commercially on nearly 5 million acres worldwide.

1998 The first complete animal genome sequence completed – the nematode roundworm *Caenorhabditis elegans*.

Researchers clone three generations of mice using nuclei of adult ovarian cumulus cells.

2000 Celera Genomics and the public Human Genome Project announce the completion of a first draft of the DNA sequence of the human genome.

The DNA sequence of the fruit fly *Drosophila* is published.

The DNA sequence of the first plant *Arabidopsis thaliana* is completed.

Genetically engineered crops are grown on 108.9 million acres in 13 countries.

Pigs are cloned.

Golden Rice, with increased β-carotene, is produced.

2001 The draft of the human genome sequence is published in *Science* and *Nature*.

President Bush permits federal funding of research using existing stem cells from nonembryonic tissues.

The DNA sequence of the rice genome is completed. This is the first crop plant to be sequenced.

2002 The first draft of a functional map of the yeast proteome is completed (its DNA sequence was published in 1996).

Scientists make progress in understanding the control factors involved in stem cell differentiation.

Genetically engineered crops are grown on 145 million acres in 16 countries.

Scientists discover how small pieces of RNA (microRNAs) control numerous cell functions.

2003 The final draft of the DNA sequence of the human genome is completed (99.9% of the gene-containing portion of the human genome sequenced to 99.9% accuracy). Other genome sequences have been completed: *Escherichia coli*; the yeast *Saccharomyces cerevisiae*; the worm *C. elegans*; and the fruitfly *D. melanogaster*. Drafts of the mouse and rat genomes completed.

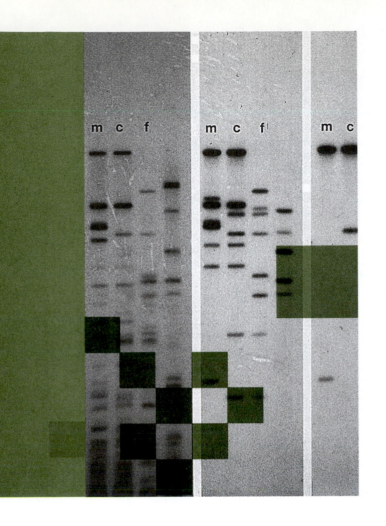

BIOTECHNOLOGY
An Introduction
Updated Second Edition

SUSAN R. BARNUM
Miami University

BROOKS/COLE
CENGAGE Learning

Australia • Brazil • Japan • Korea • Mexico • Singapore • Spain • United Kingdom • United States

**Biotechnology: An Introduction,
Updated Second Edition**
Susan R. Barnum

Acquisitions Editor: Nedah Rose

Development Editor: Kari Hopperstead

Editorial Assistant: Sarah Lowe

Technology Project Manager: Travis Metz

Marketing Manager: Ann Caven

Marketing Assistant: Leyla Jowza

Advertising Project Manager: Linda Yip

Project Manager, Editorial Production:
    Belinda Krohmer

Art Director: Rob Hugel

Print/Media Buyer: Kris Waller

Permissions Editor: Kiely Sexton

Production Service: Graphic World

Text Designer: Andrew Ogus

Photo Researcher: Kathleen Olson

Copy Editor: Maryann Short

Cover Designer: Bill Stanton

Cover Image: ©Image Source/Superstock

Compositor: Graphic World

For product information and technology assistance, contact us at
Cengage Learning Customer & Sales Support, 1-800-354-9706
For permission to use material from this text or product,
submit all requests online at **www.cengage.com/permissions**
Further permissions questions can be emailed to
permissionrequest@cengage.com

Library of Congress Control Number: 2005936742

ISBN-13: 978-0-495-11205-1

ISBN-10: 0-495-11205-4

Brooks/Cole
10 Davis Drive
Belmont, CA 94002-3098
USA

Cengage Learning is a leading provider of customized learning solutions
with office locations around the globe, including Singapore, the United
Kingdom, Australia, Mexico, Brazil, and Japan. Locate your local office at:
www.cengage.com/global

Cengage Learning products are represented in Canada by
Nelson Education, Ltd.

To learn more about Brooks/Cole, visit **www.cengage.com/brookscole**

Purchase any of our products at your local college store or at our
preferred online store **www.ichapters.com**

Printed in the United States of America
3 4 5 6 7 8 11 10 09 08

CONTENTS

4   BASIC PRINCIPLES OF IMMUNOLOGY   93

5   MICROBIAL BIOTECHNOLOGY   123

6   PLANT BIOTECHNOLOGY   147

PREFACE

Biotechnology—the use of living organisms or their products to enhance our lives and our environment—is a broad and increasingly complex discipline that encompasses many specialized areas. Our understanding of living organisms has been immensely deepened by the contributions that molecular biology and recombinant DNA methods have made to basic scientific research. We are able to isolate, analyze, and manipulate genes and alter the genetic makeup of living organisms. Cloning organisms has become almost routine, and stem cell research looks promising. Indeed, biotechnology is revolutionizing agriculture and medicine.

I have been teaching college biotechnology courses since 1986 and have long been aware that students need a single book that offers a reasonably comprehensive introduction to biotechnology. That is what motivated me to write the first edition of this book. However, the nature of modern biotechnology presents the writer of an introductory text with some unique challenges: The field is broad, diverse, and very fast moving. Its breadth means that many areas must be addressed; its diversity means that no two teachers of what are generally similar courses will cover the same topics at the same depth; and its fast pace means that a book is only truly up to date when the book is being written. This last challenge can be addressed by a timely schedule of revisions and a web site that provides updates. The first and second challenges are addressed by integrating information from many areas of biotechnology to give readers basic information on essential concepts and methods and an understanding of how the field is evolving and what developments are on the horizon.

Many changes have been made in the second edition. The most notable changes from the first edition are

that the material from Chapter 2 has been moved to Chapters 1 and 12; a new immunology chapter has been included; and Chapter 9 on the human genome project has been rewritten to include genomics, proteomics, and bioinformatics. In addition, the readings listed at the end of each chapter have been updated. New to each chapter in the book is a "Cause for Concern?" sidebar addressing ethical implications and a "Biotech Revolution" text box that illustrates a new development.

A brief tour of the book's chapters will illustrate how this book has been revised and further meets the challenges in a number of significant ways. The first chapter presents an in-depth historical review of the developments leading to biotechnology and of landmark experiments that led to the biotechnology revolution. Although the book is neither a methods manual nor a molecular biology text, three chapters provide a review of basic concepts and methodologies. Chapter 2 presents an overview of Central Dogma and gene expression. Chapter 3 presents a review of basic methods of recombinant DNA technology. This chapter has been extensively updated with new technologies, such as DNA microarray and RNA interference, included. A new chapter on the basic principles of immunology (Chapter 4) has been added.

Chapters 5 through 8 each focus on a single broad area of research and development to which modern methods of biotechnology are being applied with significant results and intriguing potential for the future: microorganisms, plants, animals, and marine organisms. Microbial, plant, and animal biotechnology chapters have been extensively revised and updated. Marine biotechnology is a developing area that is rarely covered in textbooks, although marine organisms are making important contributions to biotechnology. In each chapter, examples of discoveries and new applications are described.

Chapter 9 has been completely rewritten and updated, incorporating the new areas of genomics, proteomics, and bioinformatics. Chapter 10 has been extensively revised to include the latest developments in the area of medical biotechnology, while Chapter 11 describes the contributions to forensic science and paternity testing, paying particular attention to DNA profiling. This chapter has been revised to include new technologies. Finally, Chapter 12 focuses on regulatory and patent issues. The chapter has been extensively updated and revised, giving more attention to patents and ethical issues, and including an historical treatment of regulatory issues during the DNA revolution of the 1970s.

In addition to the breadth of coverage, other features enhance the book's usefulness:

In Chapter 9, Genomics, readers are referred to World Wide Web sites for the most up-to-date information in a rapidly moving area. Because web sites are sometimes changed, I recommend either looking addresses up on the book web site or conducting a keyword search to find the site.

The generous use of cross-references to other chapters allows teachers and students to integrate information from different areas of biotechnology and different stages in its development to achieve a better understanding of this complex and broad field.

The inclusion of several appendices to supplement some chapters (for example, summary of NIH guidelines).

Although no book can completely replace primary research papers and review articles in a rapidly advancing field, it can serve as a springboard for further study in a specific area or provide an overview for students not familiar with biotechnology. Therefore, at the end of each chapter, I list several books and many articles—both general reviews and research papers—for further reading.

A glossary gives definitions of key terms.

End papers present a list of abbreviations and acronyms, a chronological listing of the historical development of molecular biology and recombinant DNA technology, and a year- by-year account of the milestones in the Human Genome Project.

As an additional resource for instructors and students using this book, a web site may be found at www.cengage.com/biology. This site includes quizzes, glossary terms, InfoTrac exercises, guided internet activities, and Web links.

Who can use this book? Undergraduates and graduate students alike will find it valuable. Undergraduate science majors and nonmajors with some science background can use it as a stand-alone text for biotechnology courses, or as a supplement in other courses such as cell biology, molecular biology, and genetics. Selected chapters can be used in courses geared more toward a particular group of organisms (for example, plant cell biology or microbial genetics) or in a general biology or botany course for coverage of a particular topic. In graduate courses the book can provide a comprehensive overview of biotechnology, as well as background information with primary papers and review articles assigned as additional reading along with each chapter. Medical students, professionals in industry and academia, and technical writers who are interested in learning about this exciting discipline will also find the book of value.

This book could not have been written without the help of many people. I am indebted to my editor, Nedah Rose, and my assistant editors, Christopher Delgado and Kari Hopperstead, for their support, encouragement, guidance, and insight during the revision of this book. Many thanks also to Kelly Mabie of Graphic World Publishing Services and to Belinda Krohmer and the rest of the production team at Wadsworth. Finally, I would like to thank the reviewers who took the time to read and comment on this second edition. Without the thoughtful input of these individuals, all talented scholars and instructors themselves, the book as it is today would not have been possible: Olukemi Adewusi, Ferris State University; Eric Chang, Baylor College of Medicine; Denise Clark, Mesa Community College; Henry Daniell, University of Central Florida; Samuel Fan, Bradley University; Paul Farnsworth, University of Texas at San Antonio; Craig W. Fenn, Housatonic Community College; Claire Leonard, William Paterson University; Jonathan Morris, Middlesex Community College; Kurt M. Regner, Community College of Southern Nevada; Dennis M. Smith, Wellesley College; Barbara L. Stewart, J. Sargeant Reynolds Community College.

To Dan, Emily,

and Julia

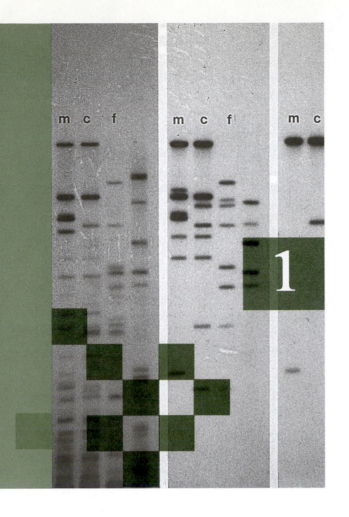

BIOTECHNOLOGY: OLD AND NEW

WHAT IS BIOTECHNOLOGY?

We are witnessing a revolution in biotechnology that has far-reaching implications for medicine and agriculture and consequently for society beyond anything previously imaginable in the history of science. The Office of Technology Assessment of the United States Congress (dismantled in 1995) defined **biotechnology** as "any technique that uses living organisms or substances from those organisms, to make or modify a product, to improve plants or animals, or to develop microorganisms for specific uses." Thus, biotechnology encompasses tools and techniques, including those of **recombinant DNA** technology; the living organisms to be improved can be plants, animals, or **microorganisms;** and the products from these organisms can be new or rare—that is, not having existed before naturally or being less abundant than we would like for certain needs or purposes. Biotechnology is multidisciplinary, involving a variety of the natural sciences—cell and molecular biology, microbiology, genetics, physiology, and biochemistry, to name only the major areas—as well as engineering and computer science. Biotechnology also encompasses recombinant DNA technology—indeed, the two terms are sometimes used synonymously. Recombinant DNA technology encompasses methods for genetically manipulating organisms ranging from bacteria and fungi to plants and animals. Scientists can change organisms and obtain products in ways not thought possible just 40 years ago.

The many applications of biotechnology include the production of new and improved foods, industrial chemicals, pharmaceuticals, and livestock. Biotechnology has contributed developments such as virus-resistant crop plants and livestock; diagnostics for detecting genetic diseases, such as Huntington's chorea and hereditary cancers, and acquired diseases such as AIDS; therapies that use **genes** to cure diseases such as adenosine deaminase (ADA) deficiency; and recombinant vaccines to prevent diseases such as malaria.

Biotechnological strategies also offer hope for restoring the environment and protecting endangered species. Microorganisms are used to clean up toxic wastes from industrial and oil spills. Conservation biologists use genetic methods to identify particular populations of endangered or threatened species (they can even use a part of an animal—a whale, for example, or an elephant killed by poachers—for information that traces the animal to its population). By determining the genetic diversity of various plant and animal populations, genetic analysis can help zoos and field biologists improve conservation practices.

What events ushered in the era of modern biotechnology that began in the 1980s and plays such a significant role in our lives today? Note the word *modern* in the previous sentence. Biotechnology is broadly considered to be modern, but by strict definition—the manipulation of living organisms—biotechnology is in fact ancient (although only very recently known by the term). In this chapter, we investigate the origins of biotechnology and review major contributory discoveries leading up to the state of biotechnology today.

ANCIENT BIOTECHNOLOGY

History of Domestication and Agriculture

For two million years Paleolithic peoples lived in mobile camps and survived by hunting wild animals and collecting wild plants. The lives of these hunter-gatherers were dictated by the migratory habits of the animals they hunted and by the distribution of edible plants, especially of grasses such as wild wheat and barley. Then, some ten thousand years ago, something remarkable occurred that changed the course of civilization. Abandoning their nomadic ways, people settled and began domesticating plants and animals: they established agrarian societies. The archaeological record indicates that early farmers in the Near East (present-day Iran, Iraq, Turkey, Syria, Jordan, Lebanon, and Israel) cultivated wheat, barley, and possibly rye. Sheep and goats provided milk, cheese, butter, and meat. Approximately seven thousand years ago in Africa, pastoralists roamed the Sahara (not then a desert) with large flocks of sheep, goats, and cattle. They also hunted and used grinding stones in food preparation. Exactly when these ancient people began to grow plants for food is not known. Archaeological evidence suggests that early farmers arrived in Egypt approximately six thousand years ago. They brought sheep, goats, and cattle and such crops as barley, emmer (an ancient wheat), flax, lentil, and chick-pea. Archaeologists also have found ancient farming sites in the New World, the Far East, and Europe. Evidence suggests that agriculture developed independently in several areas of the world at approximately the same time; there appears to have been no one region of origin. We will never be certain what prompted this apparently sudden shift to a more sedentary lifestyle; it may have been a response to population increases and the increasing demand for food, to shifts in climate, or to the dwindling of the herds of migratory animals. Whatever the reasons, this change in lifestyle allowed civilizations and, with them, new technologies to flourish. Early farmers were able to control their environment in ways that nomadic people could not.

The origins of biotechnology thus date back some ten thousand years, to these early agrarian societies in which people collected the seeds of wild plants for cultivation and domesticated some species of wild animals living around them. Each year at harvest time, farmers collected the seeds of the plants with the most desirable traits and set them aside for planting the next year. Similarly, they bred only the most prized animals. By this practice of artificial selection, farmers gradually produced new varieties of plants and ani-

Inches

FIGURE 1.1 Corn cobs demonstrating the evolutionary changes from about 5000 BC to about AD 1500. The far left ear is wild; the others are early cultivated ears. The scale shows that the far right ear is more than five times the length of the wild ear. All samples were found during excavations in the Tehuacan Valley, Mexico, under the auspices of the Robert S. Peabody Foundation for Archaeology.

mals that retained the desirable traits found in the wild species but were modified in other ways that were beneficial to humans.

Abundant evidence survives of ancient domestication and selection (Figure 1.1). The Babylonians and Egyptians left pictorial evidence that dogs, sheep, and cattle had been domesticated by 2000 BC (Figure 1.2). The Romans left written accounts of their selective breeding practices for improving livestock.

Ancient Plant Germplasm

For thousands of years, farmers at harvesttime have selected seeds, cuttings, or tubers from superior plants to save for the next planting. Farmers often protected the stored seeds from insects by sealing them in clay pots or burying them in baskets covered with ash. They stored tubers in cool areas and either replanted cuttings immediately or kept them dry until the next planting time. Farmers thus saved their genetic stocks from season to season. They could exchange surplus stocks with neighbors or barter them in the local market, with the result that the plant variety became widespread in a region. (Large-scale organized seed production did not begin until the early 1900s.)

Ancient peoples began collecting plants several thousand years ago. Records show that around 2500 BC Sumerian collectors traveled west to Asia Minor to acquire vines, rose plants, and fig trees. The Egyptians collected plants such as East African incense trees, as was depicted on a temple built during the reign of Queen Hatshepsut (in the fifteenth century BC). By the AD 1500s, plant-collecting expeditions became commonplace and collectors traversed the globe.

One of the most influential collectors of the twentieth century was Nikolai I. Vavilov (1887–1943), a Russian plant geneticist and agronomist who collected and cataloged thousands of ancient crop plants and their wild relatives. Between 1923 and 1931, he traveled extensively in the Soviet Union and in more than 50 countries collecting economically important plant varieties—

FIGURE 1.2 Stone amulets from Iraq in the shape of sheep (pierced for suspension). Some amulets were used as stamp seals and had designs drilled on the bottom. Protoliterate period, before 2900 BC.

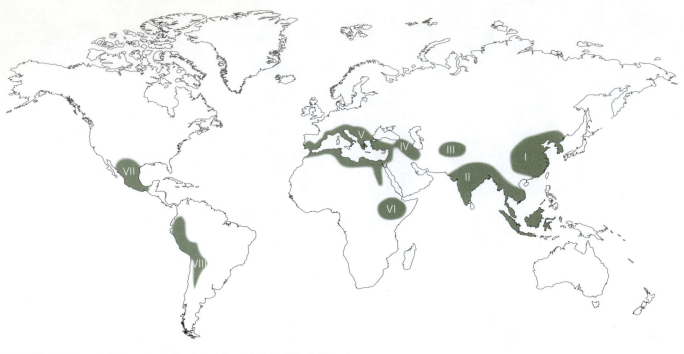

FIGURE 1.3 The eight centers of origin identified by Nikolai Vavilov.

including beans, peas, chick-peas, maize, lentils, oats, rye, and wheat (26,000 varieties of wheat alone). Vavilov identified eight primary centers of origin for plants, often referred to as centers of diversity (Figure 1.3 and Table 1.1). These centers also overlap with the centers of origin and diversity of most domesticated animals and where agrarian cultures were established and flourished. Vavilov was instrumental in establishing, at the Institute of Plant Industry in Leningrad, one of the first gene banks for long-term storage of important plant **germplasm.** He demonstrated to scientists worldwide that germplasm collections had economic value and moreover were important to the disease resistance of crop plants then in use. As president of the Lenin Academy of Agricultural Sciences and director of the Institute of Applied Botany (now the N.I. Vavilov Institute of Plant Industry), Vavilov initiated a comprehensive research and breeding program. His was the first organized, logical plan for crop genetic resource management.

Vavilov was arrested in 1940 on charges of espionage and died in prison from malnutrition in January 1943. Although charged with espionage, Vavilov had in fact become the victim of ideological forces in the Soviet Union. Under the encouragement of Trofim Lysenko (1898–1976), a rising leader in Soviet science under Stalin, the principles of Mendelian genetics were rejected in favor of Lamarckism, the theory that an organism can acquire physical traits in response to its environment and pass them on to its offspring, which also change in response to the environment. As opposed to Mendelian inheritance, these traits are acquired

through, as it were, a desire or will to change. Lysenko argued that because plants acquire characteristics through environmental influence, selective breeding is not needed. Stalin banned experimentation and scientific inquiry, and Soviet geneticists could no longer continue their research.

As the Soviet government was suppressing Mendelian genetics, the United States was establishing centers for the preservation, study, and distribution of germplasm (in New York State in 1948 and in Georgia and Washington State in 1949). In 1959, the U.S. Department of Agriculture established a national gene bank, the National Seed Storage Laboratory, at Fort Collins, Colorado.

Much of the world's rich crop germplasm resides in the developing world because of its ancient agricultural past. Unfortunately, as farmers worldwide embrace modern agriculture with its high yield, high-tech crops, ancient varieties and their wild relatives are lost, and genetic diversity among crops erodes. Wild varieties previously safe in remote, uncultivated areas are being destroyed by agricultural expansion and the widespread use of herbicides. There is now a worldwide effort to salvage remaining germplasm for preservation in gene banks.

The establishment of a global network of plant gene banks has been a significant achievement for world agriculture. Since 1971 this network has been coordinated by the Consultative Group on International Agricultural Research (CGIAR), a broadly based international consortium that works to strengthen national

TABLE 1.1 Origins of Some Domesticated Plants and Animals

Species	Center of Origin	Vavilov's Center*	Species	Center of Origin	Vavilov's Center*
Plants			**Plants**		
Alfalfa	Middle East	IV	Orange	South Asia	II
Apple	Central Asia	III	Papaya	Middle America	VII
Apricot	Central Asia	III	Pea	Middle East	IV
Avocado	Middle America	VII	Peach	China	I
Banana	South Asia	II	Peanut	South America	VIII
Barley	Middle East	IV	Pear	Central Asia	III
Bean	Middle and South America	VII, VIII	Pecan	North and middle America	VII
Beet	Mediterranean	V	Peppers (bell, chili)	Middle America	VII
Blueberry	North America	None	Pineapple	Brazil	None
Broccoli	Mediterranean	V	Plum	Central Asia	III
Brussels sprouts	Mediterranean	V	Potato	South America	VIII
Cabbage	Mediterranean	V	Pumpkin	North and middle America	VII
Cantaloupe	East Africa	VI			
Carrot	Mediterranean	V	Radish	Mediterranean	V
Cauliflower	Mediterranean	V	Rice	Southeast Asia	II
Celery	Mediterranean	V	Rye	Middle East	IV
Cherry	Central Asia	III	Sorghum	East Africa	VI
Coffee	East Africa	VI	Soybean	China	I
Cotton	Middle East, South America	IV, VIII	Spinach	Middle East	IV
			Squashes	Middle America	VII
Cranberry	North America	None	Sugarcane	South Asia	II
Cucumber	South Asia	II	Sunflower	North America	None
Date	South Asia	II	Sweeet potato	Middle America	VII
Eggplant	South Asia	II	Tangerine	Southeast Asia	II
Fig	Middle East	IV	Tobacco	South America	VIII
Flax	Mediterranean to South Asia	II through V	Tomato	Middle America	VII
			Watermelon	East Africa	VI
Garlic	Central Asia	III	Wheat	Middle East	IV
Grape	Middle East	IV	Yam	South-Southeast Asia, Africa	II, VI
Grapefruit	West Indies	None	**Animals**		
Guava	West Indies	None	Alpaca	South America	VIII
Jerusalem artichoke (sunflower root)	North America	None	Camel	Middle East, North Africa	IV, V
Lemon	South Asia	II	Chicken	South Asia	II
Lettuce	Mediterranean	V	Cow	Middle East	IV
Lime	South Asia, China	I, II	Goat	Middle East	IV
Maize	Middle America	VII	Horse	Middle East	IV
Mango	South Asia	II	Llama	South America	VIII
Millet	East Africa	VI	Pig	Middle East	IV
Oats	Northwestern Europe	None	Rabbit	Europe	V
Olive	Mediterranean	V	Sheep	Middle East	IV
Onion	Central Asia	III	Turkey	Middle America	VII

TABLE 1.2 Research Centers of the Consultative Group on International Research (CGIAR)

Center*	Location	Agricultural Programs	Focus Area
International Rice Research Institute (IRRI)	Philippines	Rice	Global
Centro Internacional de Mejoramiento de Maiz y Trigo (International Maize and Wheat Improvement Center) (CIMMYT)	Mexico	Wheat, maize, barley, triticale	Global
International Institute of Tropical Agriculture (IITA)	Nigeria	Maize, groundnut, sweet potato, yam, cassava, cowpea, soybean, plantain, agroforestry species	Tropical Africa Global
Centro Internacionalde Agricultura Tropical (International Center for Tropical Agriculture) (CIAT)	Columbia	Cassava and beans Tropical pastures	Global Central and South America
West Africa Rice Development Association (WARDA)	Côte d'Ivoire	Rice	West Africa
Centro Internacional de la Papa (International Potato Center) (CIP)	Peru	Potato, sweet potato	Global, Latin America
International Center for Agricultural Research in Dry Areas (ICARDA)	Syria	Barley, lentil, chick-pea, fava bean, bread wheat, durum wheat, forage crops	Dry regions of North Africa and West Asia
International Crops Research Institute for Semiarid Tropics (ICRISAT)	India	Pearl millet, other millets, sorghum, groundnut, chick-pea, pigeon pea	Global, semiarid regions
International Network for the Improvement of Banana and Plantain (INIBAP)	France (germplasm stored in Belgium)	Banana and plantain	Global
International Plant Genetic Resources Institute (IPGRI)	Italy	Plant genetic resources General conservation of plant germplasm	Global
International Livestock Center for Africa (ILCA)	Ethiopia	Livestock production systems Forage germplasm	Tropical Africa
International Council for Research in Agroforestry (ICRAF)	Kenya	Agroforestry	Africa
International Service for International Agricultural Research (ISNIAR)	Netherlands	National agricultural research	Global
International Food Policy Research Institutes (IFPRI)	United States	Food policy	Global
Center for International Forestry Research (CIFOR)	Indonesia	Agroforestry Tropical trees	Global

*Additional nonplant research centers: International Irrigation Management Institute (IIMI), Sri Lanka
International Center for Living Aquatic Resources Management (ICLARM), Philippines
International Laboratory for Research on Animal Diseases (ILRAD), Kenya

agricultural research programs in developing countries. The CGIAR supports 17 international agricultural research centers distributed around the world (Table 1.2) and ensures the conservation of plant species in approximately 450 non-CGIAR institutions in more than 90 countries. Germplasm storage banks house plant material such as seeds, plant cuttings, and tubers for future use and study. Their holdings are divided among short-, intermediate-, and long-term collections. Plant material in short-term storage at ambient temperatures must be used within a short period of time. Seeds in intermediate-term storage at −5 to 0°C can last up to 30 years before they must be grown out. Dried seeds stored in sealed containers at −20°C can last 100 years or longer. Periodic germination tests must be conducted to determine viability. Other types of material such as cuttings and tubers can be preserved in fields associated with the gene banks or even grown in tissue culture (described in Chapter 6, Plant Biotechnology). The research at CGIAR centers helps to preserve the world's plant genetic resources, increase the supply of basic foods to developing countries, increase productivity through genetic improvements of plants and livestock, and strengthen research programs in developing countries. CGIAR supports research that ranges from the development and introduction of integrated pest-management programs and biological control methods that encourage farmers to reduce the use of chemical pesticides, to the production of disease-resistant, high yield varieties of potato, beans, sorghum, plantain, cassava, rice, and wheat and forages, to name just a few. In addition, CGIAR is committed to strengthening national agricultural research in developing countries. Expert consultation and training programs provide thousands of scientists in developing countries with the technical expertise required to improve plants and livestock and to preserve their genetic resources.

History of Fermented Foods and Beverages

Fermented Foods   Once people settled in villages, their diets became more varied with the introduction of new foods. Many of these new foods were produced quite by accident. Microorganisms have always affected human food. Bread, yogurt, cheese, wine, and beer are produced by **fermentation**—that is, a microbial process in which enzymatically controlled transformations of organic compounds occur. (Fermentation comes from the Latin *fervere*, "to boil": the addition of yeast to fruit juices during wine making or to cereal grains during bread baking or beer brewing produces a bubbling, from the production of carbon dioxide.) The aroma of baking bread comes from the alcohol that is produced, and trapped carbon dioxide causes the bread to rise. (In beer the carbon dioxide forms the frothy head and the alcohol is present in the malt beverage.)

The existence of microorganisms and their role in contaminating foods are quite recent discoveries, dating back only some 200 years. For millennia the fermentation of foods and beverages was an art, practiced without scientific knowledge for guidance. Before methods of preserving food were devised, moist foods often became contaminated with bacteria, yeast, and molds. People learned to live with microbe-infected foods and encouraged contamination when they discovered that flavor and texture were improved.

Bread was one of the earliest foods; it predates the earliest agriculture. At some point, wild cereal grains were found to be edible. Dried in the sun, they would store for years without spoiling. Early humans probably chewed the raw grains and only later produced flour

FIGURE 1.4 A model of a bakery, Asyut, Egypt, Middle Kingdom (2040–1782 BC).

and dough for baking. Egyptian models and paintings found in tombs show them first grinding grain in a mortar and then on a sloping stone (saddle quern) with a rubbing stone (Figure 1.4). They sifted the flour with sieves made of rushes, mixed the milled flour with water to form a paste, and after adding salt, molded it into loaves. The earliest loaves, being unleavened, were flat and dense (like pita bread); they were cooked on a flat stone over a fire or baked in a clay oven. Early bread was made from an ancient cultivated wheat called *emmer* (for example, *Triticum turgidum* subspecies *dicoccum*, a principal wheat of the Old World used in beer making), from einchorn (for example, *Triticum monococcum*), or from barley (*Hordeum vulgare*).

Fermented dough was almost certainly discovered quite by accident when some dough was not baked immediately, underwent spontaneous fermentation, and when baked, produced a lighter, expanded, and more palatable bread. Some time around 1800 BC, the Egyptians and Babylonians learned that old, uncooked fermented dough could be used to ferment a new batch of dough. Bakers no longer had to depend on chance contamination. A baker would remove some flour from a batch in the mixing vessel and add it to fresh flour; the resulting paste, or starter, was used in the next day's dough. Records such as tomb paintings and reliefs and evidence such as carbonized remains of breads in tombs indicate that the Egyptians used fermentation to make bread. A wide variety of loaf shapes have been found— ovals, indented squares, triangles, cones, and animal shapes. Analysis of bread (and beer) samples has demonstrated that the Egyptians used a pure yeast strain, *Saccharomyces winlocki*, as early as approximately 1500 BC. Because at that time bakeries and breweries were usually set up in proximity, the bakers probably obtained *S. winlocki* from the brewery barm (liquid yeast skimmed from the surface of fermented beer).

Egypt and Mesopotamia exported bread making to Greece and Rome. The Romans improved the technology, producing a lighter, leavened bread by using yeast skimmed from grain-malt wort (a liquid prepared with malt that when fermented produces alcohol). This method was used for nearly 2000 years, although dough starters were also still in use. Roman society held bakers in high regard and considered bread making an art. The Romans took their bread-making methods to the lands they conquered, where they were readily adopted. A bread-specific yeast was unavailable from Roman times through the Middle Ages. In fact, not until Pasteur's experiments between 1857 and 1863 was the connection made between the role of yeast and fermentation. Finally, between 1915 and 1920 the modern production of **baker's yeast** began. Today, almost all bakers use a pure strain of *Saccharomyces cerevisiae*, which comes in a dried and often compressed form.

Bread was not the only fermented food in the ancient world. By 4000 BC the Chinese were using lactic acid–producing bacteria for making yogurt, molds for making cheese, and acetic acid bacteria for making wine vinegar. They also made soy sauce and other sauces by fermentation; the oldest known reference to soybean is Shen Nan's *Materia Medica* (2838 BC). Fermented rice was eaten not only in the Orient but also in the Andes region of what is now Ecuador. Vegetables and fruits were fermented for preservation by being packed in vessels with salt or brine. This process probably originated in the Orient, in prehistoric times, and is still used today; olives, pickles, and sauerkraut are examples of fermented produce, and in the Orient beets, turnips, lettuce, radishes, and vegetable mixtures are fermented.

Milk from domesticated animals has been a dietary staple probably since the earliest times of animal domestication. Rock drawings in the Libyan desert, dating back to almost 9000 BC, depict a cow being milked. Mammalian milk, an emulsion of oil and water, is a stable product because the surface of fat globules adsorbs phospholipids and **proteins**. However, in the absence of preservation methods, milk readily becomes contaminated with lactic acid bacteria and is soured by heat.

Milk undergoes fermentation when bacterial action causes casein, the main protein in milk, to coagulate into a curd that separates from the thin, watery whey. Cheese curd from milk was first made between 5000 and 9000 years ago. Early cheese makers discovered that using milk from a variety of animals—cows, goats, sheep, and so on—yielded cheeses with different textures, aromas, and flavors. Milk was probably first stored in animal skins or in bladders made from animal stomachs. **Enzymes** in the stomach worked with the natural bacteria in the milk to cause the casein to precipitate and form curds, which were subsequently dried. Early records report that cream, buttermilk, yogurt, sour cream, and butter also were consumed. As in early bread making, results were probably inconsistent, because people relied on spontaneous souring. The process was refined and improved through much experimentation.

Modern cheese makers inoculate the milk with lactic acid bacteria (for lactic acid fermentation) and add enzymes (such as rennet) to curdle the casein (Figure 1.5). (Rennet occurs in the gastric juices produced in the fourth stomach of nursing calves and other animals.) After the whey is separated from the curd, salt is added to the curd before ripening. The essential steps in modern cheese manufacture comprise heating, separating the curd from the whey, draining the whey (mostly water with a little protein, lipid, lactose, and lactic acid), salting, pressing the curd, and ripening. New cheeses have continued to be discovered, often by accident. Camembert and Roquefort are two French cheeses that were created by accident. In the late eighteenth century, Camembert was first made when curd was inoculated with a mold (now known as *Penicillium camemberti*). At the end of the 1800s, Camembert was produced commercially, and its flavor was refined by use of a variety of yeasts for fermentation. Roquefort was first produced when curd was contaminated by a mold now called *Penicillium glaucum roqueforti*, found in French caves. The flavor and aroma particular to a variety of cheese result from the action of different microorganisms.

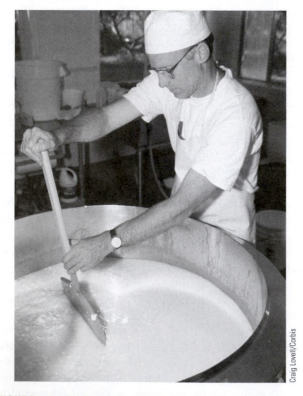

Craig Lovell/Corbis

FIGURE 1.5 Curdled milk stirred for cheese making.

Fermented Beverages Beer making may have begun in Egypt between 6000 and 5000 BC (both bread and beer were Egyptian dietary staples). Early Babylonian records recommend the use of particular varieties of barley for making particular beers. Archaeological evidence indicates that the earliest brewers made beer by partially baking dough from barley just long enough to dry it and form a crust but not long enough to break down the enzymes involved in fermentation. The dry dough was soaked in water until fermentation was complete. The resulting acidic beer was strained and poured into jars for storage (Figure 1.6). Barley was placed in earthenware vessels that were buried until the barley germinated. This germinated barley, called malt, was removed, crushed, and made into a brewer's dough. Malt contains starch and fermentable sugars, coloring, and aromatic compounds that give the beverage a particular taste and aroma. In other regions, a variety of cereal grains, such as sorghum, corn, rice, millet, and wheat, were used in ancient brewing. Yeast sediments found in ancient beer urns from somewhat later periods indicate that brewing methods improved with time.

Brewing was considered an art until the fourteenth century AD, when it was recognized as a separate trade using specialized skills. For several centuries, monasteries were the major brewers. Although brewers refined and improved their techniques, they knew nothing about the microbial basis of fermentation. In 1680 the Dutch biologist and microscopist Anton van Leeuwenhoek examined samples of fermenting beer through a microscope. He submitted a description and drawings of the yeast cells he observed to the Royal Society of London, but his discovery was forgotten. Not until 1837 was a connection made between yeast cell activity and alcoholic fermentation, although the findings, derived from microscopic examination of various fermenting liquids, were criticized. The renowned French chemist Louis Pasteur eventually established that yeast and other microbes are directly linked to fermentation. In two published manuscripts, *Études sur le Vin* (1866) and *Études sur la Bière* (1876), he described the experiments that led him to conclude that during the process of fermentation yeast converts sugar into ethanol and carbon dioxide in the absence of air.

Wine was almost certainly first made accidentally when the juices from grapes became contaminated with yeast and other microbes and fermented naturally. Some authorities believe that wine making originated in the valley of the River Tigris, in present-day Iraq. However, the exact date is unknown. The ancient Egyptians, the Greeks, and the Romans made wine. Pottery wine jars, often labeled, have been found at archaeological sites. Today, wine is aged in large wooden barrels.

FIGURE 1.6 Servant bottling beer and sealing the pottery jar with clay. Model found in a tomb, Giza, Egypt, fifth or sixth dynasty of the Old Kingdom (2498–2181 BC).

CLASSICAL BIOTECHNOLOGY

The term *classical biotechnology* describes the course of development that fermentation has taken from ancient times (that is, ancient biotechnology) to the present. The accumulation of scientific and applied knowledge from experiments and discoveries of the past has provided a solid foundation for the many industrial processes that today provide us with a plethora of products and services. From the mid-nineteenth century to the present, classical biotechnology has exploited our knowledge of cell processes to refine fermentation technology, just as have developments in modern biotechnology (described in subsequent chapters). Knowledge of fermentation has increased to such a level that a large number of important industrial compounds can be readily produced.

Brewers began producing alcohol on a large scale in the early 1700s. To produce what are known as the

English, Dutch, Belgian, and red beers, brewers introduced top fermentation, a process in which the yeast rises to the surface of the liquid as fermentation progresses. In 1833, they introduced bottom fermentation; in this process, the yeast remains at the bottom of the vat. Most beers in the United States and Europe, as well as pale ales, are made by bottom fermentation. By the 1800s, brewers had accumulated enough knowledge to begin using pure yeast cultures in the fermentation process. The equipment designed by E.C. Hansen in 1886 for producing brewer's yeast is still in use today. In 1911, brewers adopted a method for measuring the amount of acid during mashing to better control the quality of their beers.

Vinegar is another fermentation product that illustrates progress in expertise and equipment. Wine was allowed to sit in shallow barrels until it was oxidized to vinegar by the action of microorganisms. Early vinegars were most likely the product of accident, but eventually producers realized that air (that is, oxygen) enhanced the transformation. Experimentation improved vinegar production by leading to the use of a fermentation chamber packed with a material like charcoal, through which the wine or some other alcohol slowly moved for aeration; modern vinegar production uses **fermenters.** By the turn of the century, vinegar producers had mastered the variables in the control of vinegar quality. For example, they prepare a starter solution by inoculating the alcohol with vinegar (Figure 1.7).

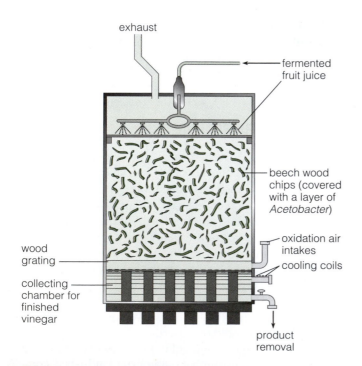

exhaust

fermented fruit juice

beech wood chips (covered with a layer of *Acetobacter*)

oxidation air intakes

cooling coils

wood grating

collecting chamber for finished vinegar

product removal

FIGURE 1.7 Large quantities of vinegar are produced by *Acetobacter* on a substrate of wood chips. Fermented fruit juice is introduced at the top of the column, and the column is oxygenated from the bottom.

From 1900 to 1940 the number of commercial fermentation products expanded to include glycerol, acetone, butanol, lactic acid, citric acid, and yeast biomass for baker's yeast. Indeed, industrial fermentation was established during World War I because Germany needed large quantities of the fermentation product glycerol for explosives. The Germans produced these quantities of glycerol by adding sodium bisulfite to the **substrates** of alcoholic fermentation. Acetone and butanol also were produced by fermentation for explosives during World War I. Later, an anaerobic acetanobutylic fermentation using pure cell culture was developed to eliminate contamination by unwanted microorganisms.

The fermentation of organic solvents began during World War I and by the 1940s had been improved by the establishment of aseptic techniques and use of fermenters that could be steam-sterilized to prevent microbial contamination. Improvements in fermenter design and in control of nutrients and aeration, methods of introducing and maintaining sterility, and methods of product purification and isolation have made it possible to produce rare and valuable chemicals.

World War II ushered in the age of the modern fermenter, or bioreactor, and antibiotics. **Antibiotics** were the first compounds to be produced primarily because of a need for drugs to combat bacteria during the war. Fermentation technology could not have been adapted to commercial antibiotic production without the development of strain isolation methods. The antibiotic penicillin was produced by fermentation of cultured *Penicillium*. Penicillin-producing strains were improved, and effective methods of sterile aeration and culture mixing were introduced. Large-scale penicillin recovery from cultures also was developed. The fermentation of other antibiotics quickly followed (Table 1.3).

Classical biotechnology has introduced chemical transformations that yield products with important therapeutic value. In the 1950s biotransformation technology was developed to convert cholesterol to other steroids such as cortisone and sex hormones. Microorganisms that are fed the proper substrate transform it into the desired compound (see Chapter 5, Microbial Biotechnology). For example, cholesterol can be converted to steroids such as estrogen and progesterone by a microbial hydroxylation reaction (that is, addition of an –OH group to the cholesterol ring). (Microorganisms can readily carry out the hydroxylation and dehydroxylation reactions that are essential steps in the conversion of steroids.) Plant steroids can even be used in the chemical transformation as substrates. This ability of microorganisms to synthesize a wide variety of compounds and use unusual substrates has been exploited commercially. By the mid-1950s, **amino acids** and other primary **metabolites** (those used for cell growth) were being

TABLE 1.3 Some Commercially Produced Antibiotics

Antibiotic	Producing Microorganism	Class
Produced by Fungi		
Cephalosporin	*Cephalosporium acremonium*	Broad spectrum
Griseofulvin	*Penicillium griseofulvum*	Fungi
Penicillin	*Penicillium chrysogenum*	Gram-positive bacteria
Produced by Gram-Positive, Spore-Forming Bacteria		
Bacitracin	*Bacillus subtilis*	Gram-positive bacteria
Polymyxin B	*Bacillus polymyxa*	Gram-negative bacteria
Produced by Gram-Positive Bacterium, Actinomycete		
Amphotericin B	*Streptomyces nodosus*	Fungi
Chloramphenicol	*Streptomyces venezuelae* (now chemical synthesis)	Broad spectrum
Cycloheximide	*Streptomyces griseus*	Pathogenic yeasts
Cycloserine	*Streptomyces orchidaceus*	Broad spectrum
Erythromycin	*Streptomyces erythreus*	Mostly Gram-positive bacteria
Kanamycin	*Streptomyces kanomyceticus*	Gram-positive bacteria
Lincomycin	*Streptomyces lincolnensis*	Gram-positive bacteria
Neomycin	*Streptomyces fradiae*	Broad spectrum
Nystatin	*Streptomyces noursei*	Fungi
Streptomycin	*Streptomyces griseus*	Gram-negative bacteria (*Mycobacterium tuberculosis*)
Tetracycline	*Streptomyces rimosus*	Broad spectrum

J. Ingraham and C. Ingraham, *Introduction to Microbiology*, Table 29.4. Copyright © 1995 Wadsworth Publishing Co.

produced. Large amounts of a specific metabolite can be produced by shifting the direction of cell metabolism during the fermentation process. Other fermentation products such as enzymes and vitamins were also produced at this time.

By the 1960s microbial cells were being produced on a large scale as a source of protein. Aeration methods were developed to new levels of sophistication, and continuous culturing replaced batch culturing. Eventually computers were used to control fermenter operations. In the 1960s and 1970s, many secondary metabolites (those compounds that are not used for cell reproduction and growth) began to be produced by fermentation and screened for therapeutic activity.

Today, primary metabolites such as amino acids (Table 1.4), pharmaceutical compounds, and a variety of chemicals, hormones, and pigments are produced by industrial fermentation for commercial use. Many antibiotics are commercially produced: Antibiotic-producing microorganisms are grown in large fermenters, and the antibiotic end product is collected. Enzymes with a variety of uses (Table 1.5) have been commercially produced from microorganisms and from animal and plant cells. Biomass was first used for food in Germany during World War I, and commercial production for animal and human consumption (such as **single-cell protein**) continues today, as does the mass production of baker's yeast, which began early in the twentieth century.

Table 1.4 Produced Amino Acids and Their Uses

Amino Acid	Use
Alanine	Added to fruit juice to improve taste
Aspartate	Added to fruit juice to improve taste
Cysteine	Added to bread and fruit juice to enhance flavor
Glutamate	Added to many foods to enhance flavor (as monosodium glutamate [MSG])
Glycine	Enhances flavor of sweetened foods
Histidine + tryptophan	Prevents rancidity in various foods
Lysine	Used in Japan to make bread a more complete protein
Methionine	Makes soybean products a more complete protein

J. Ingraham and C. Ingraham, *Introduction to Microbiology*, Table 29.5. Copyright © 1995 Wadsworth Publishing Co.

TABLE 1.5 Commercially Important Enzymes Produced by Microorganisms

Enzyme	Activity	Producing Microorganism	Use
Cellulase	Hydrolyzes cellulose	*Trichoderma konigi*	Digestive aid
Collagenase	Hydrolyzes collagen	*Clostridium histolyticum*	Promotes wound/burn healing
Diastase	Hydrolyzes starch	*Aspergillus oryzae*	Digestive aid
Glucose isomerase	Converts glucose to fructose	*Streptomyces phaeochromogenes*	Converts glucose from hydrolyzed cornstarch to a sweetener
Invertase	Hydrolyzes sucrose	*Saccharomyces cerevisiae*	Candy manufacture
Lipase	Hydrolyzes lipids	*Rhizopus spp.*	Digestive aid
Pectinase	Hydrolyzes pectin	*Sclerotina libertina*	Clarifies fruit juice
Protease	Hydrolyzes protein	*Bacillus subtilis*	Used in detergents

J. Ingraham and C. Ingraham, *Introduction to Microbiology,* Table 29.6. Copyright © 1995 Wadsworth Publishing Co.

FOUNDATIONS OF MODERN BIOTECHNOLOGY

The ability to manipulate living organisms with today's precision requires intricate knowledge of cell structure, the biochemical reactions that take place within the cell, and the genetic makeup of cells. Modern, multidisciplinary biotechnology is the result of scientific discoveries and technological developments that span the more than 300 years from the first microscopes to the first molecular cloning experiments. What were the major discoveries that led to today's sophisticated biotechnologies?

Early Microscopy and Observations

The microscope revolutionized science. Until its advent, natural scientists could describe only what they could see with the naked eye. In 1590, the Dutch spectacle-maker Zacharias Janssen made the first compound microscope (that is, one having more than one lens to magnify the image). This primitive two-lens microscope magnified an image 30 times. Microscopes were widely used by 1665, when Robert Hooke, a physicist and curator of instruments for the Royal Society of London, examined the structure of thinly sliced cork under the microscope. In his treatise *Micrographia,* he described small rectangular compartments, which he called *cellulae* (Latin for "small chambers"); Figure 1.8 shows his microscope and his drawing of what he observed. These structures were actually the walls of dead cells. A Dutch shopkeeper, Anton van Leeuwenhoek, ground lenses as a hobby; his single lenses magnified an image 200 times. In 1676, he exam-

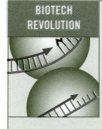

BIOTECH REVOLUTION

Old Meets New

Fermentation and the methods of genetic engineering have been applied to the production of food since the mid to late 1980s. The products of genetically engineered bacteria and fungi are being used in processed foods and as a source of enzymes to improve the foods we eat. Genetically engineered organisms are cultured in fermenters and have been modified to produce large quantities of desirable enzymes that are extracted from them and purified. These enzymes are used in the production of foods such as milk, beer, breads, cheese, and candy. Vitamins and mineral supplements also are produced. The fermentation industry has used genetic engineering to increase the amount of production and purification of enzymes, to improve an enzyme's function in food processing, and to provide a less expensive and more reliable source of enzymes from fermentation technology. Chymosin is an example of how genetic engineering has enhanced fermentation technology. In the past, chymosin, used to produce cheese by clotting milk proteins to form cheese and whey, was produced by extracting it (as rennet) from the stomach of calves. Today, chymosin is produced by genetically engineered bacteria through fermentation. More than 70% of U.S. cheeses are produced using this enzyme.

FIGURE 1.8 Robert Hooke's compound microscope and the cork tissue he sketched.

ined pond water samples and saw small living organisms (protozoa and fungi), which he called *animalcules* (for "small animals"), that were visible only through his lenses (Figure 1.9). In 1683 he saw still smaller creatures—bacteria. Nevertheless, the low resolution and magnification of these crudely constructed microscopes severely limited the understanding of cells.

Development of Cell Theory

As their tools of inquiry improved, scientists began to realize that tissues were composed of cells, and that these cells could divide to generate more cells. Therefore, each cell was a living, functioning unit. In 1838 the German botanist Matthias Schleiden determined that all living plant tissue was composed of cells and that each plant arose from a single cell. A year later, the German physiologist Theodor Schwann came to a similar conclusion for animals. The cell theory was refined by the German pathologist Rudolf Virchow, who concluded in 1858 that "all cells arise from cells" and that the cell is the basic unit of life. Until this time, the prevailing theory was vitalism: only the complete organism, rather than its individual parts, possessed life. After the cell theory was introduced, vitalism gradually fell from favor. By the early 1880s, with the improvement of microscopes, tissue preservation techniques, and stains, scientists made significant advances in the understanding of cell structure and function.

Role of Biochemistry and Genetics in Elucidating Cell Function

While the secrets of cell structure, organization, and reproduction were unfolding, scientists were elucidating the biochemical and genetic nature of organisms. Most early-nineteenth-century researchers believed that the organic and inorganic worlds (that is, the living and nonliving) were distinctly separate; the laws of chemistry applied only to the inorganic world and not to the biochemical processes of living organisms. Therefore, only living tissues could synthesize organic molecules. However, in 1828, the German chemist Friedrich Wohler obtained crystallized urea, a waste product in mammalian urine, from ammonium cyanate in the lab-

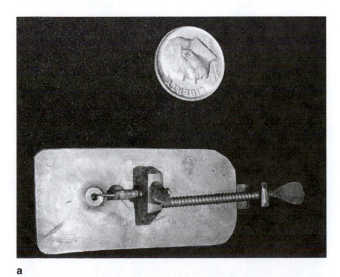

a

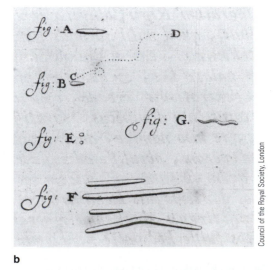

b

FIGURE 1.9 (a) Small hand–held microscope used by Anton van Leeuwenhoek and (b) animalcules from his drawings published in 1684.

oratory. This was an important discovery: The experiment demonstrated that an organic compound made by a living organism could be synthesized from inorganic compounds in the laboratory. This chemical synthesis encouraged chemists to synthesize other organic compounds.

Between 1850 and 1880, Louis Pasteur made significant contributions to our knowledge of living processes. Observing that wine sometimes became sour as it aged, Pasteur discovered that yeast cells present in the wine contributed to spoilage. He determined that wine was preserved if heated during the interval after the alcohol was made and before lactic acid was produced. This heating, called pasteurization, plays an important role in food preservation today.

Scientists had been engaged in an ongoing controversy about whether life could arise spontaneously. Experiments performed in 1668 and later in 1768 suggested that life must come from life. Although these experiments provided convincing evidence that higher life forms did not arise spontaneously, scientists thought that perhaps microorganisms could so arise. In 1860 Pasteur conducted an experiment to demonstrate that spontaneous generation of microorganisms indeed did not occur. He placed sterilized broth of boiled meat extracts in a swan-neck flask and a straight-neck flask (Figure 1.10). After some time only the broth in the straight-neck flask became contaminated. Pasteur argued that airborne microbes inoculated the broth after entering through the straight neck but were trapped in the trough made by the bend in the swan neck. Therefore, microorganisms gave rise to microorganisms. These results also supported the cell theory proposed earlier in the century. By the end of the nineteenth century, the cell theory was widely accepted, knowledge about the biochemical basis of organisms was rapidly accumulating, and the biochemical analysis of cell components was well underway.

In 1896 Eduard Buchner converted sugar to ethyl alcohol using yeast extracts rather than intact yeast cells. These *ferments* (as he called the extracts) were eventually found to consist of enzymes (enzyme means "in yeast") or biological catalysts. Now these important chemical transformations could be conducted outside the cell.

The chemical structure of protein was of great interest; scientists believed that proteins held the key to heredity and biochemical processes, and they were intrigued by the complexity and variety of proteins. In the 1920s and 1930s the biochemical reactions of many important metabolic pathways were elucidated. By 1935 all 20 amino acids were isolated. Researchers were greatly aided by advances in instrumentation. In the late 1920s the ultracentrifuge was developed, and by the early 1940s ultracentrifugation methods had been perfected for separating cell organelles and macromolecules by size, shape, and density. In 1932 the German electrical engineer Ernst Ruska built the first electron microscope (400× magnification). Although primitive at first, the instrument was greatly improved in resolution and magnification and in the 1940s and 1950s was used routinely to elucidate cell ultrastructure. Now researchers could integrate what they learned about cell organization with their knowledge of biochemical processes within the cell to create a more complete picture of the cell.

The study of the genetic nature of living organisms, which helped integrate prior knowledge of cells with what was known of the principles of heredity, is rooted in the mid-nineteenth century. In 1857, Austrian botanist and Augustinian monk, Gregor Mendel, began to experiment with peas grown in the monastery garden. He systematically cross-pollinated plants to examine the patterns of inheritance of seven traits (petal color, seed color, and seed texture among them) that existed in two alternative forms such as green versus yellow seeds, smooth versus wrinkled seeds, tall versus dwarf plants. Mendel determined that each parent pea plant contributed to its progeny one unit of heredity for each trait, in either a dominant or recessive form. From his experiments with peas, Mendel formulated the principles of inheritance that later came to be known as Mendelian genetics. However, his findings, which he published in 1865, were not understood and remained in obscurity until 1900, when three botanists independently rediscovered his paper and recognized the significance of his work.

In 1869, Johann Friedrich Miescher, a Swiss biochemist, isolated from the nuclei of white blood cells a substance that he called *nuclein* and that is now called nucleic acid. At the time, nuclein was not known to be the hereditary material, but shortly afterward an important series of observations led to the identification of chromosomes as the carriers of genetic material. In 1882, the German cytologist Walter Flemming described threadlike bodies that were visible during cell division and the equal distribution of their material to

FIGURE 1.10 Swan-neck flask used by Louis Pasteur to demonstrate that microorganisms did not spontaneously form in sterilized broth.

daughter cells. Although he did not know the significance of what he saw, the threadlike bodies were chromosomes (chromosome means "colored body," because chromosomes stain intensely; the term was coined in 1888 by W. Waldeyer), and sister **chromatids** were being equally distributed to the daughter cells. Shortly after the rediscovery of Mendel's experiments and conclusions, Walter Sutton, an American cytologist, determined in 1903 that chromosomes were the carriers of Mendel's units of heredity—or *genes*, as the Danish botanist Wilhelm Johannsen named them in 1909. Sutton observed that during meiosis (the reductive division that produces **haploid** egg and sperm cells), the gametes produced receive only one chromosome of each morphologic type. Thus, he reasoned that meiosis was the mechanism by which the hereditary units are distributed. Through meiosis, many different combinations of traits can be obtained according to how the different chromosomes are oriented during distribution to the **germ cells.**

NATURE OF THE GENE

Between 1930 and 1952, researchers focused much effort on determining the relationship of genes and proteins. They established an important connection between the two when specific mutated genes were demonstrated to produce changes in certain enzymes. The first experiments were conducted by George W. Beadle and Boris Euphrussi with mutants in the fruit fly *Drosophila* and by Beadle and Edward L. Tatum with the bread mold *Neurospora*. Charles Yanofsky and others conducted experiments with the bacterium *Escherichia coli*. Yanofsky demonstrated that there was a direct relationship, or colinearity, between the ordering of mutant sites within a gene and the linear sequence of amino acids in a protein. He and his colleagues determined the colinearity by introducing **mutations** along the tryptophan synthetase gene *(trpA)* and correlating the locations with changes in specific amino acids of the protein. Thus, genes determine the structure of proteins.

Several elegant experiments conducted in the 1940s increased our understanding of the chemical nature of the gene. Two strains of *Streptococcus pneumoniae*, a bacterium responsible for a form of pneumonia, were used: a **virulent** *smooth* strain (S) with a gelatinous coat, and a less virulent *rough* strain (R) that lacks the coat. In 1928 the British physician Fred Griffith had hypothesized that a transforming principle from the S strain of bacterial cells was responsible for the conversion of R bacterial colonies to the S form. The S strain was lethal to mice, but R cells were not (Figure 1.11). However, when Griffith injected mice with a mixture of heat-killed S and live R bacteria, the mice died, and the bacterial colonies that were isolated and plated were of the S type. In 1944 Avery, MacLeod, and

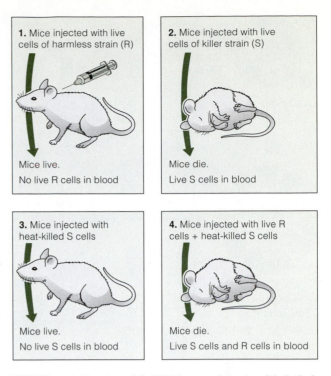

1. Mice injected with live cells of harmless strain (R)

Mice live.
No live R cells in blood

2. Mice injected with live cells of killer strain (S)

Mice die.
Live S cells in blood

3. Mice injected with heat-killed S cells

Mice live.
No live S cells in blood

4. Mice injected with live R cells + heat-killed S cells

Mice die.
Live S cells and R cells in blood

FIGURE 1.11 Results of Griffith's experiments with lethal and nonlethal strains of the bacterium *Streptococcus pneumoniae*.

McCarty extended Griffith's investigations to identify the transforming principle. They mixed the R strain with **DNA (deoxyribonucleic acid)** extracted from S-type bacteria and isolated S colonies after plating onto media (Figure 1.12). When added to the mixture, the enzyme deoxyribonuclease (DNase), which digests, or breaks down, DNA, abolished this transformation of R to S colonies. Proteases (enzymes that digest proteins) did not prevent transformation. Thus, DNA was determined to be Griffith's transforming principle.

Unfortunately, such an unexpected conclusion often meets resistance, despite the evidence. What was the source of resistance in this case? DNA was known to be chemically a much simpler molecule than the diverse array of proteins found in living organisms. Not until 1952 was DNA accepted as the genetic material. In that year Alfred Hershey and Martha Chase conducted an ingenious set of experiments using T2 **bacteriophage** (a virus that infects a bacterial host). They demonstrated that only the DNA of the T2 bacteriophage, and not its protein coat, enters the cell. Infection of the host cell and subsequent synthesis and assembly of virus progeny was therefore the result of T2 DNA infection. To identify the phage material that enters the host cell, they used radiolabeled viral proteins (sulfur-35, or ^{35}S) and **nucleic acids** (phosphorus-32, or ^{32}P) to follow these molecules during viral infection of *E. coli* (Figure 1.13). Phages were grown in medium

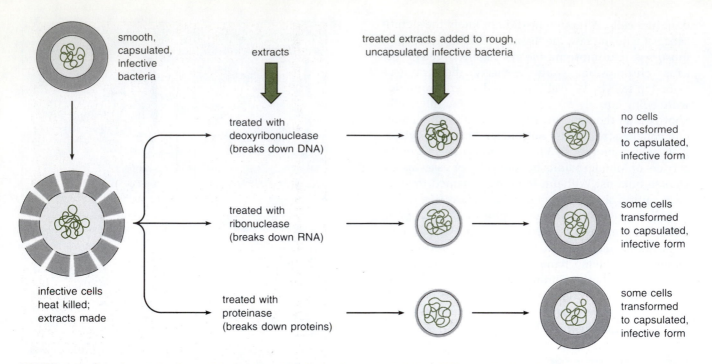

FIGURE 1.12 Experiments by Avery, MacLeod, and McCarty demonstrated that DNA was the transforming principle responsible for the transformation of *Streptococcus pneumoniae* from nonvirulent to virulent.

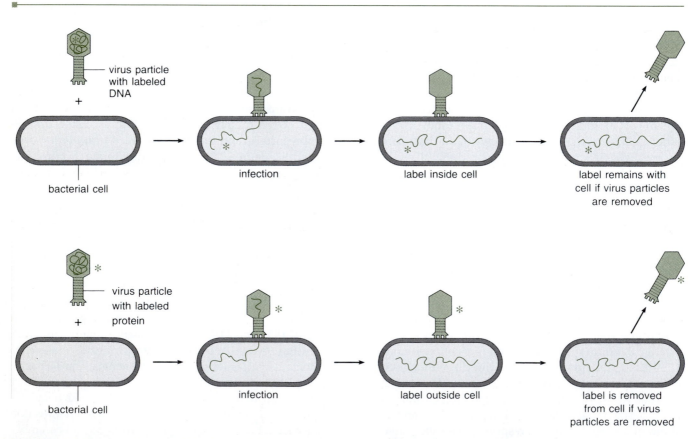

FIGURE 1.13 The experiments of Hershey and Chase demonstrated that DNA is the material that contains the genetic information required for the production of viral progeny. Growing infected cells in the presence of radioactive sulfur or phosphorus had the effect of radiolabeling the viral proteins or DNA, respectively. After infection, the viral particles were removed by shaking the cells, and the location of the radioactivity was determined. (a) Radioactivity of DNA-labeled viral particles was detected in bacterial cells. (b) Radioactivity of protein-labeled viral particles was removed when bacterial cells were shaken.

containing bacterial host cells, and bacteriophage protein and nucleic acid were labeled with ^{35}S or ^{32}P, respectively. During viral replication within host cells, phage progeny incorporated ^{35}S into their proteins or ^{32}P into the nucleic acid. Unlabeled cells were infected with the radioactively labeled phages. After infection, cells were put in a blender to remove the attached phage particles. Analysis of the host cells and culture medium showed that the ^{32}P-labeled nucleic acid was inside the bacterial cells and was used to synthesize new viral progeny. The ^{35}S-labeled proteins remained in the medium. This was strong experimental evidence that DNA was the genetic material.

Fifty years ago, the final threshold to modern molecular biology was crossed when James Watson and Francis Crick elucidated the structure of DNA, in 1953. Rosalind Franklin, an expert x-ray crystallographer, and Maurice Wilkins (who shared the 1962 Nobel Prize in chemistry with Watson and Crick) had produced x-ray diffraction patterns of DNA, and Erwin Chargaff had established the DNA base ratios (**pyrimidine** and **purine** bases were in a ratio of 1:1, adenine equaled thymine, and cytosine equaled guanine). By studying the x-rays with the knowledge of Chargaff's findings, Watson and Crick were able to develop a structural model of DNA. Their model described the width of the molecule, the number of bases per helical turn, and the spacing and location of the bases, as well as how the molecule could replicate according to complementary base-pairing rules (Figure 1.14).

Experiments by many other scientists rapidly followed to determine how the information in the gene is decoded and how gene expression is regulated in both prokaryotic and eukaryotic organisms. The scientific foundation of modern biotechnology was provided not only by the discovery and understanding of nucleic acids, but also by the elucidation of the enzymatic tools required for DNA manipulation *in vitro* (in the test tube). By studying **DNA replication, DNA repair,** and the immunity of cells to viral infection, scientists purified the enzymes that later allowed them to synthesize DNA, to cut DNA into specific fragments, and to rejoin them in different combinations *in vitro*.

Connections between structures and functions were made years later by insightful scientists who re-examined the conclusions of early experiments. The accumulated knowledge of cell structure, biochemistry, and heredity opened the door to modern molecular biology and biotechnology. Today scientists are applying what is known about cellular and molecular processes of living organisms to enhance our quality of life.

By the mid-1970s the scientific foundation was in place. However, many people were unprepared for the DNA revolution that followed and the important scientific achievements that were made during this time. The road was not a smooth one in the beginning.

FIGURE 1.14 James D. Watson (left) and Francis H.C. Crick (right), demonstrating their model of DNA structure deduced from x-ray diffraction data obtained by Wilkins and Franklin.

Rapid advances in the field of **molecular biology** have brought revolutionary change in human and veterinary medicine, agronomy, animal science, environmental science, bioethics, and patent law. We can anticipate still more new treatments for diseases, improved crops and livestock that are disease resistant and more productive, and a cleaner, safer environment.

Methods for manipulating DNA developed out of basic biological research directed toward understanding how genes are activated in living organisms. These methods were developed when the conventional genetic and biochemical techniques that had been so successful for studying bacteria proved inadequate for studying complex **eukaryotic cells** (for example, from mouse, fruit flies, and yeast) and their **genomes.** Further progress in understanding the molecular genetics of higher organisms required new techniques. **Recombinant DNA** (the recombining of DNA from different sources or organisms) technology revolutionized the study of molecular biology. A gene could now be cut and separated from the chromosome, the DNA sequence determined, and the gene placed in a different host (often in a bacterial host whose biology was better understood) for further study.

Early Years of Molecular Biology

The field of molecular biology arose from important discoveries about the cell. Genes were made of DNA and not protein as previously believed. Thus, the genetic information is encoded in the relatively simple chemical and molecular structure of DNA. These discoveries inspired researchers to seek the answers to questions about the nature of genetic information, how hereditary information in DNA is stored, and how it is retrieved and utilized.

In the 1950s and 1960s research focused on two main questions: (1) How does the DNA sequence of the gene relate to the sequence of amino acids that make up the protein? (2) What is the cell decoding process that produces a protein from the information encoded by the gene? After Watson and Crick elucidated the structure of DNA in 1953, progress was rapid. Genetic experiments in 1956 supported the hypothesis that the sequence of **deoxyribonucleotides** in DNA specified the information, or messages, of that DNA. In 1957 Matthew Meselson and Frank Stahl demonstrated how DNA is replicated (duplicated) in cells. In that same year, Francis Crick and colleagues hypothesized that the

DNA bases specify the linear amino acid sequence of a protein, and that each amino acid is specified by a triplet of bases. There was speculation that in eukaryotes a **ribonucleic acid** (RNA) served as a messenger between DNA in the nucleus and **ribosomes** (where proteins are synthesized) in the cytoplasm. That is, genetic information must flow from DNA to a **messenger RNA** (mRNA) to protein. The existence of mRNA was confirmed in 1960. By 1966, the complete 64-triplet code used by all living organisms had been elucidated (Figure 1.15).

First Recombinant DNA Experiments

Researchers in the late 1960s, although lacking the requisite methods, were nevertheless discussing the mixing of unrelated DNAs and the moving of hybrid molecules from one species to another. By the 1970s, they were using **plasmids** (replicating, extrachromosomal circular DNAs found in bacteria) and viruses as vehicles to transfer DNA into cells. Some enzymes to manipulate DNA had been isolated from bacteria—for example, **DNA polymerase** (an enzyme involved in duplicating DNA), **DNA ligase** (acts like glue to seal DNA fragments together), and **restriction endonucleases** (act like a scissors to cut DNA). This technology showed promise not only for studying gene structure and regulation but also for producing mammalian proteins in bacterial cells for commercial use. Many scientists at that time also saw the potential for human gene therapy once methods had been developed for transferring DNA into mammalian cells.

The experiments of Paul Berg, Herbert Boyer, Stanley Cohen, Janet Mertz, Ronald Davis, and their colleagues, published more than 30 years ago, ushered in the era of modern biotechnology and genetic engineering. They discovered ways to join DNA molecules from different sources (recombinant DNA) and to transfer DNA into host cells. This dramatically changed the way cells could now be studied and paved the way for cells and organisms to be genetically modified.

In late 1971, a medical student in Cohen's laboratory found that, in a test tube, bacteria treated with calcium chloride took up plasmid DNA that contained antibiotic resistance genes. Bacterial cells and their plasmids replicated (duplicated) and all cells harboring these plasmids were resistant to the antibiotic. Cohen then was able to use plasmids to transfer genes into host bacterial cells. He knew from discoveries in the mid-1950s that bacteria could become resistant to antibiotic drugs from plasmid-encoded resistance genes and that antibiotic resistance could be transferred to related bacteria. Cohen studied how plasmids reproduced and were transferred to other host cells. He was able to use restriction endonucleases to cut DNA into fragments to identify the important DNA regions required for replication and antibiotic resistance (antibiotic resistance genes).

First Base	Second Base				Third Base
	U	C	A	G	
U	phenylalanine	serine	tyrosine	cysteine	U
	phenylalanine	serine	tyrosine	cysteine	C
	leucine	serine	**stop**	**stop**	A
	leucine	serine	**stop**	**tryptophan**	G
C	leucine	proline	histidine	arginine	U
	leucine	proline	histidine	arginine	C
	leucine	proline	glutamine	arginine	A
	leucine	proline	glutamine	arginine	G
A	isoleucine	threonine	asparagine	serine	U
	isoleucine	threonine	asparagine	serine	C
	isoleucine	threonine	lysine	arginine	A
	(start) methionine	threonine	lysine	arginine	G
G	valine	alanine	aspartate	glycine	U
	valine	alanine	aspartate	glycine	C
	valine	alanine	glutamate	glycine	A
	valine	alanine	glutamate	glycine	G

FIGURE 1.15 The genetic code includes three stop codons and 61 amino acid–coding nucleotide triplets.

Breaking the Genetic Code

In 1961 Marshall Nirenberg and J.H. Matthei made the first attempt to break the genetic code by using synthetic mRNA. For their *in vitro* experiment with *E. coli*, they made a mixture of cell components that included all machinery necessary for making proteins—*E. coli* ribosomes, amino acid–transfer RNAs (tRNAs), GTP (guanosine triphosphate—provides energy), magnesium, and translation factors. They did not include nucleic acids and the cell membrane. Thus, the test tube contained all the components necessary for protein synthesis except the mRNA, which was added last as a synthetic mRNA (see Chapter 2, From DNA to Proteins). In this way, they identified the code words, or triplet ribonucleotides, that specify the individual amino acids (called codons) incorporated into protein from simple synthetic messages. These scientists were able to synthesize mRNA with a specific triplet code such as UUU, AAA, and CCC. The first codon identified was UUU (called poly-U mRNA) for phenylalanine. Nirenberg and Severo Ochoa continued this work using artificial mRNAs such as AAA for lysine, CCC for proline, and GGG for glycine. By mixing nucleotides in specific ratios (for example, uracil and guanine in a ratio of 5:1), they determined the amino acids coded by the synthetic mRNA, but they could not determine the order of the bases. For example, although they established that leucine contains two Us and one G, they did not know that the order was UGU.

In 1964 Nirenberg and Phillip Leder developed an assay that enabled them to determine which codons in the mRNA specified which amino acids (in other words, the order of the bases in the codon—for example, UGU). They synthesized specific mRNA sequences comprising different codon combinations and used these sequences in a binding assay designed to identify the aminoacyl-tRNA (a transfer RNA with its bound amino acid) that was bound to a particular codon-ribosome complex. A synthetic mRNA was added to *E. coli* ribosomes and radiolabeled amino acids bound to their appropriate tRNAs. The tRNAs (adapter molecules that properly position amino acids during

protein synthesis) bound to the appropriate codon of the mRNA-ribosome complex by complementary base pairing between the mRNA codon and the anticodon of the tRNA (Figure 1.16). Those aminoacyl-tRNAs that bound to the ribosome complex were trapped on a filter. However, tRNAs with noncomplementary anticodons passed through, and no radioactivity was detected on the filter. The amino acid bound to the tRNA was identified and associated with a specific codon. Not long after the development of this binding assay, all 61 codons that encoded amino acids and three stops (that is, stop codons that terminated protein synthesis) were elucidated.

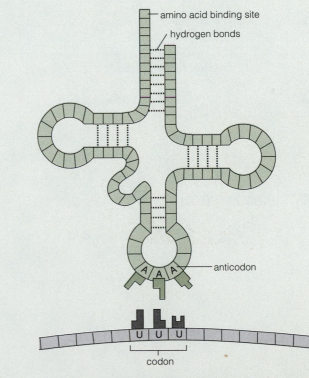

FIGURE 1.16 The folded shape of aminoacyl–transfer RNA resembles a cloverleaf. The two functional regions are the three–base anticodon region that hydrogen bonds to the codon of the messenger RNA and the end of the molecule that binds to the amino acid specified by the anticodon. The codon UUU specifies the amino acid phenylalanine.

Two significant 1972 studies conducted at Stanford University by Paul Berg and his colleagues David Jackson and Robert Symons and by Janet Mertz and Ronald Davis hinted at the exciting future potential of recombinant DNA technology. These scientists joined two DNA molecules from different sources. From these results, Berg and his colleagues speculated that mammalian cells could be transformed (the uptake of DNA) with these recombinant DNAs and then tested for activation of foreign genes.

Mertz and Davis demonstrated that the restriction endonuclease, EcoRI, could in one step generate **cohesive** (or staggered) **ends** (Figure 1.17). The cohesive ends of the two DNA strands, cut with EcoRI, spontaneously annealed (that is, came together) through hydrogen bonding of the nitrogenous bases of the deoxyribonucleotides. The DNA backbones of the two DNA fragments were covalently bonded by DNA ligase. Mertz and Davis had found an easy method of *in vitro* recombination for generating specifically oriented hybrid DNAs and allowing the original DNA molecules to be retrieved by cutting with EcoRI.

Information about how to cut plasmid DNA at a specific site and insert a new fragment was presented in November 1972 at a meeting in Honolulu, Hawaii. Mertz and Davis had just published their work showing that EcoRI produces cohesive ends, and Cohen presented his method of transferring plasmid DNA to a bacterial host. At the same meeting, Herbert Boyer, of the University of California, San Francisco, presented a paper about the purification of EcoRI from E. coli. Boyer and his colleagues hoped to use restriction endonucleases to generate specific plasmid fragments for experimental purposes. In Hawaii, Cohen and Boyer discussed a collaboration in which EcoRI would be used to generate DNA fragments for insertion into Cohen's plasmids. These recombinant molecules would be used to transform an E. coli host. At the time, no one knew if molecules engineered in the laboratory could be propagated in living organisms.

After Hawaii, Boyer went to Cold Spring Harbor Laboratories, in New York, and learned how molecular biologists were using a method called **gel electrophoresis** to separate DNA fragments cut with EcoRI and to visualize the DNA on the gel using a stain called **ethidium bromide** (Figure 1.18). Boyer immediately realized that gel electrophoresis was a powerful analytical tool for monitoring DNA digestions with restriction endonucleases.

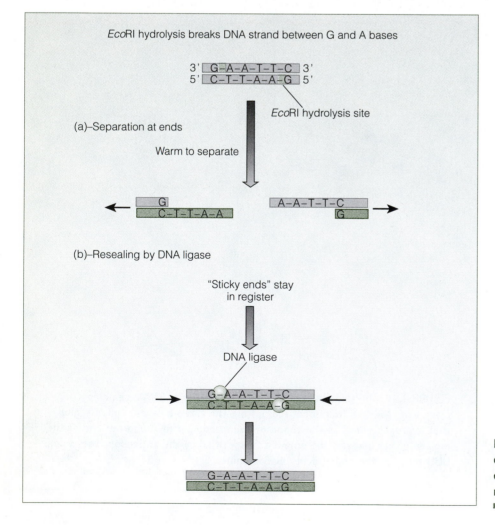

FIGURE 1.17 The generation of cohesive ends by the restriction enzyme EcoRI. The ends can be resealed by DNA ligase.

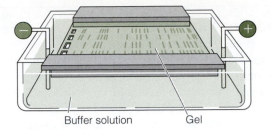

Buffer solution   Gel

FIGURE 1.18 DNA fragments are separated by gel electrophoresis. A current is applied so that the negatively charged DNA molecules migrate to the positive electrode. DNA is separated by fragment length.

First DNA Cloning Experiment

Herbert Boyer, Robert Helling, Stanley Cohen, and Annie Chang began a collaborative research project to join specific DNA fragments in a vector and transform an *E. coli* host (results were published in 1973). This collaborative group used a small bacterial plasmid, pSC101 (with a tetracycline resistance gene and one *Eco*RI restriction site), to make other useful cloning vectors (each one was named pSC102, pSC105, pSC109, and so on) with different antibiotic resistance genes. By March 1973 the cloning experiment showed much promise. *Eco*RI was used to (1) produce fragments of a large plasmid isolated from bacteria and (2) linearize the cloning vector (harboring tetracycline and kanamycin antibiotic resistance genes). The cloning vector and plasmid fragments were mixed in a test tube. The enzyme DNA ligase was added to join the ends of the DNAs (Figure 1.19). Finally, bacterial host cells were transformed with the recombinant DNA. The bacteria were plated onto solidified agar medium supplemented with tetracycline and kanamycin antibiotics to select for those bacteria that had taken up the recombinant DNA molecules. The experiments were successful: recombinant DNA could be isolated from the bacteria hosts! This method of cutting and rejoining DNA to produce a recombinant DNA that could replicate in a host cell became known as DNA cloning. The research culminated in a paper, "Construction of Biologically Functional Bacterial Plasmids *in Vitro*," published in the *Proceedings of the National Academy of Sciences*. This publication formed the basis for a patent ("Process for Reproducing Biologically Functional Molecular Chimeras"), of which Stanford University was the beneficiary. The method provided the groundwork for much of modern biotechnology. The practical applications of the method and its implications were immediately apparent.

Following this ground-breaking work, Cohen and Chang discovered that they could transfer bacterial DNA to unrelated bacteria. In a 1974 issue of the *Proceedings of the National Academy of Sciences*, they demonstrated that DNA from *Salmonella* and *Staphylococcus* could be transferred and expressed in *E. coli*. By August 1973 Cohen, Boyer, Berg, Helling, Chang, Howard Goodman (Stanford University), and John Morrow, Berg's graduate student, successfully transferred RNA genes (genes that encode information for ribosomal RNAs that make up the ribosome) from the African clawed frog *Xenopus laevis* to *E. coli*. The frog

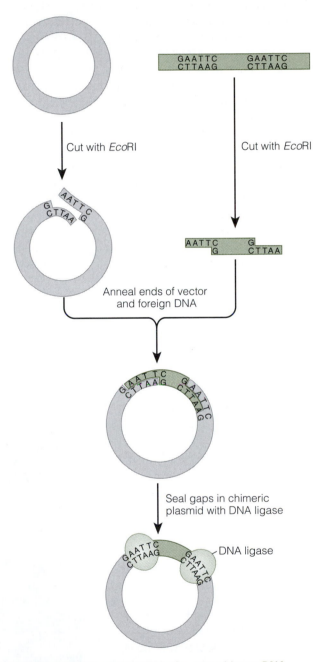

FIGURE 1.19 The production of a recombinant DNA molecule: the DNAs are cut with a restriction endonuclease to produce complementary cohesive ends, the cut plasmid and DNA molecule of interest are mixed in a test tube, and the DNAs are ligated with the enzyme DNA ligase. The product is a DNA molecule composed of two different DNAs called a recombinant DNA.

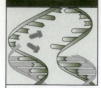

Public Reactions to Recombinant DNA Technology

Recombinant DNA cloning methods are now used routinely in many laboratories worldwide—practical applications developed out of the pursuit of basic scientific knowledge. However, more than 30 years ago, these experiments sparked a recombinant DNA revolution with implications that provoked heated debate among scientists, ethicists, the media, venture capitalists, lawyers, and many others (see Chapter 12, Regulation, Patents, and Society). Concerns in the 1970s and early 1980s focused primarily on the question of whether scientists are tampering with nature, the escape of genetically engineered bacteria from a controlled laboratory environment, and the development of a cancer epidemic caused by a cloned virus DNA in *E. coli* that is transferred to humans. As time passed, there were no disasters that occurred as a result of recombinant DNA technology, and it was concluded by most scientists that the technology did not pose any risk to human health or the environment. Concerns about the actual safety of the technology subsided.

Today, as biotechnological advances have escalated and biotechnology has influenced many diverse areas, concerns have focused on both applications and ethical implications. For example, in the medical field, gene therapy experiments raise the question of eugenics (artificial human selection) as well as testing for the presence of diseases with no cure (such as Huntington's chorea). An area that has generated heated debate is the use of embryonic stem cells for medical purposes. The reproductive field has witnessed many new developments, and one offshoot of this work has

been the generation of animal clones. Fears have been expressed that one day, in the not-too-distant future, human clones will be produced.

In agriculture there is concern that the spread of genes from transgenic crop plants to weeds may cause problems (such as herbicide-resistant weeds or weeds that produce pesticides), or that genetically engineered organisms released into the environment may adversely affect the environment and become difficult to control or outcompete native populations. One of the first occasions for public objection to the release of an organism into the environment arose from the intended use of ice-minus bacteria, an engineered bacterium, *Pseudomonas syringae*, designed to protect crop plants against frost. In August 1984, social ac-

FIGURE 1.20 A moon suit was used to protect the sprayer against unknown biohazards when applying ice-minus bacteria to strawberry plants.

tivists filed suit to block an attempt to use ice-minus bacteria and a federal judge issued an injunction against the tests. In April 1987, after a costly delay, testing was able to begin on strawberry plants. Initially, those spraying the plants with bacteria took the precaution of wearing protective clothing (Figure 1.20).

Today, fears have focused on genetically engineered foods in the marketplace. Although hundreds of genetically engineered food crops are in field trials, public concerns have hampered their commercial production. People have vigorously protested the marketing of these foods or processed foods that contain even small amounts of genetically engineered components. This has forced some companies to place a hold on crops ready for production and has resulted in the rapid growth of the organic food industry.

Despite opposition, progress continues in many areas. Today, as a result of recombinant DNA technology, hundreds of genetically engineered disease-, pest-, and herbicide-resistant plants are awaiting approval for commercialization, and many in commercial production have better yields and have reduced chemical usage in the field. Rapid progress is being made in identifying genes involved in human disease and understanding the human genome (genetic material), and new medical treatments are being developed. A whole new area, "molecular pharming" (similar to farming with plants and animals, but the goal is the production of pharmaceuticals), has emerged where plants and animals are producing valuable human proteins to treat diseases. The legal and regulatory maze has been simplified, and positive impacts on society are being realized.

genes were active in *E. coli* and were transcribed (the first step in making a product from information encoded in a gene) into RNA. The collaborators demonstrated that even animal DNA could be propagated in bacteria. When Cohen and collaborators at Stanford University isolated functional mouse protein from bacteria that were transformed with mouse DNA, they realized that bacteria could be used as microscopic factories.

In November 1980 a patent on the basic methods of cloning and transformation was awarded to Herbert Boyer and Stanley Cohen. A second patent granted the rights to any organisms that were engineered using the patented methods. Herbert Boyer also became co-founder of the biotechnology company Genentech, which offered shares of stock to the public in September 1980.

General Readings

C.W. Cowan and P.J. Watson, eds. 1992. *The Origins of Agriculture: An International Perspective*. Smithsonian Institution Press, Washington, D.C.

M.J. Crispeels and D.E. Sadava. 2002. *Plants, Genes, and Crop Biotechnology*, 2nd ed. Jones and Bartlett Publishers, Boston, Massachusetts.

R.S. MacNeish. 1992. *The Origins of Agriculture and Settled Life*. University of Oklahoma Press, Norman, Oklahoma.

E. Oura, H. Suomalainen, and R. Viskari. 1982. Breadmaking. In A.H. Rose, ed. *Fermented Foods*. Academic Press, New York, pp. 87–146.

J.D. Watson and J. Tooze. 1981. *The DNA Story*. W.H. Freeman and Company, San Francisco.

H. Wilson. 1988. *Egyptian Food and Drink*. Shire Egyptology, Bucks, United Kingdom.

D. Zohary and M. Hopf. 1993. *Domestication of Plants in the Old World*, 2nd ed. Clarendon Press, Oxford.

Additional Readings

O.T. Avery, C.M. MacLeod, and M. McCarty. 1944. Studies on the chemical nature of the substance inducing transformation of *Pneumococcal* types. *J. Exp. Med.* 79:137–158.

G.W. Beadle and E.L. Tatum. 1941. Genetic control of biochemical reactions in *Neurospora*. *Proc. Natl. Acad. Sci. USA* 27:499–506.

R.W. Beck. 2000. *A Chronology of Microbiology in Historical Context*. ASM Press, Washington, D.C.

B. Bracegirdle. 1989. Microscopy and comprehension: The development of understanding of the nature of the cell. *Trends Biochem. Sci.* 14:464–468.

T.D. Brock. 1999. *Milestones in Microbiology*. ASM Press, Washington, D.C.

A. Claude. 1975. The coming of age of the cell. *Science* 189:433–435.

D. Fredrickson. 2001. *The Recombinant DNA Controversy: A Memoir*. ASM Press, Washington, D.C.

J.S. Fruton. 1976. The emergence of biochemistry. *Science* 192:327–334.

F. Griffith. 1928. Significance of pneumococcal types. *J. Hygiene* 27:113–159.

J.R. Harlan. 1971. Agriculture origins: Centers and noncenters. *Science* 174:468–474.

J.R. Harlan and J.M.J. De Wet. 1973. On the quality of evidence for origin and dispersal of cultivated plants. *Curr. Archaeol.* 14:51–62.

J.R. Harlan and D. Zohary. 1966. Distribution of wild wheats and barleys. *Science* 153:1074–1080.

A.D. Hershey, A. Chase, and M. Chase. 1952. Independent functions of viral protein and nucleic acid in growth of bacteriophage. *J. Gen. Physiol.* 36:39–56.

G.C. Hillman and M.S. Davies. 1990. Measured domestication rates in wild wheats and barley under primitive cultivation and their archaeological implications. *J. World Prehist.* 4:157–222.

B.A. Law. 1982. Cheeses. In A.H. Rose, ed. *Fermented Foods*. Academic Press, New York.

M. McCarty and O.T. Avery. 1946. Studies on the chemical nature of the substance inducing transformation of pneumococcal types. II. Effect of deoxyribonuclease on the biological activity of the transforming substance. *J. Exp. Med.* 83:89–96.

J.H. Quastel. 1984. The development of biochemistry in the 20th century. *Can. J. Cell Biol.* 62:1103–1110.

D. Rindos. 1980. Symbiosis, instability, and the origins and spread of agriculture: A new model. *Curr. Anthropol.* 21:751–772.

W.S. Sutton. 1903. The chromosomes in heredity. *Biol. Bull.* 4:231–251.

A. Tannahill. 1973. *Food in History*. Stein and Day, New York.

E.R. Vedamuthu. 1982. Fermented Milks. In A.H. Rose, ed. *Fermented Foods*. Academic Press, New York, pp. 199–226.

J.D. Watson and F.H.C. Crick. 1953. A structure for deoxyribose nucleic acid. *Nature* 171:737–738.

J.D. Watson and F.H.C. Crick. 1953. General implications of the structure of deoxyribonucleic acid. *Nature* 171:964–967.

First Recombinant DNA Experiments

A.C.Y. Chang and S.N. Cohen. 1974. Genome construction between bacterial species *in vitro*: Replication and expression of *Staphylococcus* plasmid genes in *Escherichia coli*. *Proc. Natl. Acad. Sci. USA* 71:1030–1034.

S.N. Cohen, A.C.Y. Chang, H.W. Boyer, and R.B. Helling. 1973. Construction of biologically functional bacterial plasmids *in vitro*. *Proc. Natl. Acad. Sci. USA* 70:3240–3244.

J. Hedgpeth, H.M. Goodman, and H.W. Boyer. 1972. DNA nucleotide sequence restricted by the RI endonuclease. *Proc. Natl. Acad. Sci. USA* 69:3448–3452.

D.A. Jackson, R.H. Symons, and P. Berg. 1972. Biochemical method for inserting new genetic information into DNA of Simian Virus 40: Circular SV40 DNA molecules containing lambda phage genes and the galactose operon of *Escherichia coli. Proc. Natl. Acad. Sci. USA* 69:2904–2909.

J.L. Marx. 1976. Molecular cloning: Powerful tool for studying genes. *Science* 191:1160–1162.

J.E. Mertz and R.W. Davis. 1972. Cleavage of DNA by RI restriction endonuclease generates cohesive ends. *Proc. Natl. Acad. Sci. USA* 69:3370–3374.

J.F. Morrow, S.N. Cohen, A.C.Y. Chang, H.W. Boyer, H.M. Goodman, and R.B. Helling. 1974. Replication and transcription of eukaryotic DNA in *Escherichia coli. Proc. Natl. Acad. Sci. USA* 71:1743–1747.

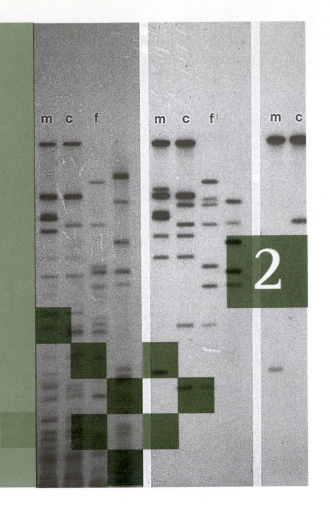

FROM DNA TO PROTEINS

2

Many scientists working in the area of biotechnology manipulate the genetic material of organisms. To do this, they must have an understanding of gene structure and function and how genes are controlled. In addition to knowing the cell- and tissue-specific genes that encode important traits, they must have also identified and studied the regulatory sequences that control gene activity. This chapter presents a brief review of the basic principles of molecular biology, paying particular attention to DNA structure, the replication of DNA, the transmission of information from DNA for protein synthesis, and the regulation of product synthesis. For a more detailed review of genes and their regulation, consult a comprehensive molecular biology textbook; a few are listed in the readings at the end of the chapter.

BASIC UNIT OF LIFE: THE CELL

The cell is the basic unit of a living organism. All organisms are composed of one or more cells, each carrying out many different functions. To manipulate living cells and organisms, scientists must have a comprehensive understanding of cell structure and function. Sci-

entists who study cell structure and function are called cell biologists. Many scientific discoveries have contributed to the advancement of cell biology. Three fields have contributed to the field of modern cell biology: cytology (the study of cell structure using microscopy), biochemistry (the study of the chemistry of living cells and organisms), and genetics (the study of the transmission of genetic information).

Prokaryotes and Eukaryotes

Two different general groups of organisms exist based on the characteristics of their cells: prokaryotes (eubacteria, archaea) and eukaryotes (nonbacterial groups). The primary distinction between the two organisms is the membrane-bound nucleus in eukaryotes (*eu* is Greek for "true," and *karyon* means "nucleus") and the lack of a membrane in prokaryotes (*pro* means "before"). Cells of prokaryotic organisms are called prokaryotic cells (Figure 2.1a), whereas those of eukaryotes are referred to as eukaryotic cells (Figures 2.1b and c). The following key features differentiate prokaryotic and eukaryotic cells. Additional distinguishing characteristics can be found in a cell biology text and in Appendix E.

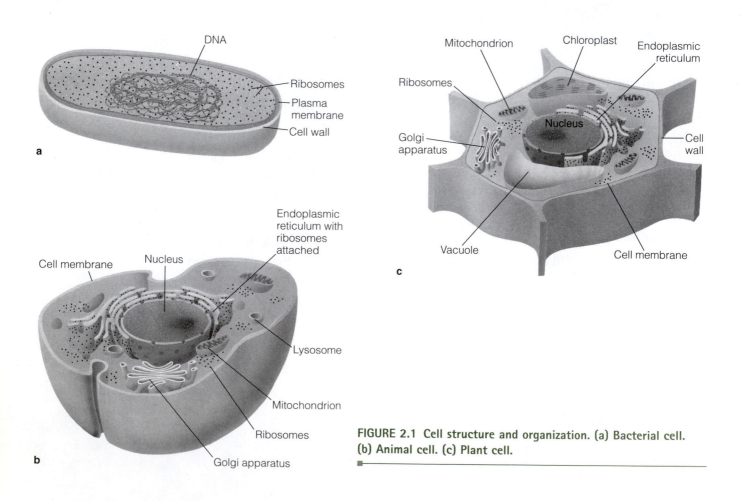

FIGURE 2.1 Cell structure and organization. (a) Bacterial cell. (b) Animal cell. (c) Plant cell.

Characteristic	Prokaryotic Cells	Eukaryotic Cells
True nucleus	No, nucleoid	Yes
Organelles	No	Yes
DNA	Circular, bound with few proteins	Linear chromosomes, complexed with many proteins

Common features of eukaryotic cells include

Organelles and their functions

Nucleus	Location of the DNA, chromosomes
Endoplasmic reticulum	Routes and modifies certain newly synthesized polypeptides, synthesizes lipids
Golgi body (Dictyosome in plants)	Modifies polypeptides, sorts and ships proteins and lipids for either secretion or for use inside the cell.
Mitochondria	Produces ATP (chemical form of energy cells use)
Chloroplasts	Site of photosynthesis in plants and algae

Other structures and their functions

Vesicles	Many functions, for example, transports or stores various substances, digestion
Ribosomes	Aids in the assembly of polypeptides during protein synthesis
Cytoskeleton	Confers shape to cells and aids in the internal organization of the cell, part of the movement mechanism in motile cells and motile structures of cells (for example, flagella); aids in the movement of internal structures (for example, chromosomes during cell division)

Macromolecules

The organelles of eukaryotes (membrane-bound structures in cells that conduct specific cell functions, for example, mitochondria) and other structures of both eukaryotic (for example, plant cell wall) and prokaryotic cells (cytoplasm) are composed of different types of polymers called macromolecules. Macromolecules are formed by the chemical bonding of monomers (single units of a molecule such as glucose). Important cellular macromolecules include polysaccharides, proteins, and nucleic acids. Lipids are generally thought of as macromolecules; however, they are constructed differently (not from repeating monomeric subunits). Lipids, polysaccharides, proteins, and nucleic acids are assembled into larger sub-cellular structures (for example, chromosomes—DNA and proteins) and enable the cell to function as a well-coordinated unit.

Lipids   Although lipids are not synthesized through the step-wise addition of monomeric subunits to form a polymer as are polysaccharides, proteins, and nucleic acids, they are macromolecules in terms of their large size. Lipids define a broad group of hydrophobic (generally insoluble in aqueous solutions, water-fearing) organic compounds. Because of this characteristic they play important roles as cellular membranes and define the boundaries of cells, organelles, and a variety of other membrane-bound structures that include vesicles and photosynthetic membranes. Lipids are chemically diverse and, therefore, play other roles in cells. For example, some lipids store energy. Typically, several classes of lipids are recognized, each with different structures and functions: triglycerides (Figure 2.2a), phospholipids (Figure 2.2b), glycolipids (for example, component of membranes of nerve tissue), steroids (for example, cholesterol), and terpenes (for example, a derivative is a component of vitamin A).

Fatty acids are the primary components of many lipids and are characterized by having a long, unbranched chain of carbon, hydrogen, and oxygen atoms (called a hydrocarbon chain) (Figure 2.3). Fatty acids are named by their hydrocarbon chain length ($C_nH_{2n}O_n$). For example, an 18-carbon chain (C_{18}) is called stearic acid. Fatty acids can be saturated (all carbons have all available sites bound with a hydrogen atom) or unsaturated (one or more double bonds reduces the number of hydrogen atoms bound to a particular carbon atom). An 18-carbon chain with one double bond is called oleic acid, whereas linolenic acid is an 18-carbon fatty acid with three double bonds (Figure 2.3). Double bonds affect the overall shape of the fatty acid. When a double bond is introduced in the chain, the straight chain kinks so that the hydrocarbon chains cannot pack tightly together. This influences the overall characteristics of the lipid. For example, lipids are liquid at room temperature if they have desaturated fatty acids, whereas solids such as butter are highly saturated. An example of a highly saturated lipid is a triglyceride (Figure 2.2a).

Polysaccharides   Polysaccharides are composed of long, repeating monomeric units called simple sugars, or monosaccharides (Figure 2.4), that function either as structure (for example, cellulose in plant cell walls) or storage (for example, starch and glycogen that can be broken down for energy). Generally, a polysaccharide is made of one type of monosaccharide or two that alternate. For example, cellulose is made of long, repeating units of glucose (Figure 2.5), whereas insect exo-

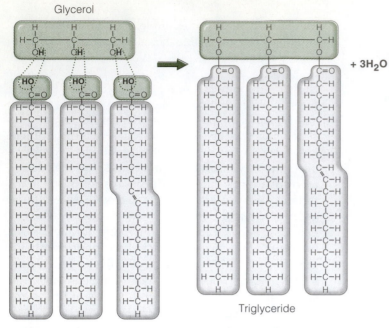

a   Three fatty acid tails

+ 3H₂O

Triglyceride

FIGURE 2.2 Two different classes of lipids have fatty acids in their structure. (a) Triglycerides are characterized by having a three-carbon–OH backbone called glycerol with a fatty acid chain attached to each carbon. These lipids store energy and are the main component of fat, and (b) phospholipids are structurally similar to triglycerides except one fatty acid chain is replaced by a phosphate. A hydrophilic (water-associating) group is bound to the phosphate. This group can vary—this example is choline and the lipid is phosphatidylcholine. Phospholipids are the main component of membranes (c) and form a lipid bilayer (for example, cell membrane) in which the hydrophilic heads associate with water and the hydrophobic (water-fearing) hydrocarbon tails associate with each other within the interior of the lipid bilayer.

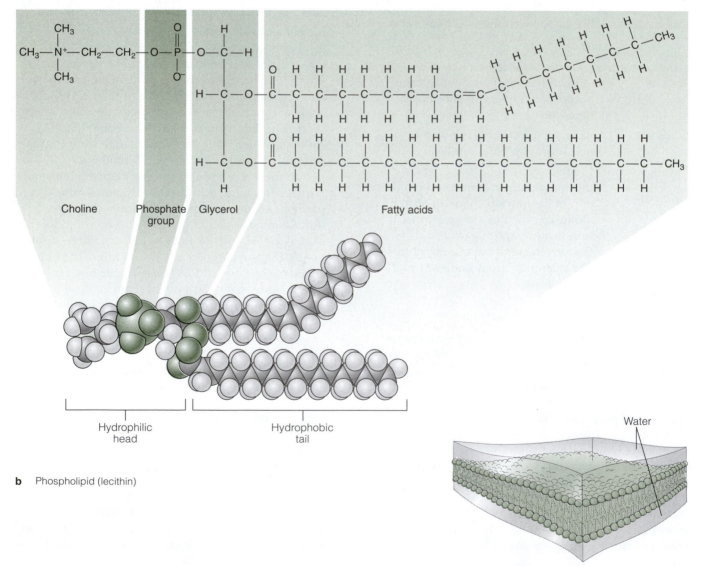

Choline   Phosphate group   Glycerol   Fatty acids

Hydrophilic head   Hydrophobic tail

b   Phospholipid (lecithin)

Water

c   Phospholipid bilayer

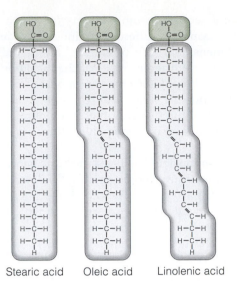

Stearic acid   Oleic acid   Linolenic acid

FIGURE 2.3 Structure of fatty acids showing the carbon backbone. In stearic acid, the carbon chain is completely saturated with hydrogen atoms. Oleic acid has one double bond in the chain and is called an unsaturated fatty acid. Linolenic acid has three double bonds in the carbon chain and is a polyunsaturated fatty acid.

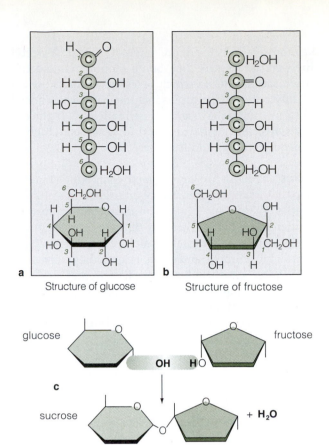

a   Structure of glucose   b   Structure of fructose

glucose   fructose

c

sucrose $+ H_2O$

FIGURE 2.4 Monosaccharides, (a) glucose and (b) fructose shown as both straight–chain and ring forms. The carbon atoms are numbered. (c) The formation of a disaccharide from two joined monosaccharides.

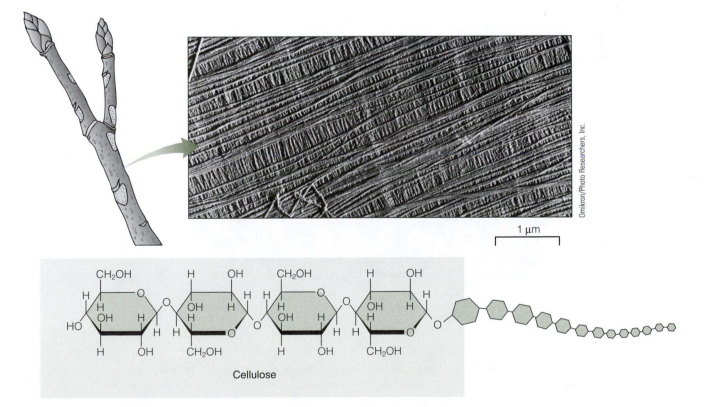

1 μm

Cellulose

Omikron/Photo Researchers, Inc.

FIGURE 2.5 Cellulose, a structural polysaccharide composed of glucose monomers, found in plant cell walls.

skeletons (called chitin) are made of units of *N*-acetyl glucosamine (Figure 2.6). Some monomers also can form disaccharides in which two simple sugars are covalently bonded. Examples include lactose composed of glucose and galactose, maltose made of two glucose molecules, and the disaccharide sucrose, which is formed by glucose and fructose.

Proteins Proteins are large organic compounds that are major cellular determinants of an organism's characteristics. Within a cell may be hundreds to thousands of different proteins. The many types of proteins found in cells include enzymes that catalyze cellular reactions for both biosynthetic and catabolic activities, hormones that help regulate cellular metabolic activities, antibodies involved in immune response, transcription factors that help control transcriptional (RNA synthesis) activity, and structural proteins such as tubulin and actin that determine cell shape (for example, the cytoskeleton) and provide motility to cells and organisms (for example, flagella). Other classes of proteins include regulatory proteins that control various processes within the cell.

Protein-encoding DNA sequences (genes) make up only about 10% of the human genome; the rest consists of sequences such as introns (noncoding regions within genes), short repetitive DNA sequences, gene sequences for RNA such as ribosomal (rRNA) and transfer RNAs (tRNA) for protein synthesis, and regulatory or control sequences. The specific functions of many repetitive sequences are unknown.

Proteins are composed of monomers called amino acids that are joined by covalent bonds to form a protein polymer (called a polypeptide). Twenty different types of amino acids are used in the synthesis of proteins (Figure 2.7). The great diversity of proteins in organisms is due to the variability of polypeptide length,

the many sequence combinations (or linear order) possible from 20 different amino acids, and the three-dimensional conformations after folding of the molecules (different types of chemical bonds and interactions help determine how a protein will fold). Some functional proteins are composed of a single polypeptide chain (for example, myoglobin), whereas others are made up of many polypeptides (for example, hemoglobin).

Each amino acid has the same basic backbone; its unique side group (represented by *R*) determines its chemical characteristics. The chemical properties of an amino acid's side group determine to which of four main classes the amino acid is assigned: nonpolar, uncharged polar, negatively charged (acidic) polar, or positively charged (basic) polar. Amino acids are joined by covalent bonds **(peptide bonds)** between the carbon atom in the carboxyl group of one amino acid and the nitrogen atom of the amino group of the adjacent amino acid (Figure 2.8). The resulting polypeptide has directionality: one end terminates with a free amino group ($-NH_2$), the amino terminus, and the other terminates with a free carboxyl group ($-COOH$), the carboxyl terminus. The combined chemical properties of the amino acid side groups (and three-dimensional shape based on folding) determine the chemical nature of the protein.

Nucleic Acids Nucleic acids are macromolecules that are involved in the storage and transmission of genetic information within the cell. They consist of long polymers that are composed of repeating units of **nucleotides.** Four different nucleotides make up a poly-

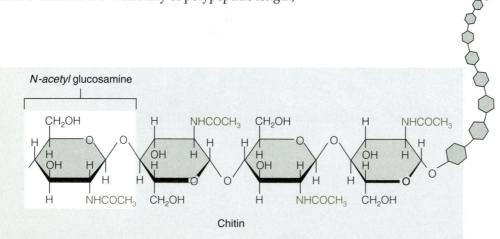

FIGURE 2.6 Chitin, a structural polysaccharide composed of *N*–acetyl glucosamine monomers, is found in the exoskeletons of insects.

Amino acid

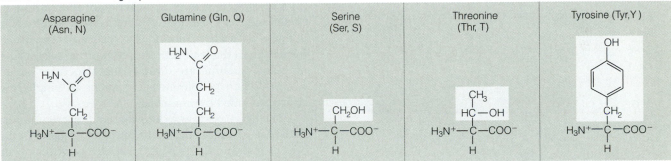

Amino acids with basic side chains

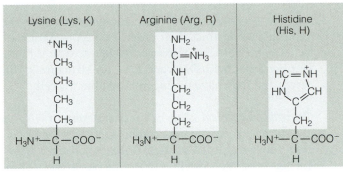

Amino acids with acidic side chains

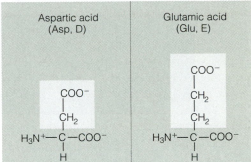

Amino acids with uncharged polar side chains

Amino acids with nonpolar side chains

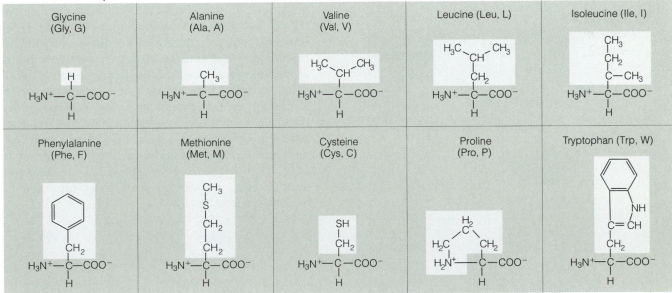

FIGURE 2.7 The 20 amino acids used in protein synthesis. The shaded boxes highlight the side chains that determine the chemical characteristics of the amino acid (and ultimately the polypeptide). The three- and one-letter abbreviations are shown. The basic structure of an amino acid is shown with the R group representing the different side chains.

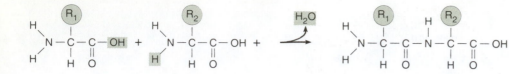

FIGURE 2.8 Peptide bond formation between two amino acids. R represents the side chain of each amino acid.

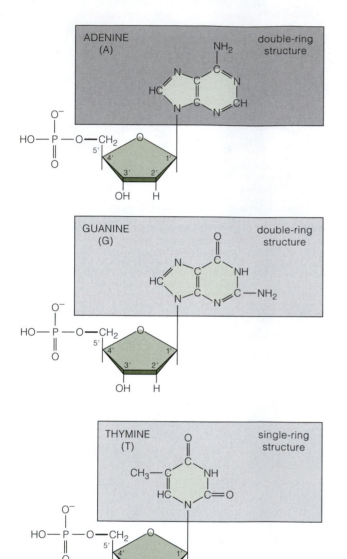

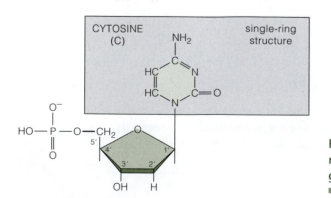

FIGURE 2.9 The four nucleotides of DNA. The numbers 1′–5′ refer to the carbon atoms on the pentose sugar to which other groups are attached.

mer; the specific genetic information stored within the molecule determines their order. Two types of nucleic acids exist: DNA, or deoxyribonucleic acid, and RNA, or ribonucleic acid. DNA and RNA are structurally different and they have distinct roles within the cell.

Deoxyribonucleic Acid The primary function of DNA is the storage of genetic information. DNA is a long polymer consisting of repeating units called deoxyribonucleotides. A deoxyribonucleotide has three components: (1) a pentose sugar (five carbons), or deoxyribose; (2) a phosphate group; and (3) one of four nitrogen-containing bases (Figure 2.9): adenine (A), guanine (G), thymine (T), and cytosine (C). DNA stores its genetic information in the four nitrogen-containing bases; adenine and guanine are double-ring structures called purines, whereas thymine and cytosine are single-ring structures called pyrimidines.

The DNA exists as a double-strand macromolecule with base pairing between the nitrogenous bases of the opposite strands. The bases project inward from the sugar–phosphate backbone, and hydrogen bonding between opposite bases (that is, one on each DNA strand) holds the two strands of the DNA molecule together (Figure 2.10). The **x-ray diffraction pattern** of B-DNA (the most common conformation) shows a helical configuration, with both strands of the molecule winding around a common central axis to form a spiral (like a spiral staircase). Two grooves are formed by this conformation: a minor groove and a wider, major groove (Figure 2.11). In this most common DNA conformation, a deoxyribonucleotide is located every 3.4 Å, with approximately 10.5 nucleotides for every helical turn of 34 Å. Each B-DNA double helix is 20 Å in diameter. Other isomers of DNA are present *in vitro*; however, most do not exist in cells.

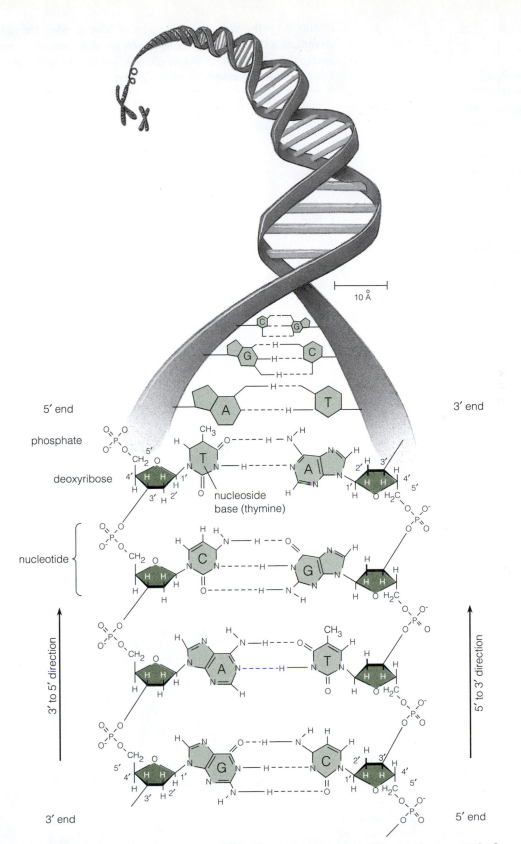

FIGURE 2.10 The double-strand structure of DNA illustrating the nucleotides, each composed of a deoxyribose sugar, a phosphate, and a nitrogen-containing base. The two strands are held together by hydrogen bonds (dashed lines) between pairs of bases. Guanine (G) and cytosine (C) are held together by three hydrogen bonds, and adenine (A) and thymine (T) by two. The two strands are antiparallel, because one strand is 5′ to 3′ in one direction and the complementary strand is 3′ to 5′.

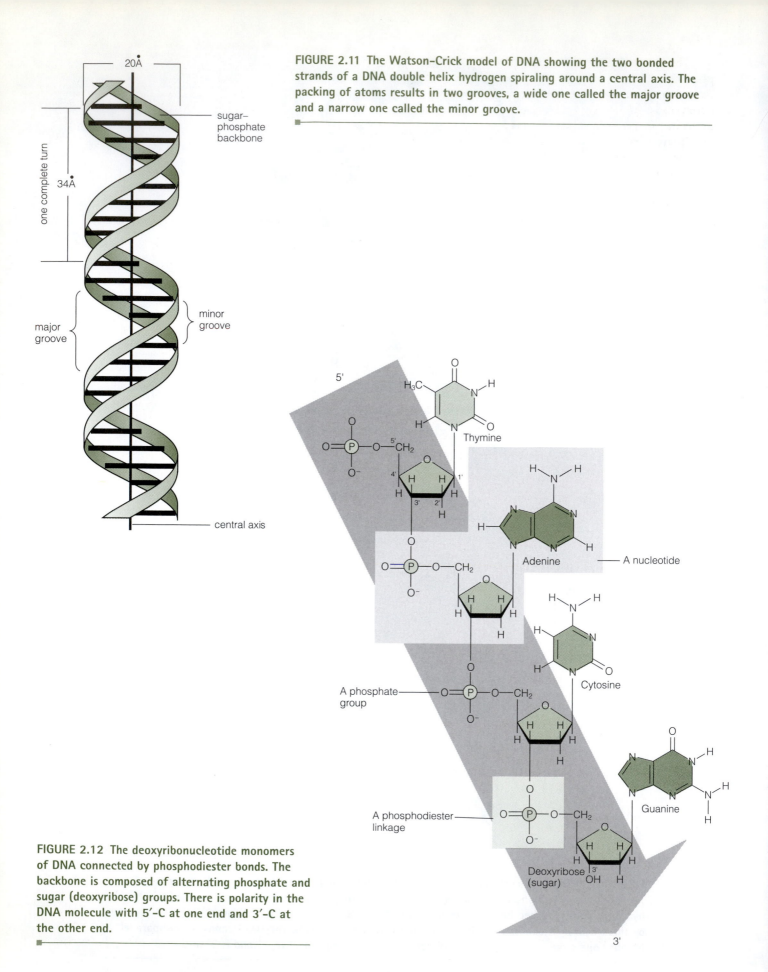

FIGURE 2.11 The Watson–Crick model of DNA showing the two bonded strands of a DNA double helix hydrogen spiraling around a central axis. The packing of atoms results in two grooves, a wide one called the major groove and a narrow one called the minor groove.

20Å

one complete turn

34Å

sugar–phosphate backbone

major groove

minor groove

central axis

5'

H₃C

Thymine

Adenine — A nucleotide

A phosphate group

Cytosine

A phosphodiester linkage

Guanine

Deoxyribose (sugar)

3'

FIGURE 2.12 The deoxyribonucleotide monomers of DNA connected by phosphodiester bonds. The backbone is composed of alternating phosphate and sugar (deoxyribose) groups. There is polarity in the DNA molecule with 5'-C at one end and 3'-C at the other end.

The deoxyribonucleotides are linked by 3', 5' phosphodiester bonds; that is, the phosphate group attached to the number 5 carbon atom (5' carbon) of the deoxyribose of one deoxyribonucleotide is connected to the number 3 carbon (3' carbon) of the adjacent deoxyribonucleotide (Figure 2.12). Alternating deoxyribose and phosphate make up the backbone of DNA. The end of the nucleic acid strand that terminates with the 5' carbon atom is called the 5' end and the end with the 3' carbon atom is the 3' end. Thus, there is a 5' end and a 3' end of the DNA polymer. Nucleotide sequences are always written from the 5' end to the 3' end of the DNA (and RNA) polymer.

Complementary Base Pairing   Nucleotides base pair by hydrogen bonding. Because a purine always pairs with a pyrimidine, a constant helix diameter is maintained. Adenine always base pairs with thymine through two hydrogen bonds, and cytosine always base pairs with guanine through three hydrogen bonds. Because each strand of the helix is a complement of the other, such base pairing is called *complementary pairing*. For hydrogen bonding of the bases, the strands must be antiparallel to one another so that one strand goes from 5' to 3' and the other from 3' to 5'. Thus, the DNA molecule has polarity: the ends differ from one another.

The two strands do not spontaneously separate under physiological conditions because the many hydrogen bonds keep the base pairs together. However, high temperatures (near boiling) or pH extremes (pH < 3 or > 10) can break hydrogen bonds so that the two strands separate, or are denatured. If the temperature is gradually lowered (for example, to 65° C), the complementary strands can recombine or reanneal. Proteins also can separate the strands of a DNA helix (see the following discussion on DNA replication).

Ribonucleic Acid   RNA has roles in the transmission of genetic information stored within the genes of DNA: (a) the expression of genetic information as messenger RNA (mRNA) that encodes the amino acid sequences of polypeptides in a process referred to as **transcription** and (b) roles in the synthesis of protein in a process called **translation**.

RNA resembles DNA in all ways except the following structural features (Figure 2.13):

1. The base uracil (U) substitutes for thymine (T), and pairs with adenine (A) (Figure 2.14).

2. The pentose sugars of the backbone are ribose molecules instead of deoxyribose.

3. RNA is a single strand.

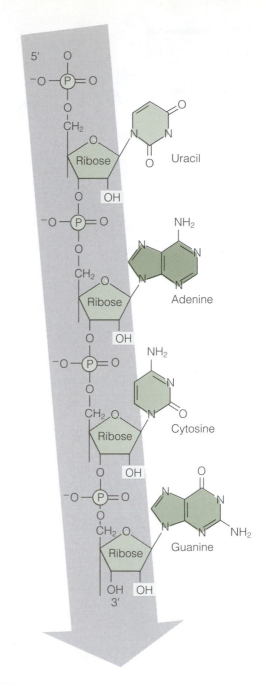

FIGURE 2.13 The nucleotide structure of RNA. Each nucleotide is connected by a phosphodiester bond. In RNA uracil replaces DNA's thymine.

There are, however, exceptions to the last statement: sometimes short nucleotide sequences that are complementary to each other within the RNA molecule base pair to form short double-strand regions. This intramolecular base pairing may be needed to maintain the integrity of the molecule or to permit some RNA functions (described later). Thus, RNA undergoes complex folding with secondary and tertiary structures

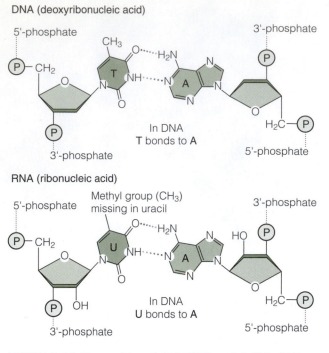

FIGURE 2.14 Base pairing, A–T in DNA and A–U in RNA.

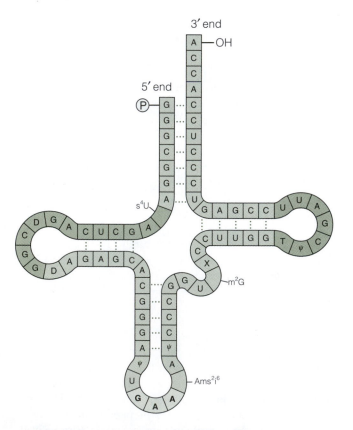

FIGURE 2.15 A mature transfer RNA showing the anticodon and 3′ amino acid binding regions.

characteristic of the type of RNA. Most RNA molecules are much shorter than DNA molecules, and although DNA is always present within the cell, RNA molecules are present only transiently and are degraded after a specific period of time.

Different genes encode functionally and structurally distinct types of RNAs. Some genes encode mRNA; others encode tRNA and **ribosomal RNA** (rRNA). Although mRNA encodes the amino acid sequence of the protein or polypeptide to be synthesized, both tRNA and rRNA molecules (which do not code for protein) are required for protein synthesis. tRNA, a small nucleic acid of approximately 75 ribonucleotides, folds into a secondary structure that resembles a cloverleaf (Figure 2.15). In the cell, tRNAs serve as adapter molecules by carrying the appropriate amino acid to the site of protein synthesis.

There are several lengths, or sizes, of rRNA. These are given by their S value, or Svedberg unit—their relative sedimentation rates during centrifugation: 5 S, 16 S, and 23 S in prokaryotes, and 5 S, 5.8 S, 18 S, and 28 S in eukaryotes. Ribosomes are large cytoplasmic structures formed when rRNAs combine with a large number of proteins. A ribosome is the workbench of protein synthesis, where amino acids covalently bond to form polypeptides. Ribosomes consist of two subunits, a large and a small, which come together in the cytoplasm to form the mature ribosome (see the following discussion on translation).

CENTRAL DOGMA OF MOLECULAR BIOLOGY

As we have discussed, genetic information is stored within DNA and retrieved through an RNA intermediate. Specifically, a discrete segment of DNA called a **gene** is a template for the synthesis of a complementary RNA. If the gene encodes the amino acid sequence of a specific polypeptide, the RNA intermediate is mRNA. (Sometimes a gene encodes information for rRNA or tRNA. This is the final gene product.) The information in a gene cannot be translated directly into a polypeptide. The process takes two steps— **transcription** and **translation**. Thus, the direction of flow of genetic information is from DNA to RNA to polypeptide. This flow of information is known as the central dogma of molecular biology (Figure 2.16).

DNA REPLICATION

When a cell divides to yield two daughter cells, the genetic material must be reproduced accurately, or replicated, so that each daughter cell contains identical DNA copies. Accuracy in replicating is essential, because the DNA stores the genetic information that the cell uses. During mitosis, each chromosome moves to opposite sides of the dividing cell, and cell division

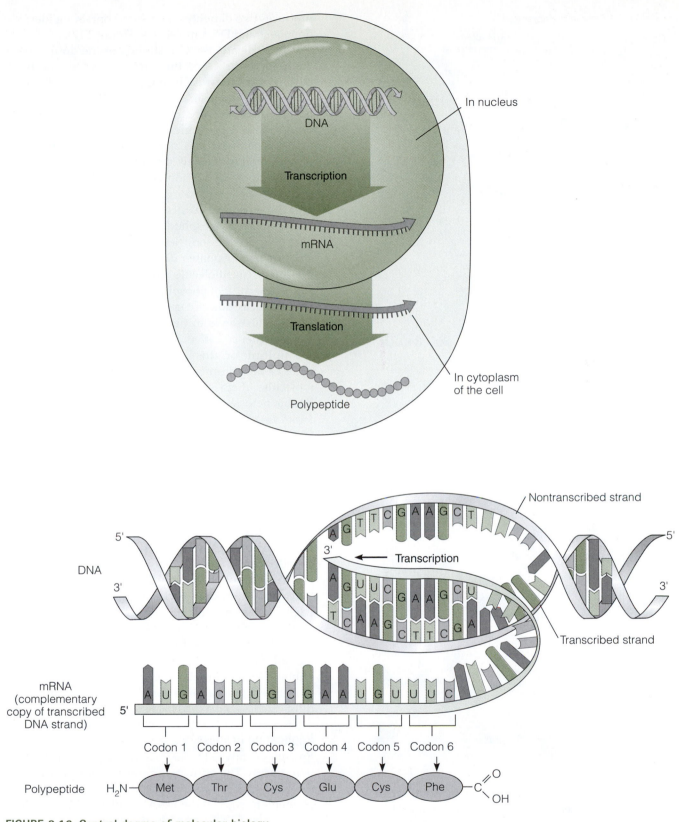

FIGURE 2.16 Central dogma of molecular biology.

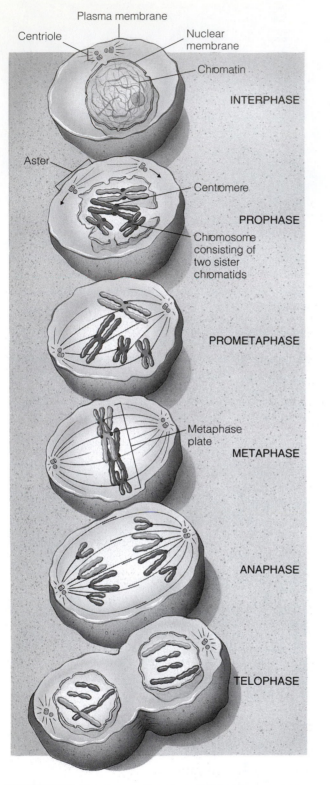

Plasma membrane

Centriole

Nuclear membrane

Chromatin

INTERPHASE

Aster

Centromere

PROPHASE

Chromosome consisting of two sister chromatids

PROMETAPHASE

Metaphase plate

METAPHASE

ANAPHASE

TELOPHASE

FIGURE 2.17 The stages of nuclear division in a process called mitosis, and cell division. In prophase, two DNA molecules make up each chromosome (each one of the pair is called a sister chromatid). During mitosis, the DNA molecules (sister chromatids) of each chromosome separate (during anaphase) and each one (now called a chromosome) ends up in one of the two resulting cells.

produces two daughter cells, each containing identical double-strand DNA molecules (Figure 2.17).

To replicate, the two strands of the double helix must separate. Once they are separated, the base-pairing rules allow each single-strand molecule to act, during replication, as a template for the formation of a new complementary strand. Each of the two identical double-strand molecules thus formed consists of one original, parent strand and one new, daughter strand. Because half of the new molecule is original material, this mode of replication is called semiconservative (Figure 2.18).

The process of DNA synthesis requires energy. The energy to form phosphodiester bonds comes from the incoming deoxyribonucleotide supplied as a deoxynucleoside triphosphate rather than a nucleoside monophosphate (the molecule actually incorporated into the DNA strand). Two phosphates are cleaved from the deoxyribonucleotides for energy.

Replication is initiated at a specific point in the DNA sequence called an origin of replication. There are multiple origins in linear eukaryotic DNA molecules of the chromosomes. In circular bacterial DNA molecules, there is one origin of replication. The DNA double helix separates at this site and two **replication forks** are created by the separation of the two strands (Figure 2.19). When replication proceeds from the two replication forks, it is called bidirectional replication. In bacterial DNA replication, a single origin proceeds around a circle bidirectionally until two double-strand DNA molecules are produced. This is called *theta* replication because the shape during replication resembles the Greek letter theta (θ). This is also true for the circular DNA molecules of the organelles, mitochondria and chloroplast.

In eukaryotic DNA, multiple origins produce multiple replication sites, each replicating bidirectionally. Each replicating unit of DNA, called a replicon, extends outward until it meets a neighboring replicon. These two replicons then fuse. Each replicon resembles a bubble and is often referred to as a replication bubble. When all replication bubbles of a linear DNA have fused, replication is complete. The DNA has been duplicated.

Replication Process

Enzymes catalyze the successive stages of the replication process. The primary enzyme is **DNA polymerase**, a large enzyme that catalyzes the synthesis of a complementary copy of the DNA template during replication. During DNA synthesis, deoxyribonucleotides are added to the 3'-OH end of the growing daughter strand. Therefore, the new strand grows at the 3' end (that is, in the 5' to 3' direction). Each incoming nucleotide (dATP, dCTP, dTTP, or dGTP) must be a complement of the template nucleotide at that site (for ex-

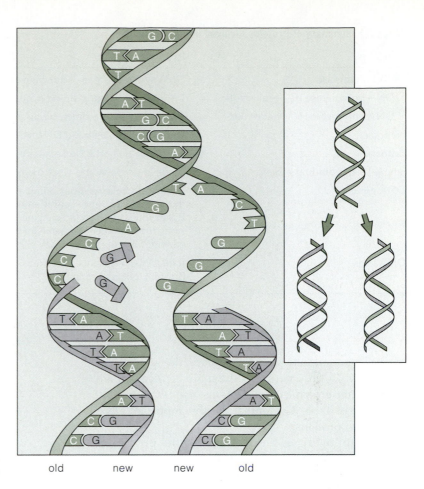

FIGURE 2.18 Semiconservative replication of DNA. Each newly replicated DNA molecule is composed of one parent (old) and one daughter (new) strand.

old   new   new   old

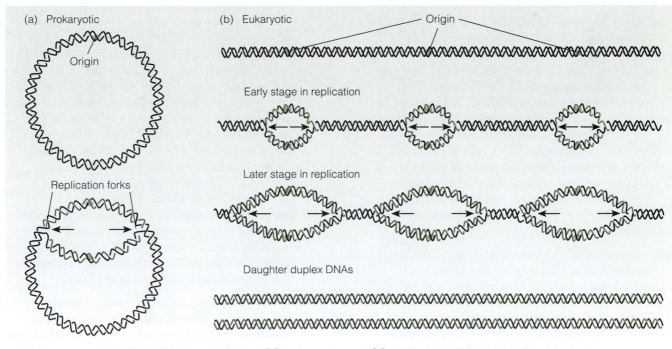

FIGURE 2.19 Bidirectional replication of DNA in (a) prokaryotes and (b) eukaryotes. The origins of replication are shown. In prokaryotes there is one origin and two replication forks. In eukaryotes, there are many origins along the DNA of a chromosome, each with two replication forks.

TABLE 2.1 Summary of Proteins Involved in DNA Replication

Protein	Functions
DNA Polymerase	DNA synthesis, 5′ → 3′ direction of new strand; requires RNA primer
DNA polymerase III—prokaryotic	Leading and lagging strand synthesis
DNA polymerase I—prokaryotic	Removal of RNA primer; filling the resulting gaps with nucleotides
Helicase	Unwinds double-strand DNA at the replication fork
Primase	Synthesis of short RNA sequence; primer for DNA synthesis
Single-strand binding protein	Binds to single-strand DNA to keep strands from base pairing
DNA gyrase	In bacteria; relaxes supercoiling ahead of the DNA replication fork that is caused by the separation of the DNA helix during replication. Eukaryotic cells have an enzyme that is similar and also serves the same function (general name for both prokaryotic and eukaryotic enzymes is a topoisomerase).
DNA ligase	Joins DNA fragments and Okazaki fragments of the lagging strand during DNA replication

ample, A in template strand, T in daughter strand). Although DNA polymerase is the main replication enzyme, other enzymes and proteins also play a role in this process. Table 2.1 summarizes the important DNA replication proteins.

The following is an overview of the process of replication in bacteria (Figure 2.20):

1. Initiation of replication. A small area of the DNA double helix is unwound at the origin of replication.

2. Unwinding of the DNA helix. Two enzymes, **DNA helicase** and a **DNA topoisomerase** called DNA **gyrase**, bind to the unwound region. The helicase continues to unwind the DNA, while DNA gyrase aids in separating the two DNA strands. Single-strand binding proteins (SSB) attach to each separated strand to stabilize them and to keep them from reannealing. The juncture where the two DNA strands separate forms a replication fork. DNA gyrase is present just ahead of the fork and relaxes any overwinding of the helix (supercoiling) caused by the unwinding of the DNA at the replication fork. Two replication forks move in opposite directions.

3. Synthesis of **RNA primers**. Although DNA polymerase is the main replication enzyme, this enzyme must initiate replication from an RNA sequence. A short complementary RNA sequence (called an RNA primer) must bind to the DNA. The RNA primer is synthesized by the enzyme **primase**. DNA polymerase (DNA polymerase I in bacteria) removes the RNA primers. The appropriate nucleotides are added where gaps are left by primer removal.

4. The enzyme **DNA polymerase** (referred to as DNA polymerase III in bacteria; DNA polymerase δ in eukaryotes) binds to each single strand and moves along the strand from the RNA primer, using information in the template DNA to mediate the formation of the new DNA strand. The appropriate nucleotides are added to the 3′ growing ends of the complementary daughter strands. The phosphate group of the 5′ end of the incoming nucleotide is added to the 3′ end of the growing DNA strand.

   DNA polymerase is a unidirectional enzyme, always moving along the template and reading the template DNA in a 3′ to 5′ direction, while synthesizing the new strand 5′ to 3′. Therefore, the DNA polymerase moves toward the replication fork on one strand (sometimes called the **leading strand**) and away from the fork on the other strand (the **lagging strand**). When DNA polymerase moves away from the replication fork, DNA synthesis must occur in short, discrete stretches called Okazaki fragments (Figure 2.21). (This process is sometimes called discontinuous synthesis.) The leading strand requires only one RNA primer because replication is a continuous process. However, because the lagging strand is synthesized in short stretches, an RNA primer must be made at the beginning of each Okazaki fragment.

5. Connection of Okazaki fragments to form one continuous DNA molecule. The enzyme **DNA ligase** connects the ends of the Okazaki fragments of the newly synthesized DNA.

Steps 1–5 continue until the DNA templates are replicated.

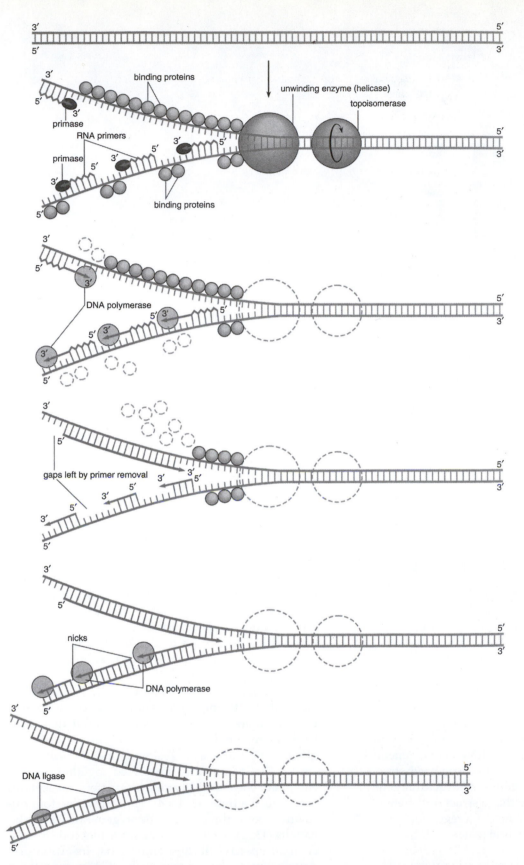

FIGURE 2.20 Replication of DNA at a replication fork in bacteria. The template is denatured, DNA helicase unwinds the DNA strands in front of the replication fork, and single-strand binding proteins bind to the single strands and keep them denatured. One strand serves as a template for discontinuous synthesis of the lagging strand and continuous DNA synthesis of the leading strand. Replication of the DNA strands proceeds in the opposite direction with leading strand replication moving in the direction of the fork. Primase adds short RNA primers to the lagging strand to prime replication, and DNA ligase produces a continuous DNA strand from discontinuous replication.

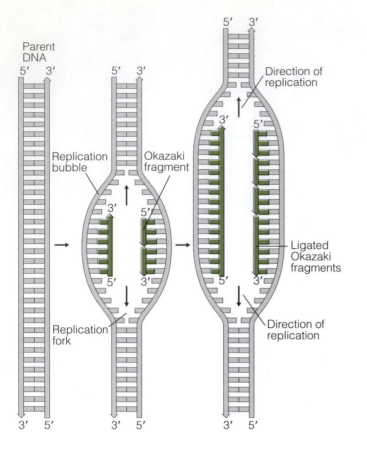

Parent
DNA

Replication
bubble

Okazaki
fragment

Replication
fork

Direction of
replication

Ligated
Okazaki
fragments

Direction of
replication

FIGURE 2.21 A DNA replication bubble showing the direction of replication in the two replication forks, leading and lagging strand synthesis, and Okazaki fragments in the lagging strand.

GENE: BASIC UNIT OF GENETIC INFORMATION

A gene is a discrete stretch of nucleotide bases on a strand of DNA that serves as a unit of information. The gene can reside on either strand and can range from a few hundred to many thousand nucleotide bases A, T, G, and C. (Noncoding regions separate the genes on a strand.) The number and sequence of the bases within a particular gene determine the information that the gene carries—that is, the amino acid sequence of the specific **polypeptide** or protein it will encode. Sometimes a gene encodes the information for RNA—and not a polypeptide (for example, tRNA). Two general classes of protein-encoding genes are recognized: structural genes and regulatory genes. A structural gene codes for a protein that has either a structural or enzymatic function, while the product of a regulatory gene controls the activity of a structural gene.

GENETIC CODE

According to the central dogma of molecular biology, the flow of genetic information is from DNA to RNA to protein. The nucleotide sequence of DNA and mRNA

specifies the amino acid sequence of a polypeptide. The genetic language of the gene is the nucleotide; however, the language of proteins is the amino acid. Because the linear order of the nucleotides is congruent with the order of amino acids in a protein, there is a direct translation from nucleotides to amino acids. The relationship of nucleotides to amino acids is known as the **genetic code**.

Although there are 20 amino acids, the four bases are sufficient to encode the amino acid sequence of a protein. Years ago, scientists discovered that three bases are required to encode the information for a single, specific amino acid. A simple calculation demonstrates that two nucleotides used together are not enough to encode all 20 amino acids (four nucleotides used two at a time: 42: $4 \times 4 = 16$). However, four nucleotides used three at a time can generate 64 combinations (43: $4 \times 4 \times 4 = 64$). Thus, a triplet code is sufficient to specify all 20 amino acids. Insertion and deletion mutagenesis experiments demonstrated a triplet code by determining the **reading frame** of the sequence. One, two, and three nucleotides were inserted or deleted to determine the effect on the reading frame of the encoded information. Addition or deletion of one or two nucleotides produced a subsequent change in the reading frame. The addition or deletion of three nucleotides close together restored the original reading frame. Therefore, nucleotides are read in threes for each amino acid. These triplets in the mRNA are called codons.

The genetic code is degenerate, because more than one codon can specify a particular amino acid (Figure 2.22 shows the amino acids arranged according to the extent of degeneracy). The protein synthesis (that is, translation) machinery reads these codons so that the appropriate amino acids are bonded to generate the correct protein encoded by the mRNA. Three stop codons (UAA, UAG, UGA) in the genetic code signal the end of the polypeptide and terminate translation. A start codon signifies the beginning of the polypeptide in translation. Methionine, specified by the codon AUG, is usually the first amino acid.

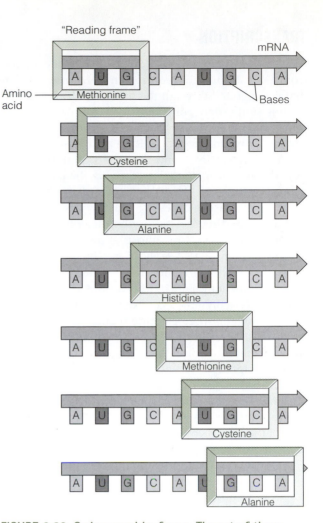

FIGURE 2.23 Codons read in-frame. The set of three nucleotides specifying an amino acid is called the reading frame. A ribosome moves along the mRNA and the specific nucleotide at which translation begins determines the amino acid sequence.

amino acids with one codon

AUG	UGG
Met	Trp

amino acids with two codons

AAA	AAC	CAA	CAC	GAA
AAG	AAU	CAG	CAU	GAG
Lys	Asn	Gln	His	Glu

GAC	UAC	UGC	UUC
GAU	UAU	UGU	UUU
Asp	Tyr	Cys	Phe

amino acids with three codons

AUA
AUC
AUU
Ile

amino acids with four codons

ACA	CCA	GCA	GGA	GUA
ACC	CCC	GCC	GGC	GUC
ACG	CCG	GCG	GGG	GUG
ACU	CCU	GCU	GGU	GUU
Thr	Pro	Ala	Gly	Val

amino acids with six codons

CGA	AGA	CUA	UUA	UCA	AGC
CGC	AGG	CUC	UUG	UCC	AGU
CGG		CUG		UCG	
CGU		CUU		UCU	
Arg		Leu		Ser	

FIGURE 2.22 The amino acids arranged according to the extent of degeneracy.

The correct protein sequence is synthesized only if the message is read in the appropriate reading frame (Figure 2.23). In-frame translation can occur only if the translational apparatus (that is, ribosome–aminoacyl-tRNA complex) initially binds to the correct codon at the translational start site. For example, suppose the following three-letter words represent codons:

THE BIG RED DOG WAS SAD

When read in-frame, the message makes sense. If reading is shifted one letter (or nucleotide), it becomes meaningless:

HEB IGR EDD OGW ASS AD

Thus, to generate the correct amino acid sequence for each polypeptide synthesized, the translation machinery must initiate translation at the correct start site and continue reading triplets in-frame.

TRANSCRIPTION

As we have discussed, the flow of genetic information from DNA to protein requires an RNA intermediate that is generated through the process of transcription. RNA is synthesized, or transcribed, from a DNA template by the main replication enzyme **RNA polymerase**. Three main types of RNA are transcribed from DNA: mRNA, rRNA, and tRNA. Through transcription, the genetic information stored in the DNA (gene) is used to make an RNA that is complementary. The RNA polymerase synthesizes the new RNA molecule in the 5′ to 3′ direction using ribonucleotides that are added to the 3′ end by phosphodiester bonds in a process similar to DNA replication. The DNA template is read by the enzyme in a 3′ to 5′ direction. Although the language of nucleotides is maintained, uracil replaces thymine as one of the nitrogenous bases.

By convention, nucleotide sequences of a gene or mRNA are referred to as *upstream* or *downstream* of a specific reference point. For example, in a gene region, the transcriptional start point of a gene is referred to as +1. Everything before that point is referred to as upstream of the gene and is denoted by negative numbers (for example, −10 refers to 10 nucleotides before the start of the gene). Downstream refers to the nucleotides after the reference point, and in the case of the transcriptional start point, the numbers would be positive (for example, +50).

In both eukaryotic and prokaryotic genes, RNA polymerase recognizes nucleotide sequences usually upstream, just before the start of the coding region, or gene. These sequences, or **promoters,** allow the RNA polymerase to be placed correctly on the DNA strand near the gene to be transcribed. Most promoters define the beginning of a gene and signal the start of transcription. Exceptions do exist; some eukaryotic gene promoters extend within the gene or are even completely within a gene, and, in prokaryotes, several genes are contiguous with one promoter located at the beginning of the transcription unit. The promoter is not transcribed and the promoter sequences of different types of genes or organisms (for example, plant versus bacteria) may vary.

As Figure 2.24 shows, the RNA polymerase binds to the promoter, the DNA strands separate, and the polymerase moves along the DNA template strand while reading the coding portion (sometimes called the antisense strand). In both eukaryotes and prokaryotes, proteins called transcription factors interact with the RNA polymerase to help regulate or promote transcription.

Transcription involves four stages:

1. Binding of the RNA polymerase to a specific sequence, called a promoter. The DNA helix unwinds in this region.

2. Initiation of transcription. RNA polymerase begins synthesizing RNA from the template strand of the DNA as the DNA helix unwinds farther.

3. Elongation of the RNA. As the RNA polymerase moves along the DNA strand in the 3′ to 5′ direction, the DNA helix unwinds and the two strands separate. The RNA elongates by the addition of ribonucleotides to the 3′ end of the newly synthesized RNA.

4. Termination of transcription. After the end of the gene is reached, termination occurs: The RNA polymerase disengages the DNA, and the new RNA molecule is released. At some termination sites in bacteria, specific transcription termination proteins, or *rho* proteins, are involved; at other sites, termination is signaled by the formation of a stem-loop structure at the end of the new RNA that forces the RNA polymerase to fall off the DNA template (Figure 2.25).

Promoters and RNA Polymerase

Bacterial promoters are approximately 30 nucleotides in length and are present near the beginning of the gene (Figure 2.26a). The most common bacterial promoter is composed of two highly conserved (among different genes) blocks of sequences approximately −10 and −35 nucleotides upstream from the beginning of the gene (transcriptional start site). These segments are called the −10 region (also called the Pribnow box) and −35 region. The *E. coli* RNA polymerase most likely recognizes and binds to the −35 region of the promoter sequence. The RNA polymerase also binds to the −10 region, which then denatures and becomes single stranded so that transcription can begin.

In bacteria, one type of RNA polymerase is present within the cell. A small transiently bound protein determines which type of promoter the enzyme will bind and thus which gene will be transcribed. This protein is called a sigma factor (σ) and ensures that the RNA polymerase binds at the appropriate site within the DNA. Thus, sigma factors provide the binding specificity of RNA polymerase to specific promoters. A single sigma factor binds to the RNA polymerase before transcription of a gene is initiated and is released immediately after transcription begins.

Promoters of genes in eukaryotic organisms are more complex than prokaryotic promoters and differ from them in sequence and location. Unlike the single RNA polymerase in prokaryotes, three types are present in eukaryotes, each one recognizing a different promoter.

1. RNA polymerase I transcribes ribosomal RNA genes for 5.8 S, 18 S, and 28 S.

2. RNA polymerase II transcribes genes encoding proteins (and most small nuclear RNAs). For eukaryotic genes encoding mRNA, the promoter is the TATA box, located approximately 25 base pairs upstream from the transcriptional start site. Many genes also have a CAAT box, approximately 75 base pairs upstream (Figure 2.26b).

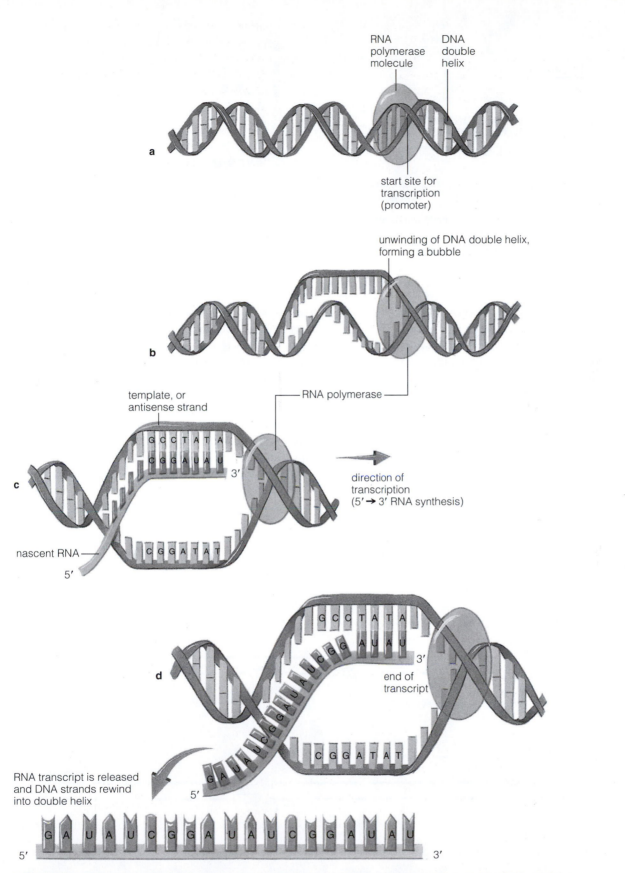

FIGURE 2.24 RNA synthesis. (a) Transcription is initiated when RNA polymerase binds to the DNA at the promoter region. **(b)** The double-strand DNA unwinds. **(c)** As the RNA polymerase travels along the DNA template, nucleotides are added to the growing RNA strand. **(d)** When RNA polymerase reaches a terminator, the RNA transcript is released and transcription is terminated.

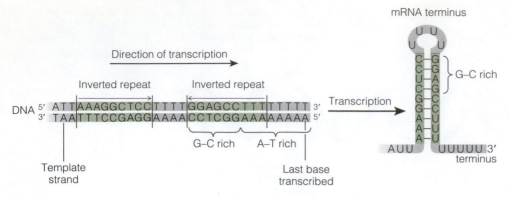

FIGURE 2.25 A stem–loop structure in the newly synthesized RNA is one way that transcription is terminated.

3. RNA polymerase III transcribes tRNA, 5S rRNA, and small nuclear and cytoplasmic RNAs. Eukaryotic promoters are not always located just in front of the transcriptional start site of the gene coding sequence. For example, the 5S rRNA and tRNA genes have internal promoters.

In addition to RNA polymerase, other proteins called *transcription factors* help initiate transcription by increasing the binding affinity of RNA polymerase for the promoter. Other proteins called regulatory proteins can modulate transcription by interacting either directly with RNA polymerase or with one of the transcription factors. Other DNA sequences in addition to

promoters also affect transcription. Short DNA sequences called **enhancers**, usually 50–100 base pairs in length, influence the level of transcription from great distances, sometimes thousands of base pairs from the gene. Enhancers are usually located either upstream or downstream from the coding region, but sometimes they are within a gene. Although we do not know how enhancers function, they most likely bind to regulatory proteins that also interact with the RNA polymerase or transcription factors at the promoter. The DNA must loop out to enable distant DNA regions to interact with the enhancer. This looping out may result from the interaction between the regulatory protein and proteins associated with the promoter (Figure 2.27).

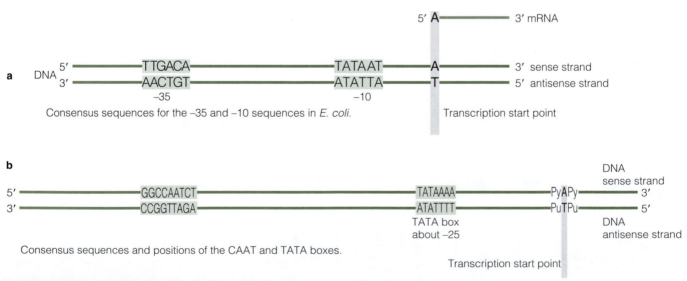

FIGURE 2.26 Conserved sequences in (a) bacterial promoters and (b) eukaryotic promoters. The transcription start site is shown. The antisense DNA strand is the gene that is transcribed into the mRNA. The sense strand is the complementary DNA that is mRNA-like.

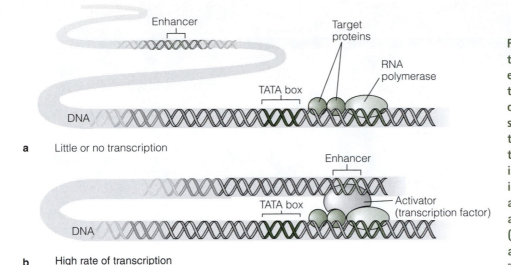

a   Little or no transcription

b   High rate of transcription

FIGURE 2.27 Enhancement of transcription by an enhancer element. (a) There is a low level of transcription without the influence of an enhancer. (b) An enhancer sequence increases the level of transcription. A regulatory protein that is a transcriptional activator is bound to the enhancer. The intervening DNA forms a loop that allows the activator protein to associate with target proteins (transcription factors that are associated with RNA polymerase).

RNA Processing

In eukaryotic cells, most newly synthesized RNA—primary transcripts called **heterogeneous nuclear RNA** (hnRNA)—must be modified before it is fully functional. Three different modifications occur in RNA that becomes mature mRNA (Figure 2.28).

1. Addition of a 5′ cap structure. A cap structure (**5′ cap**) of a modified (usually methylated) guanine base is often observed at the 5′ end.

2. Addition of a 3′ poly-A tail. Usually a string of adenine ribonucleotides, called a poly-A tail, is added to the 3′ end of eukaryotic mRNA. This process is called polyadenylation.

3. In the primary transcript of most eukaryotic genes, noncoding sequences, called **introns**, intervene between the coding sequences, the exons. The introns are spliced out of the newly synthesized RNA so that the exons are adjacent to one another.

The 5′ and 3′ modifications (1) facilitate transport out of the nucleus, (2) protect the mRNA from degradation in the cytoplasm, and (3) maintain stability for translation. Other RNAs also undergo processing. Primary transcripts of ribosomal and tRNAs undergo similar processes to become mature RNAs. Introns are spliced from tRNAs, and both the 5′ and 3′ ends are processed, including the addition at the 3′ end of CCA where the amino acid binds. In eukaryotes, a large rRNA transcript is synthesized and cut in a stepwise process to generate 28S, 18S, and 5.8S rRNAs.

TRANSLATION

In eukaryotes, transcription occurs in the nucleus where newly synthesized RNA is processed and then transported into the cytoplasm, the site of translation or protein synthesis. The major components required for protein synthesis are (Figure 2.29):

1. mRNA that encodes the linear sequence of the amino acids of a polypeptide. Thus, mRNA is the carrier of the genetic message (that is, protein) encoded in the DNA.

2. tRNA molecules that carry amino acids to the site of protein synthesis in the appropriate order specified by the linear sequence of amino acids encoded within the mRNA. Thus, tRNAs serve as adapter molecules or interpreters that convert the

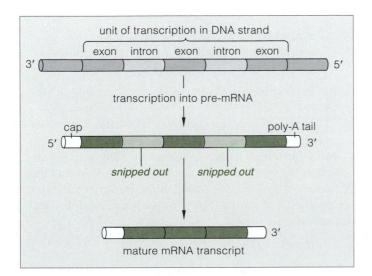

FIGURE 2.28 After transcription and formation of a pre-mRNA in eukaryotic cells, the transcript is processed—a nucleotide with functional groups is added to the 5′ end (5′ cap) and adenine nucleotides are added to the 3′ end (poly–A tail). After processing, the mature mRNA is transported to the cytoplasm for translation.

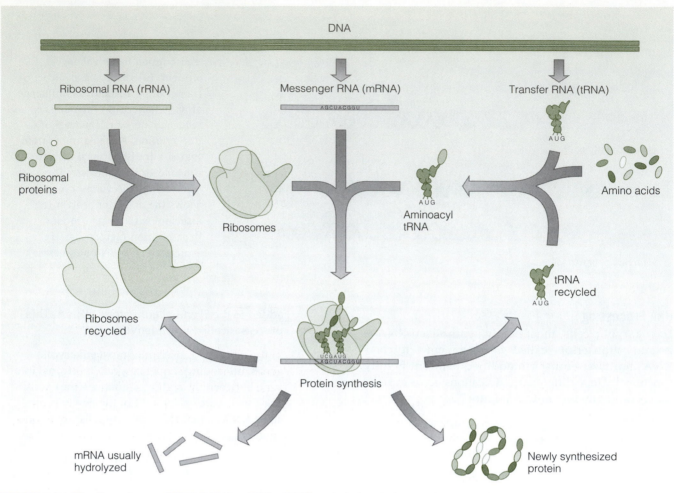

DNA

Ribosomal RNA (rRNA)

Messenger RNA (mRNA)

AGCUACGGU

Transfer RNA (tRNA)

AUG

Ribosomal proteins

Ribosomes

Amino acids

Aminoacyl tRNA

AUG

tRNA recycled

AUG

Ribosomes recycled

UCGAUG
AGCUACGGU

Protein synthesis

mRNA usually hydrolyzed

Newly synthesized protein

FIGURE 2.29 The three types of RNA (rRNA, mRNA, tRNA) and their role in protein synthesis.

RNA code or codons into the language of amino acids. Attached to the tRNA is the amino acid that is specified by the codon to which it has base paired. In this way, the tRNA carries the appropriate amino acid to the site of protein synthesis.

3. Ribosomes, composed of rRNA and proteins, that help carry out protein synthesis (Figure 2.30). The ribosome is composed of two subunits, large and small. Within the ribosome are five important sites: (1) mRNA-binding site, (2) A (aminoacyl) site where a tRNA with its bound amino acid enters the ribosome, (3) P (peptidyl) site where the growing polypeptide is bound to a tRNA, (4) the E (exit) site where tRNAs that no longer have a bound amino acid exit from the ribosome, and (5) catalytic site that forms a covalent (peptide) bond between two amino acids.

The ribosome binds to the mRNA, and the placement on the mRNA determines which bases will be read in triplets as codons. This important step ensures that the message will be read in-frame for synthesis of the correct protein. During protein

synthesis many ribosomes bind to the mRNA, resulting in the synthesis of many polypeptides.

4. Protein factors that help carry out the different steps of protein synthesis.

Translation is the conversion of information encoded in the mRNA sequence into a sequence of amino acids forming a polypeptide chain. The process involves the binding of ribosomes to the mRNA and the joining of amino acids (covalent bonding) in the order specified by the mRNA. Translation is a four-step process:

1. Initiation (Figure 2.31)

During initiation in eukaryotes, the small ribosomal subunit first binds to the initiator tRNA (the first tRNA with its amino acid, methionine; formylmethionine in prokaryotes) and its amino acid. This ribosome subunit–tRNA–methionine complex binds to the end of the mRNA near the 5' cap site. The small subunit then moves along the mRNA until the initiator AUG (first methionine of the mRNA) is recognized. The large ribosomal subunit

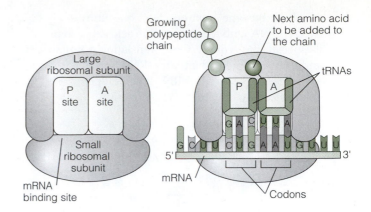

FIGURE 2.30 Ribosome structure. The ribosome is composed of two subunits. The mRNA moves through a groove between the two subunits. Two tRNA binding sites allow for the alignment of tRNA anticodons with the codons of mRNA. The A site binds a tRNA that has an amino acid attached (aminoacyl-tRNA) and the P site binds the tRNA that has the growing polypeptide attached.

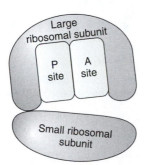

Before translation begins, the ribosomes are dissociated into small and large subunits.

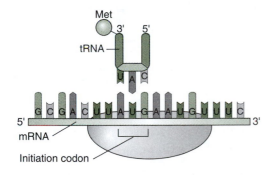

An initiation complex forms, consisting of the small ribosomal subunit, mRNA, and methionine (the initiator tRNA).

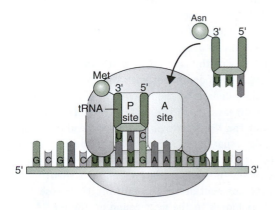

The large ribosomal subunit binds to the initiation complex. The process of peptide elongation begins with the addition of the second tRNA with its amino acid.

FIGURE 2.31 Translation initiation.

is then added. Numerous initiation protein factors play important roles in the initiation process.

In prokaryotes (for example, *E. coli*) the process is similar to that of eukaryotes, although translation begins while the gene is being transcribed because there is no membrane-bound nucleus to serve as a barrier (eukaryotic mRNAs must be transported to the cytoplasm before translation can begin), fewer protein factors are involved in initiating translation, and a special sequence in the 5' region of the mRNA (Shine-Delgarno sequence) is complementary to a sequence of the 16-S rRNA portion of the ribosome, which establishes the correct alignment of the ribosome on the mRNA.

The **anticodon** of the initiator tRNA with the amino acid base pairs with the codon AUG of the mRNA. The anticodon and codon base pairing occurs within the ribosome—antiparallel in codon–anticodon base pairing. The first two bases in an anticodon–codon pairing are specific and exactly follow base-pairing rules. However, the third position is not so restrictive and nonstandard base, or wobble, pairing, can occur (Table 2.2). (Thus, there is not always a specific type of tRNA for each of the 61 codons that correspond to amino acids. Many tRNAs recognize more than one codon for a particular amino acid.)

TABLE 2.2 Codon–Anticodon Pairing at the Third Base of the Codon According to the Wobble Hypothesis

Anticodon	Codon
U (or ψ)	U, G, or A
C	G
A	U
G	U or C
I	U, C, or A

ψ, the modified base pseudouridine; I, the modified base inosine. From F.H.C. Crick, *J. Mol. Biol.* 19:548 (1966) as used in S.L. Wolfe, *Molecular and Cellular Biology*, Table 16.3. Copyright © 1993 Wadsworth, Inc.

The ribosome accommodates one aminoacyl-tRNA (tRNA with a bound amino acid) at each of its special sites A and P. Once translation is initiated and the ribosome and initiator tRNA with its amino acid are in place on the mRNA, the polypeptide is synthesized by the process of elongation.

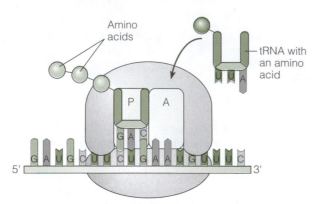

The polypeptide chain is attached to the tRNA that carries the amino acid most recently added to the chain. This tRNA is in the P site of the ribosome.

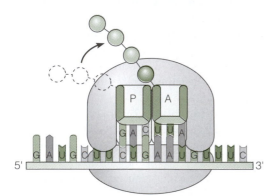

The growing polypeptide chain is detached from the tRNA molecule in the P site and joined by a peptide bond to the amino acid linked to the tRNA at the A site.

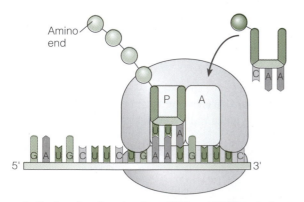

In the translocation step the mRNA and tRNA move in one direction and the ribosome moves in the opposite direction. In this way the growing polypeptide chain is transferred to the P site.

2. Elongation (Figure 2.32)

Amino acids are added stepwise to the growing polypeptide chain. The amino acid specified by the second codon of the mRNA is brought to the ribosome by a second tRNA whose anticodon is complementary to this codon. If complementary

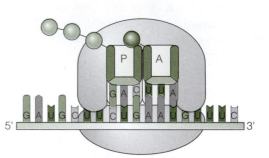

A tRNA with its specific amino acid attached has bound to the A site. Base pairs have formed between the anticodon of tRNA and the codon of mRNA.

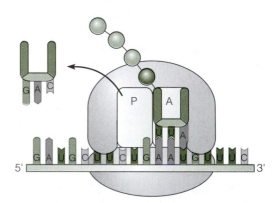

The released tRNA joins the cytoplasmic pool of tRNA and can bind with another amino acid.

FIGURE 2.32 Elongation during translation. Elongation begins with the addition of the second tRNA with its bound amino acid. Each elongation step adds one more amino acid to the polypeptide chain. The round circles represent amino acids. At the ribosome, the tRNAs transfer their amino acids to the growing polypeptide chain. The bases of the anticodons of the tRNAs form transient hydrogen bonds with the bases of the complementary codons of the mRNA. Here, the amino acid is brought into the A site by the tRNA. The UUA anticodon of the tRNA base pairs with the codon, AAU, that specifies asparagine. The polypeptide in the P site is transferred to the asparagine bound to its tRNA in the A site. A peptide bond is established and the tRNA that lost the polypeptide is released from the ribosome. The ribosome moves to the next codon on the mRNA (translocation), GUU, which codes for valine, and the process continues with the anticodon CAA of tRNA–valine base pairing with the codon.

Translation in the Nucleus of Eukaryotic Cells?

Ever since the process of translation has been elucidated, it has been thought proteins were synthesized in the cytoplasm in eukaryotic cells—that RNAs were transported out of the nucleus and into the cytoplasm for translation. Recently, however, experiments have suggested that translation may occur in the nucleus. All the components necessary for protein synthesis appear to be present (ribosomes, mRNA, tRNA, and protein factors), and in experiments, proteins are able to be translated in the nucleus. It has been estimated that perhaps 10–15% of a eukaryotic cell's protein synthesis may take place in the nucleus.

---

base pairing occurs, the amino acid attached to the first tRNA is transferred to the amino acid on the second aminoacyl-tRNA. A peptide bond is made between, for example, methionine and the second amino acid. The first tRNA, now without its amino acid, is released from the ribosome and mRNA. The elongating polypeptide chain is transferred to the incoming amino acid.

The transfer of the growing polypeptide chain to the incoming amino acid is catalyzed by one of the ribosomal RNAs of the ribosome (23S rRNA in bacteria). Thus, in this case, RNA has the catalytic activity rather than a protein. The catalytic site (where the activity occurs) of the RNA is in only one of the nucleotides of the rRNA—an adenine. Catalytic RNAs, those that have enzymatic activity, are referred to as ribozymes.

3. Translocation

Once the growing polypeptide chain in the P site of the ribosome has been transferred to the incoming amino acid in the A site, the P site contains a tRNA without a bound amino acid. The ribosome complex moves by one codon on the mRNA, in a process known as translocation. The tRNA with the growing polypeptide (called a peptidyl tRNA) is now in the P site. A new codon is exposed in the A site, and another tRNA with the appropriate amino acid moves into this site. The translation process continues as the mRNA is "read" in the 5' to 3' direction and the polypeptide is elongated one amino acid at a time.

4. Termination

Elongation of the polypeptide chain continues until a stop codon is reached (UAA, UAG, or UGA) and a protein release factor interacts with the stop codon to terminate translation. The ribosomal subunits dissociate, the messenger and transfer RNAs are released and can be used again, and the newly synthesized polypeptide is used by the cell.

REGULATION OF GENE EXPRESSION

In both prokaryotic and eukaryotic cells, complex mechanisms regulate the amount of proteins synthesized. The cell usually synthesizes only what is required and thus highly regulates the production of products. In both prokaryotes and eukaryotes, transcription initiation (turning a gene on or off) is a major control point, and proteins bind to specific DNA sequences to control transcription. However, in eukaryotes there are more regulatory proteins and DNA control sequences, as well as many more control points after transcription (that is, post-transcriptional controls). In prokaryotes, regulation is most commonly observed at the level of transcription initiation; however, other modes of regulation also occur, such as translational control where the translation rate of mRNAs can be increased or decreased. In eukaryotes, cellular control can occur at many levels, and may be quite complex (that is, regulation occurs during and after transcription and during and after translation).

Prokaryotic Gene Expression

Microorganisms must respond rapidly to sudden fluctuations in the environment. Proteins or enzymes may be required for only a very brief time, and when conditions change they might not be needed. Bacteria must rely on inducers that transmit signals so that genes can be turned on or off in response to environmental cues (such as nitrogen starvation, heat and salt stress, light intensity).

In bacteria, a single promoter often controls several structural genes or coding regions, called *cistrons*. This arrangement is called an *operon*. The genes that are part of the same operon have related functions within the cell and, therefore, the genes are turned on and off together. Some *E. coli* operons are large, such as the histidine biosynthesis operon (*his* operon), which has 11 genes; others are small, such as the lactose operon (*lac* operon), which has three genes (Figure 2.33). An operon consists of structural genes, a promoter, and a repressor binding site, called an **operator**, that overlaps

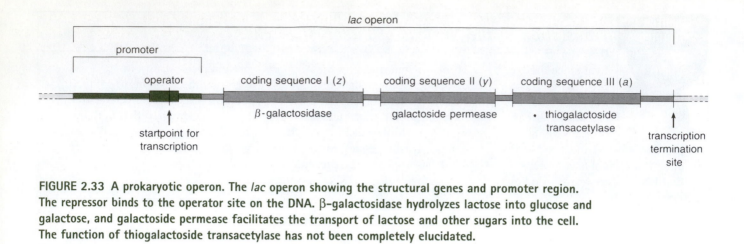

FIGURE 2.33 A prokaryotic operon. The *lac* operon showing the structural genes and promoter region. The repressor binds to the operator site on the DNA. β–galactosidase hydrolyzes lactose into glucose and galactose, and galactoside permease facilitates the transport of lactose and other sugars into the cell. The function of thiogalactoside transacetylase has not been completely elucidated.

the promoter. **Repressor proteins**, encoded by repressor genes, are synthesized to regulate gene expression. These proteins bind to the operator site to block transcription by RNA polymerase.

The *lac* Operon   The well-studied *lac* operon illustrates bacterial gene regulation. The three operon genes, *lacZ* (β-galactosidase), *lacY* (permease), and *lacA* (acetylase), encode enzymes that transport and break down the milk sugar lactose into glucose and galactose for energy. In the absence of lactose, the enzymes are not needed, and the genes in the operon are not transcribed. In an inactive operon, the *lac* repressor, encoded by the *lacI* gene

(not part of the operon), binds to the operator that overlaps the promoter region (Figure 2.34a). Transcription is prevented because the presence of the repressor prevents the RNA polymerase from binding to the promoter. The lactose repressor has two binding sites: one for lactose compounds and the other for the DNA operator site. When lactose (sometimes called an inducer) is present, the operon is transcribed and the enzymes are synthesized. Lactose binds to the repressor to form a lactose–repressor complex that cannot bind to the operator region of the DNA. The promoter region is available to the RNA polymerase, which can bind and initiate transcription (Figure 2.34b). Ribosomes immediately

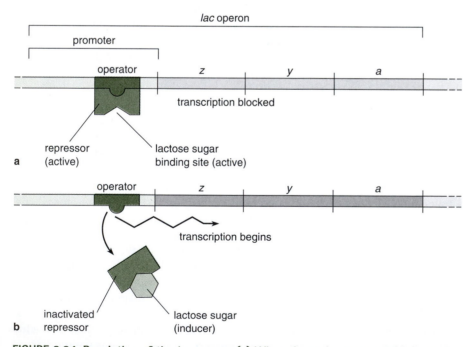

FIGURE 2.34 Regulation of the *lac* operon. (a) When the active repressor binds to the operator in the absence of lactose, RNA polymerase cannot bind to the promoter and transcription is blocked. (b) In the presence of lactose, a derivative of the sugar binds to the repressor and inactivates it so that the repressor can no longer bind to the operator, thus allowing RNA polymerase to bind to the promoter and transcription to proceed.

attach to the newly synthesized RNA, and translation begins.

The regulation of an operon by a repressor is called negative control because genes are not transcribed when the repressor is bound. The *lac* operon can also be positively regulated so that the genes are transcribed at a higher rate than they would be without this type of regulation. Specific small molecules increase the rate of transcription by facilitating both the binding of RNA polymerase to the promoter and the separation of DNA strands. The preferred substrate for *E. coli* is not lactose but glucose; however, when glucose is absent, the concentration of **cyclic AMP** (cAMP) increases in the cell. Cyclic AMP binds to a DNA-binding protein, **catabolite activator protein (CAP)**. The cAMP–CAP complex binds to a CAP binding site near the promoter region but not overlapping the region. As a result, *lac* operon transcription is greatly enhanced. When glucose is present, cAMP is low, and the cAMP–CAP complex does not form. Thus, the rate of transcription is not increased, or if lactose is absent, the *lac* operon is not expressed.

The *trp* Operon The operon that regulates tryptophan biosynthesis is another example of negative control. The synthesis of amino acids must be highly regulated to maintain the correct concentrations in the cell. The tryptophan (*trp*) operon consists of promoter and operator regions in addition to five genes encoding enzymes that catalyze the last steps of tryptophan biosynthesis.

In contrast to the repressor of the *lac* operon, the *trp* operon repressor is not active unless tryptophan binds to it. Tryptophan activates the repressor so that the repressor–tryptophan complex can bind to the operator region and block transcription (Figure 2.35). In this way, tryptophan acts as a co-repressor to turn off

transcription. Because an inactive repressor cannot bind to the operator when tryptophan is absent, the promoter remains available to RNA polymerase for transcription.

Eukaryotic Gene Expression

The regulation of gene expression in eukaryotes is much more intricate and variable than in prokaryotes. In addition to controlling transcription, eukaryotes regulate their genes by controlling mRNA processing and transport of mRNAs to the cytoplasm, the rate of translation and the availability of mRNA (controlled through stability and the masking of mRNA by bound proteins), and protein processing. The complexity of eukaryotic cells enable these cells to regulate product synthesis at many levels.

There are many reasons for this complexity. Some of them are listed here.

1. Larger genome size (that is, more DNA than prokaryotes) with extensive noncoding regions such as repetitive sequences, and other DNA sequences that do not code known information. Prokaryotic chromosomes do not have much noncoding DNA.

2. Compartmentalization within the cell (such as the nucleus and other organelles), requiring that nuclear-encoded gene products (proteins made in the cytoplasm) be transported to organelles. This means that there must be coordination between the needs of the organelle and the nucleus harboring the genes.

3. More extensive transcript processing. For example, introns must be removed from RNA and a poly-A tail to the 3′ end and a 5′ cap added.

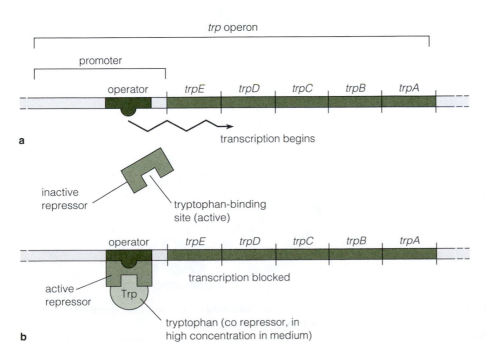

a

b

FIGURE 2.35 The *trp* operon showing the five structural genes and the promoter region. (a) In the absence of tryptophan, the repressor is inactive and cannot bind to the operator, thus allowing RNA polymerase to bind to the promoter and transcription to proceed. (b) When tryptophan is present it binds to and activates the repressor, thus allowing the activated repressor to bind to the operator and block transcription.

4. Genes are scattered around the genome, each with their own promoter–regulatory region (that is, not organized into operons to more easily regulate genes involved in the same metabolic pathway). Genes that produce products that work together (for example, polypeptides that come together to make a large enzyme such as the carbon-fixation enzyme in plants called ribulose-1,5-bisphosphate carboxylase or oxygenase) must be coordinately regulated—sometimes in different compartments within a cell. For example, a gene in the nucleus and a gene in the chloroplast must be coordinately regulated.

5. Regulation from a distance. For example, enhancer and silencer sequences can be great distances from the gene they regulate.

6. Cell- and tissue-specific gene expression. In multi-cellular eukaryotes, specific sets of genes are activated and inactivated in different cell types (for example, liver versus kidney cells). This requires complex control of sets of genes.

The amount of final product from a gene is influenced by cellular controls at many levels. Although the details of eukaryotic gene expression are beyond the scope of this book, an overview will demonstrate the many levels of control. The flow of genetic information begins in the nucleus and ends with the production of functional proteins (Figure 2.36). Thus, there are many control points (Figure 2.37).

1. Transcriptional control

Promoters and other DNA sequences that regulate gene expression are more complex and varied in eukaryotic cells. To control transcription, a complex array of general transcription factors interact with the promoter and RNA polymerase. Gene-

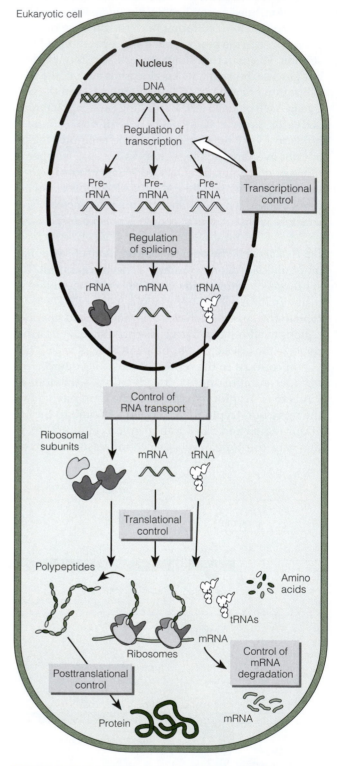

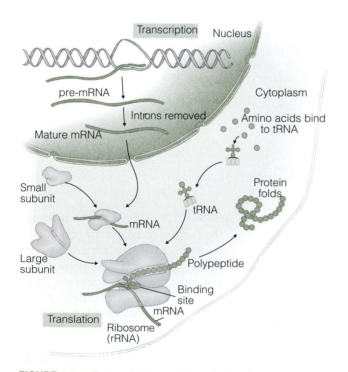

FIGURE 2.36 Transcription and translation in a eukaryotic cell.

FIGURE 2.37 Summary of the multiple levels of eukaryotic gene regulation.

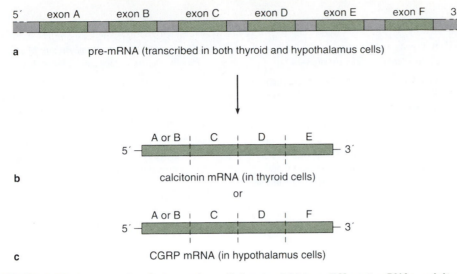

a pre-mRNA (transcribed in both thyroid and hypothalamus cells)

b calcitonin mRNA (in thyroid cells)

or

c CGRP mRNA (in hypothalamus cells)

FIGURE 2.38 An example of alternative splicing to yield two different mRNAs, calcitonin mRNA in thyroid cells and calcitonin gene-related peptide (CGRP) mRNA in hypothalamus cells. (a) Pre-mRNA comprises six exons (A–F) and five introns. (b) A or B, C, D, and E exons are retained in the mRNA that encodes the hormone calcitonin while (c) exons A or B, C, D, and F are kept in CGRP hypothalamus mRNA, thought to play a role in the function of taste receptors. Exon F is an intron in calcitonin RNA and an exon in CGRP, while exon E is an exon for calcitonin RNA and an intron for CGRP. Exon A or B, located in the 5' untranslated region, can be present in either protein.

specific regulatory proteins bind to special DNA control sequences that contain regulatory protein binding sites. These interactions also regulate gene activity. Control sequences can be adjacent to or distant from structural genes, as are enhancer sequences. Also, because operons are not found in eukaryotes, individual or distant genes often must be regulated in scattered groups or networks.

2. Regulation of RNA processing and transport out of the nucleus to the cytoplasm

Two different cell types may process the same RNA differently, by a process called alternative splicing, to yield two different proteins. That is, different introns and exons (or portions of them) in some intron-containing primary RNA transcripts are selectively removed or retained. An example is the hormone calcitonin, which is expressed in both the thyroid and hypothalamus. As Figure 2.38 shows, the pre-mRNA contains five introns separating six exons. The transcript can be processed in either of two ways to generate calcitonin mRNA in the thyroid cells or calcitonin gene-related peptide (CGRP) mRNA in the hypothalamus cells.

3. Translational control

This type of control influences the synthesis of a protein product. Translation is controlled by protein initiation factors and proteins that repress (inhibit) translation. The stability of mRNA is also important, because the rate of mRNA degradation determines how long the message is available for translation. For example, the mRNA for the protein tubulin has a half-life of 4–12 hours, whereas the insulin receptor and the enzyme pyruvate kinase mRNAs have half-lives of 9 hours and 30 hours, respectively.

4. Posttranslational control

Posttranslational control involves the alteration of protein products after they are synthesized. Alteration includes protein folding and assembly with other proteins after synthesis; processing such as removal of amino acids; cleavage of the molecule or modification (for example, different groups added such as sugar or phosphate) to activate a newly synthesized protein; import into organelles such as chloroplasts and mitochondria; and protein degradation.

General Readings

W.M. Becker, L.J. Kleinsmith, and J. Hardin. 2000. *World of the Cell.* 4th ed. Addison Wesley Longman, Inc., San Francisco.

J.E. Dahlberg, E. Lund, and E.B. Goodwin. 2003. Nuclear translation: What is the evidence? *RNA* 9:1–8.

J. Darnell. 2004. *Molecular Cell Biology.* 5th ed. W.H. Freeman and Company, New York.

J. Frank. 1998. How the ribosome works. *American Scientist* 86:428–439.

B. Hayes. 1998. The invention of the genetic code. *American Scientist* 86:8–14.

D.S. Latchman. 1999. *Gene Regulation: A Eukaryotic Perspective.* 3rd ed. Stanley Thornes, Cheltenham, UK.

H. Lodish, A. Berk, A. Matsudaira, P. Kaiser, C.A. Krieger, M. Scott, L. Zipursky, and A.B. Sach and S. Buratowski. 1997. Common themes in translational and transcriptional regulation. *Trends Biochem. Sci.* 22:189–192.

R.J. White. 2001. *Gene Transcription: Mechanism and Control.* Blackwell Science, Oxford.

M.F. Wilkenson and A.-B. Shuu. 2002. RNA surveillance by nuclear scanning? *Nature Cell. Biol.* 4:E144–E147.

S.L. Wolfe. 1995. *An Introduction to Cell and Molecular Biology.* Wadsworth Publishing Company, Belmont, California.

Additional Readings

D. Beckett. 2001. Regulated Assembly of Transcription Factors and Control of Transcription Initiation. *J. Mol. Biol.* 314:335–352.

D. Black. 2000. Protein diversity from alternative splicing: A challenge for bioinformatics and post-genome biology. *Cell* 103:367–370.

T.R. Cech. 2000. The ribosome is a ribozyme. *Science* 289:878–879.

P. Cramer et al. 2000. Architecture of RNA polymerase II and implications for the transcription mechanism. *Science* 288:640–649.

M.R. Culbertson. 1999. RNA surveillance: Unforeseen consequences of gene expression, inherited genetic disorders and cancer. *Trends Genet.* 15:74–80.

A. Das and C.C. Richardson. 1993. Control of transcription termination by RNA-binding proteins. *Annu. Rev. Biochem.* 62:893–930.

N.V. Federoff. 2002. RNA-binding proteins in plants: The tip of an iceberg? *Curr. Opin. Plant Biol.* 5:452–459.

J.W. Fickett and W. Wasserman. 2000. Discovery and modeling of transcriptional regulatory regions. *Curr. Opin. Biotech.* 11:19–24.

D.M. Gilbert. 2001. Making sense of eukaryotic DNA replication origins. *Science* 294:96–100.

B.R. Graveley. 2001. Alternative splicing: Increasing diversity in the proteomic world. *Trends Genet.* 17:100–107.

M.S. Hastings and A.R. Krainer. 2001. Pre-mRNA splicing in the new millennium. *Curr. Opin. Cell Biol.* 13:302–309.

M.W. Hentze. 2001. Believe it or not: Translation in the nucleus. *Science* 275:500–501.

U. Hubscher, H.-P. Nasheuer, and J.E. Syvaoja. 2000. Eukaryotic DNA polymerases, a growing family. *Trends Biochem. Sci.* 25:143–147.

F.J. Iborra, D.A. Jackson, and P.R. Cook. 2001. Coupled transcription and translation within nuclei of mammalian cells. *Science* 293:1139–1142.

A.B. Khodursky and J.A. Bernstein. 2003. Life after transcription: Revisiting the fate of messenger RNA. *Trends Genet.* 19:113–115.

D.L.J. Lafontaine and D. Tollervey. 2001. The function and synthesis of ribosomes. *Nature Reviews Mol. Cell Biol.* 2:514–520.

J.G. Lawrence. 2002. Shared strategies in gene organization among prokaryotes and eukaryotes. *Cell* 110:407–413.

B. Madan and S.A. Teichmann. 2003. Evolution of transcription factors and the gene regulatory network in *Escherichia coli. Nucleic Acids Res.* 31:1234–1244.

E. Martinez. 2002. Multi-protein complexes in eukaryotic gene transcription. *Plant Mol. Biol.* 50:925–947.

L. Minvielle-Sebastia and W. Keller. 1999. mRNA polyadenylation and its coupling to other RNA processing reactions and to transcription. *Curr. Opin. Cell Biol.* 11:352–357.

P. Mitchell and D. Tollervey. 2001. mRNA turnover. *Curr. Opin. Cell Biol.* 13:320–325.

K. Ogata, K. Sato, and T. Tahirov. 2003. Eukaryotic transcriptional regulatory complexes: Cooperativity from near and afar. *Curr. Opin. Struct. Biol.* 13:40–48.

A.V. Philips and A. Cooper. 2000. RNA processing and human disease. *CMLS Cellular and Molecular Life Sciences* 57:235–249.

G. Roberts, C. Gavin, and C. Smith. 2002. Alternative splicing: Combinatorial output from the genome. *Curr. Opin. Chem. Biol.* 6:375–383.

C.E. Samuel. 2003. RNA editing minireview series. *J. Biol. Chem.* 278:1389–1390.

N. Sonenberg and T.E. Dever. 2003. Eukaryotic translation initiation factors and regulators. *Curr. Opin. Struct. Biol.* 13:56–63.

S. Waga and B. Stillman. 1998. The DNA replication fork in eukaryotic cells. *Annu. Rev. Biochem.* 67:721–751.

J.D. Watson and F.H.C. Crick. 1953. Molecular structure of nucleic acid. A structure for deoxyribose nucleic acid. *Nature* 171:737–738.

B. Weisblum. 1999. Back to Camelot: Defining the specific role of tRNA in protein synthesis. *Trends Biochem. Sci.* 24:247–250.

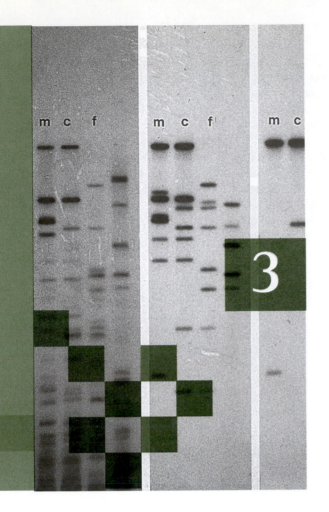

BASIC PRINCIPLES OF RECOMBINANT DNA TECHNOLOGY

3

CAUSE FOR CONCERN?

Recombinant DNA Technology: Promise and Controversy

The cloning experiments of Herbert Boyer, Stanley Cohen, Paul Berg, and their colleagues in the early 1970s ushered in the era of recombinant DNA technology (see Chapter 1, Biotechnology: Old and New). A gene can now be separated from the rest of the chromosomal DNA and transferred to a well-studied foreign host cell for further study. Genetic material introduced into foreign cells is replicated and passed on to progeny cells. Today many methods are available for isolating and characterizing genes and proteins. Selected genes are being transferred to organisms such as plants, animals, bacteria, and fungi for a variety of reasons: commercially desirable products can be efficiently produced in host cells,

genes and their proteins can be studied in ways not possible before, and new medical biotechnologies are being explored. With these technologies have come both promise and controversy. Applications include the development of new medical diagnostics and treatments, better vaccines, the development of stress-resistant crops, more nutritious foods, healthier livestock, and perhaps in the future, a cleaner environment. Opponents of recombinant DNA technology have voiced many concerns.

They fear that we will not know where to draw the line. For example, we now have sophisticated cloning technologies. Will we one day be free to clone humans? Will we be able to use them as organ donors? Are we able to select characteristics in our engineered children? At what point will these

technologies become acceptable? Other fears include the effect of genetically engineered foods on our health, hidden dangers in transgenic organisms, pollution of the environment by genetically engineered organisms (for example, displacing other organisms), and manipulating (and as some say, tampering with) the natural world.

Proponents and critics will most certainly never agree on all aspects of biotechnology; however, steps must be taken to ensure that there is open communication between scientists and lay people. New developments in biotechnology will continue, powerful DNA and protein technologies will continue to emerge, and progress will be made. We have an obligation to channel these in a positive direction.

At the heart of most molecular genetic technologies is the gene. A gene must be isolated and well characterized before it can be used in genetic manipulations. One method of isolating and amplifying a DNA of interest is to clone the gene by inserting it into a DNA molecule that serves as a vehicle or vector. When these two DNAs of different origin are combined, the result is a recombinant DNA molecule. The molecule is moved into a host—*Escherichia coli*, for example—where it can be reproduced. When the cells divide, each cell, or **clone**, in the colony contains one or more identical copies of the recombinant DNA molecule. Thus, the DNA contained within the recombinant molecule is cloned. Gene cloning has many uses. To name just a few, DNA can be amplified in host cells to obtain many copies for further study, a gene can be **expressed** to obtain a valuable protein product, and a gene and its expression can be studied in a living cell. In this chapter the basic methods of isolating and characterizing DNA are described, with emphasis on the preparation and use of recombinant DNA.

CUTTING AND JOINING DNA

Two major categories of enzymes are important tools in the isolation of DNA and the preparation of recombinant DNA (in which two DNA molecules are combined). A specific DNA or gene is removed from the DNA by cutting the sugar–phosphate backbone, and the DNA from two different sources are mixed and recombined. Recombinant DNA molecules cannot be easily generated without two types of enzymes: restriction endonucleases that act as scissors to cut DNA at specific sites and DNA ligase that is the glue that joins two DNA molecules in the test tube. Restriction endonucleases cut both strands of the DNA sugar–phosphate backbone. Restriction endonucleases recognize specific sequences within the DNA molecule (Figure 3.1). The recognized sequences are usually four to six base pairs and are palindromic (Table 3.1): the sequence of both DNA strands are the same when read in the same direction, the 5′ to 3′ or 3′ to 5′ direction.

5′GCCAATTGGC3′
3′CGGTTAACCG5′

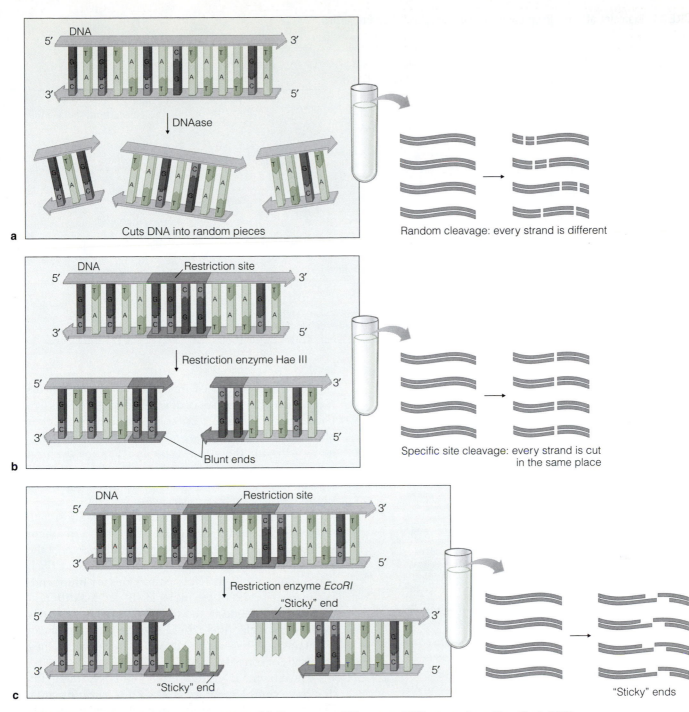

FIGURE 3.1 Cutting double-stranded DNA. (a) The enzyme DNase cuts DNA at random sites. Each DNA molecule in the test tube can be cut at different sites. **(b)** and **(c)** A specific restriction enzyme cuts each DNA molecule at the same sequence site. Some restriction enzymes make a blunt cut (for example, *HaeIII*) and others make a staggered cut (for example, *EcoRI*).

Each restriction enzyme recognizes a specific sequence and cuts at a particular place within that sequence. The enzyme cuts the double strand of DNA by breaking covalent bonds between the phosphate of one deoxyribonucleotide and the sugar of an adjacent deoxyribonucleotide.

Restriction endonucleases are found primarily in bacteria, where they cut, or fragment, foreign DNA

of bacteriophage before the invading DNA can replicate within the host bacterial cell to produce new phage that would ultimately destroy the host. The bacterial cells are resistant because their DNA is chemically modified, primarily by the addition of methyl groups, to mask most of the restriction endonuclease recognition sites so that these will not be cut.

TABLE 3.1 Examples of Some Restriction Endonucleases and Their Properties

Restriction Endonuclease	Source (Bacterial Species)	Target Site (Cuts at Arrow)	Characteristics Recognizes (No. Base Pairs)	Product
Eco RI	*Escherichia coli* R13	↓ G-A-A-T-T-C C-T-T-A-A-G ↑	6	4-base-long sticky ends
Hha I	*Haemophilus haemolyticus*	G-C-G-C C-G-C-G	4	2-base-long sticky ends
Sma I	*Serratia marcescens*	↓ C-C-C-G-G-G G-G-G-C-C-C ↑	6	Blunt ends
Hae III	*Haemophilus aegyptius*	↓ G-G-C-C C-C-G-G ↑	4	Blunt ends

J. Ingraham and C. Ingraham, *Introduction to Microbiology*. Copyright © 1995 Wadsworth Publishing Co.

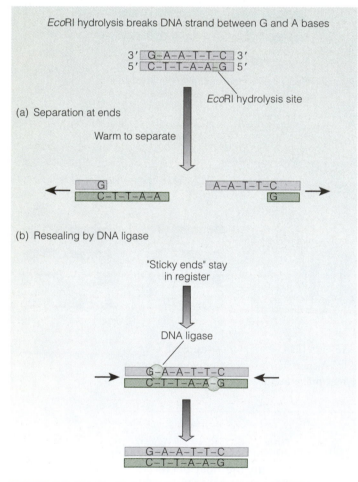

FIGURE 3.2 The ligation of two different pieces of DNA. The overlapping, cohesive ends of each DNA anneal and DNA ligase form phosphodiester bonds.

Restriction enzymes are named for the organisms from which they are isolated. For example, *Eco*RI is isolated from *E. coli* RY13: *Eco* comes from the first letter of the genus name and the first two letters of the species name; R is for the strain type and I is for the first enzyme of that type. Thus, *Bam*HI is isolated from *Bacillus amyloliquefaciens* strain H, and *Sau*3A is isolated from *Staphylococcus aureus* strain 3A.

Restriction enzyme cleavage of a sugar–phosphate backbone can produce a double-strand DNA fragment with blunt or staggered ends. When both strands of the molecule are cut at the same position, the ends are flush and no nucleotides are left unpaired; this is a **blunt end**. When each strand of the molecule is cut at a different position so that one strand (the 5' or the 3') overhangs by several nucleotides, these single-strand ends can spontaneously base pair with each other—that is, they are sticky, or cohesive.

The enzyme DNA ligase can join DNA fragments that have complementary **sticky ends** or blunt ends. Ligase catalyzes the formation of covalent bonds between the sugar and the phosphate of the adjacent nucleotides, requiring only that one nucleotide have a free 5' phosphate and the adjacent one have a 3' hydroxyl group. Ligase does not discriminate between DNAs with different origins. Thus, two DNA fragments cut from the chromosomes of two different organisms by restriction enzymes are joined by DNA ligase (Figure 3.2). The two fragments are now one DNA molecule. This cut-and-paste technique produces a recombinant DNA molecule. It is then transferred to a host cell where it is amplified (replicated) for further study. This process is called *DNA cloning* (making multiple, identical DNA copies).

SEPARATING RESTRICTION FRAGMENTS AND VISUALIZING DNA

Restriction enzyme digestions and other manipulations of DNA enable the results to be directly visualized. Agarose gel electrophoresis is a technique for separating DNA fragments by size and visualizing them after staining (Figure 3.3). Gelatinous agarose is a mixture of a powder of purified agar (isolated from seaweed) and buffer, which is boiled and poured into a mold where the agarose (generally 0.7 to 2.0%) gels and solidifies into a slab. A toothed comb forms wells in the molten agarose, and samples are loaded into the wells after the agarose solidifies. The agarose slab is submerged in a buffer solution and an electric current is applied to electrodes at opposite ends of the slab to establish an electric field in the gel and the buffer. Because the sugar–phosphate backbone is negatively charged, the DNA fragments migrate toward the positive electrode.

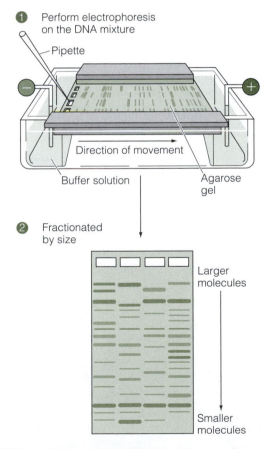

FIGURE 3.3 Agarose gel electrophoresis is used to separate DNA (and RNA) molecules according to size. The negatively charged nucleic acid moves toward the positive electrode. Larger molecules move more slowly than smaller molecules.

Pores between the agarose molecules act like a sieve that separates molecules by size. Larger molecules move more slowly than smaller molecules. Increasing the percentage of agarose produces smaller pores; these increase the resolution of smaller fragments by impeding all but the smaller molecules. Lowering the percentage (amount) of agarose produces larger pores; these increase the resolution of larger fragments by allowing greater separation among them in the gel.

Because DNA by itself is not visible in the gel, ethidium bromide is usually added to make the DNA bands visible. Ethidium bromide molecules intercalate between the bases causing the DNA to fluoresce orange when the gel is illuminated with ultraviolet light. Other dyes (nonhazardous) that stain DNA include methylene blue; however, most research laboratories use ethidium bromide because it is very sensitive and allows the detection of very small amounts of DNA. The lengths of the DNA fragments can be determined by comparing their position in the gel to reference DNAs of known lengths also in the gel. A DNA fragment migrates a distance that is inversely proportional to the logarithm of the fragment length in base pairs over a limited range in the gel. Thus, agarose gel electrophoresis allows the restriction fragment lengths to be determined.

DNA CLONING

There are several steps involved in cloning a gene. The specific methodology used in each step may vary depending on the type of DNA used, the host cell type (for example, bacterial versus plant), and the ultimate goal of DNA cloning. For example, the type of vector used for cloning will depend on whether the cloned DNA will remain in the vector or be inserted into the chromosome of the host cell (by recombination). A scientist also may want to retrieve a product from the gene (for example, human insulin or human growth hormone) and will use a vector that allows the gene to be expressed in the host cell. The vector must then have the important promoter sequences for transcription to take place and other sequences for translation.

The steps in gene cloning include

1. Isolation of DNA

2. Ligating the DNA into a vector (Figure 3.4)

3. Transformation of a host cell with the recombinant DNA (vector DNA with DNA insert) (Figure 3.5)

4. Selection of host cells harboring the recombinant DNA

5. Screening of cells for those harboring the recombinant DNA or producing the appropriate protein product

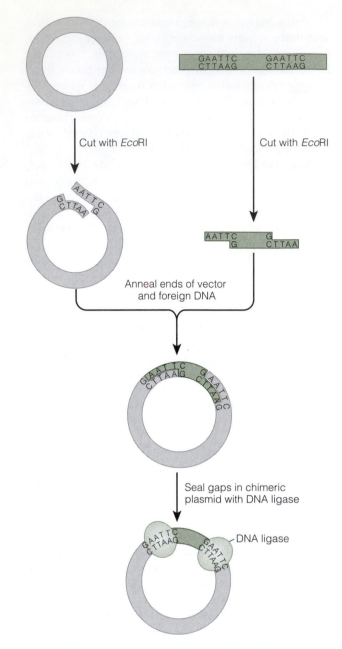

FIGURE 3.4 Ligation of a DNA fragment into a vector using the restriction endonuclease *EcoRI* and DNA ligase.

Cut with *Eco*RI

Cut with *Eco*RI

Anneal ends of vector and foreign DNA

Seal gaps in chimeric plasmid with DNA ligase

DNA ligase

CLONING VECTORS

As discussed earlier, restriction endonucleases and DNA ligase are important enzymes in the production of recombinant DNA. Often a foreign DNA is introduced into a host cell by inserting it into a cloning vehicle or vector that transports it into the host cell. Many of these vectors replicate independently in a host cell. A cloning vector must

1. Have an origin of replication so that the DNA can be replicated within a host cell

2. Be small enough to be isolated without undergoing degradation during purification

3. Have several unique restriction sites for cloning a DNA fragment so that the vector will be cut only once and several restriction sites for insertion will be available

4. Have selectable markers for determining whether the cloning vehicle has been transferred into cells and to indicate whether the foreign DNA has been inserted into the vector

Bacterial Vectors

The greatest variety of cloning vectors has been developed for *E. coli* because of the major role they have played in recombinant DNA experiments since the 1970s. Other cloning vectors are available for bacteria, such as *Bacillus subtilis*, as well as for yeast, fungi, animals, and plants.

Plasmids Bacteria harbor plasmids: circular double strands of DNA that are **extrachromosomal;** that is, they are not part of the bacterial chromosome. Often multiple copies of plasmids are present in the cell. Plasmids have diverse functions. Some encode substances for antibiotic resistance and bacteriocins—agents that kill or inhibit similar bacterial strains or species. Others perform physiological functions, such as pigment production, degradation of compounds, and dinitrogen fixation. Toxin-producing and virulence plasmids en-

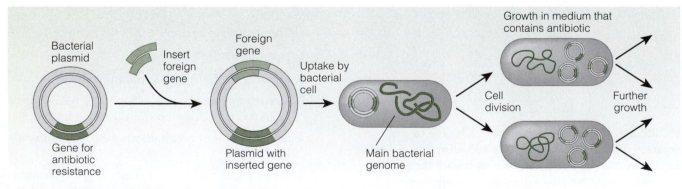

Bacterial plasmid

Insert foreign gene

Foreign gene

Uptake by bacterial cell

Growth in medium that contains antibiotic

Cell division

Further growth

Gene for antibiotic resistance

Plasmid with inserted gene

Main bacterial genome

FIGURE 3.5 The transformation of bacteria and the selection of cells harboring the recombinant vector.

Used by permission of University Science Books, from Berg and Singer, "Dealing With Genes," copyright © 1992.

code endotoxins and hemolysins, whereas some plasmids confer resistance to metals such as mercury, cadmium, nickel, and zinc. Because plasmids are relatively small and easy to manipulate, they have been engineered as cloning vectors with unique restriction sites for insertion of foreign DNA fragments—up to approximately 10 kilobases long (1 kilobase, or kb, equals 1000 nucleotides). When transported into bacterial cells, recombinant plasmid vectors are readily replicated, often in great numbers, or high copy number.

An early vector, pSC101, used by Cohen, Boyer, and their colleagues (discussed in Chapter 1, Biotechnology: Old and New) replicated to yield only one or two copies in each cell (low copy number). Co1 E1, a high copy plasmid developed in 1974, produced several thousand copies in one cell. Co1 E1 was further modified so that several unique restriction sites were available for the insertion of DNA. One of these modified plasmids was pBR322, constructed in Boyer's laboratory (Figure 3.6). The name pBR322 is derived as follows: p identifies the molecule as a plasmid; BR identifies the original constructors of the vector (Bolivar and Rodriquez); 322 is the identification number of the specific plasmid (other examples are pBR325, pBR327, and pBR328).

pBR322 is derived from three naturally occurring plasmids: the ampicillin-resistance gene from the plasmid R1, the tetracycline resistance gene from pSC101, and the replication region from pMB1. Although many engineered plasmid vectors now have similar and even more powerful features, until recently pBR322 was one of the most commonly used plasmids for several reasons:

1. The molecule is small, having only 4363 base pairs, and can be isolated easily. Consequently, this vector can accommodate DNA of up to 5 to 10 kb.

2. pBR322 has several unique restriction sites where the plasmid can be linearized and opened for inserting a DNA fragment.

3. The genes encoding resistance to ampicillin (amp^r) and tetracycline (tet^r) are used for plasmid and DNA insert selection. The DNA fragment of interest can be inserted into one of the antibiotic resistance genes, inactivating that gene. The other antibiotic gene remains active and can be used to select for bacteria carrying the plasmid. Therefore, if a foreign DNA is inserted into the tetracycline gene, the bacteria containing such a plasmid would be ampicillin resistant but tetracycline sensitive ($amp^r tet^s$) and will die if treated with tetracyline. This process is referred to as *insertional inactivation of a selectable marker.* A gene that gives a characteristic phenotype (physical characteristic) such as resistance to an antibiotic is inactivated, thus making the cells sensitive. One can easily screen cells for antibiotic resistance.

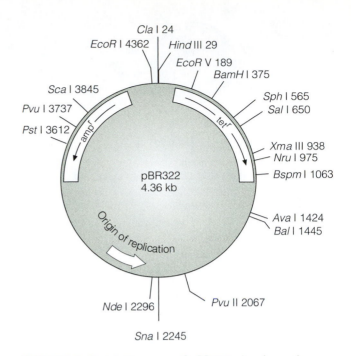

FIGURE 3.6 Restriction map of pBR322 showing unique restriction sites, the ampicillin and tetracycline resistance genes, and the origin of replication.

Insertional inactivation is a powerful way to determine whether the vector contains a DNA insert. The processes of restriction endonuclease digestion, ligation, and transformation are inefficient processes—that is, not all molecules are cut, ligated, or transferred to host cells. Consequently, in a population of bacterial cells that have been transformed, some cells have (1) received a recombinant vector, (2) received a vector not containing a DNA insert (vector only), or (3) not received a vector. A screening process allows the detection of cells that receive a recombinant DNA (vector with DNA insert). This process selectively kills cells that contain a recombinant DNA by exposure to an antibiotic because the DNA inserted into the vector inactivated an antibiotic resistance gene (Figure 3.7).

Selectively killing cells with antibiotics and identifying host cells harboring recombinant DNA involves the following:

1. After transformation of host cells with recombinant molecules produced in the laboratory, **colonies** must first be plated from the original, master plates to identical replica plates. Different antibiotics are added to the medium to identify (that is, select) cells with recombinant molecules.

2. To maintain the original bacterial colonies so that recombinant colonies from the master plates can later be further analyzed, exact copies are made by

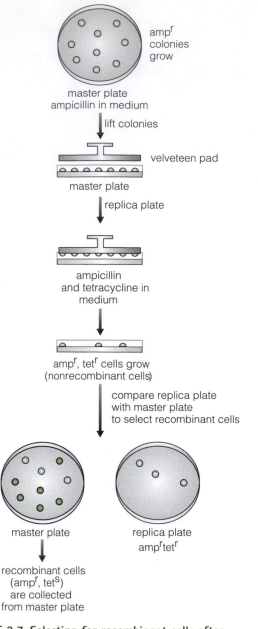

master plate
ampicillin in medium

amp^r colonies grow

lift colonies

velveteen pad

master plate

replica plate

ampicillin and tetracycline in medium

amp^r, tet^r cells grow (nonrecombinant cells)

compare replica plate with master plate to select recombinant cells

master plate

replica plate amp^r tet^r

recombinant cells (amp^r, tet^S) are collected from master plate

FIGURE 3.7 Selecting for recombinant cells after transformation.

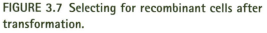

replica plating. The technique is conducted as follows:

a. After colonies of bacterial cells have grown in a medium such as agar, a sterile velveteen pad is pressed against the master plate (the dish holding the colonies). Cells from the colonies adhere to the velveteen (other transfer media can be used, such as nylon filter paper). The location of the cells on the velveteen is a mirror image of the locations of the original colonies on the master plate.

b. The velveteen pad is then pressed against media in a second and third plate, transferring cells to them. The locations of these cells will now be identical to the original colonies on the master plate.

c. Recombinant colonies are identified (and therefore selected) by the addition of different antibiotics to the replica plates and a comparison with the colony growth on the different replica plates.

3. When cells are placed onto ampicillin-containing agar medium, only cells with pBR322 will be able to survive; untransformed cells will not survive. To distinguish recombinant plasmids, bacterial colonies are replica plated onto a medium with tetracycline and ampicillin. Cells that survive do not have the DNA insert in pBR322 vectors, and cells that die have the DNA insert in pBR322. Thus, these cells are ampicillin resistant–tetracycline resistant (amp^rtet^r) and ampicillin resistant–tetracycline sensitive (amp^rtet^s), respectively.

4. The master plate is examined to select those colonies that contain recombinant plasmids.

Today, pBR322 is seldom used because of its limitations—screening for cloned inserts is time consuming, and a limited number of restriction sites for cloning are available. However, many plasmid vectors in use are derived from this early vector. More commonly used are small plasmids containing selection functions that allow a more direct screening approach, such as the pUC plasmids developed in 1982. These small vectors (less than 4 kb) contain a polycloning site made up of multiple restriction sites, where foreign DNA can be inserted. A series of pUC vectors containing the ampicillin resistance gene for plasmid selection has been constructed (for example, pUC8, pUC9, pUC18, pUC19). The multiple cloning sites differ in the type of restriction sites and their orientation (Figure 3.8). Selectable markers for insert selection are not limited to antibiotic resistance genes; the genes may encode enzymes that catalyze metabolic reactions. A chemical indicator aids in the identification of recombinant plasmids.

One type of selection, called α-complementation, allows the detection of DNA inserts in pUC plasmids. These vectors contain a portion of the *lacZ* gene (called *lacZ'*) that encodes the first 146 amino acids for β-galactosidase. The polycloning site resides in this coding region. If the *lacZ'* region is not interrupted by inserted DNA, the amino terminal portion of the β-galactosidase polypeptide (lacZ) is synthesized. An *E. coli* deletion mutant called *lacZ'*M15 harbors a chromosomal mutant of *lacZ* that encodes only the

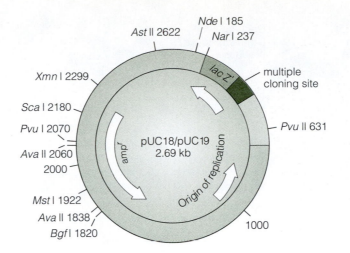

Multiple cloning sites
pUC18

Thr	Met	Ile	Thr	Asn	Ser	Ser	Ser	Val	Pro	Gly	Asp	Pro	Leu	Glu	Ser	Thr	Cys	Arg	His	Ala	Ser	Lau	Ala	Leu	Ala
ATG	ACC	ATG	ATT	ACG	AAT	TCG	AGC	TCG	GTA	CCC	GGG	GAT	CCT	CTA	GAG	TCG	ACC	TGC	AGG	CAT	GCA	AGC	TTG	GCA	CTG	GCC

Ecor I   Sac I   Kpn I   Sma I   Bamh I   Xba I   Sal I   Pst I   Sph I   Hind III
                                  Xma I   Acc I
                                                       Hinc II

pUC19

Thr	Met	Ile	Thr	Pro	Ser	Leu	His	Ala	Cys	Arg	Ser	Thr	Leu	Glu	Asp	Pro	Arg	Val	Pro	Ser	Ser	Asn	Ser	Leu	Ala
ATG	ACC	ATG	ATT	ACG	CCA	AGC	TTG	CAT	GCC	TGC	AGG	TCG	ACT	CTA	GAG	GAT	CCC	CGG	GTA	CCG	AGC	TCG	AAT	TCA	CTG	GCC

Hind III   Sph I   Pst I   Sal I   Xba I   Bamh I   Sma I   Kpn I   Sac I   EcoR I
                                 Acc I   Xma I
                                 Hinc II

* In pUC18, *EcoR* I
   In pUC19, *Hind* III

FIGURE 3.8 Restriction map of pUC18/pUC19 showing the ampicillin resistance gene, the origin of replication, and the multiple cloning site within the *lacZ* gene. The orientation is reversed in pUC18 and pUC19 with the *EcoRI* site and the *Hind*III site of the multiple cloning site immediately downstream from the lac promoter in pUC18 and pUC19, respectively.

carboxyl end of the β-galactosidase. Both the plasmid and chromosomal *lacZ* fragments encode nonfunctional proteins. However, by α-complementation the two partial proteins can associate and form a functional β-galactosidase. In cells, β-galactosidase normally hydrolyzes the sugar lactose into glucose and galactose. The α-complementation is indicated by the presence of blue colonies on special media. The blue indicates that the plasmid-encoded *lacZ'* has combined with the partial lacZ from the **complementary** fragment of the chromosomal *lacZ* gene residing on the chromosome of *E. coli*, generating an enzymati-

cally active β-galactosidase that turns a chromogenic agent blue. The chromogenic lactose analog, X-gal, or 5-bromo-4-chloro-3-indoyl-β-D-galactopyranoside, when added to the medium is broken down by β-galactosidase and produces a blue color. When the plasmid *lacZ'* gene fragment is interrupted by the insertion of a foreign DNA fragment, colonies appear white on media (Figure 3.9).

Bacteriophage A virus that infects a bacterium is called a bacteriophage. Viral DNA can be engineered for use as a cloning vector; the first to be used was

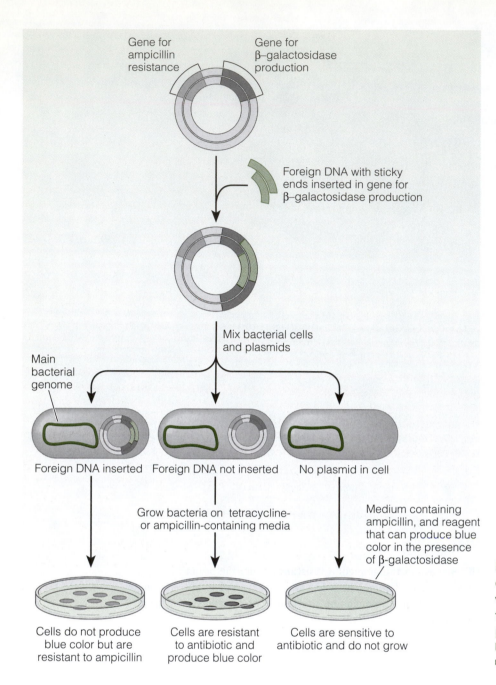

Gene for
ampicillin
resistance

Gene for
β–galactosidase
production

Foreign DNA with sticky
ends inserted in gene for
β–galactosidase production

Mix bacterial cells
and plasmids

Main
bacterial
genome

Foreign DNA inserted

Foreign DNA not inserted

No plasmid in cell

Grow bacteria on tetracycline-
or ampicillin-containing media

Medium containing
ampicillin, and reagent
that can produce blue
color in the presence
of β-galactosidase

Cells do not produce
blue color but are
resistant to ampicillin

Cells are resistant
to antibiotic and
produce blue color

Cells are sensitive to
antibiotic and do not grow

FIGURE 3.9 The selection of bacterial colonies containing a recombinant vector (pUC vector + DNA insert) by the selection of white colonies. The β–galactosidase gene is interrupted by a foreign DNA insert.

from a lambda bacteriophage, in 1974. Today many variations of lambda exist. Lambda phage vectors are derived from the 50-kb wild-type double-strand genome that has single-strand complementary ends of 12 nucleotides (**cohesive termini,** or *cos*) that can base pair. The *cos* ends (important for the **lytic** pathways) base pair, forming a circular DNA molecule once the phage DNA is inside the host cell. DNA replication then occurs from the circular molecules, producing linear lambda DNA made up of several

50-kb phage DNA end on end. In the lytic pathway (cycle), the host cell lyses after phage reproduction, releasing progeny virus (Figure 3.10). Lytic phages are used to clone and amplify a DNA of interest. The resulting **plaques**—cleared areas on the medium where host cells have lysed—contain millions of recombinant phage particles that can be isolated (Figure 3.11). In the **lysogenic** pathway, the bacteriophage genome is integrated into the host cell, no cell lysis occurs, and the phage genome replicates along

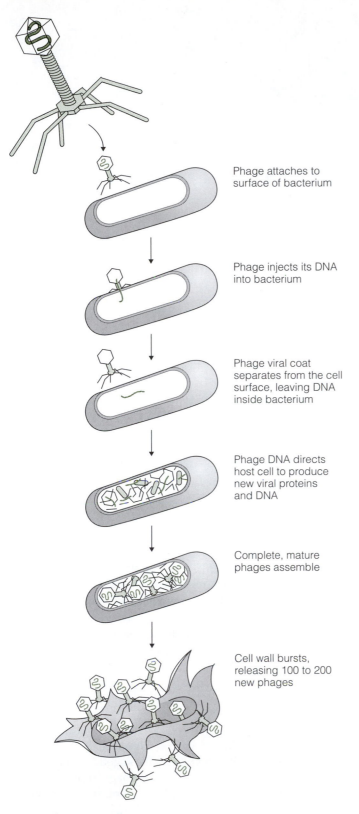

Phage attaches to surface of bacterium

Phage injects its DNA into bacterium

Phage viral coat separates from the cell surface, leaving DNA inside bacterium

Phage DNA directs host cell to produce new viral proteins and DNA

Complete, mature phages assemble

Cell wall bursts, releasing 100 to 200 new phages

FIGURE 3.10 The lytic life cycle of a bacteriophage.

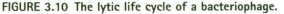

with the host genome. For example, in gene therapy, animal viruses are used to integrate a cloned therapeutic gene into the genome.

All phage vectors used as cloning vectors have been disarmed for safety and can function only in special laboratory conditions. Recombinant DNA containing the viral DNA and the DNA of interest are packaged into viral particles in the test tube. Host bacterial cells are infected with recombinant phage DNA, the DNA replicates within the host cells, and progeny phage are produced when the host cell undergoes lysis. Plaques then become visible on plates where cells in the colony have lysed.

The DNA of lambda-type phage can accommodate only an additional 3 kb, or 5%, of its genome, for a total size of 52 kb. DNA that is too large or too small cannot be packaged into the head of the viral particle. Fortunately, a significant portion of the viral genome can be deleted without adversely affecting packaging and infection of *E. coli* host cells. Removal of one-third of nonessential DNA from the central portion of the phage allows a DNA insert of up to about 20 kb to be ligated into the phage DNA. The deletion contains most of the DNA necessary for integration and excision from the *E. coli* genome and is not necessary for use as a cloning vector. Vectors that have a segment of nonessential DNA removed, making room for foreign DNA, are called replacement vectors. Another type of phage vector, called an insertion vector, has one restriction site for the insertion of DNA of 5 to 10 kb.

Cosmids A significant disadvantage of using plasmids and bacteriophage as vectors is that only relatively small DNA fragments can be inserted. Larger DNA fragments can be cloned using engineered hybrids of phage DNA and plasmids called **cosmids.** These are plasmids with a small portion of lambda DNA, the *cos* sites. A cosmid vector not only is composed of *cos* sites for packaging into phage particles but also contains a plasmid replication origin for replication in bacterial hosts and genes for plasmid selection (for example, antibiotic resistance genes to confer resistance to antibiotics such as ampicillin or kanamycin). The cosmid vector is packaged in a protein coat *in vitro* as with a bacteriophage vector; however, after the packaged DNA infects *E. coli* host cells, the DNA replicates as a plasmid rather than as bacteriophage DNA and the cells are not lysed. Bacterial colonies rather than plaques are formed on petri plates. Cosmid vectors are small (some only 2.5 kb), and because they are packaged for infection of host cells, if *cos* sites are separated by 37 to 52 kb, they accommodate large inserts of foreign DNA. Typically, 35 to 45 kb can be cloned into cosmid vectors.

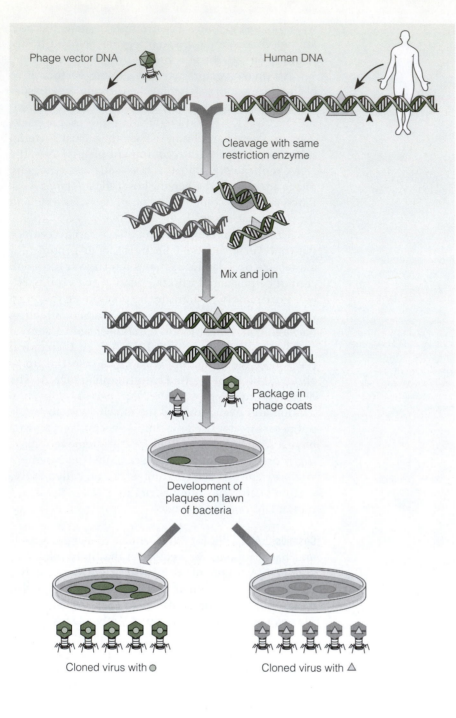

Phage vector DNA

Human DNA

Cleavage with same restriction enzyme

Mix and join

Package in phage coats

Development of plaques on lawn of bacteria

Cloned virus with ●

Cloned virus with △

FIGURE 3.11 The cloning of human DNA restriction fragments (● and △) into a bacteriophage vector.
Used by permission of University Science Books, from Berg and Singer, "Dealing With Genes," copyright © 1992.

The following compares available vectors for bacterial host cells:

Vector	Insert Size	Characteristics
Plasmids	≤10 kb	Autonomously replicates in host cell, high or low copy number plasmids
Bacteriophage	5–20 kb	Packaged into protein coat, kills host cells, variable insert size
Cosmids	35–45 kb	Packaged like a bacteriophage, replicates like a plasmid without killing host cells

Vectors for Other Organisms

Although many cloning experiments are carried out using *E. coli* as the host cells, other organisms also are used that require different types of cloning vectors.

When the goal is to obtain a protein product such as insulin or growth hormone or to modify the properties of a specific organism, such as to introduce pest resistance into soybean, the cloning vector must be compatible with the organism used. Cloning vectors are available for yeast and other fungi, plants, insects, fish, and mammals.

Yeast Artificial Chromosomes

Yeast artificial chromosomes (YACs) are useful for eukaryotic molecular studies. The yeast chromosome has the following necessary components:

1. A **centromere** distributes the chromosome to the daughter cells during cell division

2. A **telomere** at the end of the yeast chromosome ensures that the end is correctly replicated and protects against degradation

3. An **autonomously replicating sequence** (ARS) consists of specific DNA sequences that enable the molecule to replicate. YACs also have a gene that provides a way to detect an inserted DNA fragment. These components are joined to make a YAC that can replicate in yeast host cells.

YACs are especially useful for cloning large DNA fragments. Many animal genes being studied can be 200 kb or more, requiring a cloning vector that accommodates very large fragments. As noted earlier, many vectors accommodate only rather small DNA inserts. YAC vectors accept fragments of between 200 and 1500 kb, allowing complete genes as well as gene clusters to be cloned for study. YAC libraries are available and are very useful for large genomes, such as the human genome. They enable researchers to isolate and sequence specific regions of the genome.

Bacterial Artificial Chromosomes

Bacterial artificial chromosomes (BACs) are synthetic vectors and have been the most widely used DNA cloning system for large genome sequencing projects. They have been used to clone very large fragments of eukaryotic genomes—100 to 300 kb of DNA, with the average size being 150 kb. BACs are constructed using a very low copy *E. coli* plasmid vector—the naturally occurring fertility factor plasmid—called the F factor. The F factor of *E. coli* is a circular plasmid of approximately 100 kb that encodes proteins that allow the molecule to replicate. BAC vectors have been engineered to be approximately 7.4 kb and contain cloning sites and selectable markers of different types (for example, color-based *lacZ* selection). The benefit of using F factor plasmids as a part of the vector is that they have the ability to carry up to 25% of the bacterial chromosome. In addition, unlike bacteriophage and cosmid vectors, there is no packaging limitation—the recombinant vector is introduced

into cells by electroporation (discussed later). The BAC cloning system is more stable than YACs (cloned DNA inserts are less likely to undergo rearrangements). This type of vector is useful for analyzing large portions of complex genomes, whole genes, and constructing physical maps of genomic regions.

Plant Cloning Vectors

DNA is being cloned into plants for several purposes: to generate resistance to disease, pests, and herbicides; to improve crop yields, quality, and nutritional value; to develop new ornamental plant characteristics; and to increase the shelf life of many common fruits and vegetables. Several cloning vectors have been constructed that allow efficient transfer of DNA to plant cells. The most commonly used are plant viruses, such as tobacco mosaic virus (TMV), and the **Ti**, or tumor inducing, **plasmid** of the soil bacterium *Agrobacterium tumefaciens* (Figure 3.12; see Chapter 6, Plant Biotechnology).

A. tumefaciens is a soil microorganism that induces crown gall formation in many species of dicotyledonous plants. This bacterium infects tissue wounds in plants (such as in the stem) and induces plant cells to proliferate, resulting in a cancerous tissue mass, or crown gall, near the infection site. The bacterium's large Ti plasmid—greater than 200 kb—is what induces crown gall formation: Ti plasmid genes are involved in infection and induce plant cell division that leads to the tumorlike growth. A special region on the plasmid, the **T-DNA** (transferred DNA), containing approximately eight genes that encode the disease characteristics, is incorporated into the plant's genome. Some of these genes direct the synthesis of unusual compounds, called opines, that the bacterial cells use as nutrients for growth. Thus, *A. tumefaciens* redirects transcription and translation within the plant host cell for its own use.

The properties of the Ti plasmid make it an efficient cloning vector in certain plants. The T-DNA, which integrates into the plant chromosomes, can be used to transfer foreign genes into plants. For use in biotechnology, the engineered Ti plasmid lacks some of the genes that contribute to the cancerous properties.

One of the challenges facing scientists is the development of efficient methods of transferring foreign DNA into all the cells of a plant. One way is to integrate the DNA of interest into a few cells, which, once transformed, divide and give rise to the whole plant (see plant tissue culture in Chapter 6, Plant Biotechnology). Another way is to transfer DNA to cells in the embryo by soaking seeds or buds of plants in a solution containing recombinant *A. tumefaciens* bacteria. A selectable marker gene enables scientists to identify plants that harbor T-DNA with the foreign gene or complementary DNA (**cDNA**).

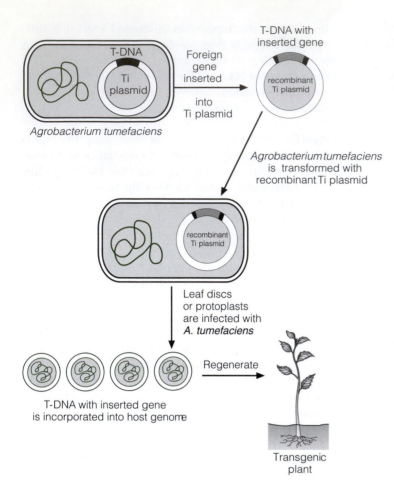

T-DNA with inserted gene is incorporated into host genome

Transgenic plant

FIGURE 3.12 The use of *Agrobacterium tumefaciens* to transfer DNA into plants. A gene of interest is cloned into the T–DNA of the *A. tumefaciens* plasmid (along with a selectable marker gene), and *A. tumefaciens* is transformed with the recombinant Ti plasmid. Transformed *A. tumefaciens* is used to infect plant cells in culture; the T–DNA, including the foreign gene, is transferred into plant chromosomal DNA. Plants regenerated from cultured cells are tested for the presence of the foreign gene.

Mammalian Cell Vectors The first eukaryote-infecting virus to be used for cloning, and one of the most studied viruses, was simian virus 40, or SV40, a small, circular, double-strand DNA tumor virus. Although the virus can infect the cells of several mammalian species, including monkey cells, its host range is limited. Moreover, only a limited amount of foreign DNA can be inserted because the DNA must be packaged into a viral coat, or capsid. To provide space for inserted foreign DNA, both nonessential and essential genes have been deleted in vectors derived from SV40. Thus, a helper virus containing the necessary genes is required to propagate and package recombinant SV40 virus. Finally, only transient expression of an inserted gene has been achieved.

Other vectors include retrovirus and adenovirus. **Retroviruses** are single-strand RNA viruses that show much promise for use as vectors in a wide variety of animal cells, including human. The viral genome contains two single-strand RNA molecules that are held together by hydrogen bonds at the 5' ends. To replicate, the virus uses the enzyme **reverse transcriptase** to make a double-strand DNA molecule from the RNA template. This DNA integrates stably into the chromo-

somes of dividing host cells, where transcription and translation of the provirus (that is, the integrated DNA) occur. Some of the synthesized RNAs are packaged into virus particles.

The adenovirus is a double-strand DNA virus. Like the retrovirus, it can infect cells with high efficiency and has a broad host range. Because the pathogenicity in humans is low, the adenovirus makes a desirable vector for gene therapy. Moreover, unlike the retrovirus, adenovirus does not have to infect dividing cells.

The synthesis of complex eukaryotic gene products requires eukaryotic hosts (such as human or mouse cells). Often, eukaryotic gene products must be modified in ways not possible with bacterial cells, such as by posttranslational cleavage, glycosylation, and amino acid modifications such as phosphorylation, acetylation, sulfation, and acylation. Mammalian host cells often are used if a gene encoding a protein product contains introns and the mRNA must be processed or if the protein requires processing after synthesis (although cDNAs can be used with prokaryotic host cells). Many animal proteins can be produced only in eukaryotic hosts, because only these cells have the capability to process RNA and protein. Many important pharmaceutical compounds for therapeutic use are produced in this way; among them are factor VIII used by hemophiliacs to aid in blood clot formation, β-interferon, growth hormone, and erythropoietin. Vectors specific for animal hosts have been developed for a variety of uses (for example, gene therapy; see Chapter 10, Medical Biotechnology).

CELL TRANSFORMATION

To introduce new DNA into organisms or to use hosts for gene cloning, methods of gene transfer are required. A variety of methods are available for transforming bacteria, yeast, plants, and animals with foreign DNA. Some bacteria are naturally **competent**—that is, they have the ability to take up extracellular DNA. However, many bacteria must be treated chemically to become competent. Exposing bacteria to a salt solution such as calcium chloride and applying a heat shock in the presence of DNA is usually sufficient for transferring DNA into cells; that is, **transformation.** However, eukaryotes require more complicated methods.

In organisms that have cell walls, like fungi, algae, and plants, enzymes are used to degrade the walls, producing **protoplasts**, or cells without walls. Foreign DNA then can be introduced into protoplasts by **electroporation:** Protoplasts are exposed to a brief electrical pulse, which is thought to introduce transient openings in the cell membrane through which the DNA molecules enter. After transformation, cells are washed and cultured so that the cell wall re-forms and cell division begins. Transformed plants can be regenerated from cell culture (see Chapter 6, Plant Biotechnology).

Another method of transforming cells is microprojectile bombardment, or **biolistics** (Figure 3.13). Very small (4 μm) microprojectiles made of gold or tungsten are coated with DNA and shot at high velocity from a particle gun into cells or tissue. Because the projectiles penetrate the cell, the walls do not have to be removed. Biolistics shows much promise for use in gene therapy, whereby cells of organs in living animals can be transformed with a corrective gene.

Lacking walls, animal cells are easily transformed by electroporation, as well as by a method that precipitates DNA onto the cell surface with a calcium phosphate solution. Viruses often are used to transfer DNA into animal cells for the stable integration into the chromosome. (Viruses are also an important tool in the transformation of plant and bacterial cells.) However, the introduction of genes into all the cells of a multicellular animal requires a different method, called **microinjection.** Unlike plant cells, animal cells cannot regenerate into an entire organism. To express a gene in all the cells of an animal, thus generating a transgenic organism, fertilized eggs or very early embryos are transformed by microinjection (Figure 3.14). DNA is injected directly into the nucleus of animal cells with an extremely fine pipette. After DNA is transferred into the cell, it is integrated into the chromosome and the transformed fertilized egg is implanted into an animal for the completion of development.

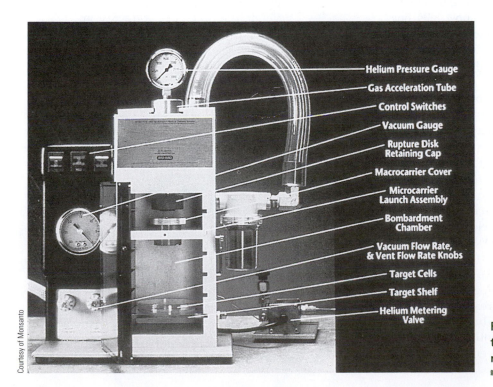

Courtesy of Monsanto

Helium Pressure Gauge
Gas Acceleration Tube
Control Switches
Vacuum Gauge
Rupture Disk Retaining Cap
Macrocarrier Cover
Microcarrier Launch Assembly
Bombardment Chamber
Vacuum Flow Rate, & Vent Flow Rate Knobs
Target Cells
Target Shelf
Helium Metering Valve

FIGURE 3.13 Particle gun used for transforming cells with DNA–coated microprojectiles.

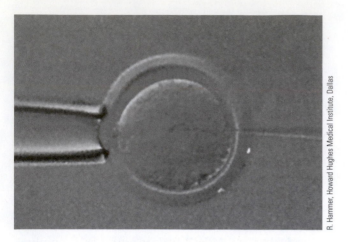

FIGURE 3.14 Microinjection of DNA into the pronucleus of a fertilized animal egg.

CONSTRUCTING AND SCREENING A DNA LIBRARY

Vectors are used to compile a library of DNA fragments that have been isolated from the genomes of a variety of organisms. This collection of fragments is used to isolate specific genes and other DNAs. The relationship of a single DNA fragment to a **DNA library** is analogous to that of one book to a library containing thousands of books. Searching for one specific piece of DNA is similar to looking for one book in a library.

DNA fragments are generated by cutting the DNA with a specific restriction enzyme. These fragments are ligated into vector molecules, and the collection of recombinant molecules is transferred into host cells, one molecule in each cell. The total number of all DNA molecules make up the library. This library is searched, that is, screened, with a molecular **probe** that specifically identifies the target DNA.

To allow scientists to learn more about the organization and regulation of genes, libraries for a variety of organisms have been constructed to **map** and sequence their genomes. For example, libraries are available for the bacteria *Bacillus subtilis* and *E. coli,* the yeast *Saccharomyces cerevisiae,* the nematode *Caenorhabditis elegans,* the fruit fly *Drosophila,* and the plant *Arabidopsis.* To achieve such ambitious goals, researchers must construct ordered, overlapping genomic clones representing the entire genome; that is, all the DNA in each chromosome.

Two main types of libraries can be used to isolate specific DNAs: genomic and cDNA. A genomic library contains DNA fragments that represent the entire genome of an organism. A **cDNA library** is derived from mRNA of a tissue and consequently contains only sequences that are expressed at a given time in a specific tissue.

Genomic Library

Soybeans will be used as an example to demonstrate how a genomic library is constructed. First, total nuclear DNA is isolated from soybean cells and cut with a specific restriction enzyme. At the same time, a cloning vector (which can be a plasmid, cosmid, or bacteriophage) is cut with the same restriction enzyme so that the vector is linearized and the ends are complementary to those of the genomic DNA fragments. The two DNAs—genomic fragments and vector—are mixed in a test tube and DNA ligase is added to form recombinant molecules (Figure 3.15).

The recombinant DNA molecules are introduced into host cells, usually *E. coli,* if plasmids, cosmids, or bacteriophage vectors are used. Transformed bacterial cells, each containing recombinant plasmids or cosmids, multiply when plated onto antibiotic-containing medium. Each colony contains a specific soybean DNA fragment cloned into the vector. Plaques, cleared areas

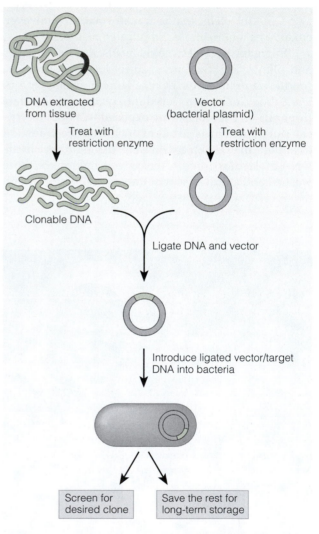

DNA extracted from tissue

Vector (bacterial plasmid)

Treat with restriction enzyme

Treat with restriction enzyme

Clonable DNA

Ligate DNA and vector

Introduce ligated vector/target DNA into bacteria

Screen for desired clone

Save the rest for long-term storage

FIGURE 3.15 The steps involved in the construction of a DNA library (genomic).

where *E. coli* cells have lysed, contain recombinant virus with soybean genome inserts.

The collection of plaques or colonies that together contain all the DNA fragments of a genome constitute the library. For example, the human genome comprises more than 100,000 cosmid clones, or approximately 10,000 YAC clones. The soybean genomic library usually includes at least one complete soybean genome cut into fragments and housed within *E. coli* cells. It is best to have clones representing four or five genomes so that all of the genetic information is represented at least once. From the average size of the DNA inserted into the vector and from the genome size of the organism, a simple calculation tells researchers how many library clones are required to represent the entire genome.

cDNA Library

Some organisms have very large genomes, and a genomic library may yield too many different clones to manipulate easily, even when a host–vector system is used that can accommodate large DNA inserts. When the library must be screened for a particular gene, especially in a plant or animal, the number of clones to be screened may be unmanageable. Thus, a library that includes only expressed genes from a certain type of cell may be more feasible (for example, leaf-specific cDNAs from soybean). This library, called a cDNA library, dramatically reduces the total amount of DNA to be cloned because only a fraction of genomic DNA is expressed by any given cell type. Genes are expressed differentially; consequently, in a complex multicellular organism, specialized cells produce cell-specific proteins. Although a cell has a full complement of genes, only a specific set of genes (in addition to general housekeeping genes) will be expressed and all the others will be silent.

To make, for example, a soybean cDNA library, RNA from leaf cells is isolated and used as a template to make DNA by a method called cDNA synthesis (Figure 3.16). Reverse transcriptase catalyzes the reverse synthesis of cDNA from the mRNA template. The mRNA then is degraded with a **ribonuclease** or an alkaline solution, and DNA polymerase is used in the synthesis of the second DNA strand. Double-strand **DNA linkers** with ends that are complementary to the cloning vector are added to the double-strand DNA molecule before ligation into a cloning vector. Recombinant clones are introduced into bacteria. Thus, the library is composed of cDNAs from the expressed mRNAs in soybean leaf cells.

The differences between genomic and cDNA libraries are significant, and they influence the choice of library for scientific research. First, the noncoding introns present within most eukaryotic genes are not included in the cDNA, because the mRNA has already undergone posttranscriptional modification before iso-

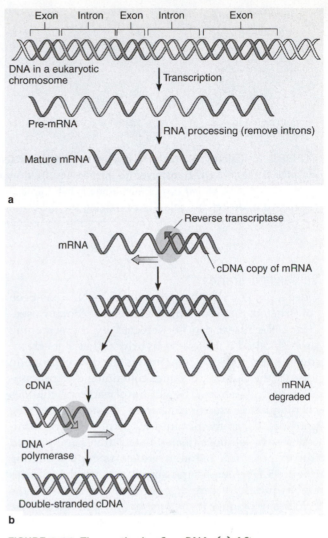

FIGURE 3.16 The synthesis of a cDNA. (a) After transcription, RNA processing in the nucleus produces a mature mRNA (minus introns, with a 5' cap and a 3' poly(A) tail). **(b)** Scientists isolate mature mRNA and use the enzyme reverse transcriptase to produce single-stranded cDNA that is complementary to the mRNA. This produces a cDNA–mRNA hybrid. The mRNA is degraded, and a second DNA strand is synthesized by DNA polymerase that is complementary to the single-stranded DNA. The result is a double–stranded cDNA.

lation for cDNA synthesis. In addition, a cDNA library does not contain regulatory elements associated with genes, such as promoters and enhancers; nor does it contain the portion of the genome that does not code for RNA—that is, noncoding DNA. Fewer clones comprise the library, making the screening process much simpler and less time consuming. If noncoding DNA (such as promoter sequences) is to be isolated from a library, a genomic library must be used.

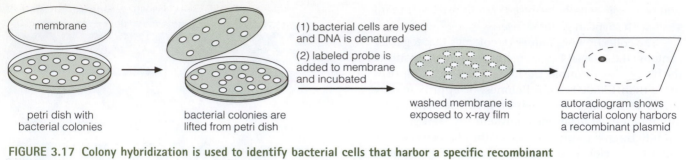

memberane

(1) bacterial cells are lysed and DNA is denatured

(2) labeled probe is added to membrane and incubated

petri dish with bacterial colonies

bacterial colonies are lifted from petri dish

washed membrane is exposed to x-ray film

autoradiogram shows bacterial colony harbors a recombinant plasmid

FIGURE 3.17 Colony hybridization is used to identify bacterial cells that harbor a specific recombinant plasmid. Bacterial cells transferred to membranes are disrupted and the DNA denatured by an alkaline solution. The DNA bound to the membrane is hybridized to a radioactively labeled DNA probe. X-ray film is exposed to detect the colonies containing the DNA of interest on the membranes.

Screening Libraries

Once a library, whether genomic or cDNA, has been constructed, the recombinant clones that contain a specific DNA insert can be detected by the screening process called nucleic acid **hybridization**. For screening, a specific DNA sequence is used as a probe to identify the clones or plaques containing the appropriate target sequence. The library of bacterial colonies or plaques is transferred from master plates (the petri dishes) to membranes made of nylon or nitrocellulose (a cellulose treated with nitrates and pressed into paper) in a manner similar to replica plating (Figure 3.17). Membranes are placed over the colonies or plaques, pressed against them to transfer the colonies or plaques to the membranes, and lifted. More than one membrane can be used for each master plate to produce multiple membranes for screening the library in duplicate or triplicate.

Within the library are perhaps one to several colonies or plaques that contain the DNA of interest. A single-strand DNA probe is designed to comple-

ment the target DNA and base pair (that is, hybridize) with it. The result is a DNA hybrid: one target DNA strand hydrogen bonded to one probe strand. The membranes first are treated so cells in colonies are lysed and cell debris is removed. The remaining DNA is denatured to obtain single-strand molecules. Single-strand DNA molecules bind tightly to the membrane by the sugar–phosphate backbones, and the unpaired nucleotides are free to base pair with a complementary DNA probe. Hybridization of single-strand probes to these membranes is used to identify specific recombinant DNA molecules from the bacterial colonies (Figure 3.18) or bacteriophage plaques transferred to the membranes. Although the probe and target DNA may not be completely complementary, enough base pairing must be established to produce a stable hybrid.

To detect whether the DNA probe and the target DNA bound to the membrane have hybridized, the probe is first made radioactive or is labeled by a nonradioactive method that results in fluorescence or stain-

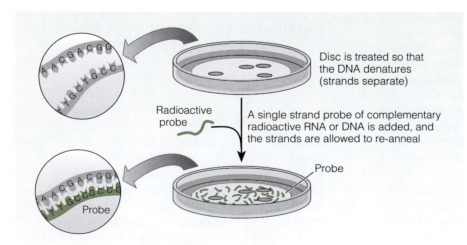

Disc is treated so that the DNA denatures (strands separate)

Radioactive probe

A single strand probe of complementary radioactive RNA or DNA is added, and the strands are allowed to re-anneal

Probe

Probe

FIGURE 3.18 Probe hybridization (cDNA or DNA) to single-stranded DNA in cloned DNA. The labeled probe base pairs with its complement.
Used by permission of University Science Books, from Berg and Singer, "Dealing With Genes," copyright © 1992.

ing after hybridization. After hybridization, the excess probe is washed from the membranes, which are then exposed to photographic film. Where hybridization has occurred, radiolabeled DNA makes an autoradiographic image.

Probes must be sufficiently complementary to the target DNA for hybridization to occur. A cDNA probe can be a piece of DNA or cDNA (for example, from another organism). For example, a **heterologous probe** (that is, the same DNA fragment from another species) from spinach might be used to hybridize to soybean DNA. For the nucleotide sequence of the probe to base pair with the target sequence, the spinach DNA must have enough sequence conservation; that is, the majority of the nucleotide sequence of the spinach DNA must be the same as the soybean DNA. Stringency (the extent of matching of base pairs) can be controlled during hybridization by adjusting temperature and salt concentration. Under conditions of highest stringency, hybridization occurs only when the bases between probe and template base pair almost perfectly. Under conditions of lowered stringency, base pairing between partially complementary sequences can occur and **mismatches** are tolerated. Thus, closely matching but nonidentical DNA can be used as probes.

The probe also can be made by a DNA synthesizer. If a portion of the amino acid sequence of the protein is known, the possible nucleic acid sequences can be deduced. A short DNA fragment called an **oligonucleotide** can be synthesized, labeled radioactively, and used as a probe in hybridizations. The genetic code is degenerate (see Chapter 2, From DNA to Proteins), and the specific codons for amino acids sometimes preferentially used by an organism (codon usage) usually are unknown. Consequently, a degenerate oligonucleotide probe must be made to account for all the different codons that may encode an amino acid. For example, to make an oligonucleotide that encodes the amino acids

ISOLEUCINE–ASPARTIC ACID–METHIONINE–
TRYPTOPHAN–GLUTAMIC ACID–GLUTAMINE

the degenerate oligonucleotide would look like this (positions 3, 6, 15, and 18 are degenerate in this example):

ATAGATATGTGGGAGCAG
  C   C   A A
  T

Expression Libraries

Expression libraries are made with a cloning vector that contains the required regulatory elements for gene expression, such as the promoter region. In an *E. coli* expression vector, an *E. coli* promoter is placed next to a unique restriction site where DNA can be inserted. When a foreign gene or cDNA is cloned into an expression vector in the proper reading frame (see Chapter 2, From DNA to Proteins), the gene is transcribed and translated in the *E. coli* host cell. When a library is made using an expression vector, only DNA—that is, genes (with introns if a eukaryotic gene) and cDNA (without genes)—inserted in-frame within the vector are expressed. Expression libraries are useful for identifying a clone containing the gene or cDNA of interest when an **antibody** (see Chapter 4, Basic Principles of Immunology) to the protein encoded by that gene or cDNA is available.

Antibodies are made by eukaryotic cells in response to proteins and other molecules that are recognized as foreign. Antibodies can be used instead of a nucleic acid as a probe to isolate a clone containing a sequence of interest that is transcribed and translated. A radioactively labeled antibody can be used to identify a specific protein made by one of the clones of the expression library (see Western blotting). Antibody binding, a technique similar to nucleic acid hybridization, identifies the clone containing the gene expressing the specific protein.

Expression vectors and a variety of host cells are used routinely to produce large quantities of a specific protein and have important commercial applications. Table 3.2 lists a few host cells and their important recombinant products.

TABLE 3.2 Examples of Host Cells Used to Produce Recombinant Proteins

Mammalian Cells	Saccharomyces cerevisiae (yeast)	Insect Baculovirus System
Growth hormone	Insulin	Adenosine deaminase
Erythropoietin	Epidermal growth factor	Erythropoietin
Interferon	α1 antitrypsin	Interferon
Interleukin-2	Granulocyte-macrophage colony stimulating factor	Poliovirus proteins
Monoclonal antibodies		Influenza virus hemagglutinin
Tissue plasminogen activator	Hepatitis B virus surface antigen	Tissue plasminogen activator
Blood-clotting factors	Hepatitis C virus protein	Rabies glycoprotein

REPORTER GENES

A **reporter** gene, connected to the DNA of interest and under the same control, is used to indicate whether the foreign gene or cDNA is being expressed. Many reporter genes are available that encode enzymes with readily assayed activities. One important reporter used today is the luciferase gene taken from either the firefly or bacteria *Vibrio harveyi*. When the firefly luciferase gene is expressed in the presence of the substrate luciferin and the energy source, ATP, the transformed cells give off detectable **bioluminescence** (Figure 3.19) that can be measured. Reporter genes enable the investigator to quantify the level of expression, test the expression level from different promoters, and identify the tissue in which the gene is being expressed. Reporter genes have been used to investigate gene expression in the cells of many different organisms (for example, plants, mammals, fish, and bacteria). Other reporter genes in use include

1. Green fluorescent protein (GFP)—This protein is produced by the jellyfish *Aequorea victoria* and interacts with a photoprotein, acquorin, resulting in a green fluorescence. GFP naturally fluoresces green when the protein is excited with blue or ultraviolet light, and thus needs no substrate for the reaction to occur. Variants of GFP have been engineered to fluoresce at different wavelengths and intensities, although because the signal is relatively weak, expression of GFP requires the use of very strong promoters or other regulatory sequences. A powerful application of GFP is the

tracking of specific proteins. When the GFP gene is fused to a gene of interest to create a fusion protein, GFP fluorescence indicates the presence of the fused protein of interest, and the transport and fate of the fusion protein is followed by fluorescence.

2. β-glucuronidase gene (GUS)—Isolated from *E. coli*, this gene encodes an enzyme that catalyzes the breakdown of a variety of β-D-glucuronides. In one type of assay, GUS generates a blue color that is formed by the breakdown of an intact, uncolored substrate. Alternatively, a more sensitive assay can be used when a different substrate is used to generate a fluorescent product. The product can be quantified to measure the amount of product formed.

SOUTHERN BLOT HYBRIDIZATION

In recombinant DNA technology, the location of specific regions within a cloned gene or DNA insert often must be known before the fragment can be sequenced or studied further. For example, the exact location of a 2-kb gene must be identified within a 15-kb insert so that it can be isolated for further analysis or sequencing. Restriction endonuclease digestion reduces the large DNA insert to DNA fragments, which then are separated by agarose gel electrophoresis. To identify the fragment that contains the gene of interest, a specific DNA probe, such as a small region of the gene of interest, can be used to hybridize to the DNA fragment.

In the mid-1970s, Edward Southern developed a simple technique, called **Southern blotting**, by which DNA fragments are transferred from gels to a special membrane that then is used for hybridization. During Southern blotting, these DNA fragments are denatured by an alkaline (high pH) buffer, and the single strands are transferred to a nylon or nitrocellulose membrane cut to the size of the agarose gel. The membrane is sandwiched between the gel on the bottom and paper towels stacked on top (Figure 3.20). Buffer in a trough moves by capillary action through a wick placed under the agarose slab, facilitating the transfer of DNA from the gel to the membrane.

The membrane is a replica of the agarose gel and is used in hybridizations as described earlier for colony or plaque hybridizations. If a probe hybridizes to fragments on the membrane, photographic film placed next to the membrane will be exposed where the probe has hybridized to a specific DNA band or bands. Southern blot hybridization is commonly used to identify a specific gene fragment from the often many DNA bands on a gel.

Although this is the traditional method of Southern blotting, efficiency, sensitivity, and reproducibility have greatly improved.

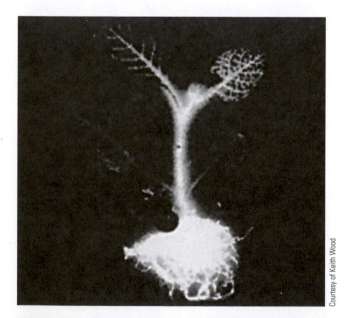

Courtesy of Keith Wood

FIGURE 3.19 A tobacco plant transformed with the firefly luciferase gene. The bioluminescence indicates the gene was incorporated into the plant DNA and is being expressed.

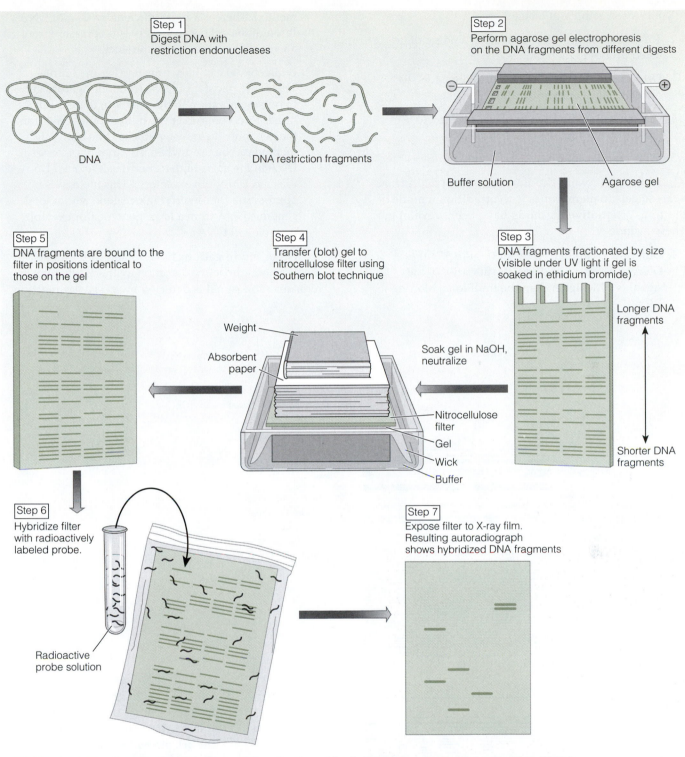

Step 1
Digest DNA with restriction endonucleases

DNA

DNA restriction fragments

Step 2
Perform agarose gel electrophoresis on the DNA fragments from different digests

Buffer solution

Agarose gel

Step 3
DNA fragments fractionated by size (visible under UV light if gel is soaked in ethidium bromide)

Longer DNA fragments

Shorter DNA fragments

Soak gel in NaOH, neutralize

Step 4
Transfer (blot) gel to nitrocellulose filter using Southern blot technique

Weight

Absorbent paper

Nitrocellulose filter

Gel

Wick

Buffer

Step 5
DNA fragments are bound to the filter in positions identical to those on the gel

Step 6
Hybridize filter with radioactively labeled probe.

Radioactive probe solution

Step 7
Expose filter to X-ray film. Resulting autoradiograph shows hybridized DNA fragments

FIGURE 3.20 The steps involved in conducting a Southern blot hybridization. In Southern blotting, DNA is transferred from an agarose gel to a membrane for hybridization by making a sandwich of the gel, membrane, filter paper, and absorbent paper. A salt solution that moves by capillary action facilitates the transfer of DNA from the gel to the membrane.

1. Increased activity of labeled probes

2. Treatments to prevent nonspecific binding of labeled probes to membranes

3. Use of sensitive detectors (phosphoimagers) to obtain images of hybridization signals

4. Use of vacuum blotting or electroblotting to greatly reduce the DNA transfer time

5. DNA transfer set up

Although upward capillary transfer of DNA from agarose gels to membranes is the traditional mode of blotting, alternative methods have been developed. These include

1. Downward capillary transfer (Figure 3.21a)—DNA transfer is rapid and the intensity of the signal is 30% higher than the traditional blotting method. There is a more efficient transfer of DNA through the gel because there is no pressure from a weight placed on top of the stack.

2. Bidirectional capillary transfer (Figure 3.21b)—This is the transfer of DNA from one agarose gel to two membranes. In this method, there is not a buffer reservoir, rather the buffer comes only from the gel; thus, the transfer efficiency is low. This method can be used when the amount of target DNA in the gel is high (for example, cloned DNA, plasmid DNA, other vectors). This method is not appropriate for identifying one gene within total genomic DNA from a large genome (for example, mammalian DNA).

Other modifications have included the development of nonradioactive detection methods whereby the hybridization signal is detected by a (1) color reaction

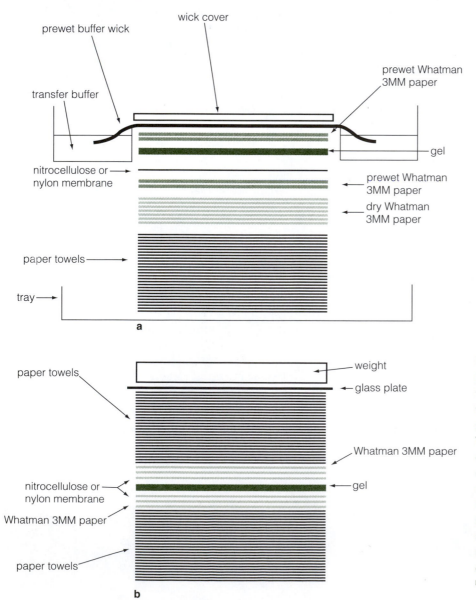

prewet buffer wick
wick cover
transfer buffer
prewet Whatman 3MM paper
nitrocellulose or nylon membrane
gel
prewet Whatman 3MM paper
dry Whatman 3MM paper
paper towels
tray

a

paper towels
weight
glass plate
Whatman 3MM paper
gel
nitrocellulose or nylon membrane
Whatman 3MM paper
paper towels

b

FIGURE 3.21 Southern blot capillary DNA transfer. (a) Downward capillary transfer. The DNA is transferred from the gel, placed toward the top of the stack, to the membrane beneath it. The transfer buffer moves down by capillary action. (b) Bidirectional capillary transfer. This procedure is useful for analyzing cloned DNAs but is not sensitive enough to analyze large, complex genomes. In this method, DNA is transferred to two membranes using only the liquid contained in the gel.

on the membrane (least sensitive), (2) fluorescence on the membrane detected by a transilluminator that emits light at 290 nm, and (3) chemiluminescence that is detected by exposure to x-ray film for 1 hour (fast and sensitive). Although more than 50 different nonradioactive detection kits are available commercially, they do not match the sensitivity and reproducibility of radioactive methods.

Southern blot hybridizations are used to detect differences in banding patterns between organisms or within DNA regions. Differences in restriction fragment lengths of DNA, called **restriction fragment length polymorphisms**, or RFLPs, are used to generate individual DNA fingerprints. DNA **fingerprinting** is a powerful tool in forensic analysis (see Chapter 11, DNA Profiling, Forensics, and Other Applications). Individuals can be identified by their DNA fingerprints, and because RFLPs are inherited, such fingerprinting can identify maternally and paternally inherited DNAs to settle paternity disputes. Another important application is in medical **diagnostics**, where differences in RFLPs are used to identify genetically inherited diseases. Some defective genes differ from their normal counterparts by only one nucleotide; however, if the mutation is in a restriction enzyme recognition site, the enzyme may not be able to cut the DNA, thereby generating a different DNA banding pattern after hybridization with a probe specific to the DNA region. A comparison of the hybridization patterns enables medical clinicians to detect normal and variant genes in an individual.

Southern blot hybridization differs from colony hybridization in two distinct ways. In Southern blot hybridization (1) the blotted DNA can be cloned or un- cloned (for example, total digested genomic DNA) and (2) DNA is first separated on an agarose gel and then transferred to a membrane before hybridization. In colony hybridization (1) the DNA to be probed is cloned into vectors and (2) the colonies of bacteria are blotted onto membrane and then lysed to release the DNA to the membrane before hybridization. Complete genomic libraries can be screened by colony hybridization. From the results of hybridization, the colony harboring the DNA of interest is identified.

NORTHERN BLOT HYBRIDIZATION

A similar blotting and hybridization method, called Northern blotting, can be used to probe RNA molecules separated by denaturing agarose gel electrophoresis. Instead of DNA, RNA is transferred from the gel to a membrane in a similar way to that described for DNA transfer. The probes are DNA. Northern blot hybridization is used to measure the quantity and determine the size of specific transcribed RNAs. Thus, this method is useful for studying the expression of specific genes.

POLYMERASE CHAIN REACTION

Traditional DNA isolation methods rely on the construction and screening of genomic or cDNA libraries, a process that takes many steps. A method developed in 1985 enables a researcher to rapidly isolate a specific DNA without building and screening a library. The method, called the **polymerase chain reaction** (PCR), uses the components of DNA replication to replicate a specific DNA fragment in the test tube (Figure 3.22).

FIGURE 3.22 Polymerase chain reaction is a powerful method for multiplying specific sequences of DNA.

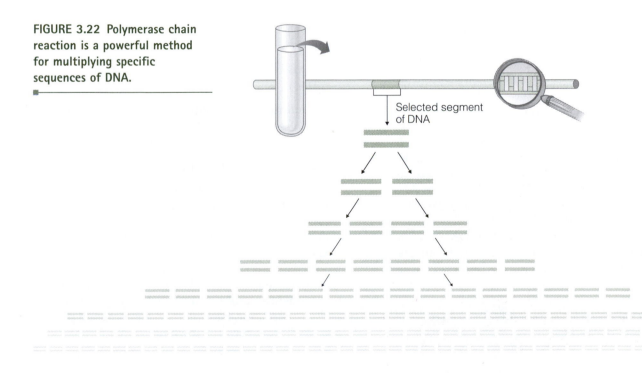

Selected segment of DNA

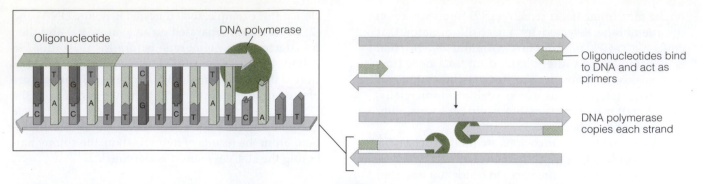

FIGURE 3.23 Short fragments of DNA called oligonucleotides act as primers to allow a specific DNA to be replicated by DNA polymerase. The approximately 18–21 nucleotide primer base pairs with the template DNA to initiate replication in the test tube. In this illustration, one primer is shown.

PCR enables scientists to selectively amplify specific DNA sequences without using hybridization. Any DNA sequence can be isolated from the total DNA of an organism. Generally, however, the sequence of the region that flanks the DNA to be amplified must be known so that **primers** used in amplification can be synthesized.

Two short oligonucleotide primers are used that flank the DNA region to be amplified. The primers anneal, or hybridize, to the target sequence (Figure 3.23), one on each strand of the double-strand DNA molecule. The oligonucleotides define the limit of the region to be amplified, and the DNA polymerase replicates the DNA between the primers using all four of the deoxyribonucleotides (dGTP, dATP, dCTP, dTTP) provided in the test tube (Figure 3.24). In an amplification (that is, replication) cycle, the template is denatured by high temperature, the primers are annealed by lowering the temperature, and the DNA polymerase extends the DNA from the primers.

Repeated cycles of denaturation, primer annealing, and DNA synthesis result in the exponential amplification of DNA. About 25 to 40 cycles are generally conducted using a thermal cycler, an instrument that automatically controls temperatures and times. A special DNA polymerase—*Taq* polymerase, isolated from a thermophilic bacterium, *Thermus aquaticus*, living in hot springs—is stable at the high temperature used to denature the template DNA. The product generated from the polymerase chain reaction is analyzed by agarose gel electrophoresis.

PCR allows a specific gene or other DNA region to be rapidly isolated from total DNA without the time-consuming task of screening a library. PCR is used to

1. Rapidly isolate specific sequences for further analysis or for cloning

2. Identify specific genetic loci for diagnostic or medical purposes

3. Generate DNA fingerprints to determine genetic relationships or to establish identity in forensics

4. Rapidly sequence DNA

DNA SEQUENCING

Today, laboratories routinely determine the order of deoxyribonucleotide sequences in DNA. DNA sequencing methods are valuable tools; they

1. Confirm the identity of genes isolated from DNA libraries by hybridization or amplified by the polymerase chain reaction

2. Determine the DNA sequence of promoters and other regulatory DNA elements that control expression

3. Reveal the fine structure of genes and other DNA

4. Confirm the DNA sequence of cDNA and other DNA synthesized in the test tube

5. Help scientists deduce the amino acid sequence of a gene or cDNA from the DNA sequence

Today, large genome projects, such as the Human Genome Project, mouse genome project, and rice genome project, are yielding information about the evolution of genomes; the location of genes, regulatory elements, and other sequences; and the presence of mutations that give rise to genetic diseases. Computers are routinely used to find coding sequences, predict protein sequences, and rapidly search databases (EMBL and Genbank, for example) for homologous protein and DNA sequences of known function.

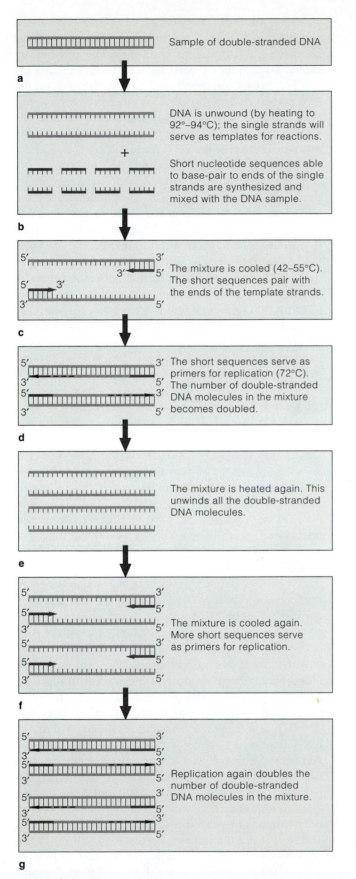

a Sample of double-stranded DNA

b DNA is unwound (by heating to 92°–94°C); the single strands will serve as templates for reactions.

+

Short nucleotide sequences able to base-pair to ends of the single strands are synthesized and mixed with the DNA sample.

c 5′　　　　　　　3′
3′　　　　5′
5′　　3′
3′　　　　　　　5′

The mixture is cooled (42–55°C). The short sequences pair with the ends of the template strands.

d 5′　　　　　　　3′
3′　　　　　　　5′
5′　　　　　　　3′
3′　　　　　　　5′

The short sequences serve as primers for replication (72°C). The number of double-stranded DNA molecules in the mixture becomes doubled.

e

The mixture is heated again. This unwinds all the double-stranded DNA molecules.

f 5′　　　　　　　3′
5′　　　　　5′
3′　　　　　5′
5′　　　　　3′
5′　　　　　5′
3′　　　　　5′

The mixture is cooled again. More short sequences serve as primers for replication.

g 5′　　　　　　　3′
3′　　　　　5′
5′　　　　　3′
3′　　　　　5′
5′　　　　　3′
3′　　　　　5′
5′　　　　　3′
3′　　　　　5′

Replication again doubles the number of double-stranded DNA molecules in the mixture.

FIGURE 3.24 The polymerase chain reaction (PCR) is an *in vitro* DNA replication method.

FIGURE 3.25 The structure of a dideoxynucleotide shows the substitution of an H instead of an OH on the 3′ carbon of the deoxyribose (shaded area).

Two DNA sequencing methods were first reported in 1977. Allan Maxam and Walter Gilbert at Harvard University developed a chemical method that uses selective degradation of bases, and (2) Frederick Sanger developed a DNA synthesis method (that is, test tube replication) that involves the termination of newly synthesized DNA strands at specific sites. In both methods, the DNA is labeled and gel electrophoresis separates DNA fragments. Most investigators now use the Sanger sequencing method.

In the Sanger method, a DNA primer is annealed to the denatured template DNA, and DNA polymerase extends the sequence from the primer. During DNA synthesis, labeled nucleotides are incorporated (for example, ^{32}P-labeled dATP) when the nucleotide is specified. Alternatively, instead of the label being incorporated into the growing DNA sequence, the primer is labeled. At random points during the synthesis, 2′,3′- **dideoxyribonucleotides** (or dideoxynucleoside triphosphates, ddNTPs) are incorporated into the growing strands that then terminate DNA synthesis (Figure 3.25). Four different ddNTPs are used: dideoxyadenosine, dideoxycytidine, dideoxyguanidine, and dideoxythymidine triphosphates—one for each of the nucleotide termination reactions. DNA synthesis cannot proceed when, for example, the dideoxyribonucleotide ddATP is incorporated into the DNA strand instead of dATP, because the absence of a 3′-OH group prevents the addition of a new deoxyribonucleotide.

The DNA to be sequenced is put into four test tubes, one for each of the four nucleotide bases; the tubes contain all four nucleotides (dATP, dTTP, dGTP, and dCTP, one of which is radioactively labeled), DNA polymerase, and a small amount of one of the four ddNTPs. For example, in the A tube (the other tubes being C, T, and G) all the normal nucleotides are added, but the only ddNTP is ddATP, which terminates DNA synthesis at adenine (Figure 3.26). The A tube contains many identical DNA molecules that are being sequenced; however, termination occurs only at A sites. The resulting strands are of different lengths, because termination is a random process. If all the

DNA template to be sequenced:

3' GGGCCTAAGCCTTAAACTGAAGGTTATGCCCCTTAGCC 5'

The sequencing primer is annealed to the denatured template to initiate DNA sequencing:

3' GGGCCTAAGCCTTAAACTGAAGGTTATGCCCCTTAGCC 5'
   | | | | | | | | | | | |
   5' CCCGGATTCGG3'

Deoxyribonucleotides used:

dCTP, dATP, dTTP, dGTP

Dideoyribonucleotide used to teminate DNA sequencing in the "A" tube:

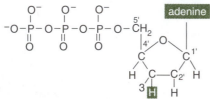

Sequencing Reactions in the tube (A* indicates sequencing was terminated by a dideoxy ATP):

3' GGGCCTAAGCCTTAAACTGAAGGTTATGCCCCTTAGCC 5'
5' CCCGGATTCGGA*

3' GGGCCTAAGCCTTAAACTGAAGGTTATGCCCCTTAGCC 5'
5' CCCGGATTCGGAA*

3' GGGCCTAAGCCTTAAACTGAAGGTTATGCCCCTTAGCC 5'
5' CCCGGATTCGGAATTTGA*

3' GGGCCTAAGCCTTAAACTGAAGGTTATGCCCCTTAGCC 5'
5' CCCGGATTCGGAATTTGACTTCCA*

3' GGGCCTAAGCCTTAAACTGAAGGTTATGCCCCTTAGCC 5'
5' CCCGGATTCGGAATTTGACTTCCAA*

3' GGGCCTAAGCCTTAAACTGAAGGTTATGCCCCTTAGCC 5'
5' CCCGGATTCGGAATTTGACTTCCAA*

3' GGGCCTAAGCCTTAAACTGAAGGTTATGCCCCTTAGCC 5'
5' CCCGGATTCGGAATTTGACTTCCAATA*

3' GGGCCTAAGCCTTAAACTGAAGGTTATGCCCCTTAGCC 5'
5' CCCGGATTCGGAATTTGACTTCCAATACGGGGA*

3' GGGCCTAAGCCTTAAACTGAAGGTTATGCCCCTTAGCC 5'
5' CCCGGATTCGGAATTTGACTTCCAATACGGGGAA*

FIGURE 3.26 DNA sequencing: the "A" tube reaction. A hypothetical DNA sequence is used to illustrate how dideoxy sequencing is conducted. Either the primer or one of the deoxyribonucleotides is radiolabeled to allow detection of the banding pattern on x-ray film. Three other reaction tubes for "G," "C," and "T" would be used to obtain the DNA sequence.

strands are pooled, a termination will have occurred at each A. In other words, all possible A sites in the sequence will have an incorporated ddATP. The same process occurs in the C, T, and G tubes with their respective dideoxynucleotides. After termination, the DNAs in all four tubes are denatured by heating and the DNA strands are then separated by differences in length through electrophoresis on a denaturing polyacrylamide gel (a chemical polymer that keeps the single-strand DNAs from hydrogen bonding to one another). The band pattern is detected by **autoradiography.** X-ray film is placed against the sequencing gel and is exposed and developed. The dark bands on the film correspond to the DNA fragments that have migrated to a certain point according to their length. The base sequence is read directly from the film, beginning from the bottom of the gel (Figure 3.27).

The Maxam and Gilbert method of sequencing uses double-strand DNA molecules that are radioactively labeled at the 5' end. The DNA is denatured and added to four tubes, one for each of the four nucleotides. To each tube is added a certain chemical that modifies and then removes a specific base from the DNA strand. For example, a chemical is added to the C tube that removes cytosine. After the chemical reaction, the DNA backbone is cut where the base is missing. This process generates many radioactively labeled fragments of varying lengths, each ending where a specific base was eliminated. As with the Sanger method, DNA is separated by size on a denaturing gel, a researcher detects the radiolabeled DNA fragments on the gel by autoradiography, and the sequence is read.

DNA sequencing has been automated to rapidly sequence thousands of base pairs—a requirement in laboratories that daily sequence long stretches of DNA. A variety of methods are available; some protocols rely on labeling the nitrogenous base of each of the four dideoxynucleotides (ddATP, ddGTP, ddTTP, and ddCTP) with fluorescent dye (a BigDye-terminator method), each emitting a different wavelength (color) upon fluorescence. Unlike conventional sequencing and gel electrophoresis, in this automated method all four reactions (G, A, C, and T) occur in one tube and the fragments are separated in one lane of the gel (often a capillary tube with a very small diameter—50–100 μm). A laser beam scans the lane or capillary tube and excites the dye molecules. The wavelengths are detected by a special camera (CCD camera) or, in older sequencing models, a photomultiplier tube (Figure 3.28). Results are presented as an electropherogram, with each base peak printed in a different color.

PCR also is used in a variation of Sanger dideoxy DNA sequencing, a method called cycle sequencing. Annealing of the sequencing primer, primer extension (that is, DNA synthesis), and chain termination all can be conducted in the thermal cycler in one tube.

The efficiency and accuracy of automated DNA sequencing has increased dramatically in recent years. Today, it is not uncommon for some machines to run 96 samples or more at one time and, with an automated sample loader, handle more than 4000 samples each week with an average nucleotide reading of 600–650 per single run.

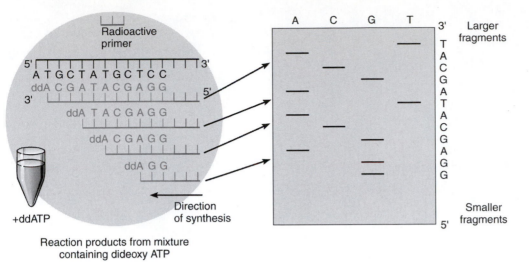

The random incorporation of dideoxy ATP into the growing chain generates a series of smaller DNA fragments ending at all the possible positions where adenine is found in the newly synthesized fragments. These correspond to positions where thymine occurs in the original template strand.

a

The radioactive products of each reaction mixture are separated by gel electrophoresis and located by exposing the gel to x-ray film. The nucleotide sequence of the newly synthesized DNA is read directly from the film (5' → 3'). The sequence in the original template strand is its complement (3' → 5').

b

FIGURE 3.27 A Sanger dideoxy DNA sequencing showing (a) the random incorporation of dideoxy ATP into the growing DNA chain. Direction of synthesis is from 5' to 3'. Fragments end at all the possible positions where adenine is found in the newly synthesized DNA. These correspond to thymine in the complementary template DNA. (b) The radiolabeled products of each reaction mixture are separated by gel electrophoresis and located by exposure to x-ray film. The nucleotide sequence is read in the 5' to 3' direction, beginning at the bottom of the film (called an autoradiogram). The sequence of the template strand is the complement (3' → 5') of a sequencing gel showing the bands that are read to obtain the sequence. Sequence is read from bottom to top. (c) An exposed x-ray film showing the sequence in all four lanes. The DNA sequence is read from the bottom to top, incorporating sequence from all four lanes.

An exposed x-ray film of a DNA sequencing gel. The four lanes represent A, C, G, and T dideoxy reaction mixes, respectively.

c

PROTEIN METHODS

The study of proteins is an important component of recombinant DNA technology. Proteins are the molecules scientists ultimately seek to modify quantity and sequence. The structural and functional characteristics of a protein must be elucidated before the protein's structure is altered. The important amino acids must be identified and the ultimate effect of the modification on, for example, enzyme activity or protein stability must be determined. A few of the methods used to study proteins are discussed in the following sections.

Protein Gel Electrophoresis

Electrophoresis is used to separate individual proteins by size or overall charge. The proteins are separated by one- or two-dimensional polyacrylamide gel electrophoresis; polyacrylamide is used instead of agarose because it gives better resolution. Usually the one-dimensional gel is a vertical slab, unlike the horizontal slab used for separating nucleic acids. Proteins are visualized as bands (Figure 3.29) in a one-dimensional gel by a dye that binds to the proteins in the gel or by radiolabeling and autoradiography.

Proteins are visible as spots rather than bands in a two-dimensional separation (Figure 3.30). First, the proteins are separated by charge in a slender tube gel (a pH gradient is established in the gel) in a process called isoelectric focusing. The tube is then placed horizontally along the top of a vertical slab gel, and proteins move by electrophoresis out of the tube and into the gel, which separates them by size. A detergent called *sodium dodecyl sulfate* (SDS) that is used in the second dimension binds to the hydrophobic regions of proteins, giving the proteins an overall negative charge. Thus, in the second dimension, proteins are resolved by differences in size rather than charge. Because all proteins are negatively charged, proteins move toward the positive end of the gel, with large

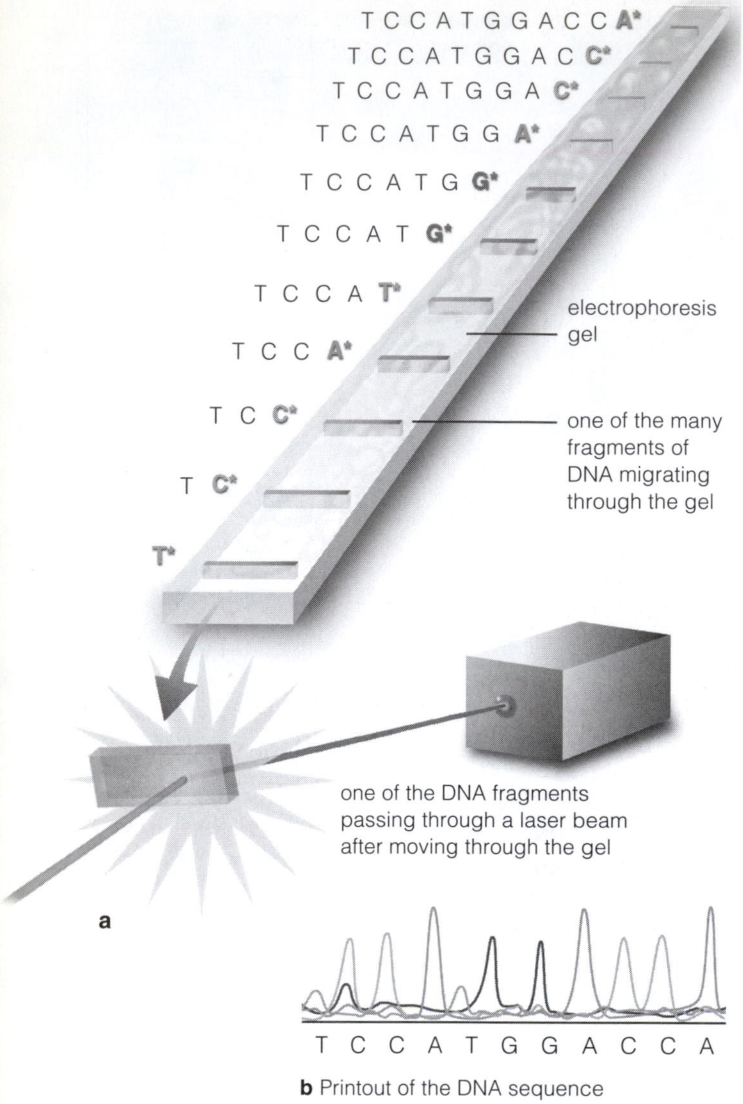

TCCATGGACC**A***
TCCATGGAC**C***
TCCATGGA**C***
TCCATGG**A***
TCCATG**G***
TCCAT**G***
TCCA**T***
TCC**A***
TCC**C***
TC**C***
T**C***
T*

electrophoresis gel

one of the many fragments of DNA migrating through the gel

one of the DNA fragments passing through a laser beam after moving through the gel

a

T C C A T G G A C C A

b Printout of the DNA sequence

FIGURE 3.28 Automated DNA sequencing. (a) A dideoxynucleotide at the end of each synthesized DNA fragment is labeled with a fluorescent dye. Each modified nucleotide (A*, G*, C*, T*) fluoresces a different color when it passes through a laser beam. Whenever a modified nucleotide is incorporated into the growing DNA chain, DNA synthesis is terminated. (b) Printout of the DNA sequence. Each peak indicates the absorbance of a particular labeled nucleotide.

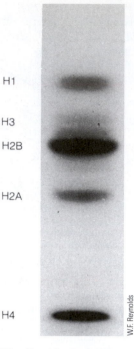

H1
H3
H2B
H2A
H4

W.F. Reynolds

FIGURE 3.29 The five histone proteins that associate with eukaryotic DNA are shown. Proteins are separated into discrete bands by gel electrophoresis and stained or radiolabeled for visualization.

molecules moving more slowly than smaller ones. Separating proteins in two dimensions achieves greater resolution of the molecules.

Protein Engineering

Protein engineering is an exciting area that enables researchers to improve specific protein characteristics. By modifying the protein's gene sequence, they can

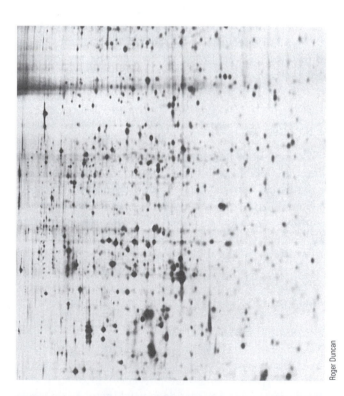

Roger Duncan

FIGURE 3.30 Human HeLa cell proteins separated by two-dimensional gel electrophoresis are visualized as spots.

change protein structure to enhance protein function in specific ways. Changes can include increased stability, such as resistance to degradation, pH change, temperature, oxidation, and contamination; enhanced enzyme activity; a change in substrate specificity or enzyme activity in extreme or less-than-optimum conditions; and increased nutritional value. The researcher must know the amino acid sequence (often deduced from the DNA sequence), the chemical properties of the protein, the predicted three-dimensional structure, which is usually generated by a computer (sometimes from analysis of the purified protein), and in some cases the specific amino acids that contribute to the catalytic function of the enzyme. The properties of proteins are modified by changing certain amino acids.

Several methods are available for changing the amino acids of proteins by mutating the corresponding nucleotides of the DNA. Frequently, only one nucleotide must be mutated to produce the desired amino acid change. One method, oligonucleotide-directed mutagenesis, uses a short, single-strand DNA, an oligonucleotide 15 to 20 bases long, that is complementary to the gene region to be mutated except for the individual bases to be changed. The mutagenesis process is outlined in the following:

1. The gene to be mutated must first be cloned. A viral vector, the bacteriophage M13, is used in the cloning process. M13 has both a single-strand DNA stage that is packaged within a viral coat and a double-strand phase, or **replicative form** (RF), that is synthesized after cell infection. The RF, which is structurally similar to a plasmid, is used for cloning the DNA to be mutated. From these recombinant molecules, single-strand molecules are then generated for the mutagenesis process (Figure 3.31).

2. The oligonucleotide containing a specific change in sequence is then annealed to the single-strand circular DNA molecule, except at the specific mismatch site that cannot base pair with it.

3. The oligonucleotide serves as a primer for replication by DNA polymerase, and DNA ligase seals the ends of the newly synthesized circular DNA. The resulting molecule is a double-strand, circular DNA with a mismatch region where the oligonucleotide mutation is located.

4. The DNA is transferred to *E. coli*, where it is replicated. One strand of the original DNA molecule has the mutation from the oligonucleotide and the other does not. Thus, some replicated RF double-strand molecules will have the mutation in their DNA sequence and others will not; the latter are called *wild type*.

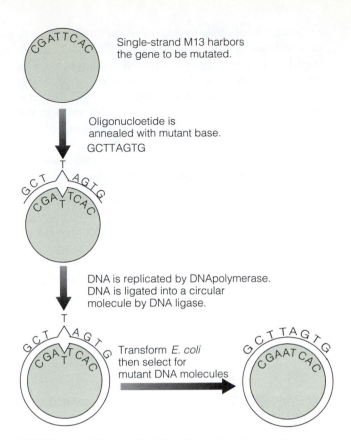

FIGURE 3.31 Oligonucleotide–directed mutagenesis involves the introduction of a desired base change into the target DNA during replication of the DNA.

5. Progeny phage are produced within the bacterial host cells and are released as a single-strand DNA packaged in a viral protein coat.

6. Mutant plaques are determined by hybridizing the oligonucleotide (used to generate the mutation) to the resulting plaques containing either wild-type or mutant DNA.

Because this mutagenesis procedure is inefficient and generates far fewer mutant molecules than expected, methods for enriching mutated DNA are used. Once obtained, the double-strand mutant DNA (gene or cDNA) is isolated, placed in an expression vector, transferred to appropriate host cells such as bacteria or yeast, and expressed as a mutant protein for further study.

Another way to mutate proteins is to substitute a DNA fragment, or *cassette*, containing the selected nucleotide mutations for the same DNA fragment in the organism. The DNA region containing the nucleotides to be changed is removed with a restriction enzyme and replaced with the cassette. This replacement is conducted *in vitro* in a vector that is then transferred to *E. coli* for replication.

A variety of other mutagenesis methods are used, some technically complex and using PCR, such as PCR-amplified oligonucleotide-directed mutagenesis. The method used depends on the researcher's goal: to obtain random, multiple-point mutations by producing single nucleotide changes at different intervals in the DNA; to introduce deletions or insertions of nucleotides; or, if the exact nucleotides to be changed are unknown, to generate a series of DNAs with different mutations in a specific area for further study.

Protein Sequencing

Experiments sometimes require that the amino acid sequence be determined for a protein. For example, if a protein has been purified and a DNA probe is unavailable for hybridization and gene isolation, the protein can be partially sequenced. Oligonucleotide probes, generated from the amino acid sequence, can be used for PCR or hybridization to isolate the gene.

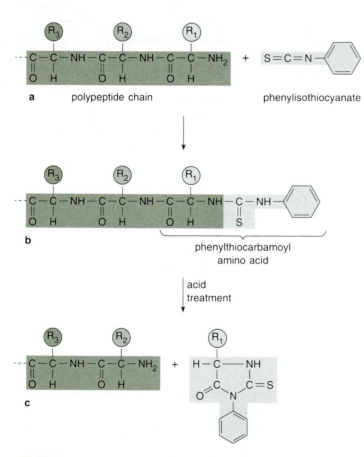

FIGURE 3.32 The Edman degradation method for determining the amino acid sequence of a protein. (a and b) The NH_2–terminal end of the polypeptide is chemically modified to phenylthiocarbamoyl amino acid by the addition of phenylisothicyanate. (c) Acid treatment removes the phenylthiocarbamoyl amino acid that can be identified by analytical methods. The process, repeated for each amino acid at the NH_2–terminus, continues until the polypeptide is sequenced.

The linear order of amino acids in a protein can be elucidated by a process called Edman degradation. In this method, the amino acids that make up the amino terminal end (the NH_2 terminus) of the polypeptide strand are determined one at a time by chemical cleavage (Figure 3.32). The first amino acid is chemically modified to a phenylthiocarbamoyl amino acid and removed by acid treatment for identification. The next amino acid is now at the beginning of the chain (NH_2 terminus), and the process is repeated until all the amino acids of the protein chain are identified. This method has been automated. More recently, state-of-the-art mass spectrometry methods are being used because it is a more sensitive method. While Edman degradation sequencing requires approximately 10 picomoles of protein, mass spectrometry can be at the picomole scale or smaller.

DNA MICROARRAY TECHNOLOGY

As the number of sequences obtained from a variety of organisms such as bacteria, fungi, fruit fly (*Drosophila*), mice, plants, and humans increases with rapid sequencing methods, there is a growing need for new, powerful methods to conduct functional analyses on sequences. Older single-gene methods such as Northern blot hybridization and RT-PCR (reverse transcriptase–PCR), a version of PCR that allows the detection of RNA being expressed, are not as efficient or sensitive. One emerging technology is DNA microarray, which allows the analysis of an entire genome in a single experiment. For example, a single yeast DNA microarray (approximately 6200 genes in the genome) is equivalent to approximately 6200 Southern blots. DNA microarray is being used to

1. Analyze gene activity (expression). Scientists are able to determine which genes are active, for example, in a particular cell, tissue, or organ. They are able to correlate changes in gene expression with changes in a physiological or biochemical process such as photosynthesis in high and low light conditions. Other examples include studying the genes expressed during a particular stage in the early development of an organism such as a mouse or genes expressed in different tissues. Microarray technology also can be used to compare gene expression patterns in normal versus diseased cells (for example, cancer). The effect of various therapeutic drugs on gene expression patterns also can be evaluated.

2. Follow changes in genomic DNA. Several laboratories are using DNA microarrays for the analysis of changes in genomic DNA. For example, changes in DNA occur in cancerous cells. Frequently, there are genomic modifications in the form of DNA amplification, point mutations, translocations, and deletions. Loss of gene function can occur or oncogenes

(genes that cause uncontrolled cell growth; can cause cancer) can become activated. Interestingly, single nucleotide polymorphisms (nucleotide differences in only one nucleotide) can be detected using microarrays that allow the detection of sequence variants in a specific gene. Oligonucleotides are used as probes, and hybridization is sensitive enough to detect one-nucleotide mismatches between target and probe.

Microarrays are used to study gene expression (that is, transcription) *in vivo*. The genes that are transcribed at any one point in time are referred to as the transcriptome. Known DNA sequences (for example, genomic sequences) placed on a substrate in an ordered array (a gridlike pattern) are used as probes to hybridize to unknown DNA or cDNA molecules labeled (from mRNA—transcription products) with a fluorescent tag of different colors. The locations where binding occurred (and the colors) are used to identify the DNA being expressed and to quantify the level of binding. Changes in the position of bound labeled DNA to the microarray DNAs correlate with changes in the specific genes being transcribed.

To make a microarray, typically, numerous single-strand molecules of DNA samples representing known expressed gene sequences as cDNAs or synthesized short oligonucleotides (10–25 nucleotides, from a portion of cDNA sometimes called EST, or expressed sequence tag) are immobilized, each with a defined location, onto a small substrate that can be a glass microscope slide or membrane filter. Sometimes DNA is permanently fixed (printed) onto a small chip (DNA chip technology—high density DNA microarrays). Chips are manufactured by high speed, precision robotics that can fix thousands of DNA samples on a glass slide in a 1-cm² area. The diameter of one sample is approximately 100 μm. Each spot consists of many single-strand copies of a single gene or oligonucleotide (Figure 3.33). Thousands of sequences placed very close together are bound to the substrate. This allows the screening of genes expressed in an entire organism, organ, or tissue—both healthy and diseased states. Examples of chips available commercially for research and diagnostic purposes include the HIV gene chip, breast cancer BRCA-1 gene chip, and the p53 chip.

The ordered microarray is then hybridized with a solution of unknown DNA sequences that are fluorescently labeled. Hybridization produces a specific pattern on the microarray that can be analyzed or compared with other patterns. A high-resolution laser scanner is used to detect the fluorescence (where hybridization occurred) on the substrate surface. By analyzing the signal from each spot with digital imaging software, the hybridization pattern can be qualitatively and quantitatively analyzed.

In much of the literature (and here), the probe refers to the immobilized cDNA or oligonucleotide sequences of the microarray, while the target refers to the mixture of fluorescently labeled nucleic acids (from expressed gene products, that is, cDNAs from isolated RNA).

The following outlines the steps in conducting a DNA microarray analysis (Figure 3.34):

Sequence of one gene

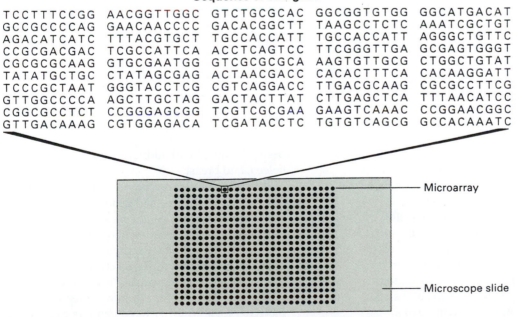

```
TCCTTTCCGG   AACGGTTGGC   GTCTGCGCAC   GGCGGTGTGG   GGCATGACAT
GCCGCCCCAG   GAACAACCCC   GACACGGCTT   TAAGCCTCTC   AAATCGCTGT
AGACATCATC   TTTACGTGCT   TGCCACCATT   TGCCACCATT   AGGGCTGTTC
CCGCGACGAC   TCGCCATTCA   ACCTCAGTCC   TTCGGGTTGA   GCGAGTGGGT
CGCGCGCAAG   GTGCGAATGG   GTCGCGCGCA   AAGTGTTGCG   CTGGCTGTAT
TATATGCTGC   CTATAGCGAG   ACTAACGACC   CACACTTTCA   CACAAGGATT
TCCCGCTAAT   GGGTACCTCG   CGTCAGGACC   TTGACGCAAG   CGCGCCTTCG
GTTGGCCCCA   AGCTTGCTAG   GACTACTTAT   CTTGAGCTCA   TTTAACATCC
CGGCGCCTCT   CCGGGAGCGG   TCGTCGCGAA   GAAGTCAAAC   CCGGAACGGC
GTTGACAAAG   CGTGGAGACA   TCGATACCTC   TGTGTCAGCG   GCCACAAATC
```

— Microarray

— Microscope slide

FIGURE 3.33 Each spot on the slide or chip represents the location of many copies of a single DNA sequence. All the spots in a grid–like pattern compose the DNA microarray. From Discovering Geonomics, Proteomics, and Bioinformatics, by Campbell and Heyer. Copyright 2003, pg. 108. Copyright © 2003, Benjamin Cummings.

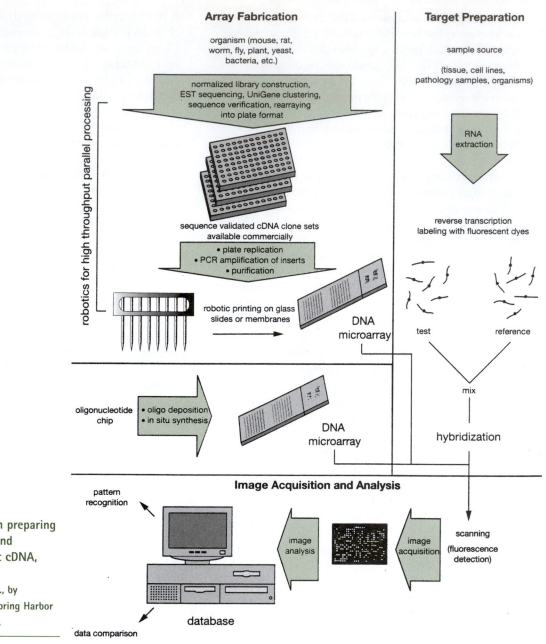

Array Fabrication

organism (mouse, rat, worm, fly, plant, yeast, bacteria, etc.)

robotics for high throughput parallel processing

normalized library construction, EST sequencing, UniGene clustering, sequence verification, rearraying into plate format

sequence validated cDNA clone sets available commercially

• plate replication
• PCR amplification of inserts
• purification

robotic printing on glass slides or membranes

DNA microarray

oligonucleotide chip

• oligo deposition
• in situ synthesis

DNA microarray

Target Preparation

sample source

(tissue, cell lines, pathology samples, organisms)

RNA extraction

reverse transcription labeling with fluorescent dyes

test   reference

mix

hybridization

Image Acquisition and Analysis

pattern recognition

image analysis

image acquisition

scanning (fluorescence detection)

database

data comparison

FIGURE 3.34 The steps in preparing a microarray, preparing and hybridizing labeled target cDNA, and analyzing the data.
From Molecular Cloning, 3 ed., by Sambrook and Russell, Cold Spring Harbor Laboratory Press, Fig. A10–11.

1. Isolating cDNA clones or preparing oligonucleotides for the microarray (probes). Custom DNA microarrays can be produced by the user (although time consuming) or purchased. For example, Affymetrix (http://www.affymetrix.com) sells gene chips designed for diagnostic purposes, such as the cytochrome P450 gene chip that is used to detect mutations of specific liver enzymes that metabolize drugs or the p53 chip that is used to detect single nucleotide polymorphisms of the p53 tumor-suppressor gene. Another example of a commercial microarray is from Operon Technologies (http://www.operon.com). Their products include tissue-specific arrays for eight human tissues—brain, heart, kidney, liver, prostate, lung, spleen, and breast.

Microarrays also can be genomic or cDNA sequences, cDNA clones, and genomic clone sets obtained from large-scale genome sequencing projects, including complete genome libraries, that have been assembled and verified. Examples are the 6000 rat cDNA clones offered by Research Genetics (http://www.resgen.com) and the compete set of PCR-amplified genes of *E. coli* from Genosys Biotechnologies (http://www.genosys.com/expression/frameset.html). Incyte Genomics (http://www.incyte.com/expression/argem1.html) has available a plant microarray that contains clones representing 7900 *Arabidopsis thaliana* genes from the major tissue categories such as roots, stems, leaves, and flowers. Other genomic sources for microarrays include sequences from humans, mice, yeast, the worm *Caenorhabditis elegans*, the fruit fly, and various bacteria.

2. Applying the cDNA or oligonucleotides to a substrate or chip.

3. Isolation of RNA—mRNA or total RNA. RNA is isolated from a particular cell or tissue state. RNA can be isolated, for example, from specific tissues, developmental stages, cell types, diseased tissue states, and cells in different physiological states (for example, after exposure to stress such as heat shock, cold shock, or nutrient deprivation).

4. Preparation of the target—cDNAs are synthesized from the RNA templates using the enzyme reverse transcriptase. The cDNAs are subsequently fluorescently labeled. Usually, control and test target sequences are prepared and labeled with different fluorescent dyes. For example, the test sequences may emit green (for example, cDNAs prepared from a disease state of a tissue), while the control sequences (normal state) emit red. Microarray methods usually use two fluorophores (color labels)—Cy3 (green) and Cy5 (red).

5. Hybridization of the fluorescently labeled target sample to the DNA microarray. Both Cy3- and Cy5-labeled cDNA samples are hybridized to the microarray.

As an example, the mRNA from plants grown in two different conditions, light and dark, is isolated. Plant cells are isolated from each plant treatment, and both mRNA groups are made into cDNAs—one labeled with Cy3 (say, the plants grown in light) and the other with Cy5 (in the dark). Each population of labeled mRNAs (either red or green) represents the transcriptome from one treatment (light or dark).

The two cDNA populations are mixed (green and red) and incubated with the DNA microarray or chip. After incubation overnight, the unbound cDNAs are washed away and the chip is dried in the dark to prevent light from bleaching the fluorescent dyes.

6. Data acquisition and analysis. After hybridization, the DNA microarray is scanned with lasers (each detecting specific wavelengths) to measure the amount of fluorescence in the hybridized target and probe. Images that show the locations and intensities for each fluorophore, green and red, are generated. The red image and green image are stored in a computer, which also superimposes the two images. Yellow spots indicate that the specific gene is transcribed in both populations of transcripts (red + green = yellow).

For quantification of the hybridized image, the intensities of the fluorescent spots—red, green, and the ratio of red to green are converted to numbers (Table 3.3).

TABLE 3.3 The Fluorescence Intensity for Each Dye, Red and Green, and the Calculated Ratio of Red to Green

Block	Column	Row	Gene name	Red	Green	Red/Green Ratio
1	1	1	*tub1*	2,345	2,467	0.95
1	1	2	*tub2*	3,589	2,158	1.66
1	1	3	*sec1*	4,109	1,469	2.80
1	1	4	*sec2*	1,500	3,589	0.42
1	1	5	*sec3*	1,246	1,258	0.99
1	1	6	*act1*	1,937	2,104	0.92
1	1	7	*act2*	2,561	1,562	1.64
1	1	8	*fus1*	2,962	3,012	0.98
1	1	9	*idp2*	3,585	1,209	2.97
1	1	10	*idp1*	2,170	1,005	2.78
1	1	11	*idh1*	1,896	4,245	0.51
1	1	12	*idh2*	1,896	2,996	0.63
1	1	13	*erd1*	1,023	3,354	0.31
1	1	14	*erd2*	1,698	2,896	0.59

The data is for only a portion of a microarray hybridization result. The location of each spot is shown as block, column, and row of the chip. The gene name of the immobilized DNA is given.

RNA Interference Technology: Gene Silencing

In 1998, a researcher at the Carnegie Institute of Embryology in Baltimore, Maryland, discovered that if an experimental worm was injected with double-strand RNA that was complementary to a specific gene, the worm could not produce the protein. Amazingly, inhibition of protein synthesis was observed even when the worms ate the double-strand RNA. The process of inhibiting a protein product by double-strand RNA is called RNA interference, or RNAi.

This method of gene silencing has been found to occur naturally in organisms such as earthworms, trypanosomes, fruit flies, some fish, mice, and plants. Although the exact role of RNA interference is unknown, it is thought to protect animals and plants from viruses or from the insertion of DNA sequences like transposons (mobile DNA sequences; they may multiply and insert into the genome by cDNA synthesis from the transcript).

During RNAi, long double-strand RNAs (dsRNAs) silence the expression of target genes to which they are complementary (Figure 3.35). During this process

1. Double-strand RNAs are processed into approximately 21–23 nucleotide RNAs with two nucleotide overhangs on the 3′ end, called small interfering RNAs (siRNAs). An RNase enzyme called Dicer cuts the dsRNA molecules (from a virus, transposon, or through transformation) into short siRNAs.

2. Each siRNA complexes with ribonucleases (distinct from Dicer) to form an RNA-induced silencing complex (RISC).

3. The siRNA unwinds and RISC is activated.

4. The activated RISC targets complementary mRNA molecules. The siRNA strands act as guides where the RISCs cut the transcripts in an area where the siRNA binds to the mRNA. This destroys the mRNA.

In experiments, cells can be transformed with siRNAs to reduce the level of gene expression dramatically. RNA interference is a powerful method for reducing the expression of specific genes up to 90% in living cells by specifically reducing the amount of mRNA. RNAi may become an effective therapeutic method for reducing the presence of unwanted or faulty proteins in humans and animals. This technology may provide an alternative approach (other methods include the use of ribozymes and antisense; see Chapter 10, Medical Biotechnology) to the treatment of disease.

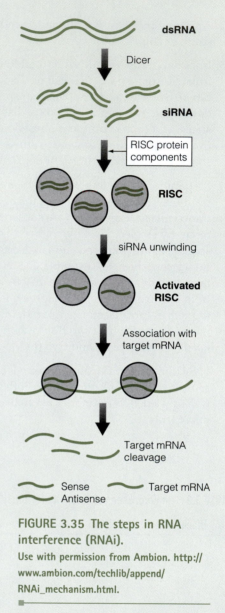

FIGURE 3.35 The steps in RNA interference (RNAi).
Use with permission from Ambion. http://www.ambion.com/techlib/append/RNAi_mechanism.html.

The same microarray technology is also used in the production of protein chips where thousands of proteins are bound in an ordered array. These chips are used to study protein–protein interactions.

APPLICATIONS OF RECOMBINANT DNA TECHNOLOGY

Recombinant DNA technology has revolutionized methods of conducting research in a variety of fields, not only molecular and cell biology and biochemistry but also in fields such as ecology and anthropology. Because of the complexity of eukaryotic genomes, genes and their regulation can be studied only when a gene or several genes have been isolated from the genome and are *in vitro* or in host cells. Recombinant DNA technology also plays a major role in modern biotechnology. Applications of recombinant DNA are found in many areas such as agriculture and medicine (Table 3.4). For example, the manufacture of important products (therapeutic human proteins, for example) from cloned genes; the analysis, diagnosis, and repair of genetic disorders; and the production of

TABLE 3.4 Some Applications of Recombinant DNA Technology

Field	Application	Importance
Basic biology	DNA sequencing	Answers questions about gene structure, gene function, and relatedness of genes and organisms
	Directed mutagenesis	Answers questions about gene function
Medicine	Therapeutic proteins	Makes human proteins for treating diseases such as diabetes, pituitary dwarfism, hemophilia
	Gene therapy	Treats genetic diseases such as cystic fibrosis
	Improved vaccines	Produces more effective vaccines with fewer side effects
	Diagnosis	Allows rapid, accurate diagnosis of infections and other diseases
	Veterinary medicine	Better diagnosis, prevention, and treatment of disease
Industry	Altering microorganisms	Improved production of antibiotics, amino acids, vitamins, and enzymes; also improved disposal of waste, including persistent toxic chemicals
Agriculture	Altering plants	More rapid breeding of disease-resistant and improved plants (e.g., tomatoes that stay fresh-tasting longer)
	Altering farm animals	More rapid development of superior breeds
Criminal investigation	DNA fingerprinting	Can determine if a biological sample such as blood, semen, or tissue is from a particular person

J. Ingraham and C. Ingraham, *Introduction to Microbiology*, Table 7.2. Copyright © 1995 Wadsworth Publishing Co.

transgenic plants, animals, and bacteria can be achieved only with recombinant DNA methods. Exciting new applications have the potential to provide us with new medical treatments, to augment current methods of removing toxic wastes and pollutants from the environment, and to increase agricultural productivity.

The following chapters review new advances in various fields of biotechnology, including the applications of recombinant DNA technology.

General Readings

F.M. Ausubel, R. Brent, and R.E. Kingston, eds. 1994. *Current Protocols in Molecular Biology.* Wiley, New York.

K. Drlica. 2004. *Understanding DNA and Gene Cloning: A Guide for the Curious.* 4th ed. Wiley, Hoboken, New Jersey.

B.R. Glick and J.J. Pasternak. 2003. *Molecular Biotechnology: Principles and Applications of Recombinant DNA.* 3rd ed. ASM Press, Washington, D.C.

M.A. Innis, D.H. Gelfand, J.J. Sninsky, and T.J. White. 1990. *PCR Protocols: A Guide to Methods and Applications.* Academic Press. San Diego, California.

H. Kreuzer and A. Massey. 2001. *Recombinant DNA and Biotechnology: A Guide for Students.* 2nd ed. ASM Press, Washington, D.C.

K.B. Mullis, F. Ferre, and R.A. Gibbs, eds. 1994. *The Polymerase Chain Reaction.* Birkhauser Verlag, Basel, Switzerland.

P. Peters. 1993. *Biotechnology: A Guide to Genetic Engineering.* Wm. C. Brown, Dubuque, Iowa.

J. Sambrook and D.W. Russel. 2001. *Molecular Cloning: A Laboratory Manual.* Vol. 1–3, 3rd ed. Cold Spring Harbor Laboratory Press, New York.

Additional Readings

J. Alam and J.L. Cook. 1990. Reporter genes: Application to the study of mammalian gene transcription. *Anal. Biochem.* 188:245–254.

L. Alphey. 1997. *DNA Sequencing: From Experimental Methods to Bioinformatics.* BIOS Scientific Publishers Ltd., New York.

P. Anderle et al. 2003. Gene expression databases and data mining. *BioTechniques Suppl.* March: S36–S44.

P. Balbas and F. Bolivar. 1990. Design and construction of expression plasmid vectors in *E. coli. Methods Enzymol.* 185:14–37.

V. Benes and M. Muckenthaler. 2003. Standardization of protocols in cDNA microarray analysis. *Trends Biochem. Sci.* 28:244–249.

Best Bio Protocols on the Web. http://orbigen.com/protocols/Best_WebProtocols.html

F. Bolivar, R.L. Rodriguez, P.J Greene, M.C. Betlach, H.L. Hyneker, H.W. Boyer, J.H. Crosa, and S. Falkow. 1977. Construction and characterization of new cloning vehicles. II. A multipurpose cloning system. *Gene* 2:95–113.

D.T. Burke, G.F. Carle, and M.V. Olson. 1987. Cloning of large segments of exogenous DNA into yeast by means of yeast artificial chromosome vectors. *Science* 236:806–812.

C. Cagoni and G. Macino. 2000. Post-transcriptional gene silencing across kingdoms. *Genes Dev.* 10:638–643.

J.-T. Chi, H.Y. Chang, N.N. Wang, D.S. Dustin, N. Dunphy, and P.O. Brown. 2003. Genomewide view of gene silencing by small interfering RNAs. *Proc. Natl. Acad. Sci. USA* 100:6343–6446.

S. Choi and U.-J. Kim. 2001. Construction of a bacterial artificial chromosome library. *Genomics Protocols, Meth. Mol. Biol.* 175:57–68.

Cold Spring Harbor Laboratory Library. Web Resources Protocols. http://nucleus.cshl/CSHLib/webres/protocols.htm

M.R. Davey, E.L. Rech, and B.J. Mulligan. 1989. Direct DNA transfer to plant cells. *Plant Mol. Biol.* 13:273–285.

A.M. Denli and G.J. Hannon. 2003. RNAi: An ever-growing puzzle. *Trends Biochem. Sci.* 28:196–201.

S.M. Elbashir, W. Lendeckel, and T. Tuschi. 2001. RNA interference is mediated by 21- and 22-nucleotide RNAs. *Genes Dev.* 15:188–200.

H.A. Erlich, D. Gelfand, and J.J. Sninsky. 1991. Recent advances in the polymerase chain reaction. *Science* 252:1643–1651.

E. Frengen et al. 1999. A modular, positive selection bacterial artificial chromosome vector with multiple cloning sites. *Genomics* 58:250–253.

J. Geisselsoder, F. Whitney, and P. Yuckenberg. 1987. Efficient site-directed *in vitro* mutagenesis. *BioTechniques* 5:786–791.

D.L. Gerhold, R.V. Jensen, and S.R. Gullans. 2002. Better therapeutics through microarrays. *Nature Genetics* 32:547–552.

S.K. Goda and N.P. Minton. 1995. A simple procedure for gel electrophoresis and northern blotting of RNA. *Nucleic Acids Res.* 28:3357–3358.

H.G. Griffin and A.M. Griffin. 1993. DNA sequencing: Recent innovations and future trends. *Appl. Biochem. Biotechnol.* 38:147–159.

M. Grunstein and D.S. Hogness. 1975. Colony hybridization: A method for the isolation of cloned DNAs that contain a specific gene. *Proc. Natl. Acad. Sci. USA* 72:3961–3965.

S.M. Hammond, A.A. Cady, and R. Jorgensen. 2001. Post-transcriptional gene silencing by double-stranded RNA. *Nature Rev. Gen.* 2:110–119.

G. Hulvagner and P.D. Zamore. 2002. RNAi: Nature abhors a double-strand. *Curr. Opin. Genetics and Development* 12:225–232.

U.-J. Kim et al. 1996. Construction and characterization of a human bacterial artificial chromosome library. *Genomics* 34:213–218.

K. Kretz, W. Callen, and V. Hedden. 1994. Cycle sequencing. *PCR Methods Appl.* 3:5107–5112.

X.J. Lou, M. Schena, and F.T. Horrigan. 2001. Expression monitoring using cDNA microarrays: A general protocol. *Genomics Protocols, Meth. Mol. Biol.* 175:323–340.

O.J. Lumpkin, P. Dejardin, and B.H. Zimm. 1985. Theory of gel electrophoresis of DNA. *Biopolymers* 24:1573–1593.

J. Marx. 2000. DNA arrays reveal cancer in its many forms. *Science* 289:1670–1672.

A.M. Maxam and W. Gilbert. 1977. A new method for sequencing DNA. *Proc. Natl. Acad. Sci. USA* 74:560–564.

Y. Nosoh and T. Sekiguchi. 1990. Protein engineering for thermostability. *Trends Biotechnol.* 8:16–20.

M.P. Piechocki and R.N. Hines. 1994. Oligonucleotide design and optimized protocol for site-directed mutagenesis. *BioTechniques* 16:702–707.

L. Radnedge and H. Richards. 1999. The development of plasmid vectors. *Methods Microbiol.* 29:51–95.

N. Rosenthal. 1994. Stalking the gene—DNA libraries. *New Engl. J. Med.* 331:599–600.

F. Sanger, S. Nicklen, and A.R. Coulson. 1977. DNA sequencing with chain-terminating inhibitors. *Proc. Natl. Acad. Sci. USA* 74:5463–5467.

D.D. Shoemaker et al. 2001. Experimental annotations of the human genome using microarray technology. *Nature* 409:922–927.

T. Sijen, J. Fleenor, F. Simmer, K.L. Thijssen, S. Parrish, L. Timmons, R.H.A. Plasterk, and A. Fire. 2001. On the role of RNA amplification in dsRNA-triggered gene silencing. *Cell* 107:465–476.

L. Smith and A. Greenfield. 2003. DNA microarrays and development. *Human Mol. Gen.* 12:R1–R8.

E.M. Southern. 1975. Detection of specific sequences among DNA fragments separated by gel electrophoresis. *J. Molec. Biol.* 98:503–517.

W.R. Tolbert. 1990. Manufacture of biopharmaceutical proteins by mammalian cell culture systems. *Biotechnol. Adv.* 8:729–739.

R. Walden and J. Schell. 1990. Techniques in plant molecular biology—progress and problems. *Eur. J. Biochem.* 192:563–576.

T.A. Waldmann. 1991. Monoclonal antibodies in diagnosis and therapy. *Science* 252:1657–1662.

J.G. Wetmur. 1991. DNA probes: Applications of the principles of nucleic acid hybridization. *Crit. Rev. in Biochem. Molec. Biol.* 26:227–259.

T.J. White. 1996. The future of PCR technology: Diversification of technologies and applications. *Trends Biotechnol.* 14:478–483.

S. Wiseman, B. Singer, and D. Thomas. 2002. Applications of DNA and protein microarrays in comparative physiology. *Biotech. Adv.* 20:379–389.

T.D. Yager, J.M. Dunn, and J.K. Stevens. 1997. High speed DNA sequencing in ultrathin slab gels. *Curr. Opin. Biotechnol.* 8:107–113.

J.H. Yang, J.H. Ye, and D.C. Wallace. 1984. Computer selection of oligonucleotide probes from amino acid sequences for use in gene library screening. *Nucleic Acids Res.* 12:837–843.

R.A. Young and R.W. Davis. 1983. Efficient isolation of genes using antibody probes. *Proc. Natl. Acad. Sci. USA* 80:1194–1198.

M.J. Zoller and M. Smith. 1982. Oligonucleotide-directed mutagenesis using M13-derived vectors: an efficient and general procedure for the production of point mutations in any fragment of DNA. *Nucleic Acids Res.* 10:6487–6500.

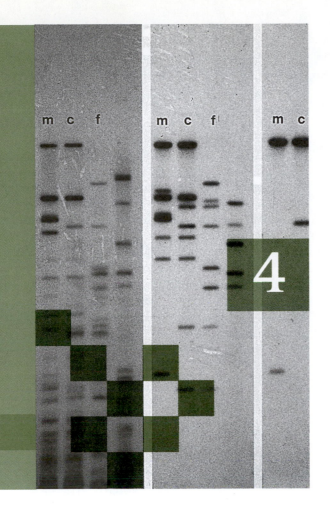

BASIC PRINCIPLES OF IMMUNOLOGY

NATURAL OR NONSPECIFIC IMMUNITY
Antimicrobial Agents
Phagocytic Cells
Nonphagocytic Cells
Natural Killer Cells
Inflammation and Fever

ACQUIRED OR ADAPTIVE IMMUNITY
Lymphocytes
    T cells
    B cells
Major Histocompatibility Complex
Antibodies

OVERVIEW OF CELL- AND ANTIBODY-MEDIATED IMMUNE RESPONSES

VACCINES

■ *Cause for Concern?*
   New Vaccines for the Developing World?

IMMUNE SYSTEM DISORDERS
Hypersensitivity
    Allergies
    Autoimmune Disorders
    Immunodeficiencies
      HIV and AIDS

MONOCLONAL ANTIBODIES

■ *Biotech Revolution*
   Future of Monoclonal Antibody Production

TOOLS OF IMMUNOLOGY
Western Blotting
Fluorescent Antibody Technique
Enzyme-Linked Immunosorbent Assay

The immune system is a complex defense system that protects us against foreign invaders such as pathogenic microorganisms, viruses, abnormal cells (for example, cancer), and tissues from another individual. It is a network of both nonspecific and specific processes that protect the body against pathogens. There are two types of immunity: (1) natural immunity that is a nonspecific, general immunity against a pathogen such as a microorganism and (2) acquired or adaptive immunity that is a response to a specific type of microorganism, pathogen, or substance. This type of immunity recognizes and reacts to specific molecules on the pathogen's surface. These two components of the immune system differ in the way they recognize an invasion by a microorganism, but work together to protect the body against pathogens and defective cells (Table 4.1).

NATURAL OR NONSPECIFIC IMMUNITY

Natural immunity is a rapid way for an organism to fight off microorganisms in a general way before the specific mechanisms of acquired immunity are activated (Figure 4.1). The body has several physical barriers to infection that act as a first line of defense. Two important physical barriers against a foreign invader are the skin and mucous membranes of the body. The skin is an effective barrier that prevents the entry of pathogens and viruses. It is a fairly tough covering that prevents the entry of pathogens into the tissues and circulatory system. Perspiration produced by sweat

glands in the skin and tears from ducts near the eyes contain the enzyme lysozyme, which degrades the cell walls of bacteria. Blinking bathes the eyes in tears that wash away potentially harmful invaders. The acidity of perspiration prevents the growth of many harmful microorganisms.

Other protective physical barriers are present in the digestive and respiratory tracts. The air we breathe contains many microorganisms, some of which cause diseases. The warm, moist environment of the respiratory tract is an excellent place for microorganisms to multiply; however, mucous traps many of these before they are able to reach the lungs, and cilia lining the respiratory tract move these microorganisms toward the mouth where they are swallowed and destroyed. Note that during a respiratory infection, more mucous is produced. Sometimes the food we eat is contaminated with microorganisms. Saliva, also containing lysozyme, is produced in the salivary glands of the mouth, and can destroy many types of microorganisms. Others that survive in the mouth cannot live in the harsh environment of the stomach and intestines (both containing very acidic digestive enzymes).

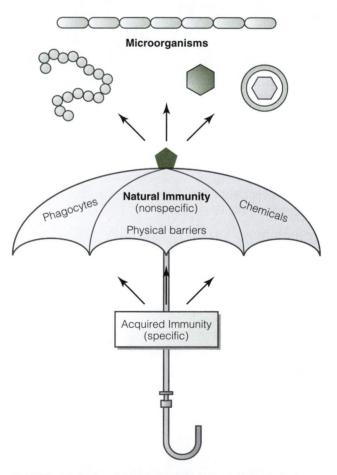

FIGURE 4.1 Natural immunity involves a general response to pathogens and serves as an umbrella before the more specific acquired immune system is able to respond.

TABLE 4.1 Three Lines of Defense Against Pathogens

Barriers at Body Surfaces (*nonspecific* targets):

1. Intact skin; mucous membranes at other body surfaces
2. Infection-fighting substances in tears, saliva, etc.
3. Normally harmless bacterial inhabitants of body surfaces that outcompete pathogenic visitors
4. Flushing effect of tears, urination, and diarrhea

Nonspecific Responses (*nonspecific* targets):

1. Inflammation
   a. Fast-acting white blood cells (neutrophils, eosinophils, basophils)
   b. Macrophages (also take part in immune responses)
   c. Complement proteins, blood-clotting proteins, other infection-fighting substances
2. Organs with phagocytic functions (e.g., lymph nodes)

Immune Responses (*specific* targets):

1. White blood cells (macrophages, T cells, B cells)
2. Communication signals (e.g., interleukins) and chemical weapons (e.g., antibodies, complement proteins)

Sources: C. Starr and R. Taggart. *Biology: The Unity and Diversity of Life,* 7th ed. Copyright © 1995 Wadsworth Publishing Co., Inc.

TABLE 4.2 Characteristics of Natural and Acquired Immunity

Characteristic	Innate Immunity	Adaptive Immunity
Speed of response	Rapid, begins to act within a few hours	Slow, takes days or weeks to act
Distribution in nature	Present in all animals	Present only in vertebrates
Variation over time	Invariant, an inherited capability that doesn't change over our lifetime	Changes with our exposure to microorganisms and foreign substances
Means of recognizing pathogens	Recognizes classes of molecules shared by groups of microorganisms (for example, peptidoglycan)	Recognizes unique molecules on particular pathogens (for example, cholera enterotoxin)
Numbers of molecules recognized	Only a few	An almost infinite number
Change during response	Constant	Improves during response

Antimicrobial Agents

Antimicrobial agents are chemicals or molecules that act to deter or destroy microorganisms. Some of these agents act in conjunction with physical barriers. For example, mucus is present in the respiratory tract (see previous discussion) along with cilia that move the trapped invaders. In addition to lysozyme, other antimicrobial agents include

1. Interferon—Interferons are a member of a large group of proteins (and peptides) called cytokines that act as signals during both the natural and acquired immune responses (Table 4.2). This molecule (glycoprotein) primarily protects the body against viral infection. There are several types of interferon that are produced early during viral infection. They are produced by cells of the immune system. Although interferon does not stop the virus from infecting cells, it does act as a signal to warn cells that an infection is coming (Figure 4.2). Thus, interferon activates and sets in motion a series of reactions that ultimately destroy the invading virus. After the cell receives the signal that a virus is present, a special protein on the surface of cells is activated, which then triggers the synthesis of antiviral proteins (AVPs). AVPs are enzymes that prevent the assembly of virus particles within the cell after infection occurs. Interferon can also prevent the spread of infection by certain viruses by inducing uninfected cells to secrete a lipoprotein (lipid-protein complex) coating outside the plasma membrane so that the virus cannot infect.

2. Interleukins are another class of cytokines (Table 4.3) that are produced by cells of the immune system and regulate interactions of these cells with other body cells. Tumor necrosis factors (TNFs) are another cytokine that stimulate cells of the immune system to initiate an inflammatory response. TNF cells also kill tumor cells.

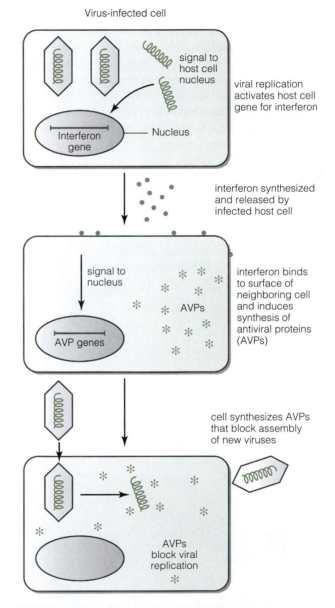

FIGURE 4.2 Interferon is an intermediary messenger molecule that is released from cells infected by viruses. Viruses usually kill cells they infect; however, the interferon that is released triggers adjacent cells to synthesize antiviral proteins that inhibit viral replication.

TABLE 4.3 Action of Certain Cytokines Released by Macrophages

Cytokine	Local Effects[a]	Systemic Effects
Interleukin-1	Tissue destruction; increases access of other leukocytes	Fever
Interleukin-6	Stimulates adaptive immune response (antibody production)	Fever
Interleukin-8	Chemotactic factor; attracts leukocytes including neutrophils to infected area	
Interleukin-12	Activates NK cells; also induces CD4 T cells to differentiate	
Tumor necrosis factor-alpha (TNF-α)	Increases permeability of blood capillaries in infected area	Fever; shock

[a]Most cytokines exert multiple effects; only the most relevant are listed here.

3. Lactoferrin and transferrin—These two proteins bind and sequester the element iron to reduce its cellular concentrations. Iron is a requirement for microbial growth, and when it becomes inaccessible, growth is dramatically reduced or inhibited. Transferrin is present in blood, while lactoferrin is produced in milk, saliva, mucus, and tears.

4. Complement—A family of more than 20 different proteins found in the serum of the blood (the fluid portion of the blood, excluding cells and clotting factors) that function together to protect the body against infection. They are called complement because they complement the actions of other components of the immune system in the fight against infection. These proteins are important in both the nonspecific and acquired parts of the immune system (Figure 4.3). Complement is generally inactive, but when the body encounters an antigen it perceives as foreign, complement is activated by a complex series of reactions. Activated complement responds in four general and

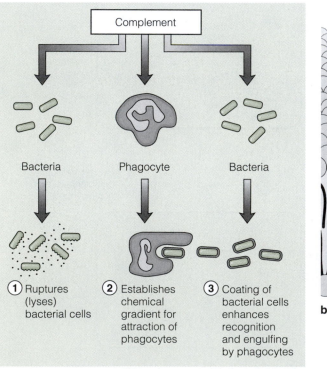

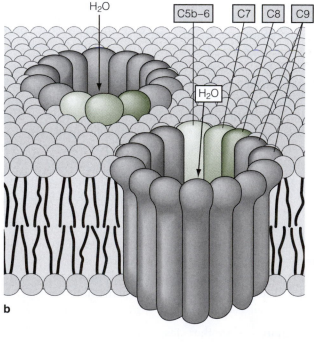

FIGURE 4.3 The complement system. (a) The complement system acts in several different ways. **(b)** The membrane attack complex (MAC). Complement proteins (c5b–6, c7, etc.) insert into the plasma membrane of a microorganism and create a pore that allows water to move into the cell and cause it to burst.

nonspecific ways to an antigen. Each mode of action requires specific complement proteins.

1. They can form a coating on the pathogen surface so that phagocytes (macrophages and neutrophils) can engulf them more readily.

2. Some complement proteins lyse the cell wall of microorganisms.

3. They can aid in the immune response by releasing substances such as histamine that increase inflammation, causing increased capillary permeability and dilated blood vessels.

4. They can attract lymphocytes (white blood cells) to an infection site.

Phagocytic Cells

Sometimes an invader is able to avoid the antimicrobial agents and cross the physical barriers. When this occurs, the microorganism can enter the bloodstream and tissues. A third line of defense is available to prevent further infection. Phagocytic cells (amoeba-like that move by cytoplasmic streaming) engulf particles, including microorganisms, in digestive vacuoles and then break down the cells, by the action of lysozyme (Figure 4.4a). The process of engulfing invading cells is referred to as phagocytosis (Figure 4.4b). The phagocyte contains many types of enzymes for digestion such as various lysozymes to break down peptidoglycans in

cell walls, proteases to break down proteins, enzymes to break down RNA and DNA, and lipases to break down lipids. In addition to digestion, these cells produce toxic substances such as hydrogen peroxide.

There are two types of phagocytes: (1) stationary phagocytes that reside along blood vessel walls and connective tissue and (2) wandering phagocytes that circulate in the blood. Both types of phagocytes are made in the bone marrow. Each type plays different roles in natural immunity.

1. Stationary phagocytes are large cells called macrophages that are made in the bone marrow. After they are made they circulate in the bloodstream for several days. At this stage they are referred to as monocytes (see wandering phagocytes, discussed later). After they stop circulating and become localized, they are called macrophages. Macrophages are like scavengers because they engulf and consume other cells such as microorganisms (Figure 4.5), as well as old and dead cells of the body, cancer cells, and cells infected with viruses. They harbor a large array of lysosomal enzymes. Macrophages' life span ranges from months to many years.

2. Wandering phagocytes are white blood cells (leukocytes) that circulate in the bloodstream and are called neutrophils and monocytes. There are

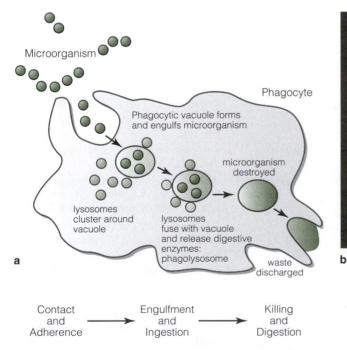

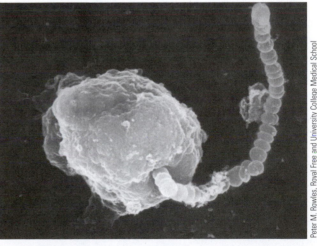

a

Microorganism

Phagocyte

Phagocytic vacuole forms and engulfs microorganism

microorganism destroyed

lysosomes cluster around vacuole

lysosomes fuse with vacuole and release digestive enzymes: phagolysosome

waste discharged

b

Peter M. Rowles, Royal Free and University College Medical School

Contact and Adherence → Engulfment and Ingestion → Killing and Digestion

FIGURE 4.4 Phagocytes break down microorganisms and other cells by a process of phagocytosis. (a) The process of phagocytosis. (b) A scanning electron micrograph showing a phagocyte ingesting streptococci bacterial cells.

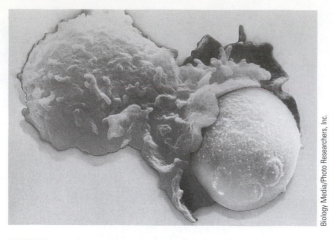

FIGURE 4.5 A scanning electron micrograph shows a macrophage (left) engulfing a yeast cell.

Biology Media/Photo Researchers, Inc.

approximately 4000–6000 neutrophils per mm³ of blood (Figure 4.6). They account for 65% of the total white blood cells in the blood. (Generally, there are approximately one or two white blood cells for every 1000 red blood cells.) When there is a bacterial infection, neutrophils increase in number, sometimes doubling. However, their life span is very short—approximately 50% die every 7 hours. Therefore, they are constantly being made. More than 100 billion neutrophils are produced every day in the bone marrow. Neutrophils are very mobile and can squeeze through capillary walls and into infected or injured tissues, as well as into cells that are infected with a bacterial pathogen such as *Salmonella typhimurium*. They are

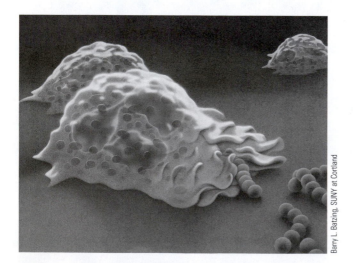

Barry L. Batzing, SUNY at Cortland

FIGURE 4.6 A neutrophil showing its amoeboid shape. This neutrophil is engulfing streptococci bacterial cells (scanning electron micrograph).

even able to enter the spinal column to fight infection, for example meningococcal meningitis.

Nonphagocytic Cells

Three types of nonphagocytic white blood cells are basophils, eosinophils, and lymphocytes (Table 4.4). Basophils and eosinophils together only make up less than 5% of the white blood cell population. Basophils play a role in inflammation and allergies, such as to pollen and mold spores. This type of white blood cell circulates in the bloodstream and also is present in tissues and localized along capillary walls (nonmobile basophils are called mast cells). Eosinophils are only somewhat phagocytic and function primarily in diseases caused by parasitic worms (helminths) such as tapeworms and pinworms. Lymphocytes are large cells that respond to specific microorganisms and play a very important role in acquired immunity. They are discussed in more detail in the acquired immunity section of this chapter.

TABLE 4.4 Summary of White Blood Cells and Their Roles in Defense

Cell Type	Main Characteristics
Macrophage	Phagocyte; acts in nonspecific and specific responses; presents antigen to T cells, and cleans up and helps repair tissue damage
Neutrophil	Fast-acting phagocyte; takes part in inflammation, not in sustained responses, and is most effective against bacteria
Eosinophil	Secretes enzymes that attack certain parasitic worms
Basophil and mast cell	Secrete histamine, other substances that act on small blood vessels to produce inflammation; also contribute to allergies
Dendritic cell	Processes and directly presents antigen to helper T cells
Lymphocytes:	*All take part in most immune responses; following antigen recognition, form clonal populations of effector and memory cells.*
B cell	Effectors secrete five types of antibodies (IgM, IgG, IgA, and IgD known to protect a host in specialized ways; also IgE)
Helper T cell	Effectors secrete interleukins that stimulate rapid divisions and differentiation of both B cells and T cells
Cytotoxic T cell	Effectors kill infected cells, tumor cells, and foreign cells by a touch-kill mechanism.
Natural killer (NK) cell	Cytotoxic cell of undetermined affiliation; kills virus-infected cells and tumor cells by a touch-kill mechanism

Natural Killer Cells

Natural killer (NK) cells are not phagocytic, but they attach to cell surfaces and produce enzymes that destroy cells that have been infected with viruses or microorganisms. They also are able to destroy cancer cells. NK cells do not destroy invaders, rather they kill body cells that have been modified or are infected. It is thought that infected or modified cells are coated with antibodies that bind to specific proteins on the cell surface (see later discussion on acquired immunity) and that NK cells recognize those altered cells that are bound with antibodies. Thus, although NK cells are nonspecific, the antibodies that recognize proteins such as a toxin produced by a microorganism or a protein on the surface of a cancer cell are specific.

Inflammation and Fever

Inflammation and fever are nonspecific antimicrobial responses to infection by microorganisms. Infection occurs when a pathogenic microorganism passes through a physical barrier (for example, when the skin is punctured or opened and tissues are damaged) (Figure 4.7).

Chemicals such as prostaglandins (which cause vasodilation and are synthesized from phospholipids of the membranes from damaged cells) and histamine (which causes increased capillary permeability and is produced by mast cells and basophils) are released that set in motion an inflammatory response, a beneficial reaction of the immune system (Table 4.5). Inflammation is characterized by swelling, redness, and warmth at the site, often accompanied by pain. Inflammation is a response by the immune system to tissue damage. The increase in redness and heat are due to increased blood flow to the injured or damaged site when there is an increase in the diameter of local blood vessels (vasodilation). Capillaries become more permeable, and an influx of phagocytes (neutrophils and macrophages) and plasma (fluid with clotting factors without blood cells) into the damaged tissue causes swelling. Swelling causes localized pain. Infection is limited by various blood proteins (for example, fibrin) that trap microorganisms by the formation of blood clots. Blood cells, dead cells from tissue, and microorganisms are trapped. Phagocytes destroy and remove invading microorganisms.

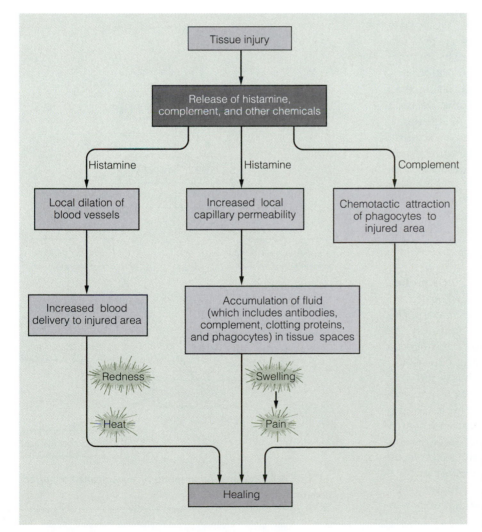

FIGURE 4.7 An injury or infection initiates an inflammatory response that involves redness, warmth, swelling, and pain in the surrounding tissue.

TABLE 4.5 Inflammatory Mediators[a]

Inflammatory Mediator	Source/Identity	Activity
Histamine	Degranulation of mast cells/a small molecule derived from the amino acid histidine	Causes vasodilation and increased blood vessel permeability; stimulates nerve endings, causing pain
Complement fragments	Formed during the activation of the complement system/small proteins	Cause vasodilation; activate and attract leukocytes
Kinins	Formed as blood clots/a family of short peptides	Cause vasodilation and increased blood vessel permeability; stimulate nerve endings causing pain
Bacterial by-products	By-products of the breakdown of bacteria by phagocytes/for example, the short peptide formyl-methionyl-leucyl-phenylalanine	Attract leukocytes to damaged area and activate them
Prostaglandins	Formed in mast cells from arachidonic acid, a fatty acid component of the cell membrane/small molecules	Cause vasodilation and increased blood vessel permeability[b]
Leukotrienes	Formed in mast cells from arachidonic acid, a fatty acid component of the cell membrane/small molecules	Have the same activity as histidine but are 100 times more active

[a]Some cytokines also act as inflammatory mediators.
[b]Aspirin reduces inflammation by inhibiting the synthesis of prostaglandin.

During some infections, fever occurs. Fever is usually induced by toxic substances produced by invading microorganisms (for example, bacterial endotoxins) and is beneficial to the infected organism. The increase in body temperature actually kills some bacterial pathogens, increases inflammation and stimulates phagocyte activity, and later, stimulates an acquired immune response. Elevated body temperature also reduces the concentration of iron in the blood and the level of iron-uptake proteins in the invading microorganism, thereby limiting the amount of iron available to the microorganism.

ACQUIRED OR ADAPTIVE IMMUNITY

Although natural immunity usually effectively protects the body against many types of potentially harmful microorganisms, there are times when these nonspecific defenses are overwhelmed. Acquired immunity is a way for organisms to effectively destroy many specific types of microorganisms, cancer cells, and viruses (Figure 4.8). It is a second line of defense that is activated when an invader is not destroyed by our nonspecific defenses. This immune response is a complex, highly organized network of defenses. Our immune system also identifies us as self and the tissues of others as nonself. Thus, if the immune system identifies a molecule or cell as foreign, it has the capacity to destroy it.

Acquired immunity is an immune response that develops over time through encounters with a wide vari-

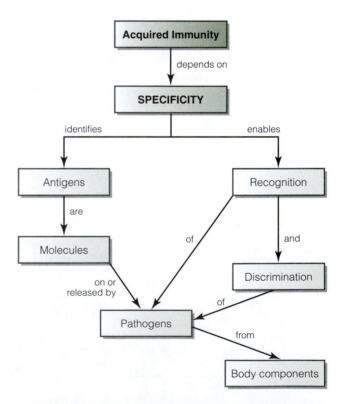

FIGURE 4.8 The specific immune response called acquired immunity.

ety of microorganisms, toxins, and many different types of molecules (for example, proteins, glycoproteins). The acquired immune system first encounters a molecule or microorganism and recognizes it as foreign—that is, different from self. Then the immune system takes action to eliminate the invader or molecule. Acquired immunity is specific and interacts only with a particular component or portion of a microorganism. For example, it might recognize a particular protein or polysaccharide component of the cell wall. Sometimes the molecule is a food component—for example, when someone develops an allergy to peanuts. A specific protein in the peanut triggers the acquired immune system.

The components or molecules that are recognized as foreign and trigger the acquired immune response are called antigens. An antigen can be a protein, glycoprotein, some polysaccharides, and even DNA or RNA. Components of the immune system specifically interact with the antigens of, for example, an invading microorganism or toxin. The portion of the antigen that triggers the immune response is called the antigenic determinant (also referred to as an epitope) (Figure 4.9). In an antigen that is a long protein, perhaps only seven amino acids might be sufficient to trigger

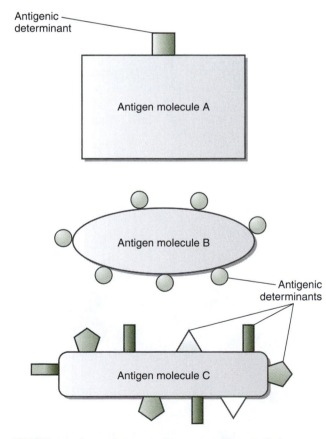

FIGURE 4.9 An antigen may have one or more of the same antigenic determinants or different antigenic determinants. Antigenic determinants are also called epitopes.

the immune response (the actual antigenic determinant). A microorganism can have many different antigens on its surface, each having one specific antigenic determinant or several different ones. Thus, each type of microorganism or molecule has a unique molecular fingerprint (the antigenic determinants) that is recognized by the acquired immune system. Because there are thousands of different microorganisms, each having different molecular fingerprints, the immune system must be able to recognize many different antigenic determinants.

During the acquired immune response, the body responds by producing antibodies that recognize and bind specific antigens. There are many types of cells and substances that are involved in the acquired immune response. In addition, some cells have an immunological memory—that is, they are able to respond more rapidly and effectively the second time they encounter a specific antigen. These cells remember a specific antigen or foreign substance that they encountered previously and are able to elicit a more aggressive attack.

Acquired immunity is based on the coordination and interaction of different types of cells and other components (Figure 4.10). There are two main types of acquired immunity. One is cell-mediated immunity. Cell-mediated immunity is a specific immunity that is mediated by cells called lymphocytes (T cells, see the section on lymphocytes). These cells attach to antigens on the surface of body cells that have been modified by infection or mutation and destroy them (Figure 4.11). The other type is antibody-mediated immunity, which involves the production of antibodies (see the section on antibodies) that are specific for the different antigenic determinants (epitopes) of antigens present on the surface of microorganisms (Figure 4.12). Cells called B cells differentiate into plasma cells and produce the antibodies (also called immunoglobulins [Igs]). Antibodies inactivate circulating viruses, toxins, and microorganisms.

During acquired immunity, cells of the immune system must be circulated through the body. The lymphatic system is a drainage system through which these cells circulate. Lymph nodes are small bean-shaped organs of the lymphatic system and are loaded with lymphocytes and macrophages (Figure 4.13). Several incoming lymph vessels merge at the node and form one large outgoing lymphatic vessel. Lymphocytes are the main cells transported in the lymph and are carried throughout the body by the lymphatic system. Lymphocytes circulate in the blood and lymph (fluid that is present in the vessels of the lymphatic system) and in tissues such as the tonsils, adenoids, thymus gland, spleen, appendix; in regions of the small intestine; and in the bone marrow (Table 4.6). The lymphatic vessels circulate the lymph throughout these lymphoid

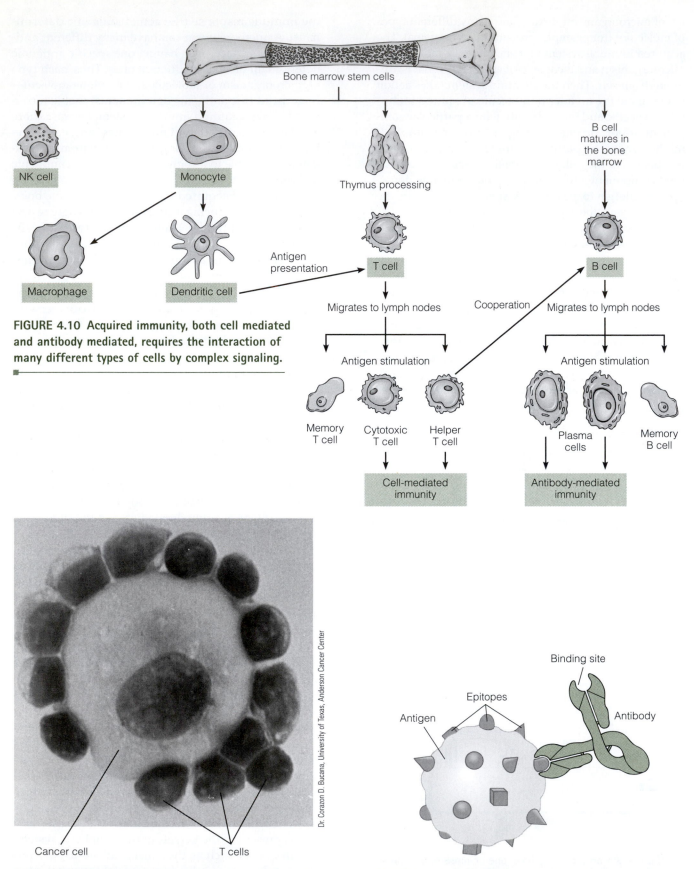

Bone marrow stem cells

NK cell

Monocyte

Macrophage

Dendritic cell

Antigen presentation

Thymus processing

T cell

Migrates to lymph nodes

Antigen stimulation

Memory T cell

Cytotoxic T cell

Helper T cell

Cell-mediated immunity

B cell matures in the bone marrow

B cell

Cooperation

Migrates to lymph nodes

Antigen stimulation

Plasma cells

Memory B cell

Antibody-mediated immunity

FIGURE 4.10 Acquired immunity, both cell mediated and antibody mediated, requires the interaction of many different types of cells by complex signaling.

Dr. Corazon D. Bucana, University of Texas, Anderson Cancer Center

Cancer cell

T cells

FIGURE 4.11 T cells react to antigens on the surface of modified body cells. This photomicrograph shows T cells directly interacting with a cancer cell.

Binding site

Epitopes

Antigen

Antibody

FIGURE 4.12 An antibody molecule binds to an antigenic determinant (epitope) of a cell that has different antigens on its surface.

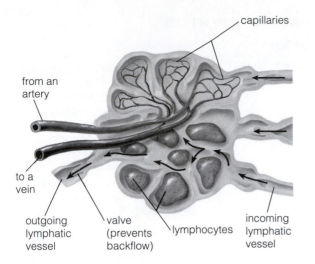

FIGURE 4.13 A lymph node is located at the junction where incoming lymph vessels come together to form an outgoing lymphatic vessel. The node is filled with macrophages and lymphocytes. Lymphocytes enter through arterial blood, pass through capillary walls, and enter storage regions.

tissues, assisting in the contact of lymphocytes and macrophages with specific antigens. Lymphocytes have specific receptor molecules on their surface that recognize specific antigenic determinants. When a particular antigenic determinant makes contact with a cell surface receptor of a lymphocyte, the microorganism or toxin is destroyed by the lymphocyte.

Once a lymphocyte has been stimulated by an antigen, it undergoes cell division and differentiation and becomes a lymphoblast.

Lymphocytes

There are three primary types of lymphocytes that play roles in immunity (see Table 4.4). T cells and B cells have important roles in acquired immunity, while NK cells are important for natural immunity (discussed earlier).

T cells   Approximately 60% of lymphocytes are T cells, which develop in the bone marrow then migrate to the thymus gland where they mature and acquire the ability to respond during infection (Figure 4.14). The *T* refers to the fact that they mature in the thymus (B cells mature in the bone marrow). T cells are involved primarily in cell-mediated immunity. That is, they attack body cells that are infected by microbial pathogens or have become cancerous (also cells from transplanted organs—for example, heart, liver, kidney). While T cells are maturing, they develop the ability to recognize specific surface antigens on microorganisms and viruses. After maturation, T cells move into the blood and lymph where they circulate throughout the body. Some types of T cells produce lymphokines when activated, which act as chemical regulatory signals that modulate the intensity of the immune response. T cells produce a variety of cytokines that are important in T, B, and NK cell development and stimulate cell-mediated and antibody-mediated immunity. To modulate the immune response, different types of T cells enhance and

TABLE 4.6 Tissues and Vessels of the Immune System

Tissue	Location	Role in Adaptive Immune System
Primary Lymphoid Tissues		
Bone marrow	Within vertebrae, ribs, sternum, long bones, pelvis	Produces all blood cells, including leukocytes and lymphocytes; site of B-cell differentiation
Thymus	Gland that lies in front of the heart	Site of T-cell differentiation
Secondary Lymphoid Tissues		
Lymph nodes	Small kidney-shaped organs located on lymphatic vessels throughout the body	Mature lymphocytes are stored, interact with one another, and move between blood and lymph vessels here
Spleen	Upper abdomen	Mature lymphocytes stored and interact here
Tonsils	Pair of lymphoid tissues in the throat	Mature lymphocytes stored and interact here
Adenoids	Ring of lymphoid tissues in the nose	Mature lymphocytes stored and interact here
Appendix	Small outpouching of the intestinal tract	Mature lymphocytes stored and interact here
Peyers's patches	Small collections of lymphoid tissue within the intestinal wall	Mature lymphocytes stored and interact here
Lymphatic Circulation		
Lymph vessels (lymphatics)	Vessels throughout the body that collect lymph and return it to the bloodstream	Transports lymphocytes
Thoracic duct	Largest lymph vessel	Empties into heart to return lymph to the bloodstream

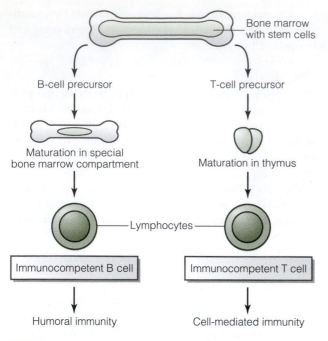

FIGURE 4.14 Lymphocytes originate from bone marrow stem cells. T cells develop in the bone marrow and differentiate in the thymus. B cells differentiate in the bone marrow.

TABLE 4.7 Summary of T Cells

Cell Type	Action
Killer T cells	Destroy body cells infected by viruses and attack and kill bacteria, fungi, parasites, and cancer cells.
Helper T cells	Produce a growth factor that stimulates B-cell proliferation and differentiation and also stimulates antibody production by plasma cells; enhance activity of cytotoxic T cells.
Suppressor T cells	May inhibit immune reaction by decreasing B- and T-cell activity and B- and T-cell division.
Memory T cells	Remain in body awaiting reintroduction of antigen, when they proliferate and differentiate into cytotoxic T cells, helper T cells, suppressor T cells, and additional memory cells.

suppress the immune system (Table 4.7). There are two primary types of T cells: CD8 T cells (cytotoxic T cells) and CD4 T cells (helper T cells).

1. CD8 T cells are cytotoxic, or killer, T cells that respond to foreign antigens on cell surfaces by directly killing them. Cancer cells, foreign cells (from a transplant or graft), and pathogen-infected and virus-infected cells can be targeted by killer T cells.

2. CD4 T cells are helper T cells that secrete substances that either activate or enhance the immune system. Two types of helper T cells each respond differently during an immune response. T helper 1 (T_h1) cells are involved primarily in cell-mediated immunity, while T helper 2 (T_h2) cells are involved in antibody-mediated immunity and induce B cells to divide and produce antibodies.

The T-cell receptor recognizes and binds both foreign antigens and antigens on a body cell that signals self (as opposed to nonself). The T-cell receptor is a glycoprotein (protein-carbohydrate complex)—two different glycoprotein chains (alpha and beta) linked by disulfide bonds. These alpha and beta chains have both constant and variable regions (see antibody section for a discussion of the formation of antibody and T-cell re-

ceptor diversity). The variable regions provide the tremendous T-cell receptor diversity necessary to recognize the many foreign antigens.

The T-cell receptor can recognize a foreign antigen (for example, proteins from another individual) that extends from an antigen-presenting cell (APC). An APC is a cell such as a macrophage or B cell that expresses the foreign antigen on its surface (Figure 4.15). In addition to macrophages and B cells, dendritic cells also are APCs. These cells are found in the internal organs and skin. Also associated with the T-cell receptor are CD4 and CD8 proteins that determine T-cell function (CD4 is associated with the T-cell receptor complex of helper T cells, CD8 is associated with the T-cell receptor of cytotoxic T cells).

B cells B cells develop and undergo maturation in the bone marrow (see Figure 4.14). The three types of B cells are based on function.

1. Naïve B cells

2. Plasma cells

3. Memory B cells

Differentiated B cells that have not yet been exposed to antigens are called naïve B cells. Each one of these cells has approximately 100,000 identical antibody molecules bound to its surface. As these cells continue to develop, they move into the lymph nodes where they are exposed to a variety of antigens (for example, on microorganisms, toxins) that have entered the body. Naïve B cells are activated, undergo cell division, and fully develop into plasma cells when they encounter a matching antigen from a foreign substance,

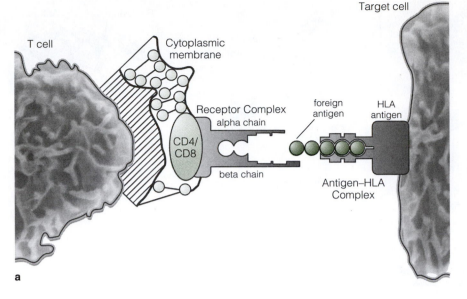

T cell

Cytoplasmic membrane

Receptor Complex
alpha chain

CD4/CD8

beta chain

foreign antigen

HLA antigen

Antigen–HLA Complex

Target cell

a

FIGURE 4.15 The T–cell receptor recognizes and binds to an antigen–MHC (major histocompatibility complex) (human leukocyte antigen [HLA] in humans) complex on a target cell. (a) Variable amino acid sequences on each of the two glycoproteins (alpha and beta) form the receptor complex groove that binds to a histocompatibility antigen (MHC or HLA) and a portion of a foreign antigen that is displayed on the target cell as a complex. (b) T cells respond to changes in histocompatibility antigens as in this virus-infected cell. When a virus infects a body cell, viral antigens complex with HLA proteins on the cell surface. T cells sense this change and kill the virus-infected cell.

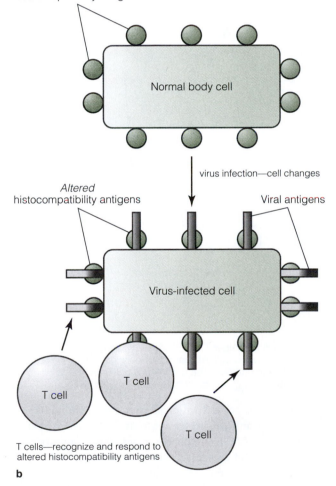

Histocompatibility antigens

Normal body cell

virus infection—cell changes

Altered histocompatibility antigens

Viral antigens

Virus-infected cell

T cell

T cell

T cell

T cells—recognize and respond to altered histocompatibility antigens

b

an invading microorganism, or virus. Each type of B cell secrets a specific antibody in response to a particular antigen (Figures 4.16 and 4.17). A plasma cell can produce more than 10 million antibody molecules in 1 hour. The B cell is activated by a complex reaction that involves helper T cells. The antibodies that are produced bind to their specific antigens that originally activated the B cells and mark them for destruction (Figure 4.18).

Some activated B cells (and T cells) do not develop into plasma cells, rather, they become memory cells (Figure 4.19). Memory cells do not die like plasma cells. They produce small amounts of antibody after an infection has been eliminated by the immune system. Memory B cells enable an organism to attack rapidly and aggressively if a reinfection occurs by the same microorganism or the immune system contacts a foreign antigen it has been exposed to previously. The antibody that is circulated in the body binds to the antigen it encounters. Memory B cells begin dividing and differentiate into plasma cells to destroy the pathogen. This is one way that clinicians can determine whether someone has been exposed to a particular pathogen. Antibodies against the pathogen can be detected long after infection (for example, Epstein-Barr virus).

Major Histocompatibility Complex

Body cell antigens are called major histocompatibility complex (MHC) proteins or antigens. (The MHC in humans is called the HLA complex, for human leukocyte antigen.) These proteins identify cells of the body as self. There can be 40 or more different forms for a given MHC gene in a human population. Therefore, in each individual, the different proteins that make up this complex are slightly different.

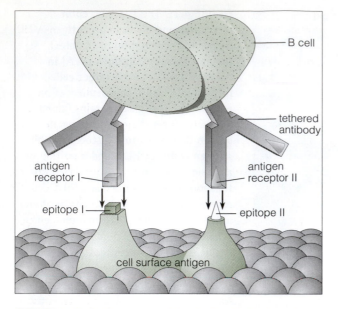

FIGURE 4.16 Antigen recognition in B cells that produce antibodies. Antibodies bound to two B cells each recognize and bind to a different epitope on a single cell surface antigen.

MHC markers are involved in the defense against foreign invaders. Some MHC markers are common to all cells and others are found only in lymphocytes and macrophages. APCs express foreign cell surface antigens as antigen-MHC complexes on their own cell surface. These antigen-presenting cells activate cell-mediated and antibody-mediated immune responses. Once activated, B and T lymphocytes rapidly divide and differentiate into effector cells, while memory cells remember the antigen that activated the immune response so that when the antigen reappears, a more rapid and stronger immune response is initiated.

MHC is one of the reasons why there must be a close match between individuals during a tissue graft—for example, in a bone marrow transplant—otherwise, the MHC antigens on the donated cells will be recognized as foreign by T cells and the cells will be destroyed.

The MHC is composed of three different classes of proteins: MHC class I, MHC class II, and MHC class III antigens. MHC class 1 antigens are found on most cells of the body and are involved in the recognition of self versus nonself. These antigens bind to foreign antigens of foreign cells (from a transplant or tissue graft), as well as viruses. The complex that results, the foreign antigen-MHC complex, is projected on the surface of the cell and is recognized by cytotoxic T cells.

MHC class II antigens are found on cells of the immune system such as macrophages, B cells, dendritic cells, and some types of T cells. These antigens regulate interactions among APCs (for example, macrophages),

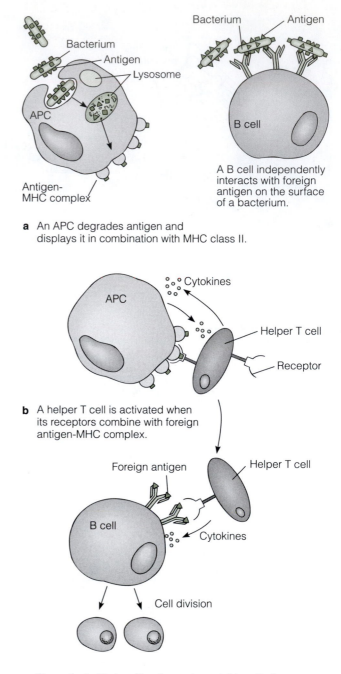

a An APC degrades antigen and displays it in combination with MHC class II.

b A helper T cell is activated when its receptors combine with foreign antigen-MHC complex.

c The activated helper T cell secretes cytokines that can activate a B cell that has previously interacted with foreign antigen. The activated B cell then divides, forming a clone of competent B cells.

FIGURE 4.17 B cell activation. B cells are activated by the action of antigen–presenting cells (APCs) and helper T cells. (a) An APC (for example, macrophage) ingests a bacterial cell and presents the foreign antigens on its cell surface, forming a foreign antigen–MHC complex. B cells associate with foreign antigens on the bacterial cell surface. (b) Helper T cells are activated when their receptors bind to foreign antigen–MHC complexes on APCs. Cytokines secreted by both cells play a role in cell signaling. (c) Activated helper T cells secrete cytokines that activate B cells. B cells divide and form B cell clones.

T cells, and B cells. MHC class II antigens bind to epitopes of antigens that are presented on the surface of APCs that have engulfed and destroyed, for example, invading bacteria. The foreign antigen-MHC complex is on the cell surface and activates helper T cells, which then stimulate other components of the immune system. The last class, MHC class III proteins, makes up some of the proteins of the complement system.

Antibodies

Antibodies, called immunoglobulins, are protein molecules produced by lymphocytes from advanced vertebrates like mammals; they protect the organism by binding to and eliminating foreign molecules by activating mechanisms of the immune system.

Antibodies are Y-shaped molecules that are composed of two antigen-binding fragments (Fab fragments)—the arms of the molecule. The stem interacts with cells of the immune system (Fc fragment, called *c* because it crystallizes in the cold) (for example, phagocytes) (Figure 4.20). Each antibody can bind with two antigen molecules and forms an antibody-antigen complex.

An antibody molecule is composed of four polypeptide chains. There are two identical light-chain short polypeptides and two identical heavy-chain long polypeptides linked by a disulfide bond. Each polypeptide chain has a constant (C) region and a variable (V) region. The V regions are located at the ends of the Y. The variable regions are able to change during differentiation and activation of the antibody-producing cells. This allows for the generation of a huge diversity of antigen-binding sites. The C regions are primarily invariable sites within the antibody that anchor the molecule to the cell.

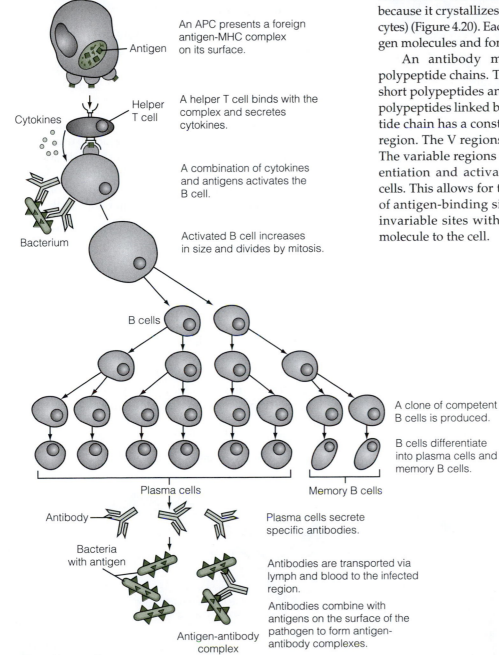

An APC presents a foreign antigen-MHC complex on its surface.

Antigen

A helper T cell binds with the complex and secretes cytokines.

Helper T cell

Cytokines

A combination of cytokines and antigens activates the B cell.

Bacterium

Activated B cell increases in size and divides by mitosis.

B cells

A clone of competent B cells is produced.

B cells differentiate into plasma cells and memory B cells.

Plasma cells

Memory B cells

Antibody

Plasma cells secrete specific antibodies.

Bacteria with antigen

Antibodies are transported via lymph and blood to the infected region.

Antibodies combine with antigens on the surface of the pathogen to form antigen-antibody complexes.

Antigen-antibody complex

FIGURE 4.18 Steps in antibody–mediated immunity. A specific B cell becomes activated when it is exposed to cytokines produced by an activated helper T cell and when it binds with a specific antigen (for example, on a bacterial cell). A clone of cells is produced when the activated B cell divides. Most of the cells differentiate into plasma cells and secrete antibodies that are transported to the infection site by the lymph or blood. A small population of cells becomes memory B cells.

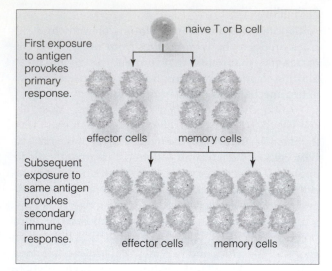

FIGURE 4.19 Both T and B cells produce both activated (effector cells) and memory cells during an immune response to antigen (primary immune response). Many cells circulate as memory cells that activate when exposed to the same antigen at a later time (secondary immune response).

The antibody tags the antigen (can be on the cell surface of a pathogen) as foreign by binding to it and thus marking it for destruction. Antibodies do not directly destroy antigens. In general, several antibodies bind to antigens, forming a large aggregate of antigen-antibody complex (each antibody can bind two antigens on different cells). This complex triggers several responses.

1. The inactivation of the pathogen; antibodies bind to a virus or bacterium and prevent a host cell from being infected

2. The activation of phagocytes that ingest pathogens

3. Antibodies (immunoglobulin G, IgG, and immunoglobulin M, IgM, described in the following) bind to specific antigens on the pathogen and activate complement proteins that then destroy the particular pathogen

4. The Fc portion of antibodies that have attached by their Fab fragments to antigens on the surface of the pathogen can bind to Fc receptors of phagocytes

Antibodies are classified based on the five amino acid sequences of the C region of the heavy chains: IgA, IgD, IgE, IgG, and IgM. Each type of antibody has a distinct function (Table 4.8, Figure 4.21). Approximately 75% of antibodies are in the IgG class.

IgA—Found in mucus, tears, milk, and saliva. Secreted into the respiratory, digestive, urinary, and reproductive tracts. Prevents bacteria and viruses from binding to epithelial surfaces; defends against pathogens that are ingested or inhaled. Passed on to newborns in breast milk.

IgD—On the surface of B cells; helps activate B cells after binding to antigens.

IgE—Binds to mast cells; when antigens bind to IgE, mast cells release histamine, triggering an inflammatory response; involved in immunity to parasitic worms.

IgG—Found in plasma; involved in long-term immunity. Long lived. This antibody type crosses the placenta during pregnancy and protects the fetus and newborn for up to 1 year after birth (called passive immunity in the newborn).

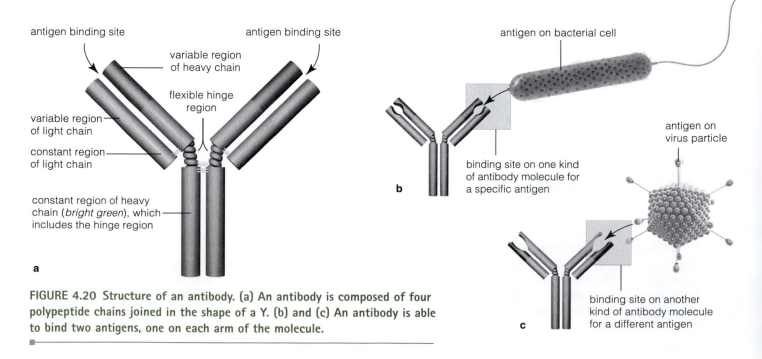

FIGURE 4.20 Structure of an antibody. (a) An antibody is composed of four polypeptide chains joined in the shape of a Y. (b) and (c) An antibody is able to bind two antigens, one on each arm of the molecule.

TABLE 4.8 Types and Functions of the Immunoglobulins

Class	Location and Function
IgD	Present on surface of many B cells, but function uncertain; may be a surface receptor for B cells; plays a role in activating B cells.
IgM	Found on surface of B cells and in plasma; acts as a B-cell surface receptor for antigens secreted early in primary response; powerful agglutinating agent.
IgG	Most abundant immunoglobulin in the blood plasma; produced during primary and secondary response; can pass through the placenta, entering fetal bloodstream, thus providing protection to fetus.
IgA	Produced by plasma cells in the digestive, respiratory, and urinary systems, where it protects the surface linings by preventing attachment of bacteria to surfaces of epithelial cells; also present in tears and breast milk; protects lining of digestive, respiratory, and urinary systems.
IgE	Produced by plasma cells in skin, tonsils, and the digestive and respiratory systems, overproduction is responsible for allergic reactions, including hay fever and asthma.

IgM—Large antibody; produced during the first encounter with a foreign antigen. For example, when someone is vaccinated against hepatitis B, IgM antibodies are primarily produced. These antibodies are found on the surface of B cells, along with IgD. This complex molecule is called a macroglobulin.

Antibodies are able to recognize and bind to millions of different antigens. This means that there must be tremendous antibody diversity. An unusual feature of antibody synthesis is the process of programmed DNA rearrangements that provide the antibody diversity required for a healthy immune system. Although there are not millions of different antibody genes, different DNA segments encoding portions of the antibody can be recombined to generate a variety of different antibodies (Figure 4.22). Antibody diversity is generated during the maturation and differentiation of B cells. In undifferentiated B cells, within the DNA, gene regions or segments exist for several hundred V regions, one or more junction (J) regions, and one or more different C regions. During B cell differentiation, the DNA regions recombine. For example, a randomly selected V region (such as V_{100}) DNA of the light chain is joined with J and C DNA segments to form a new gene. Two of these light chains combine with two heavy chains to form the antibody. Thus, within B cells different DNA regions join in a variety of combinations. At the same time, changes within the newly generated genes also may occur (mutations) that provide further diversity. The same process occurs within T cells for T-cell receptor diversity on the surface of these cells. The number of different combinations is tremendous, allowing for the potential generation of 100 million different antibody genes. Thus, in a typical adult human, there may be millions of B cells produced, with each population of B cells producing a specific antibody that recognizes and binds specific antigens. Although a person inherits genes encoding antibodies and T-cell receptors from his or her mother and father, the genes will be rearranged differently, resulting in different antibodies and T-cell receptor molecules.

In summary, the natural (nonspecific) and acquired (specific) immune responses work together to keep a person or animal healthy. Table 4.9 reviews the responses to a pathogen.

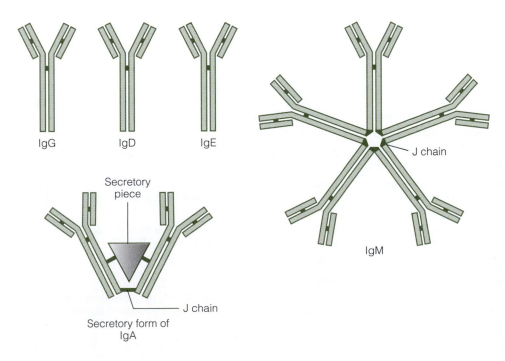

IgG   IgD   IgE

Secretory piece

J chain

Secretory form of IgA

J chain

IgM

FIGURE 4.21 Five classes of antibodies. Each class has specific functions. IgG, IgD, and IgE are monomers. IgM is composed of five identical antibody monomers, connected by a small protein called a J chain, arranged in the shape of a star. Two monomers connected by a J chain and a unique secretory piece make up IgA.

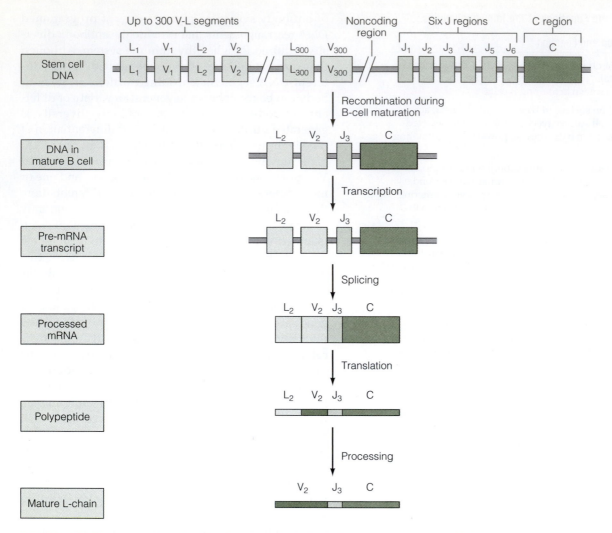

FIGURE 4.22 The generation of a functional antibody L–chain by recombination of DNA regions. In immature B cells, light–chain genes contain several hundred V–L (variable–leader) regions, six J regions, and one C region. In maturing B cells, random combinations of V–L and J regions are connected by recombination to the C region.

TABLE 4.9 Nonspecific and Specific Immune Responses to Bacterial Invasion

Nonspecific Immune Mechanisms	Specific Immune Mechanisms
Inflammation Engulfment of invading bacteria by resident tissue macrophages	Processing and presenting of bacterial antigen by macrophages
Histamine-induced vascular responses to increase blood flow to area, bringing in additional immune cells	Proliferation and differentiation of activated B-cell clone into plasma cells and memory cells
Walling off of invaded area by fibrin clot	Secretion by plasma cells of customized antibodies, which specifically bind to invading bacteria
Migration of neutrophils and monocytes/macrophages to the area to engulf and destroy foreign invaders and to remove cellular debris	Enhancement by helper T cells, which have been activated by the same bacterial antigen processed and presented to them by macrophages
Secretion by phagocytic cells of chemical mediators, which enhance both nonspecific and specific immune responses	Binding of antibodies to invading bacteria and activation of mechanisms that lead to their destruction
	Activation of lethal complement system
Nonspecific activation of the complement system Formation of hole-punching membrane attack complex that lyses bacterial cells	Stimulation of killer cells, which directly lyse bacteria
Enhancement of many steps of inflammation	Persistence of memory cells capable of responding more rapidly and more forcefully should the same bacterial strain be encountered again

OVERVIEW OF CELL- AND ANTIBODY-MEDIATED IMMUNE RESPONSES

During the cell-mediated immune response (Figure 4.23),

1. Virus invades cells of the body

2. Foreign antigens combine with MHC and form foreign antigen-MHC (class I) complex that is displayed on surfaces of antigen-presenting cells

3. Specific T cells are activated by foreign antigen-MHC complex

4. T-cell clone produced by cell division—some become cytotoxic T cells

5. Cytotoxic T cells migrate to infection site

6. Cytotoxic cells secrete proteins that destroy target cells—those infected by virus

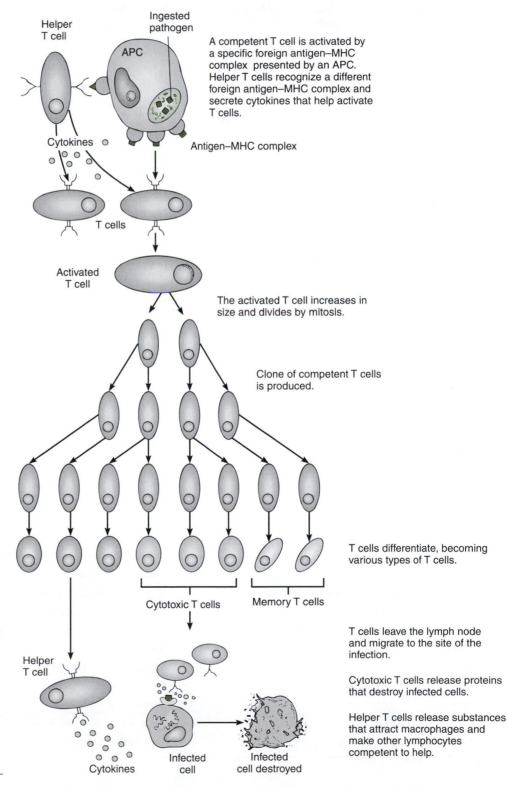

Helper T cell

Ingested pathogen

APC

A competent T cell is activated by a specific foreign antigen–MHC complex presented by an APC. Helper T cells recognize a different foreign antigen–MHC complex and secrete cytokines that help activate T cells.

Cytokines

Antigen–MHC complex

T cells

Activated T cell

The activated T cell increases in size and divides by mitosis.

Clone of competent T cells is produced.

T cells differentiate, becoming various types of T cells.

Cytotoxic T cells

Memory T cells

Helper T cell

T cells leave the lymph node and migrate to the site of the infection.

Cytotoxic T cells release proteins that destroy infected cells.

Helper T cells release substances that attract macrophages and make other lymphocytes competent to help.

Cytokines

Infected cell

Infected cell destroyed

FIGURE 4.23 Steps in cell-mediated immunity. A T cell is activated by a foreign antigen-MHC complex that is presented on the surface of an antigen-presenting cell (APC) and cytokines. Helper T cells produce cytokines that act on other T cells. Some T cells become cytotoxic T cells that release proteins that kill infecting pathogens.

7. Helper T cells secrete substances that attract macrophages and other cells to help fight infection

During the antibody-mediated immune response (see Figure 4.18),

1. A pathogen with a foreign antigen penetrates body

2. Antigen-presenting cells (APCs) (for example, macrophages) phagocytize the pathogen. That is, macrophages phagocytize bacterial cells and display the antigen on their surface so they are recognized by the immune system

3. The foreign antigen along with the MHC (antigen-MHC complex) is displayed on the surface of the phagocytes

4. Helper T cells bind with foreign antigen and become activated

5. Activated helper T cells interact with B cells that display the same antigen-MHC complex. That is, B cells display the antigen of the infecting microorganism

6. B cells are activated

7. B cell clone is produced through cell division

8. B cells differentiate and become plasma cells

9. Plasma cells secrete antibodies

10. Antibodies complex with the pathogen

11. Antibody-pathogen complexes trigger the destruction of the pathogen

VACCINES

Following exposure to an antigen, the immune system responds in a process called active immunity. Active immunity can be achieved either naturally, as when a person is exposed to a virus (for example, measles, rubella), or artificially, as when an inactivated virus or other antigen is injected into the body to activate the immune system (primary immunity) and produce long-term immunity (a process called vaccination) (Table 4.10). Memory B cells would be produced that, when the antigen was again encountered naturally, would elicit a more effective immune response (secondary immune response). Vaccine production occurs in several ways.

1. A virus can be weakened (live, attenuated vaccine) by modification so that it is unable to cause disease in humans. This includes reducing its ability

TABLE 4.10 Some Currently Used Vaccines

Vaccine	Component Antigens
MMR	
Measles	Live, attenuated virus
Mumps	Live, attenuated virus
Rubella	Live, attenuated virus
Polio	
Oral polio vaccine (OPV; Sabin)	Live, attenuated viruses (3)
Inactivated polio vaccine (IPV; Salk)	Inactivated viruses
DtaP	
Diphtheria (*Corynebacterium diphtheriae*)	Toxoid
Tetanus (*Clostridium tetani*)	Toxoid
Pertussis (whooping cough; *Bordetella pertussis*)	Inactivated pertussis toxin; acellular pertussis (aP)
Influenza	Whole inactivated or disrupted ("split") virus
Rabies	Inactivated virus
Meningococcal meningitis (*Neisseria meningitidis*)	Capsular polysaccharide
Streptococcal pneumonia (*Streptococcus pneumoniae*)	Capsular polysaccharide
Haemophilus meningitis (*H. influenzae* type B)	Capsular polysaccharide
Tuberculosis	Live, attenuated bovine strain (BCG strain of *Mycobacterium bovis*)
Smallpox	Live, attenuated *vaccinia* virus
Hepatitis B	Recombinant: surface antigen
Hepatitis A	Inactivated virus
Chickenpox (varicella)	Live, attenuated (herpes varicella zoster virus)
Lyme disease (*Borrelia burgdorferi*)	Recombinant: outer surface protein

New Vaccines for the Developing World?

The development of new types of vaccines and methods of delivery will greatly enhance healthcare in the near future, especially for children and the elderly. Vaccines that fight cancer and other diseases (for example, herpes simplex virus and HIV) will be important weapons in the medical arsenal. Biotechnology companies spend millions of dollars for research and development and expect to make up their expenses when a new vaccine enters the commercial market. Those in industrialized countries will benefit greatly; however, will these new technologies and vaccines be available to those living in developing countries (and even people without medical insurance in the United States)? Furthermore, will vaccines be developed for diseases that are primarily found in developing countries or will there not be an incentive if there is lower economic benefit to companies (for example, Dengue fever virus)? Developing countries have a great need for vaccines that target such diseases as HIV, malaria, hepatitis, yellow fever, and polio. New modes of vaccine delivery should be tailored for those regions where there is no refrigeration or rapid means to reach those requiring vaccination. If the majority of people globally are to be vaccinated against common diseases, industrialized countries must determine whether they will provide the means including funding for research, development of the technologies, and distribution to those in need.

---

to enter the body, cause disease, or reproduce. However, the pathogen is still antigenic and can produce an immune response. Examples are the measles, mumps, and Sabin polio vaccines.

2. Sometimes killed pathogens are used (inactivated vaccines). They cause an immune response because the antigens are still present to cause an immune response. Often chemical treatment is used. Examples are the typhoid, whooping cough, and rabies vaccines.

3. The toxin isolated from a pathogen can be modified to eliminate the toxic properties of the substance; however, it still remains antigenic (toxoid vaccine). The tetanus vaccine is an example.

4. Subunit vaccines are used to provide an additional safeguard against side effects. Only specific antigens from the pathogen, not the entire pathogen or virus, are used in the vaccine. For example, one coat protein from a virus will produce an immune response. Unfortunately, subunit vaccines do not stimulate the immune system as strongly as other types of vaccines.

Recombinant DNA technology enables the genes encoding the proteins (antigens) to be cloned and produced in bacteria such as *E. coli* or yeast cells and, therefore, readily mass-produced for vaccine production. An example is the hepatitis B vaccine that has been developed using a cloned viral coat protein in yeast host cells. At the same time, recombinant DNA technology allows scientists to modify the antigens (that is, the gene encoding the protein) so that they are able to increase the response of the immune system (increase the strength of antigenicity).

5. Different antigens can be mixed to create a stronger immune response and, thus, better protect the individual. These multivalent vaccines also can ensure that immunity is established against different but closely related organisms, as well as those that may have several forms or express different antigens at different times.

New vaccines are being tested that are more potent and may produce fewer side effects. These new vaccines include small synthetic peptides that elicit an immune response but lack components that cause side effects and nucleic acid vaccines (DNA or RNA) made from a pathogen's genome. The DNA is injected into a patient and the DNA is incorporated into the host genome where it can be expressed. The protein synthesized will be an antigen that stimulates the immune system. Some DNA vaccines are in clinical trials (for example, prevention of HIV infection).

Therapeutic vaccines have the potential to fight cancer cells. Cancer cells have altered antigens and MHC proteins on their surface and often hide from the immune system. Thus, the immune system is often unable to kill these cells. Generically engineered antigens, specific for tumor cells, can be used to elicit a more virulent attack against tumor cells.

Another new development is the production of edible vaccines in plants such as potato, tomato, and banana. Diseases such as malaria, dysentery, whooping cough, diphtheria, measles, polio, tuberculosis, and tetanus claim many lives in the developing world. In many cases there is neither refrigeration to preserve vaccines nor the funds to distribute them. An edible vaccine would be more readily accepted, cost less to distribute, and would not require refrigeration (see Chapter 6, Plant Biotechnology).

IMMUNE SYSTEM DISORDERS

Although the immune system protects the body against infection, there are occasions when things go awry and the immune system responds to and attacks molecules and cells produced by the body. This is called an autoimmune response. There are many autoimmune diseases, each based on the response of the immune system to different regions of the body (Table 4.11). There are most likely several possible causes of immune system breakdown. For example, a defect in T-cell function such as with T-helper or T-suppressor cells might enable the immune system to attack body cells. An example might be in the development of diabetes when insulin-producing islet cells are attacked by the immune system. It also is possible that there is cross-reactivity of self antigens and foreign antigens so that the immune system mistakes self for foreign. Latent changes in the antigen composition of some body cells may cause the body to respond to these antigens as foreign.

Hypersensitivity

Allergies Many people have allergies—diseases resulting from increased activity of the immune system. Examples of allergies include food allergies, allergies to insect stings such as from bees and wasps, drug allergies such as to the antibiotic penicillin, and allergies to plant oils (for example, poison ivy) or pollen. Antigens that elicit an allergic response are referred to as allergens. The response to an allergen varies greatly in people. Some people do not respond abnormally to an allergen, while others suffer a variety of symptoms (for example, people who have hay fever in response to ragweed pollen often have runny, itchy eyes and noses). Two types of adverse immune responses to antigens are discussed.

Cell-mediated allergy is a response mediated by T cells and macrophages. It is elicited by contact with allergens such as plant oils, organic compounds and other chemicals, cosmetics, and latex. This type of allergy (that is, contact allergy) is a delayed-hypersensitivity because it often takes several hours to days for the immune system to respond. In this allergy, compounds or chemicals bind to and modify proteins of plasma membrane of cells that make up the skin. This activates the cell-mediated immune response. An example of this is the response to a tuberculin skin test after a person is exposed to tuberculosis (TB) and harbors antibodies to TB. A positive reaction to the test is slow and takes from 24 to 72 hours.

TABLE 4.11 Some Autoimmune Diseases

Disease	Organ or Tissue Affected
Addison's disease	Adrenal glands
Autoimmune hemolytic anemia	Red blood cells
Goodpasture's syndrome	Kidneys and lungs
Hashimoto's thyroiditis (hypothyroidism)	Thyroid gland
Juvenile onset diabetes	Pancreas (islet cells)
Multiple sclerosis	Brain tissue
Myasthenia gravis	Heart and skeletal muscle
Pernicious anemia (vitamin B_{12} deficiency)	Gastrointestinal mucosa
Rheumatoid arthritis	Joints, synovial membranes
Sjogren's disease	Salivary and tear glands
Systemic lupus erythemastosus (SLE)	Systemic
Ulcerative colitis	Colon

Antibody-mediated allergy includes asthma, hay fever, and the allergic response to insect stings. During this type of an allergy, IgE is synthesized and circulates in the body. The tail region of IgE antibody binds to basophil cells and mast cells, while the antigen binding site attaches to an allergen molecule. Basophils and mast cells contain histamine, one of the molecules that elicit allergy symptoms. This type of allergy is an immediate-response hypersensitivity because, upon exposure to the allergen, an immune reaction is very rapid.

IgE-mediated allergies are usually treated with antihistamine drugs that block the release of histamine from mast cells and basophils. Steroids such as cortisone are sometimes used to reduce inflammation and generally suppress the immune system. Finally, desensitization treatments are sometimes used. For example, people who are allergic to bee venom may become desensitized by being exposed to increasing concentrations of venom through injection. This works almost 100% of the time. The low-dose exposure to antigen stimulates the production of IgG antibodies that bind to antigen, thereby preventing the antigen from binding to IgE on mast cells and eliciting an adverse immune response.

Autoimmune Disorders When the immune system malfunctions, sometimes cells, tissues, and molecules of the body are attacked. For example, in insulin-dependent diabetes mellitus, insulin-producing beta cells of the pancreas are destroyed. Autoimmune reactions can be caused by a number of factors. Antigens on a microbe may be very similar to those on cells of the body and may cause a cross-reaction. Infections may change the surface antigens of cells so that they are perceived as foreign.

Cytotoxic hypersensitivity occurs when the immune system attacks self. This occurs when IgM or IgG antibodies mistake the body's own cells for those of an invading microbial cell (this is an autoimmune disorder). Body cells that are bound by these antibodies can be destroyed by phagocytes or killer cells, by agglutination (clumping) with IgM antibody, or by being lysed through the complement system. An example of cytotoxic hypersensitivity is Graves' disease, where antibodies bind to cells that produce thyroid hormone and cause them to produce even more thyroid hormone. High levels of thyroid hormone cause a number of physical symptoms such as bulging eyes and weight loss. Another disease is Hashimoto's thyroiditis, where antibodies attack and destroy the thyroid, dramatically reducing the production of thyroxin. People with this disorder must take thyroxin supplements. If not treated, they may develop a goiter.

Immune-complex hypersensitivity is similar to cytotoxic hypersensitivity because antibodies attack self cells. However, it also is different in that the body components that are attacked are not cells, but circulating molecules. In this reaction, soluble antigen–antibody complexes that circulate in the body initiate this type of immune response. Macrophages typically remove such complexes, but when they are not eliminated, these complexes become stuck in capillaries of body tissues that then activate complement and elicit a strong inflammatory response. White blood cells that are present release enzymes that cause extensive tissue damage. An example of this type of disorder is systemic lupus erythematosus (SLE)—sometimes referred to as lupus. Although symptoms vary (as they do with any immune reaction), any organ can be affected such as the kidneys and the brain. In addition, the joints (arthritis) and skin can become inflamed. The disease can be fatal. Another type of disorder is rheumatoid arthritis. This disease is triggered by IgG antibodies that react with rheumatoid factor, a protein that people with this condition synthesize. This antibody–antigen complex causes joints in the body to be damaged, or in severe cases, destroyed. Although there is not a cure for these immune disorders, steroids are often prescribed to repress the immune system.

Immunodeficiencies Sometimes the immune system is unable to function at a normal level. Any type of immune system cell might be affected. The disorder is defined by the particular component that is faulty (Table 4.12). When the immune system is deficient, the individual is susceptible to infections. Some types of immunodeficiencies are hereditary (for example, severe combined immunodeficiency (SCID), see Chapter 10, Medical Biotechnology), while others are acquired and can be caused by infection or cancer. Sometimes immunosuppressive medications cause immunodeficiency disorders.

TABLE 4.12 Examples of Immunodeficiency Disorders

Disorder	Immune Defect	Congenital vs. Acquired	Clinical Manifestations
Severe combined immunodeficiency (SCID)	Deficiency of B and T cells	Congenital	Overwhelming infections that cause death in early infancy unless patient is maintained in germ-free environment or receives a bone marrow transplant
X-linked agammaglobulinemia	Severe deficiency to near absence of B cells	Congenital (inherited on the X chromosome, so it affects males only)	Frequent, often life-threatening bacterial infections that begin in infancy
IgA deficiency	Selective inability to produce IgA	Congenital	Repeated bacterial infections of the respiratory and gastrointestinal tracts
DiGeorge's syndrome	Abnormal development of T cells associated with abnormal thymus	Congenital	Severe immunodeficiency associated with other developmental abnormalities, often leading to death in infancy
AIDS	Severe deficiency of T-helper lymphocytes	Acquired through infection by HIV	Recurrent bacterial, viral, fungal, and protozoal infections beginning when TH cell count falls below 500 per microliter of blood
Chronic granulomatous disease	Deficiency in phagocyte function–neutrophils unable to kill ingested pathogens	Congenital; inherited on the X chromosome	Recurrent bacterial infections, both blood-borne and soft tissue, beginning during the first year of life
C5 dysfunction	Abnormal function of the C5 complement protein	Congenital	Recurrent bacterial infections

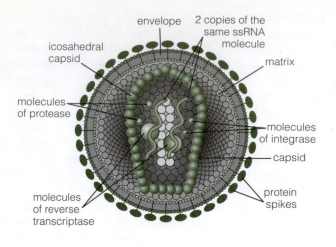

icosahedral capsid

envelope

2 copies of the same ssRNA molecule

matrix

molecules of protease

molecules of integrase

capsid

molecules of reverse transcriptase

protein spikes

FIGURE 4.24 Structure of HIV. The capsid is surrounded by a structural protein matrix and an envelope embedded with protein spikes (gp120 protein). Within the capsid are two copies of the RNA genome (called the plus-strand RNA) and the enzymes reverse transcriptase, integrase, and protease. Reverse transcriptase uses the RNA genome as a template to synthesize DNA (the final product is a double-strand DNA). Integrase integrates the DNA into the host chromosome. Protease is required for virus replication and assembly of new viruses in the host cell. As newly assembled HIV viruses leave the host cell, protease processes the proteins within the capsid (part of maturation, see Figure 4.25).

HIV and AIDS   Acquired immunodeficiency syndrome (AIDS) is a well-known global disease caused by the human immunodeficiency virus (HIV, a retrovirus) (Figure 4.24). It acts by repressing the immune system in specific ways. People who develop AIDS are susceptible to cancers, respiratory infections, and a variety of fungal, viral, and bacterial infections. Because there is no cure, the disease is fatal if viral inhibitor drugs are not taken. Once HIV enters the human body, it is circulated in the bloodstream and lymph. The viral envelope is composed of proteins—one of which, gp120, binds to CD4 receptors expressed on cells such as helper T cells and macrophages (the targets of HIV). Once the virus binds to the receptor, the envelope of the virus fuses with the cell membrane (Figure 4.25). This fusion of the two cells allows the virus to gain entry into the cell. Once the virus is inside the cells, the RNA genome of the virus and the enzyme, reverse transcriptase, are released from the envelope. Reverse transcriptase is involved in synthesizing a double-strand DNA molecule using the RNA genome as a template. This DNA integrates into the host cell genome. This provirus can remain dormant for a number of years before it becomes active. The virus only becomes active when the particular host cell becomes active. When the virus is active, it reproduces and these viruses infect large numbers of cells. T cells eventually die from HIV infection, thereby disrupting immune system function. People with active HIV have reduced numbers of T cells that results in an immune deficiency so that the body cannot mount an effective attack on foreign invaders and on cancerous cells. Thus, the symptoms of AIDS result from the dramatic reduction of helper T cells and the severe immunodeficiency, which ultimately leads to death.

The immune system cannot detect HIV because, once the virus infects host cells, they are within the cell and cannot be detected by the immune system. Although the cell-mediated immune response system can attack and kill infected cells, cytotoxic T cells must be able to recognize HIV antigens on the infected host cell surface (antigen presenting). However, when the virus is not active, the immune system cannot recognize virus-infected cells. Furthermore, HIV coat proteins undergo mutation frequently, which enables the virus to evade the immune system. Because the immune system cannot detect all the variants, the virus cannot be completely eliminated. Finally, as the immune system weakens over time, the body becomes susceptible to a variety of infections. The unusual properties of HIV are summarized and some of the treatments are indicated in Table 4.13.

MONOCLONAL ANTIBODIES

Antibodies are used extensively in modern biotechnology to detect genetic and acquired diseases (diagnostics), as therapeutic agents in medicine, and as tools in basic research. They are produced in the laboratory against a variety of selected antigens. When an antigen such as a protein is injected into an animal (usually mice, rabbits, or goats), a mixture of antibodies (produced from different lymphocyte clones)—that is, polyclonal antibodies—having different specificities to parts of the protein (antigenic determinants or epitopes) are produced and isolated. Thus, each antibody recognizes a specific epitope within the protein molecule. Some of these antibodies may even bind to other proteins, nonspecifically. The same process occurs when our immune system responds to infection by a

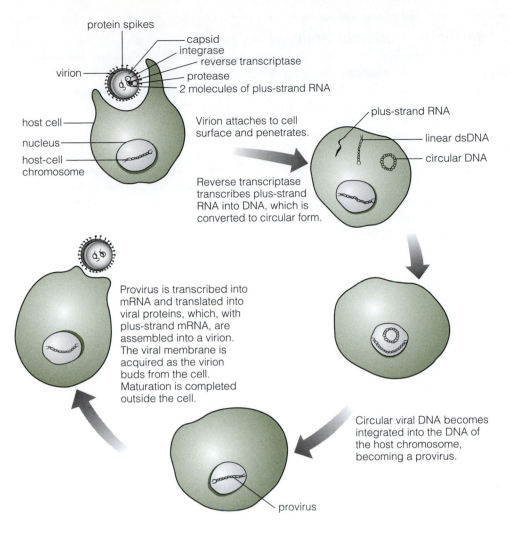

FIGURE 4.25 Replication of HIV.

TABLE 4.13 Unusual Properties of HIV

Property	Comments
RNA genome is diploid	Significance is not understood.
Intact virion contains three enzymes: reverse transcriptase, integrase, and protease	Inhibitors of two of these enzymes are powerful anti-AIDS drugs.
Plus-strand genome serves as a template for making dsDNA (action of reverse transcriptase), not as mRNA	Certain anti-AIDS drugs, including AZT, act by inhibiting this reaction.
Maturation is completed within the intact virion when protease cuts strings of connected proteins into active units	Anti-AIDS drugs called protease inhibitors act by inhibiting this reaction.
Capacity for rapid genetic change	Possibly accounts for its having become a human pathogen; may foretell its gaining further pathogenic potential.

Future of Monoclonal Antibody Production

Hybridoma cells grow slowly (because they must be cultured for several months in culture and in mice) and are expensive to produce. Cells must be screened for the desired antibody. In addition, mouse cells cannot produce human antibodies for certain types of therapeutic applications. Therefore, new methods of generating monoclonal antibodies are being investigated. For example, antibody-encoding genes have been cloned into cells of animals, plants, and bacteria such as *E. coli* using bioreactors (mass cell cultures in large volumes of liquid medium). These cells produce specific antibodies (monoclonal antibodies) that can be used therapeutically. This process is less expensive and time-consuming.

The production of genetically engineered corn that produces monoclonal antibodies in the field is also being investigated. This method will be used to mass produce specific monoclonal antibodies. Plant-generated antibodies are more cost-effective and are thought to produce fewer side effects because plant viruses and pathogens generally do not infect a human host (unlike those from an animal host). Recently, in clinical trials, a plant-generated monoclonal antibody was used to prevent infection by tooth decay–promoting bacteria.

---

bacterium. Hundreds and possibly thousands of different epitopes are involved in activating hundreds to thousands of B-cell clones, each producing a different antibody (that is, polyclonal antibodies).

By contrast, **monoclonal antibodies** (MABs, identical antibodies to a specific epitope of a protein produced by a clone originating from one cell) provide researchers with an unlimited amount of very specific antibodies. However, B cells, the precursors of the plasma cells that produce antibodies, do not divide or produce antibodies in culture. Therefore, B cells must be fused with ones that are able to divide indefinitely (a property of a cancer cell) in culture for the mass production of monoclonal antibodies (a property of the B cell). To immortalize an antibody-producing cell, it is fused with a cancerous form of B cell (for example, myeloma cell) that grows continuously in culture. Spleen cells, often obtained from mice, are fused with immortal cancer cells from the same lineage—for example, a cancerous mouse cell, often a myeloma cell that is of B cell origin—that can divide indefinitely. The fusion product, called a **hybridoma** (the fusion of two different cells to create one cell), secretes one type of antibody against a specific antigen. The cells derived from the hybridoma are all clones and thus produce the same specific monoclonal antibody. Scientists must separate the hybridomas that produce the desired antibody and clone them in cell culture. Each antibody from one clone is specific for a single epitope.

Monoclonal antibodies have many important applications in research, diagnostics, and therapeutics; a few are briefly introduced. Monoclonal antibodies have been used to identify a clone that produces a specific protein from an expression library (recall that proteins are synthesized in an expression library). Their specificity makes these antibodies important tools in medical diagnostics; they increase the sensitivity of **immunoassays** for detecting a variety of hormones (such as concentrations of progesterone and estrogen) and antigens (blood group antigens, for example).

Other sensitive diagnostics use monoclonal antibodies to detect a variety of diseases (for example, exposure to HIV; chicken pox; or hepatitis A, B, or C) or specific molecules in very low concentration. For example, home pregnancy tests use monoclonal antibody specific for human chorionic gonadotropin (hCG), which is synthesized by a developing human embryo.

Monoclonal antibodies are being studied to evaluate their therapeutic potential to help fight disease such as cancer (for example, MABs that recognize and bind specific tumor antigens), boost the immune system, and prevent organ rejection. In this last case, antibodies synthesized by the recipient's lymphocytes against the newly transplanted organ are removed from the body by monoclonal antibodies. Monoclonal antibodies also can aid in the prevention of graft-versus-host (GVH) disease, which occurs when donor T cells from a bone marrow transplant attack the host. The donor T cells are removed by monoclonal antibody binding.

Monoclonal antibodies also are used in the isolation and purification of proteins in a process called affinity purification. MABs can be synthesized to specifically bind a wide variety of proteins—even antibodies. This method enables a scientist to separate a specific component from a mixture. One way of doing this is to use affinity chromatography where specific MABs are bound (immobilized) to an inert substrate within a column. The substrate is typically an agarose gel bead. The mixture of proteins is run through the column. The specific target protein binds to the MABs within the column and unbound proteins are washed (eluted) from the column. Following this, an elution buffer releases the protein of interest from the column for isolation.

TOOLS OF IMMUNOLOGY

Many methods have been developed for both research and diagnostic purposes. A few of them are discussed in this section.

Western Blotting

Western blotting (sometimes called immunoblotting) is a method used to detect a specific protein, often present in very low concentration and mixed with other proteins (Figure 4.26). After proteins are separated on a polyacrylamide gel by gel electrophoresis, they are blotted onto a membrane (the Western blot) in a manner similar to that of Southern blotting (see Chapter 3, Basic Principles of Recombinant DNA Technology) and a labeled antibody is used as a probe to identify its target protein. This method allows an investigator to detect the presence, quantity, and size of specific proteins in a particular preparation. The following is the step-by-step process:

1. Sodium dodecyl sulfate (SDS)-polyacrylamide gel electrophoresis is conducted to separate proteins by size.

2. The proteins migrate into the gel, and the distance they move is inversely proportional to their size.

3. The gel is blotted in a manner similar to Southern blotting so that the protein bands are transferred to a membrane.

4. The membrane is bathed in a solution of monoclonal antibodies that can be detected by using a labeled secondary antibody that binds to the primary monoclonal antibody probe.

5. The membrane is washed and the places where proteins have bound with antibody are visible. Depending on the type of method, the bands are visible using autoradiography, chemiluminescence, or the enzymatic production of a colored precipitate.

Fluorescent Antibody Technique

Fluorescent antibody technique, also called immunofluorescence microscopy, is used to detect antigens in a cell or tissue. That is, using antibodies, corresponding antigens can be detected. This method uses a fluorescent tag (molecules that emit a specific visible wavelength when excited by a short wavelength, in the ultraviolet range) that is covalently bound to an antibody that can be detected using a fluorescence microscope. The most commonly used fluorescent tag is fluorescein isothiocyanate (FITC). This tag fluoresces a yellow-green light when excited with ultraviolet light. There are both direct and indirect methods (Figure 4.27). In both types, specific

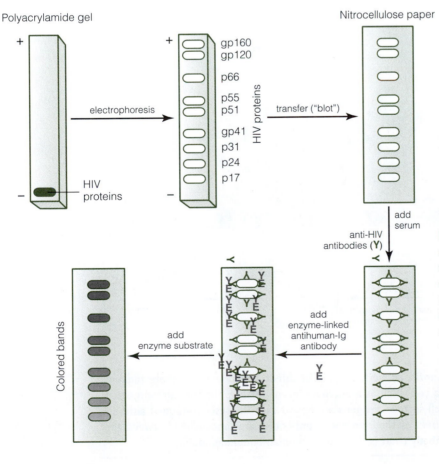

FIGURE 4.26 Western blotting for antibodies to HIV proteins. Proteins are separated by polyacrylamide gel electrophoresis and transferred to a membrane by blotting. The membrane is exposed to antibodies to HIV, and the antibodies bind to specific antigens (proteins). A secondary antibody conjugated (linked) to an enzyme is added. The antihuman–Ig antibody binds to the anti-HIV antibodies on the membrane. A substrate for the enzyme is added that yields colored bands where binding occurred.

antibodies are detected. Either tissue samples or pathogens are bound to microscope slides, and tagged antibodies are used as a probe.

1. In the direct assay, labeled fluorescent antibodies bind directly to antigens such as on the cell surface. In this way, various tissues can be examined for infection. For example, when an animal is suspected of being infected with rabies virus, tissue samples bound to microscope slides can be used to detect rabies virus antigens. Fluorescent-tagged antibodies specific to the virus are used and antigen-antibody binding is detected using fluorescence microscopy.

2. In an indirect assay, unlabeled antibodies are first used followed by a secondary fluorescent-tagged antibody that is specific to the unlabeled antibody. Like the direct assay, this method also is used to detect specific antibodies. This so-called immune sandwich serves to amplify the tagged signal. An example is the determination of infection by a human pathogen. During infection, antibodies are produced and can be detected by this method. The pathogen is bound to a microscope slide and unlabeled antibodies from human serum are added to the slide. A second antibody, fluorescent-tagged antihuman immunoglobulin, indicates whether an antibody-antigen reaction has occurred.

Enzyme-Linked Immunosorbent Assay

Immunoassays quantify antigen-antibody reactions. Either tagged antibodies or antigens allow the detection of very small quantities of antigen or antibodies to be detected, respectively. Tags can be fluorescent (immunofluorescence assay), radioactive (radioimmunoassay), or enzymatic (enzyme immunoassay). DNA probes and polymerase chain reaction

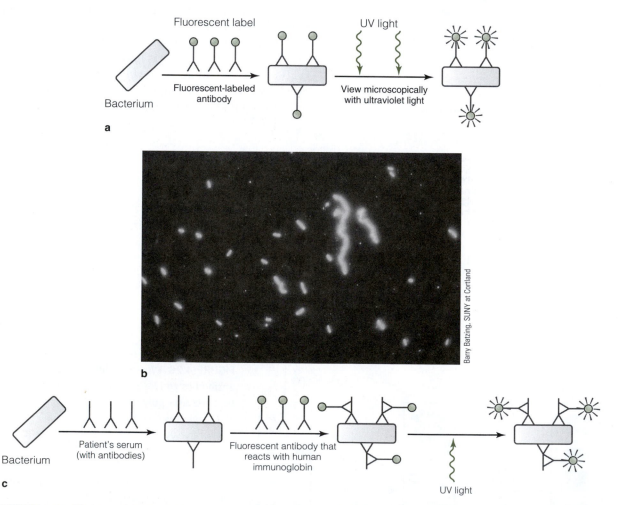

FIGURE 4.27 Direct and indirect fluorescent antibody methods. (a) The direct antibody assay where tagged antibody interacts directly with antigens on a cell (such as a microorganism) surface. (b) The microorganism *Legionella pneumophila* fluoresces after its cell surface antigens are bound with fluorescent–tagged antibody using the direct binding method. (c) In the direct method, untagged antibodies are sandwiched between the known antigen and a fluorescent–tagged antibody (that is, tagged antihuman immunoglobulin).

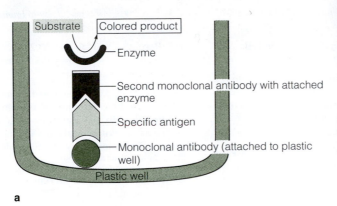

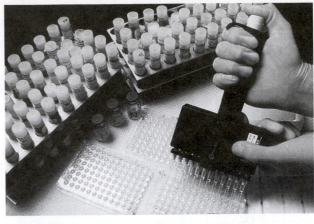

FIGURE 4.28 The ELISA method. (a) The sandwich of substrate-bound antibody, specific antigen (to be tested), and a second antibody with a bound enzyme, with the subsequent conversion of a substrate to a colored product. (b) The steps in conducting an ELISA assay. A colored product indicates that specific antibodies are present in a person's serum. The photograph shows a laboratory technician conducting an ELISA assay. An automatic pipetting instrument allows many wells to be filled with assay reagents. From Hoffman-LaRoche Inc., Nutley, New Jersey.

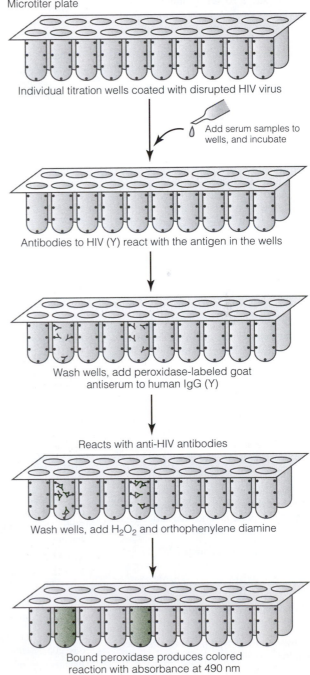

(PCR) are being more widely used clinically to diagnose disease and infection, although immunoassays are still used.

One of the most commonly used immunoassays used today is the enzyme-linked immunosorbent assay (ELISA) (Figure 4.28). ELISA is often used to detect specific antibodies in human serum. Instead of a fluorescent-tagged antibody, the detectable tag is an enzyme. Different enzymes can be used such as alkaline phosphatase, glucose oxidase, and lysozyme.

First, the target antigen (for example, from a bacterial pathogen or HIV) is bound to a solid phase such as to the bottom of a well in a plastic microplate. One type of ELISA uses a sandwich method similar to that described for indirect immunofluorescence. A patient's serum is added to the well of the microplate. If the serum contains antibodies to the antigen, they will form antibody-antigen complexes. The well is washed with antihuman immunoglobulin antibody that is bound to an enzyme that will be detected. This will bind to the person's antibodies if an antibody-antigen complex is present. To detect the enzyme-labeled antibody, a substrate for the enzyme is added. This results in a color reaction that can be quantified.

ELISA is a highly sensitive and relatively inexpensive method compared to other types of assays and is regularly used as a clinical diagnostic tool. For example, the assay is used to detect viral infections such as hepatitis infection, HIV, rubella, and herpes simplex infection.

General Readings

A. K. Abbas, A. H. Lichtman, and J. S. Pober. 2000. *Cellular and Molecular Immunology*. 4th ed. W. B. Saunders, Philadelphia.

E. Benjamini, G. Sunshine, and S. Lleskowitz. 1996. *Immunology: A Short Course*. Wiley, New York.

M. Bestagno, M. Occhino, M. V. Corrias, O. Burrone, and V. Pistia. 2003. Recombinant antibodies in the immunotherapy of neuroblastoma: Perspectives of new developments. *Cancer Lett*. 197:193–198.

S. A. Camperi, N. B. Lannucci, G. J. Albanesi, E. M. Oggero, M. Etcheverrigaray, A. Messeguer, F. Albericio, and O. Cascone. 2003. Monoclonal antibody purification of affinity chromatography with ligands derived from the screening of peptide combinatory libraries. *Biotechnol. Lett*. 25:1545–1548.

L. A. Doughty and P. Linden, eds. 2003. *Immunology and Infectious Disease*. Kluwer Academic, Norwell, Massachusetts.

R.A.B. Ezekowitz and J.A. Hoffman, eds. 2003. *Innate Immunity*. Humana Press, Totowa, New Jersey.

J. M. Jacobson, ed. 2002. *Immunotherapy for Infectious Diseases*. Humana Press, Totowa, New Jersey.

J. Kuby. 1997. *Immunology*. W.H. Freeman, New York.

L. Gordis. 2000. *Epidemiology*. 2nd ed. W.B. Saunders, Philadelphia, PA.

S. Graslund, R. Falk, E. Brundell, C. Hoog, and S. Stahl. 2003. A high-stringency proteomics concept aimed for generation of antibodies specific for cDNA-encoded proteins. *Biotechnol. Appl. Bioc*. 35:75–82.

J.L. Ingraham and C.A. Ingraham. 2004. *Introduction to Microbiology: A Case History Approach*. 3rd ed. Brooks/Cole—Thomson Learning, Inc., Pacific Grove, California.

J. Krauss. 2003. Recombinant antibodies for the diagnosis and treatment of cancer. *Mol. Biotechnol*. 25:1–17.

M. Nisnevitch and M. A. Firer. 2001. The solid phase in affinity chromatography: Strategies for antibody attachment. *J. Biochem. Biophys. Meth*. 49:1–3.

N. Ramirez, M. Ayala, D. Lorenzo, D. Palenzuela, L. Herrera, V. Doreste, M. Perez, J. V. Gavilondo, and P. Oramas. 2002. Expression of a single-chain Fv antibody fragment specific for the Hepatitis B surface antigen in transgenic tobacco plants. *Transgenic Res*. 11:61–64.

C.V. Rao. 2002. *An Introduction to Immunology*. CRC Press, Boca Raton, FL.

K.S. Rosenthal and J.S. Tan. 2002. *Microbiology and Immunology*. Mosby Publishing, St. Louis, Missouri.

M. Shimizu, A. Matsuzawa, and Y. Takeda. 2003. A novel method for modification of tumor cells with bacterial superantigen with a heterobifunctional cross-linking agent in immunotherapy of cancer. *Mol. Biotechnol*. 25:89–94.

S. Tonegawa. 1995. The molecules of the immune system. *Sci. Am*. 253:123–131.

H. Warzecha and H. S. Mason. 2003. Benefits and risks of antibody and vaccine production in transgenic plants. *J. Plant Physiol*. 160:755–764.

D.J. Wise and G.R. Carter. 2002. *Immunology: A Comprehensive Review*. Iowa State University Press, Ames, Iowa.

M. Zouali. 2000. Tracking immunoglobulin variable-gene expression in HIV infection. *Appl. Biochem. Biotechnol*. 83:1–3.

Additional Readings

A. Anderson and R.J. Ulevich. 2002. Toll-like receptors in the induction of the innate immune response. *Nature* 406:782–784.

R. Barrington, M. Zhang, M. Fisher, M. Carroll. 2001. The role of complement in inflammation and adaptive immunity. *Immunol. Rev*. 180:5–15.

B.R. Bloom and P.-H. Lambert. 2003. *The Vaccine Book*. Academic Press, Boston, Massachusetts.

C.A.K. Borrebaeck. 2000. Antibodies in diagnostics: From immunoassays to protein chips. *Immunnology Today* 21:379–382.

D.J. Chiswell and J. McCafferty. 1992. Phage antibodies: Will new "coliclonal" antibodies replace monoclonal antibodies? *Trends Biotechnol*. 10:80–84.

L.A. Doughty and P. Linden. 2003. *Immunology and Infectious Disease*. Kluwer Academic Publishers, Norwell, Masssachusetts.

C. Ezzell. 2001. Magic bullets fly again. *Sci. Am*. 285:34–41.

A. Farzaneh-Fat, J. Rudd, and P.L. Wessberg. 2001. Inflammatory mechanisms. *Br. Med. Bull*. 59:55–68.

G.F. Fischer. 2000. Molecular genetics of HLA. *Vox Sang*. 78:261–264.

M. Giese. 1998. DNA-antiviral vaccines: New developments and approaches. *Virus genes* 17:219–232.

T. Gura. 2001. Innate immunity: Ancient system gets new respect. *Science* 291:2068–2071.

H.C.J.E. Hildegund, ed. 2003. *DNA Vaccines*. Kluwer Academic Press, New York.

W.H.R. Langridge. 2000. Edible vaccines. *Sci. Am*. 283:66–71.

M. Little, S.M. Kipriyanov, F. Le Gall, and G. Moldenhauer. 2000. Of mice and men: Hybridoma and recombinant antibodies. *Immunology Today* 21:364–370.

M.J. McCluskie and H.L. Davis. 1999. Mucosal immunization with DNA vaccines. *Microbes Infect*. 1:685–698.

P. Parham. 2000. *The Immune System*. Garland Publishing, New York.

N.R. Rose, R.G. Hamilton, and B. Detrick, eds. 2002. *Manual of Clinical Laboratory Immunology*. 6th ed. ASM Press, Washington, D. C.

A.A. Salyers and D.D. Whitt. 2001. *Bacterial Pathogenesis*. 2nd ed. ASM Press, Washington, D. C.

J. Sprent and D.F. Tough. 2001. T cell death and memory. *Science* 293:245–248.

H.J. Stauss, Y. Kawakami, and G. Parmiani, eds. 2003. *Tumor Antigens Recognized by T Cells and Antibodies*. Taylor & Francis, New York.

G. Taubes. 1997. Salvation in a snippet of DNA. *Science* 278:1711–1714.

A. Vladutiu. 1993. The severe combined immunodeficient (SCID) mouse as a model for the study of autoimmune disease. *Clin. Exp. Immunol*. 93:1–8.

T.A. Waldmann. 1991. Monoclonal antibodies in diagnosis and therapy. *Science* 252:1657–1662.

R.D. Weeratna, M.J. McCluskie, L. Comanita, T. Wiu, and H.L. Davis. 2000. Optimization strategies for DNA vaccines. *Intervirol*. 43:218–226.

D.B. Weiner and R.C. Kennedy. 1999. Genetic vaccines. *Sci. Am*. 281:50–57.

F. Yamamoto. 1995. Molecular genetics of the ABO histo-blood group system. *Vox Sang*. 69:1–7.

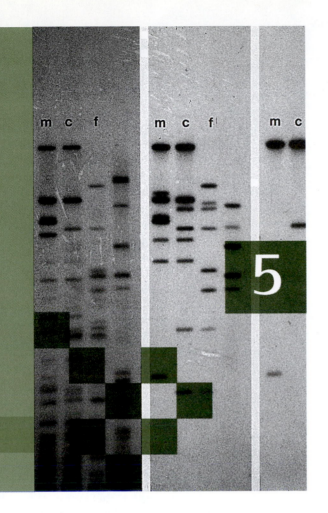

MICROBIAL BIOTECHNOLOGY

5

Microorganisms are used in industrial settings to produce many important chemicals, antibiotics, organic compounds, and pharmaceuticals. Using living organisms as chemical synthesis factories reduces many of the risks and complexities of industrial syntheses, allowing costly and polluting raw materials to be replaced by less expensive processes. The by-products of biosynthetic reactions are usually less toxic and hazardous than those of industrial chemical reactions.

Bacteria also have a potentially more prominent role in environmental cleanup of industrial spills and accidents, chemical leaching, crude oil spills, radiation leaks, toxic herbicides and pesticide residues, acid rain, chemical effluents, and accumulations of lead and mercury.

Microorganisms have been exploited for hundreds of years. Consequently, we know far more about their biochemical properties than we know about plants and animals. Because the extent of their commercial and industrial exploitation far exceeds the scope of a single chapter, we examine only a few of the most important uses of microorganisms.

COMMERCIAL PRODUCTION OF MICROORGANISMS

Biochemists define fermentation as an anaerobic process that generates energy by the breakdown of organic compounds; the end products can be microbial metabolites such as lactic acid, enzymes, the alcohols ethanol and butanol, and acetone. Industrial users of fermentation have broadened the definition to include (1) any process that produces bacteria and fungi (for example, yeast) (biomass) as the end product and (2) biotransformation (transformation by cells of a compound added to the fermentation medium of a commercially valuable compound). In large-scale fermentations, metabolites, proteins, carbohydrates, and lipids are provided to enable the propagation of microbial cells, or biomass.

Aerobic or anaerobic microorganisms are cultured under controlled conditions in large chambers, or fermenters, sometimes called bioreactors. Most fermentations, whether anaerobic or aerobic, require a number of steps: (1) sterilization of the fermentation vessel and associated equipment; (2) preparation and sterilization of the culture medium; (3) preparation of a pure cell culture for inoculation of the medium in the fermentation vessel; (4) cell growth and synthesis of the desired product under a specific set of conditions; (5) product extraction and purification or cell collection; and (6) disposal of expended medium and cells, and the cleaning of the bioreactor and equipment.

Industrial Fermenters

A goal that drives development in fermentation technology is to improve the productive performance of microorganisms by optimizing their growth conditions. Most of the microorganisms used in industrial fermentations today are aerobic, because aerobic metabolism is more efficient than anaerobic metabolism (although scientists are investigating the use of anaerobic microorganisms for potential new products). Besides requiring oxygen for rapid growth, they need a consistent pH in the growth medium, temperature control, a supply of nutrients for rapid growth, and an antifoaming agent to alleviate excess foaming during aeration of the culture. (The culture is aerated either by bubbling air into the medium or by agitating it.)

Fermenters have ports for adding the gases, nutrients, culture inoculum, and antifoaming agents. A probe continuously monitors pH, and a separate port allows acid or alkali to be added to adjust the pH. Temperature is regulated by a water jacket that circulates water around the outside of the culture vessel. An exit valve allows microbial cells, medium, or microbial products to be collected. Sterility is essential; the sterilized vessel and culture medium are kept free of contamination by the use of filters, such as for admitting sterilized gases.

Different applications dictate the use of different types of fermenters. Figure 5.1 shows the most common type, the stirred tank reactor, which relies on an agitator to circulate oxygen. Another common type is the airlift fermenter, which supplies oxygen to the culture through an intake valve in the bottom of the culture vessel. Air flow into the fermenter creates high pressure that vigorously circulates the culture medium.

Products and cells are collected from fermenters by either of two methods, continuous fermentation or batch culturing. In continuous fermentation nutrients are fed into the fermenter while an equal volume of products, cells, and medium is collected. (Sometimes the collected cells are added back to the fermenter.) Thus, continuous fermentation allows continuous culture growth to be maintained for long periods. This method is most suitable for obtaining compounds that are produced proportionally to cell growth, such as primary metabolites (vitamins and enzymes, for example). Continuous fermentation is also very useful for treating wastewater and for degrading organic industrial wastes. The waste compounds are carbon and nutrient sources for microbes.

In batch culturing, the cells, products, and medium are collected after the fermentation process has terminated, and the fermenter is set up for the next fermentation. This method is preferred when synthesis of the products does not depend on the amount of cell biomass; antibiotics are an example. The cells and liquid

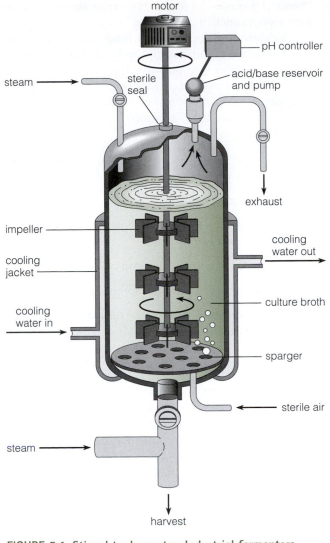

motor

pH controller

steam

sterile seal

acid/base reservoir and pump

impeller

cooling jacket

cooling water in

cooling water out

culture broth

sparger

sterile air

steam

exhaust

harvest

FIGURE 5.1 Stirred tank reactor. Industrial fermenters produce very large volumes of cell cultures.

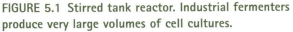

5. Enzymes

6. Surfactants

7. Ethanol

8. Biogas

9. Microbial polysaccharides

10. Secondary metabolites such as antibiotics

In solid substrate fermentation microorganisms are grown on solid substrates that are not submerged in liquid media. Rather, moisture is absorbed onto or complexed within the substrate. In this way it is thought that oxygen is more readily available to the cells because they are in direct contact with air. Other advantages are that this process requires less energy, uses fermenters that take up less space than submerged fermentation, and results in downstream processing of products that is generally not as complex. Although many microorganisms are able to grow on solid substrates, most cannot grow rapidly or to a large mass in the absence of free-flowing media. Furthermore, fermentation parameters are more difficult to control. This process has been most successfully used in the production of enzymes and secondary metabolites where stationary phase cells are used.

A recent development in fermentation is the use of ultrasound to increase the production of cells. Sonication has often been used to break cells open. In this highly controlled process, ultrasonication (exposure to a particular sound frequency inaudible to humans) enhances biochemical activities of plant, animal, and microbial cells during fermentation. Ultrasound is often used to clean surfaces (for example, jewelry), for underwater sonar, and in medical imaging; however, scientists believe it can enhance production in industrial fermentation and are investigating a variety of sono-bioreactor designs.

Single-Cell Protein

Microbes have been used as food and food supplements for several thousand years. For example, yeast cells have been used as a supplement in soups, sausages, and animal feed to contribute additional protein, minerals, and vitamins. A monoculture of algal, bacterial, or fungal cells has a protein content that is 70–80% of its dry weight. When such a monoculture is grown in large volumes for use as human or livestock feed supplements, it is called single-cell protein, or SCP. SCP is high in such nutrients as minerals, vitamins, carbohydrates, and lipids, as well as essential amino acids like lysine and methionine that are often lacking in plant protein.

SCP is produced by using inexpensive substrates to supply nitrogen and carbon—substrates such as

are separated by sedimentation or filtration, and the metabolites and enzymes are extracted from the collected liquid. Cells also can be concentrated for later use, or if intracellular products are to be obtained, cells must be lysed and the product purified and concentrated. In fed-batch fermentation, a variation of batch culturing, fresh medium is added periodically during fermentation.

In recent years, solid substrate fermentation has been investigated for the production of

1. Protein-enriched animal feed

2. Bioremediation substances

3. Biopesticides

4. Single-cell protein from waste products

methanol from natural gas, carbohydrate-containing raw materials, and wastes from cheese production and pulp mills. An early example of SCP is baker's yeast from *Saccharomyces cerevisiae*. Most recently, mycoprotein has been marketed as a meat substitute.

The cyanobacterium *Spirulina* also has been used as SCP in Israel, Mexico, and Taiwan. Photosynthetic microorganisms require only water, a nitrogen source such as ammonia or nitrate, sunlight, and minerals (often present in the water supply) and can be grown in outdoor ponds, which eliminates the need for costly culturing facilities.

In the future, SCP may be produced using wastes or by-products of industrial processes or waste-treatment facilities. Although a few SCPs are well established for human and livestock consumption, others that have potential are ruled out for the present because they contain toxic compounds. Nucleic acids are potentially toxic but can be degraded by nucleases. However, other extremely toxic compounds may also be concentrated in SCP, such as hepatotoxins produced in cyanobacteria or heavy metals absorbed from the substrate. Cost-effective methods to remove toxins from SCP must be developed if it is to be widely distributed. In a world in which most people receive inadequate nutrition, SCP may eventually provide a low-cost solution.

FOOD BIOTECHNOLOGY

On a global scale, food production has become extremely complex. Raw materials are obtained around the world, and food production and processing require new technologies. As demands increase for new, safe, and high-quality foods, producers increasingly look toward biotechnology for solutions. As discussed in Chapter 1, Biotechnology: Old and New, biotechnology has been used to produce foods and beverages for more than 8000 years through the process of fermentation.

Breads	Soy sauce	Miso	Tea
Cheeses	Cider	Kefir	Tofu
Yogurt	Coffee	Salami	Cocoa
Vinegar	Olives	Pickles	Beer
Cottage cheese	Sour cream	Buttermilk	Tamari
Distilled liquors	Sauerkraut	Wine	Tempeh

Microorganisms are a major component of the fermentation process. As we have discussed, fermentation relies on the enzymatic reactions found in a variety of microorganisms. Although the food industry today uses sophisticated methods to mass produce foods and beverages, develop new food products, improve the microorganisms that conduct food and beverage fermentations, and reduce costs of production, the bio-

chemical processes of fermentation remain the same as they were in ancient times.

Modern developments in food production have focused on the quality, safety, and nutritional value of products. Methods to improve the processing of raw materials used in food production also are being investigated. The following is a partial listing of ways that the food industry is improving food quality and production:

1. More environmentally friendly manufacturing processes

2. Better waste treatment

3. Better assessment of food safety during production and processing

4. The use of natural flavors and colors

5. The use of new enzymes and emulsifiers

6. Improved starter cultures for fermentation and the improvement of raw materials

Scientists are working on methods to improve the quality and safety of food production during fermentation. One way to improve microorganisms for food fermentations is to develop virus-resistant strains through recombinant DNA technology. Bacteria that are used to produce fermented dairy products are susceptible to virus infection that can cause significant economic losses to the food industry. Scientists are developing important bacterial strains that are resistant to infection for a variety of food fermentations.

Sometimes fermenters become contaminated with bacteria that cause foods to spoil or that produce toxins that cause food poisoning. The use of bacteria that produce chemicals that kill these contaminating bacteria are being used as an additive in fermentations to prevent contamination of fermented foods.

Microorganisms also are important sources of food additives. These additives serve several important functions, such as enhancing the flavor of foods, increasing the nutritional value, and preventing spoilage during production. Examples of microbial additives include amino acid supplements, flavor enhancers, flavorings, vitamins, guar gum, and xanthan gum. The vitamins added to breakfast cereals are produced through microbial fermentation.

Numerous microbial enzymes produced by recombinant DNA technology and fermentation are now used in food processing. For example, the first commercial food product produced by biotechnology was a microbial enzyme used in making cheese. At one time the enzyme had to be extracted from the stomach of calves and lambs—now the recombinant protein is produced by microorganisms through industrial fermentation.

The food industry now uses more than 55 different enzymes in food processing alone. The number is increasing as new microorganisms and their enzymes are identified, and recombinant DNA technology is used to manipulate these microorganisms and enzymes.

Food safety is a growing field. The protection of our food supply has been an important focus of discussion by scientists, the public, and the media. For example, methods such as food irradiation to kill harmful microorganisms have been debated for several years. Microorganisms, growing on foods and beverages after production, spoil foods and may produce harmful toxins. In addition to new methods of food processing and treatment, sensitive diagnostic methods are being developed to detect these microorganisms and their toxic products (such as aflatoxin produced by fungi that grow on crops like peanuts) to ensure that food products are safe before distribution. Examples of methods include the use of the polymerase chain reaction (PCR), monoclonal antibody tests, and specific DNA probes. Bacteria such as the potent bacterium *E. coli* 0157:H7 and others such as *Clostridium botulinum*, *Listeria,* and *Salmonella* can be detected and distinguished from harmless bacteria using sensitive assays. These assays are rapid, very sensitive, and can be used outside the laboratory setting (that is, on site). For example, results for the detection of *Salmonella* can be obtained in 36 hours using newly developed diagnostic tests (as compared with 3–4 days in the past).

PRODUCTS FROM MICROORGANISMS

The production of antibiotics through fermentation was a biotechnological breakthrough that revolutionized medicine after World War II. Today, fermentation using microorganisms yields a still wider variety of commercially important compounds. Microbial fermentation provides flavorings, nutrients, and colorings for a wide variety of foods. For example, among the many fungal varieties used in cheese manufacturing, the fungus *Penicillium roquefortii* gives blue cheese its characteristic flavor. Amino acids are used as flavor enhancers (for example, alanine, aspartic acid, and glutamic acid) and nutritional supplements (for example, threonine, methionine, and lysine), as well as in industrial biochemical syntheses. Fermentation produces pharmaceutically active compounds such as antiinflammatory agents, antidepressants, anticoagulants, and coronary vasodilators. The ability to genetically alter cells through recombinant DNA technology has expanded the list of commercially important compounds obtained by fermentation processes. Microbial cells (and other cells such as animal, plant, and yeast) can now synthesize compounds with important therapeutic value, such as interferons and in-

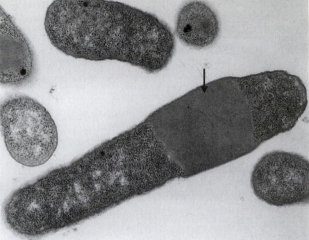

FIGURE 5.2 When the human gene for insulin is cloned into the bacterium *E. coli*, a very large inclusion of human insulin protein is formed. When a bacterium expresses a human gene, the resulting protein often accumulates within the cell.

Dr. Daniel Williams, Lilly Research Laboratories

terleukins for the treatment of a variety of cancers; factor VIII for hemophilia; erythropoietin for anemia; human insulin (Figure 5.2) for diabetes, α_1-antitrypsin for the treatment of emphysema; nerve growth factor to promote the repair of nerve damage; and human growth hormone for the treatment of pituitary dwarfism (Table 5.1).

In the following sections, a few of the important applications of commercial fermentation are examined. These include the production of important metabolites, enzymes, antibiotics, natural rubber, fuels, and plastics.

Metabolites

Two types of metabolites are synthesized by microbial cells: primary and secondary metabolites. Primary metabolites are produced during the organism's growth phase; these compounds are essential to an organism's metabolism and can be intermediary metabolites or end products. Secondary metabolites are not essential to cell growth or function and are characteristically produced quite late in the growth cycle; they are the end products of metabolism and are usually derived from primary metabolites or the intermediates of primary metabolites. Although the specific functions of secondary metabolites often are unknown, these compounds probably give the organism a selective advantage against its competitors in its natural environment. Table 5.2 shows examples of commercially important primary and secondary metabolites.

TABLE 5.1 Some Pharmaceutical Products Manufactured Using Recombinant DNA Technology

Product	Comments
Tissue plasminogen activator (t-PA)	Dissolves the fibrin of blood clots; therapy for heart attacks; produced by mammalian cell culture.
Erythropoietin (EPO)	Treatment of anemia; produced by mammalian cell culture.
Human Insulin	Therapy for diabetics; better tolerated than insulin extracted from animals; produced by E. coli.
Interleukin-2 (IL-2)	Possible treatment for cancer; stimulates the immune system; produced by E. coli.
Alpha- and gamma-interferon	Possible cancer and virus-disease therapy; produced by E. coli, S. cerevisiae.
Tumor necrosis factor (TNF)	Causes disintegration of tumor cells; produced by E. coli.
Epidermal growth factor (EGF)	Heals wounds, burns, ulcers; produced by E. coli.
Prourokinase	Anticoagulant; therapy for heart attacks; produced by E. coli and yeast.
Factor VIII	Treatment for hemophilia; improves clotting; produced by mammalian cell culture.
Colony-stimulating factor (CSF)	Counteracts effects of chemotherapy; improves resistance to infectious disease such as AIDS; treatment for leukemia; produced by E. coli, S. cerevisiae.
Superoxide dismutase	Minimizes damage caused by oxygen free radicals during reperfusion of oxygen-deprived tissues; produced by S. cerevisiae and Pichia pastoris (yeast).
Monoclonal antibodies	Possible therapy for cancer and transplant rejection; used in diagnostic tests; produced by mammalian cell culture (from fusion of cancer cell and antibody-producing cell).
Hepatitis B vaccine	Produced by S. cerevisiae that carries hepatitis-virus gene on a plasmid.
Bone morphogenic proteins	Induce new bone formation; useful in healing fractures and reconstructive surgery; produced by mammalian cell culture.
Taxol	Plant product used for treatment of ovarian cancer; produced by E. coli.

Enzymes

Enzymes isolated from microorganisms and fungi have applications in the production of foods, cosmetics, and pharmaceuticals and in the synthesis of industrial chemicals and detergents (Table 5.3). In industry, enzymes are used to convert substrates, often from plant material, into commercially valuable products. An example is the production of high-fructose corn syrup from corn starch, a highly branched polymer composed of covalently bonded glucose monomers. The synthesis uses three enzymes—bacterial β-amylase, glucoamylase, and glucose isomerase. Corn is steamed under high pressure to make it more susceptible to enzymatic action. After it has cooled, β-amylase is added to hydrolyze the glucose polymer into short-chain polysaccharides. Glucoamylase completely hydrolyzes the polysaccharides to glucose by breaking bonds of the branching cross-links. By a transformation reaction, the enzyme glucose isomerase then converts glucose into fructose. Although the product is commonly called high-fructose corn syrup, it is a mixture of glucose and fructose. Because this process yields a lower cost per unit of production than the process for producing sucrose, fructose has replaced sucrose as the major sweetener in the United States. This industrially important bioconversion uses the bacteria Bacillus amyloliquefaciens for β-amylase, Aspergillus niger for glucoamylase, and Arthrobacter sp. for glucose isomerase.

Recombinant DNA technology provides a way to mass produce commercially important enzymes. The gene encoding a newly identified enzyme can be moved into commonly used cells such as E. coli or yeast cells for mass production through fermentation or for genetic manipulation to change, for example, enzyme

TABLE 5.2 Examples of Primary and Secondary Metabolites Produced by Fermentation

Primary Metabolites	Secondary Metabolites
Amino acids	Antibiotics
Vitamins	Pigments
Nucleotides	Toxins
Polysaccharides	Alkaloids
Ethanol	Many active pharmacological compounds (e.g., the immunosuppressor cyclosporin, hypotensive compound dopastin)
Acetone	
Butanol	
Lactic acid	

TABLE 5.3 Examples of Microbial Enzymes and Their Uses

Enzyme	Uses
Lipase	Enhances flavor in cheese making
Lactase	Breaks down lactose to glucose and galactose; lactose-free milk products
Protease	Detergent additive; hydrolyzes suspended protease proteins in beer that form during brewing for a less cloudy chilled beer
α-amylase	Used in production of high fructose corn syrup
Pectinase	Degrades pectin to soluble components, reduces cloudiness in chilled wine, fruit juice
Tissue plasminogen activator (TPA)	Dissolves blood clots

activity or substrate specificity. Examples include glucose oxidase produced in fungi, which reduces food spoilage, especially in egg-based products, and beta-glucanase produced by bacteria, which improves beer filtration.

The search for new microorganisms and enzymes is ongoing. Less than 1% of the world's microorganisms have been cultured and characterized, thus they essentially represent an untapped gold mine. Because microbes have an ancient history (more than 3.5 billion years) and have adapted to practically every known habitat, they have evolved complex biochemical pathways. Scientists are discovering new enzymes from microorganisms, extremophiles, that tolerate extreme environments (heat, cold, acidity, salinity), as well as those that are found in wastewater that may be applicable to a variety of industrial processes. Once applications are identified, enzymes also can be improved (for example, enhanced catalytic properties, modified substrate recognition) through genetic engineering.

Antibiotics

Antibiotics are small metabolites with antimicrobial activity that are produced by Gram-positive and Gram-negative bacteria as well as fungi. Their role in nature is probably to enable the antibiotic producer to effectively compete for resources in the environment by killing or inhibiting the growth of competitor microorganisms. The first antibiotic used was penicillin, discovered in 1929 by Alexander Fleming. Since the widespread use of penicillin during World War II, antibiotics have been produced commercially by microbial fermentation. Table 1.3 identifies microorganisms and the important antibiotics they produce.

A wide variety of antibiotics, many of them from the Gram-positive soil bacterium *Streptomyces*, are available today to treat many bacterial infections. These antimicrobial drugs act in different ways, by (1) disrupting the plasma membrane of microbial cells, (2) inhibiting cell wall synthesis, and (3) inhibiting synthesis of important metabolites such as proteins, nucleic acid, and folic acid.

Today many bacteria are becoming resistant to the present arsenal of antibiotics used to fight infections. Scientists and clinicians are searching for new antibiotics to combat antibiotic resistance. One way to identify new antibiotics is to screen purified secondary metabolites for antimicrobial activity. This is an arduous task that may not yield new antibiotics. A second way is to feed unusual substrates and substrate analogs to microorganisms that can be used in antibiotic biosynthetic pathways. Chemical variants of existing antimicrobial compounds may be produced (an example is the synthesis of novel β-lactams, a family of antibiotics that includes penicillin). Other methods include chemically modifying preexisting antibiotics or synthesizing in the laboratory chemicals with antimicrobial activity.

Recombinant DNA technology is also playing an important role in the development of powerful new antibiotics. The manipulation of genes for antibiotic synthesis enables scientists to produce novel antibiotics that differ in structure from the original antibiotic. Methods include combining antibiotic biosynthetic pathways and the targeted genetic modification of pathways. The genes for two related but independent antibiotic biosynthetic pathways can be placed in the same microorganism. The enzymes of one pathway are able to act on the intermediates produced in the other pathway. That is, the compounds produced in one antibiotic biosynthetic pathway can be used as substrates in the other pathway. In this way, new structural variants are generated that differ from the intermediates produced in both pathways. These new compounds can have different antimicrobial activities. Novel antibiotics in the isochromanequinone family are produced by moving cloned genes for a specific isochromanequinone antibiotic biosynthetic pathway on plasmids into host organisms that already synthesize a particular antibiotic in the same antibiotic family (antibiotics include actinorhodine, granaticin, and medermycin). Variants have been made by cloning the genes for actinorhodine from *Streptomyces coelicolor* and placing them into other species of *Streptomyces* that synthesize different antibiotics.

Another way to produce new antibiotics is to feed unusual substrates to a microorganism that contains a pathway from another organism. For example, one biotechnology company has produced intermediates of cephalosporin (closely related to penicillin) by feeding adipic acid to *Penicillium chrysogenum*, which harbors the cephalosporin biosynthetic pathway genes from *Streptomyces* or *Cephalosporium*.

Fuels

The industrialized world consumes enormous quantities of fuels that are nonrenewable resources, and the nations of the developing world are burning increasing amounts as well. The Earth harbors only finite amounts of coal, oil, and natural gas; eventually supplies will run out unless alternatives are found. All produce pollution when burned, although in different amounts. We must find cleaner, renewable alternatives. One such alternative is methane, a natural gas produced by anaerobic bacteria in swamps and landfills. These bacteria may serve as a renewable resource in the future. They are inexpensive to culture, because they use domestic and agricultural sewage and industrial effluents as nutrients for growth. Landfill gas already is harnessed for energy. (Scrubbing removes the undesirable gaseous compounds, and the separated methane is burned, producing CO_2 and H_2O.)

One potential fuel that is extremely attractive with respect to its environmental impact is hydrogen, because combustion produces only energy and water. There are several primary obstacles to adopting hydrogen as a fuel of choice. First, pure hydrogen is extremely combustible (much more so than gasoline), so there are great concerns about the practicality of storing it safely in conditions of normal use, such as in the fuel tank of an automobile. Second, the high cost of extraction precludes its commercial use. Here again bacterial fermentation shows great promise, although a drawback to this is that most hydrogen-producing microorganisms also are able to utilize it for energy metabolism. The bacterial enzyme hydrogenase readily produces hydrogen gas from two hydrogen ions. Research using both *Clostridium* bacteria and the eukaryotic algae *Chlorella* looks promising, because both organisms can produce hydrogen gas for prolonged periods. Today, genomic databases are being screened to identify novel hydrogen producers. Though the production of fuel gas from natural organisms is still in the experimental phase, there is much promise for the future.

Biopolymers

A tremendous amount of plastic is used worldwide. Most of the plastic in the United States is used for products that must be durable and resist deterioration. The rest is used for packaging (wraps, lids, caps, sacks, bottles, and so on) and for coatings, sheeting, and photographic and video film. Polyethylene, polypropylene, polystyrene, polyvinyl chloride (PVC), or polyethylene terephthalate are the types most commonly used. Most of these synthetic polymers are produced from the naphtha fraction of petroleum or from natural gas and are heavy polluters because they are not biodegradable.

Most plastics end up in landfills, on our shores, or in the ocean. Several hundred thousand tons are discarded into the marine environment each year. If such disposal stopped today, plastic litter would continue to wash up on our shores for a century or longer. The heavy toll on marine mammals, turtles, and birds has been well documented. More than a million marine animals die each year by choking, suffocating, drowning, or starving after ingesting or becoming entangled in discarded nondegradable plastics. Plastic is now found in the gut of almost all seabirds examined. Plastic particles mistaken for food and ingested reduce the capacity for feeding; malnutrition and starvation follow.

The so-called biodegradable or photodegradable plastics require ultraviolet light, oxygen, and other elements in the environment to slowly break them down or partially degrade them. Although some plastics contain additives that are truly biodegradable, there is not a plastic on the market that is completely biodegradable because their usefulness as products (for example, packaging) requires that their refractory hydrocarbon-based polymers resist degradation.

Burning plastic wastes has not been a solution, because combustion releases airborne pollutants; some plastics produce hydrogen cyanide, and burning PVC produces hydrogen chloride. These harmful chemicals must be removed from the combustion products before they are released into the air.

At least a partial solution to this devastating environmental problem is the development and use of bioplastics that are readily broken down by natural elements and the action of microorganisms. Biotechnology may one day provide the tools to synthesize such natural plastics, and do so using methods of production that do not pollute the environment. Current research looks promising, and advances have been made with both bacteria and plants (see Chapter 6, Plant Biotechnology). Some of the chemicals that are required for plastic production may be produced by microorganisms. For example, alkene oxides are used extensively in plastics and polyurethane foams. The conventional method of producing these oxides generates pollutants, but a bacterium, *Methylococcus capsulatus*, found in the hot springs of Bath, United Kingdom, produces alkene oxides as a gas at 81° C. Use of this bacterium may simplify the collection of alkene oxides and eliminate pollution.

Living organisms are capable of synthesizing many complex polymers and catalyzing reactions at various temperatures. A group of very diverse microorganisms (including examples from the genera *Alcaligenes, Azospirillum, Acinetobacter, Clostridium, Halobacterium, Microcystis, Pseudomonas, Rhizobium, Spirulina, Streptomyces,* and *Vibria*) produces polymers in the poly(β-hydroxyalkanoate), or PHA, family, which are used for energy and as a storage form of cellular carbon. These microorganisms can accumulate from

30–80% of their dry weight in PHA (Figure 5.3). PHA compounds are produced and stored in intracellular inclusions during unbalanced growth conditions—conditions in which there is excess carbon, such as glucose, but too little of another nutrient, such as ammonia, or mineral, such as magnesium, sulfate, or phosphate. All PHA compounds are linear polymers or polyesters of β-hydroxyalkanoic acids containing different alkyl groups at the β-position.

The most extensively characterized PHA is poly(β-hydroxybutyrate), or PHB. Polyesters with different chemical and physical properties for a variety of uses are produced by controlling the chain lengths of repeating units. The desired polymers are produced by supplying the microorganism-defined nutrients in correct ratios or, in some cases, using raw materials from agricultural biomass (for example, corn, sugarcane). The metabolic flexibility of many microorganisms allows them to synthesize new varieties of bioplastics to replace petrochemical-based plastics in use today. Growth is initially maximized to provide a large amount of biomass; the medium is then changed to create an imbalance by depleting one nutrient, and growth slows. A polymer-forming substrate is supplied that the cells use to convert the compound into the desired polymer. Homopolymers of PHB have been produced that have chemical and physical properties almost identical to the chemically synthesized, nonbiodegradable polypropylene. Recombinant DNA technology provides a means by which genes involved in biopolymer synthesis can be transferred to microorganisms (and plants; see Chapter 6, Plant Biotechnology) to produce new, useful polymers that can bypass costly and sometimes hazardous chemical processes.

Biological polymers have been produced commercially in recent years. For example, recently Cargill Dow established a biorefinery in Nebraska to convert sugars from field corn into polylactic acid (PLA), which can be used to produce packaging, clothing, and bedding. The price and quality have been competitive with petroleum-based plastics and polyesters. In 2001, DuPont and its partners developed a polymer from corn sugar, Sorona, that is used in clothing.

Environmentally friendly production methods of biopolymers are being sought. Recently, the integrated coproduction of PHA polymers with sugar and ethanol production has been investigated at a sugarcane mill. In Brazil a large sugarcane industry produces white sugar, fuel ethanol, and their by-products. The mills generate the energy they require by burning sugarcane residue obtained after the juice has been extracted from sugarcane. The availability of energy from renewable sources (both electrical and thermal), the recycling of by-products, availability of low-cost sugar in large quantities, and the readily available large-scale fermentation technologies overcome some of the obstacles that have prevented the large-scale acceptance of biodegradable

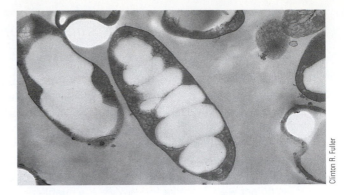

Clinton R. Fuller

FIGURE 5.3 Transmission electron micrograph of the bacterium *Pseudomonas olevorans*, showing the accumulation of electronlucent poly(β-hydroxyalkanoate) (PHA) inclusions.

polymers. Furthermore, the remaining biomass after PHB production is high in calcium, nitrogen, phosphate, and a variety of micronutrients. These can be applied to sugarcane fields. In a test mill, PHB purity has been shown to be greater than 98%. An average sugarcane mill crushes 12,000 tons of sugarcane each day, yielding juice for sugar and ethanol production. In such a mill, PHB production can reach 10,000 tons per year. Furthermore, significant numbers of low-fertility pastures are available for additional sugarcane production if PHB demand increases. This is an example of how environmentally sound biotechnology products can be successfully produced in countries other than those that are highly industrialized.

Being 100% biodegradable microbial bioplastics, PHA polymers have many important potential industrial and medical applications, for example, as biodegradable carriers for fertilizers, insecticides, herbicides, and fungicides. They could be used in the medical field for such surgical paraphernalia as sutures, pins, and staples; for artificial blood vessels; bone replacements, and capsules for pharmaceuticals; and for containers or bags for intravenous solutions, medications, and nutrients. Medical use of biopolymers would dramatically reduce the amount of nonbiodegradable waste. Bioplastics also could be used in packaging as containers, bottles, bags, and the like. PHA compounds could be used in disposable diapers, a product that is not biodegradable and the disposal of which has become a great concern as the nation's landfills reach their maximum capacity.

Other biopolymers produced through biotechnology (by microbial fermentation) include spider silk and adhesives from barnacles. Through genetic engineering, it may one day be possible to synthesize even better polymers.

Biotechnologies and Developing Countries

Most biotechnological achievements have been obtained in industrialized countries. Unfortunately, it is often costly to transfer successful biotechnologies to developing countries, and companies do not want to make the investment. Soon new products generated by biotechnology will replace many products produced by developing countries (see Chapter 6, Plant Biotechnology, for examples of plant-derived specialty oils, plant secondary metabolites). This would be economically devastating, especially because many countries are politically and economically unstable. Developing countries have many valuable resources that would be invaluable to biotechnology. Industrialized countries must ask what roles less developed countries can play in the production of new products. By using practices already in place, some technologies can be readily incorporated. One example is the potential mass production of inexpensive biopolymers using an integrated approach. Brazilian sugar mills already produce sugar and ethanol, and biopolymer production can be introduced without a significant additional investment (see earlier discussion on biopolymers).

Rubber (*cis*-1,4-polyisoprene) is one of the most valuable polymers obtained from plants. It provides the raw material for more than 40,000 products, more than 400 of which are used in the medical field. Natural rubber has many important properties; it is very elastic, resists abrasion, disperses heat efficiently, is resilient, and is impact resistant. Medical gloves made from rubber outperform the synthetic gloves that are becoming more commonly used today. They do not tear easily, provide an excellent barrier against pathogens, and yet they retain excellent tactile sensation.

At this time, the only commercial source is naturally harvested rubber from the Brazilian rubber tree (*Hevea brasiliensis*). Unfortunately, due to the chemical nature of natural rubber, it cannot readily be chemically synthesized. Commercial production of natural rubber is at risk because of the lack of tree genetic diversity and the possibility that the trees could be destroyed by a pathogen. Furthermore, the places where the tree can be grown are limited to very specific tropical regions. Extracting rubber is labor intensive because it is tapped from trees by hand and, thus, the price of rubber depends on labor costs. Finally, latex allergies, which can be life-threatening, are increasing (it is an IgE-mediated allergy caused by proteins in the latex). Unfortunately, the removal of allergy-causing proteins is not cost-effective or very effective. Thus, the biotechnology industry has been investigating alternative plant sources of hypoallergenic, natural rubber. Mass production will most likely be crop plants that have been genetically engineered to accumulate new types of isoprenoid metabolites (for example, rubber). Other effective methods of production may use genetically engineered yeast and bacteria that harbor rubber-synthesizing genes.

BIOCONVERSIONS

Bioconversions occur when microorganisms modify a given compound to a structurally related compound. Microorganisms can use substrates that are not usually present in their environment. Synthesis of these unusual substrates can result in products that are more useful than those resulting from synthesis of the naturally occurring substrate. Enzymes convert the substrate into a metabolite that is not normally a constituent of the cell. The compound may be of no value to the cell for cell function or growth, but the cell metabolizes it nevertheless and yields a product that has commercial value. These bioconversions are exploited commercially to produce a vast array of chemicals and pharmaceuticals.

Bioconversions are useful when a multistep chemical synthesis is more expensive or inefficient—or indeed is impossible to achieve—in the test tube. Industrial conversions often employ a mixture of both chemical synthesis and cellular reactions to generate a given end product.

Biotechnology research is producing bacteria that are more efficient bioconverters (for example, they might transform a substrate in fewer steps). New enzymatic pathways can be transferred into bacteria, or existing pathways can be modified to enable microorganisms to synthesize new, commercially desirable end products.

Important examples of bioconversions include the syntheses of steroids. To obtain the end product prednisone—an antiinflammatory drug used for allergic reactions to poison ivy and for some **autoimmune diseases** such as arthritis—chemical synthesis requires 37 steps, beginning with a product extracted from specific plants. Substituting microbial bioconversions at

Microbial Cell-Surface Display

Microbial cell-surface display is a new technology in which foreign proteins are displayed on the surface of microbial or yeast cells by anchoring them to cell-surface proteins or fragments of proteins (Figure 5.4). The protein to be displayed is referred to as the *passenger protein*, while the anchoring protein is called the *carrier protein*. This complex is embedded in the host cell membrane. This process has an enormous range of potentially important commercial and biotechnological applications such as

1. Live vaccine development by exposing epitopes (antigenic determinants of antigens) on disabled bacterial pathogenic cells to activate a specific antigen-antibody response

2. Antibody production by expressing surface antigens

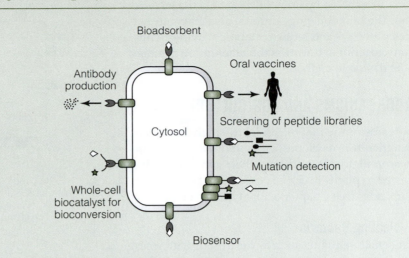

FIGURE 5.4 Microbial cell-surface display and applications.

to raise polyclonal antibodies in animals

3. Bioadsorbents to remove hazardous chemicals and heavy metals

4. Bioconversions using whole-cell biocatalysts with anchored, immobilized enzymes

5. Signal-sensitive receptors or components for diagnostic applications in medicine (for example, immunoassays) or for monitoring substances in the environment (for example, toxins, genetically engineered microorganisms)

---

specific points in the synthesis reduces the process to 11 steps. Bioconversions significantly reduce the cost and shorten the time required to synthesize this important steroid. Other examples of microbial bioconversions are the synthesis of the amino acid L-phenylalanine from phenylpyruvic acid and the conversion of D-sorbitol to L-sorbose—an important intermediate in the synthesis of vitamin C—by *Acetobacter suboxydans* and its enzyme sorbitol dehydrogenase. In the synthesis of ascorbic acid from glucose, only one bacterial bioconversion step is currently available; all other steps must be conducted chemically. If other microorganisms can provide additional enzymatic reactions, the chemical steps will be further reduced. In the near future, microorganisms may supply most if not all of the enzymatic steps. Recombinant DNA technology may allow us to give microorganisms all the enzymatic steps needed for completely converting glucose to ascorbic acid. Recently, a gene from *Corynebacterium* sp. encoding one of the enzymes in vitamin C synthesis has been transferred into *Erwinia herbicola*, which supplies an-

other step in vitamin C synthesis, thus reducing the number of chemical steps in the synthesis.

Use of Immobilized Cells

In some cases, cell immobilization is a preferred alternative to fermentation for synthesizing commercially important compounds. Cells are immobilized by chemical cross-linking or by using a polymer matrix such as natural agar or carrageenan, which comes from algae. Immobilization allows cells to be concentrated, thereby increasing cell densities that can result in an increase in product for a given reaction volume. In addition, the isolation of product is easier and less expensive because there is not the complication of cell-liquid product separation during downstream processing. Cells that are localized (that is, bound to a substrate) also may increase the efficiency of biotransformation and other synthetic reactions because the substrate and enzymes will be in close contact.

The method presents a few disadvantages that sometimes preclude its use. Because the cell contains many enzymes, undesirable biochemical reactions may occur that alter the desired product. This problem can sometimes be solved by immobilizing only the purified enzyme. The immobilizing substrate also may interfere with cell processes and reduce catalytic activities. Thus, for many reactions, traditional fermentation remains the more desirable method.

MICROORGANISMS AND AGRICULTURE

Microorganisms are playing an increasingly important role in agriculture as biopesticides and fertilizers to increase crop yields. In addition, their genes are being transferred to plants to confer resistance to insect pests and herbicides (see Chapter 6, Plant Biotechnology).

Ice-Nucleating Bacteria

In frost-sensitive plants such as deciduous fruit trees, corn, tomatoes, beans, and various subtropical plant species, frost injury results when ice crystals form within cells or between cells. Crystals enter plant cells and damage membranes by mechanical disruption. Upon thawing, plants become soggy and limp because cells have ruptured. The loss of important crops such as oranges and strawberries to frost each year is extremely costly to the grower and to the consumer, who pays dramatically higher prices for fruit and fruit juices.

At least ten species of bacteria have been identified (for example, *Pseudomonas viridiflava*, *Erwinia herbicola*, *Xanthomonas campestris*) that are active in ice nucleation at temperatures above −5° C. *Pseudomonas syringae*, one of the best characterized microorganisms for biological ice nucleation, is a microbial colonizer of plant surfaces and produces a protein that promotes the formation of ice crystals between 0 and 2° C. If these ice-nucleating bacteria are not present, frost damage does not occur until the temperature is between −6 to −8° C. *P. syringae* reduces liquid or solid media supercooling and enhances the formation of ice crystals at subzero temperatures.

The gene for ice-nucleating activity (INA) protein *(Ina)* was deleted from the chromosome of *P. syringae* by Steven Lindow and his colleagues at the University of California at Berkeley (Figure 5.5). In 1983 safety concerns voiced by the public prevented field testing of the genetically modified *P. syringae*. In 1987, these ice-minus bacteria were tested on strawberry and potato plants. When sprayed onto plants, the bacteria readily colonize plant surfaces before the wild variety can do so, thus providing frost resistance (Figure 5.6). Concerns over the use of recombinant *P. syringae* prompted scientific investigators to explore the potential of naturally occurring ice-minus bacteria. In 1992 Frost Technologies Corporation registered with the Environmental Protection Agency (EPA) a mixture of three naturally occurring bacterial strains (two *P. sy-*

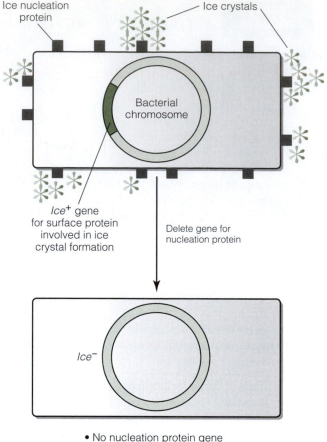

FIGURE 5.5 To engineer ice-minus bacteria, a gene (depicted here as ice⁺) encoding an ice-nucleation protein (Ina) on the cell surface of *Pseudomonas syringae* was deleted. The protein is a focus for the formation of ice crystals, causing damage to plants colonized by ice-nucleating bacteria such as *Pseudomonas*. When the gene is deleted, bacteria do not facilitate ice formation, and when they colonize plants, the mutant bacteria protect the plants from frost damage.

ringae strains and one *P. fluorescens*) for use in frost control. The trade name is Frostban B. The EPA has specified that because bacteria used for biological control, including those for frost control, decrease the presence of the wild-type bacteria, they are pesticides and must be registered as such. At this time, the cost of registration, efficacy testing, and environmental impact studies makes biological pesticides expensive to produce.

Today, research has focused on microbial ice-nucleation applications such as spray-ice technology for snowmaking, a process in the food industry to preserve the flavors and aromas of products such as coffee and fruit juices (called freeze-concentration), and the

FIGURE 5.6 The potato plant on the right has been sprayed with ice−minus *Pseudomonas syringae* and supercooled but is not damaged by frost, while the plant on the left has been frozen because of the presence of ice nucleation−active bacteria.

Steven E. Lindow

use of the ice-nucleating gene as a reporter gene (see Chapter 3, Basic Principles of Recombinant DNA Technology). An additional application is the use of bacterial INA in assays (for example, bacterial ice-nucleating assay) to detect *Salmonella* in food samples. The ice-nucleating gene is placed into a *Salmonella* host-specific bacteriophage. A green dye and the bacteriophage are added to a food sample to test for the presence of *Salmonella*. If *Salmonella* is present, the bacteriophage infects the cells. After incubation, to allow for the expression of INA proteins, the food samples are cooled to −5° C. The sample freezes and the green dye turns dark and nonfluorescent.

Microbial Pesticides

As plants evolved, they acquired a variety of insect defense mechanisms. However, when vast agricultural areas are devoted to monocultures, these resistance mechanisms do not always provide enough protection—insects evolve to become resistant to protective mechanisms. Each year, to prevent insect infestations that would devastate important crops, chemical pesticides must be sprayed several times during a growing season because crops are so susceptible to insect larvae.

Chemical pesticides have been widely used since the early 1940s. One of the first was DDT, or dichloro-diphenyltrichloroethane, a chlorinated hydrocarbon. DDT was an extremely effective and widely used pesticide that attacked insect nervous systems and muscle tissue. Other common chlorinated hydrocarbon pesticides are chlordane, lindane, dieldrin, and toxaphene. All contain arsenic, lead, or mercury, and are called persistent pesticides because they are not degraded by bacterial action or sunlight. By 1960, more than 100 million acres of agricultural fields in the United States were treated with chemical pesticides. Winds picked up dust from contaminated soils and spread it around

the world. In the 1960s, at the height of DDT use, rain distributed 40 tons of DDT each year on the United Kingdom alone.

Unfortunately, insects eventually become resistant to any pesticide, requiring that more and more of the chemical be applied. Moreover, beneficial insects—those that prey on pests—are often more susceptible to the chemicals than the target pests and their numbers decline. As organisms consumed other organisms that had absorbed DDT in their diet, the pesticide accumulated in the food chain, becoming more concentrated at successively higher trophic levels. DDT ended up in the fatty tissues of animals and birds at the top of the food chain. The populations of many birds of prey such as bald eagles, peregrine falcons, sparrow hawks, and brown pelicans declined dramatically because DDT caused their egg shells to become too thin and fragile to protect the embryos. The number of young plummeted during the 1960s; the eastern peregrine falcon became extinct and many other birds were threatened or endangered. DDT was banned in 1972, after its effect on birds was realized. However, its half-life (the time required for one-half of the chemical to degrade into other compounds) may be as long as 20 years. Even today, more than 30 years after DDT was banned, seals, penguins, and fish in the Antarctic show traces of DDT in their fat. The long-term effects on human and animal health will never be known completely.

Microorganisms and insect viruses may provide a source of insecticides that are **biodegradable** and insect specific. Although more than 100 fungal, viral, and bacterial strains that infect insects have been identified, only a few of these have been exploited commercially. Bacterial pesticides degrade rapidly in the environment because the active components are labile molecules. Unlike chemical pesticides, these organic pesticides are proteins that are produced by microorganisms such as

Bacillus thuringiensis. Once exposed to sun, rain, and other environmental elements, the proteins break down rapidly. Recombinant DNA technology can be used to increase the potency and stability of the biological insecticide, to move genes encoding toxins into other hosts such as plants, and to increase the range of specificity. Many crop plants such as corn, potatoes, cotton, tobacco, and tomatoes are being engineered to have increased insect resistance; they produce a compound that is toxic to specific insects but does not affect humans and wildlife.

Recent developments include the engineering of insect-killing viruses to increase their killing potency. However, these types of viral insecticides have caused concern among scientists and lay people alike, who ask whether these superviruses might kill beneficial insects or transfer their genes to related viruses. The use of *Bacillus thuringiensis* (Bt) and **baculovirus** as biopesticides is the focus of intense research and development. The insertion of Bt toxin genes into plants in particular has been promising. A significant proportion of genetically engineered corn, soybean, and cotton crops are Bt-containing varieties. Examples of crops that have been marketed include Rogers brand Attribute Bt Sweet Corn and NK brand YieldGard Corn developed by Syngenta Seeds, and Bollgard Insect-Protected Cotton by Monsanto (see Chapter 6, Plant Biotechnology, for a discussion on Starlink corn and additional new developments).

Bacillus thuringiensis The soil bacterium *Bacillus thuringiensis* has been used as a biopesticide for many years. Bt is a Gram-positive bacterium that forms **endospores** within a **sporangium** during adverse environmental conditions or when nutrient availability is low. Sporulating cells produce parasporal crystalline inclusions composed of aggregated precursor proteins; the crystallized proteins are lethal to insect larvae once

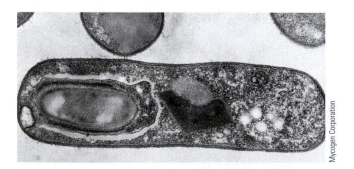

Mycogen Corporation

FIGURE 5.7 HD–1 strain of *Bacillus thuringiensis*. The bipyramidal crystal protein (dark triangular structure) has insecticidal activity and is adjacent to a spore (oblong structure) surrounded by a sporangium. The crystal is composed of three closely related Cry1A proteins, each approximately 130,000 daltons.

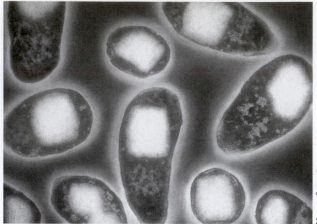

Mycogen Corporation

FIGURE 5.8 Transformed *Pseudomonas fluorescens*, a harmless leaf saprophyte, harbors a Bt crystal and expresses insecticidal activity. Production of Bt toxin is very high in *P. fluorescens* cells.

ingested (Figure 5.7). The crystal proteins, referred to as δ-endotoxins, or insecticidal crystal proteins (ICPs), are toxic to specific insect pests.

The Bt toxin protects crops from insect pests without affecting mammals, fish, or birds. It has been used by farmers and home gardeners for 30 years without negative side effects. Dormant spores are dusted on crops and later become active and cover plants with Bt cells. Now, through recombinant DNA technology, the Bt toxin genes have been inserted into *Pseudomonas* bacteria, which are Gram negative, nonpathogenic leaf colonizers (Figure 5.8). These bacteria produce ICPs and are heat killed before their use on crop plants. This method of delivery by encapsulation overcomes the problem of protein instability that results from exposure to environmental conditions in the field. Bt toxin genes also have been inserted into crops so that transgenic plants can produce their own ICPs (Figure 5.9). Unlike chemical pesticides, these toxins are effective against the larvae of specific insects, and only if the ICPs are ingested. Because of toxin selectivity, no one Bt protoxin from any subspecies of *B. thuringiensis* is effective against all insect pests. More than 50 subspecies are recognized within the *B. thuringiensis* species, each with different pesticidal activity, established by host-range testing. (Examples include the subspecies *B.t. tenebrionis*, beetles such as potato beetles and boll weevils; *B.t. kurstaki*, larvae of butterflies, moths such as gypsy moths, tent caterpillars, cabbage worms, and tobacco hornworms; and *B.t. israelensis*, dipteran larvae such as mosquitoes and black flies).

When crystals are ingested by insect larvae, protoxins are dissolved and digested, where exposure to the alkaline midgut and digestive proteases activates

FIGURE 5.9 The Bt toxin gene has been transferred to crop plants to confer insect resistance. (a) Transgenic plants are resistant to insect larvae damage (left), while plants without the Bt gene are susceptible to damage by larvae. (b) Transgenic cotton (left) after challenge with lepidopteran larvae are resistant to attack, unlike the cotton from a plant that does not contain the Bt gene (right).

them. These smaller toxic proteins bind to specific receptors on the midgut epithelial cell membranes. Within minutes, the gut is paralyzed and the insect ceases to feed. Death occurs within three to five days after a generalized paralysis. The toxins are thought to disrupt ion channels in cell membranes and may increase the flux of important ions such as potassium, thereby altering the osmotic balance. The specific mode of interaction of the ICPs with the cell membrane may depend on the type of toxin produced. Because parasporal crystals are sensitive to ultraviolet light, sunlight inactivates the ICPs.

Although toxin instability is undesirable, it may be that the transient presence of pesticide is what prevents the development of insect resistance. (Such resistance may occur through a decrease in the binding affinity of the toxin for its receptor, or perhaps through a decrease in specific receptors for a particular toxin.) In the future, we are likely to see less labile sprays, more stable toxins, increased host ranges, more effective methods of killing insect borers (which are rarely affected by Bt toxin applications), and less-expensive biopesticides.

The *B. thuringiensis* protoxin proteins and their plasmid-encoded genes have been studied extensively. The central portion of the protein determines insect specificity. Recombinant DNA technology may allow scientists to one day create hybrid or fusion proteins by introducing gene fragments encoding the central portion of different toxin proteins into recipient Bt toxin genes. A single subspecies could then synthesize several types of Bt toxins. In experiments, subspecies that have acquired additional toxin genes affected additional types of insect larvae. Fusion proteins were effective against insects that normally are not targets of either individual toxin, and the combined toxins were sometimes more potent against established target insects than either single toxin, thus demonstrating synergism. As additional toxin genes and fragments encoding core protein regions are exchanged and mutated, new and more potent toxin-producing *B. thuringiensis* strains are likely to be produced.

In the field, insect-resistant plants enable farmers to use less pesticide on their crops. Special care is required so that insecticide-resistant insects do not develop. As with chemical insecticides, insects can develop resistance rapidly because their high reproduction rates allow them to rapidly accumulate mutations in response to selection pressure. To ensure longer-lasting resistance, plants are being transformed with several types of resistance genes. Another strategy for reducing the number of resistant insects is to supply reservoirs of insect-sensitive plants in fields; by allowing wild-type, nonresistant insects to breed, these plants reduce selection for resistant insects. Second generation Bt-containing crop varieties also are being developed that reduce the likelihood of insect resistance (for example, Bollgard II Insect-Protected Cotton by Monsanto) and offer a broader spectrum of control against insects.

Baculoviruses Baculoviruses are invertebrate-specific DNA viral pathogens. They infect mostly larval stages of various insect orders such as Lepidoptera, Diptera, Hymenoptera, Coleoptera, and Homoptera. Because the larval forms of insects are the most damaging to

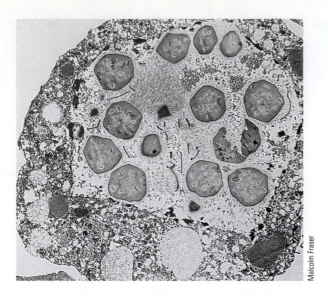

FIGURE 5.10 Transmission electron micrograph of a cultured *Spodoptera frugiperda* (a moth whose larva is called fall armyworm) cell nucleus infected with baculovirus particles (Baculovirus *Autographa californica* multicapsid nuclear polyhedrosis virus). An occlusion body is in the process of enlarging and occluding virions that are observed near the nuclear envelope. The electron dense chromatin–like material is the virogenic matrix where nucleocapsid assembly is thought to occur.

Malcolm Fraser

crop plants, baculoviruses are potentially useful as biopesticides. These viruses do not harm nontarget organisms, because they have specific host ranges. However, unless the host range can be extended through biotechnology, their usefulness as pesticides will be rather limited.

Insects are infected by ingesting plant material contaminated with the virus. The viral particles are released into the gut lumen, where digestive enzymes break down the crystalline protein polyhedron that protects the particles. Eventually, the viral DNA is transported to the nuclei of insect cells, where the virus replicates. Viral particles are produced, cells lyse, and the virus infects other cells and tissues (Figure 5.10). The progressive infection of cells results in mass destruction of tissues. By the time the insect dies, there is practically only virus within the insect cuticle. The virus is released onto plant material, ready to be consumed by other insect larvae. The virus can spread by wind and rain, as well as by birds and other animals that feed on dead insects. Animal digestive enzymes do not break down the protein coat of the virus, which is deposited with the feces in locations that may be far from the original location.

Novel engineered baculoviruses may provide an alternative to the use of chemical pesticides. The main disadvantage of using baculovirus insecticide is the length of time that the infected insect takes to die: after exposure to the virus, the insect continues to feed and damage crop plants for several days. The insertion of foreign genes may enhance the virulence of these viruses or disrupt the host insect's metabolism. For example, a useful insertion would be a gene encoding an insect hormone or enzyme that affects the life cycle and eliminates feeding. Biotechnology offers the opportunity to increase the effectiveness of baculovirus insecticides and to propagate engineered baculovirus in insect culture for commercial use.

Transferring insect toxin genes to baculoviruses may be an effective way to control insects. Although toxins tend to be insect-specific, their transfer to baculovirus may extend their range; for example, the neurotoxin gene from the North African scorpion *Androctonus australis Hector* was transferred to baculovirus and has shown extended specificities. A scorpion toxin gene has been inserted into the *Autographa californica* multicapsid nuclear polyhedrosis virus (AcMNPV), a baculovirus that kills the larvae of butterflies and moths by paralyzing them. Research has demonstrated that the AcMNPV becomes much more lethal after the scorpion gene is added.

Studies on the use and safety of recombinant viruses in the field are ongoing. Primary concerns focus on the question of whether recombination with related viruses promotes the spread of foreign genes and will negatively affect nontarget insects. Issues of virus dispersal and persistence are being addressed by extensive research. The use of disabled virus mutants will help ensure safety in the field and restrict the spread of recombinant viruses.

BIOREMEDIATION

Hydrocarbons are abundant in the environment and come from both human sources (industry, tanker accidents, transportation of petroleum products, atmospheric fallout, fires, mining) and natural sources (phytoplankton, natural seeps). **Biodegradation** is the natural process whereby bacteria and fungi are able to break down hydrocarbons and produce carbon dioxide, water, and partially oxidized biologically inert molecules as by-products of metabolism. The hydrocarbons are oxidized and provide energy for the microorganisms. Hundreds of species of bacteria, yeasts, microalgae, cyanobacteria, and filamentous fungi have been identified that can metabolize hydrocarbons. Each species is able to break down a few specific types of hydrocarbons. The types of molecules that can be degraded depend on the metabolic pathway in each species. Polycyclic aromatic hydrocarbons (PAHs) are an example of a group of complex hydrocarbons. **Bioremediation** is the process of reclaiming or cleaning up contaminated sites using microorganisms to remove or degrade toxic wastes and other pollutants from the environment (a variant process using plants is called phytoremediation). Currently, bioremediation involves two methods:

1. The use of nutrients to encourage growth and enhance the activity of bacteria already present in the soil or water. The natural breakdown of hydrocarbons is accelerated by adding fertilizer (and sometimes trace metals and other micronutrients as well as microorganisms), thus providing a source of nitrogen and phosphorus that may be limiting in the natural environment.

2. The addition of new bacteria to the polluted site. The majority of bioremediation applications use naturally occurring microorganisms to clean up wastes, although genetically engineered microorganisms are being tested. Sometimes by-products resulting from microbial breakdown of wastes can be used for certain applications. For example, methane is produced by some microorganisms during the breakdown of sulfur liquor, a waste product generated during the production of paper. The methane can be captured and used as a source of energy.

Toxic wastes such as crude oil, refined petroleum products, and heavy metals accumulate in the environment through spills and other accidents, dumping, and leaching. Because these toxic compounds do not degrade naturally, or do so extremely slowly, they persist in the environment and may poison wildlife and cause cancer in humans. Such compounds are called **xenobiotics** (*xeno* means "foreign") because they are foreign to a living organism and do not occur naturally in the environment. Examples are such volatile organic compounds as benzene, vinyl chloride, trichloroethylene, polychlorinated biphenyls, and PAHs, and such semivolatile compounds as phenol, naphthalene, benzopyrene, and benzofluorene. Chlorophenols have been widely used to preserve wood, hides, textiles, paints, and glues and as a disinfectant to keep industrial cooling waters free of organisms such as algae. Pentachlorophenols (PCPs) are used to preserve wood. Wood-treatment facilities and sawmills have been significant sources of contamination; their wastes are now treated as hazardous. Polychlorinated biphenyls (PCBs) have been widely used as cooling fluids in electrical equipment and some industrial cooling systems. They are no longer manufactured in the United States and are being removed from most uses. Polybrominated biphenyls (PBBs) are widely used as flame retardants in textiles, carpets, and plastics. PVC is a common plastic that is often abundant in municipal waste.

Many industrial processes and other activities may contribute to the production of a toxic compound. For example, PAHs come from sources as diverse as incineration of refuse and waste, forest fires, charcoal broiling, automobile exhaust, refinery and oil storage wastes, and spills from oil tankers, to name just a few.

The disposal of toxic wastes presents a formidable challenge worldwide. Incineration and chemical treatment of wastes are costly, are unavailable in many areas, and may themselves create additional environmental problems. One of the drawbacks of using microorganisms for environmental clean-up is the fact that most cannot survive in radioactive sites. Recently, experiments have been conducted to study a radiation-resistant bacterium, *Deinococcus radiodurans*, that has been genetically modified with biodegradation genes. Toluene dioxygenase from *Pseudomonas* has been shown to be functional in *D. radiodurans*.

Oil Spills

Oil is composed of a variety of hydrocarbons; its specific composition depends on whether it is unrefined—the crude oil that makes up most tanker spills—or has been distilled to separate its components by molecular weight into products such as very volatile gasolines, diesel fuel, jet fuels, and heating oils.

The world consumption of petroleum is enormous. In 1989, 1 trillion gallons per day were consumed. During transport some oil is spilled, polluting oceans and shorelines in many areas of the world. Tanker accidents, which receive much media attention, release approximately 100 million gallons of crude oil into the environment every year, yet they account for only a small portion of the oil that is spilled: The largest tanker spill occurred in 1978, when the *Amoco Cadiz* lost 67 million gallons. Although that spill was six times the size of the 1989 *Exxon Valdez* spill, it nevertheless accounted for only 8% of the oil lost at sea in 1978. Every year, some 900 million gallons are released into the sea.

Oil spills from tanker accidents and tank ruptures have serious environmental consequences and affect vast numbers of wildlife. However, people are more directly affected by small releases into the environment, such as when petroleum leaking from underground storage tanks contaminates soil and groundwater. Microorganisms are responsible primarily for the biodegradation of evaporated and dissolved molecules of crude oil and refined products in the environment. Molecules that have not evaporated or dissolved in the sea will degrade over time—although the process may require centuries.

Bioremediation of petroleum requires microorganisms with hydrocarbon-oxidizing enzymes and the ability to bind to hydrocarbons. Also required are nutrients for microbial growth; the major limiting factor in the bioremediation of petroleum-contaminated water and soil is the availability of phosphorus and nitrogen. These can be added in the form of fertilizers (Figure 5.11). Currently, this is the only enhancement

FIGURE 5.11 Bioremediation of an oil-contaminated beach after the Exxon Valdez tanker accident off the coast of Alaska. An untreated area of beach is covered with oil (left), while a section treated with fertilizer (Inipol) showed a dramatic reduction in oil on cobble sediment shortly after treatment (right).

used to clean up oil wastes and spills. In the future, biotechnology may produce microbes with enhanced oil-degrading abilities.

Wastewater Treatment

The best-known application of bioremediation technology to date is microbial aerobic oxidation of wastewater. To treat domestic sewage and industrial aqueous wastes, bacteria are introduced into wastewater in a carefully controlled environment containing nutrients for microbial growth. In most wastewater-treatment plants, microbes (often called biomass) are free floating, but sometimes they are immobilized on permeable plastic films and the wastewater flows over them. The treatment of soils and sludges requires the use of water, nutrients, and microbes, which are sometimes attached to a substrate, often the soil or sludge particles. Contaminated soils can be treated by a variety of *ex situ* methods (that is, removed for treatment off site) or *in situ* methods (that is, in position on site). Two common *in situ* methods are pumping contaminated water to the surface before adding nutrients to encourage bacterial action and percolating water and nutrients into the contaminated soil.

Artificially constructed wetlands also are effective for treating urban runoff, industrial effluents, landfill leachate, coal-pile seepage, agricultural wastes, acid mine drainage, and municipal and domestic wastewater. In these systems, soils and aquatic plants with attached microbes adsorb and degrade organic and inorganic wastes. Wastewater also can be treated *in situ* in waste treatment ponds and lagoons, as well as in oil-field production pits. Often such lagoons or ponds contain a sludge composed of hydrocarbons, which must be broken down. First, aerators agitate the sediment and liberate the hydrocarbons, then nutrients, often in the form of commercial fertilizers, are added to promote the growth of microorganisms. If the indigenous population is not effective, the lagoon or pond can be inoculated with additional microorganisms.

Often, however, indigenous microorganisms, even after inoculation, cannot efficiently degrade toxic chemicals (and the chemicals may themselves inhibit microbial growth). For example, resins and aromatics, such as penta-, tetra-, and naphtheno-aromatics, are very resistant to biodegradation. Most aromatic hydrocarbons containing more than five rings degrade slowly and some not at all. A comparison of half-lives is telling: benzopyrene, a five-ring compound, has a half-life of 200–300 weeks; pyrene, a four-ring compound, 34–90 weeks; naphthalene, a two-ring compound, only 2.4–4.4 weeks. Solvents of highly chlorinated hydrocarbons are not readily degraded by natural microbial populations. In fact, the number of halogen atoms directly affects the rate of degradation; the more halogens on the molecule, the slower the degradation. Effective treatment requires bacteria that are genetically engineered to treat these specific types of pollutants. In the future more of these microbial products of biotechnology will be available for bioremediation, offering a cost-effective and environmentally friendly means of removing toxic compounds.

Chemical Degradation

In the mid 1960s several microorganisms were discovered to have the ability to degrade pesticides, herbicides, and many organic chemicals. Today many species of bacteria and fungi are known to oxidize a wide range of compounds. Strains of *Pseudomonas*, the most common soil bacteria, degrade more than 100 organic compounds. The bacteria use the chemicals as a carbon source and metabolize the compound using enzymes of biodegradative pathways. The genes encoding the enzymes of the metabolic pathway may reside on the chromosome or plasmids or both. The plasmids encoding these enzymes are usually large, from 50 to 200 kb, and currently are being isolated and studied in the laboratory.

The most abundant components of herbicides and pesticides, such as DDT, are halogenated aromatic chemicals. Chemicals that contain the halogen elements astatine, bromine, chlorine, fluorine, or iodine are hazardous pollutants found at many toxic waste sites. Many halogens that are important industrial chemicals—for example, the dry-cleaning solvent carbon tetrachloride and the PCB insulation in electrical equipment—are carcinogenic and toxic to fish and wildlife. Halogenated compounds also occur naturally in the environment; most contain chlorine. More than 200 halogenated compounds are known to be produced by algae, bacteria, and sponges.

PAHs, nonhalogenated aromatic chemicals, are hazardous chemicals composed of two or more benzene rings in linear, clustered, or angular arrangements. Examples are naphthalene, anthracene, pyrene, and fluoranthene. These compounds occur as byproducts of industrial processing, cooking, and fossil fuel combustion. They enter the environment in chemical and sewage effluents and through leakage, aerial fallout, industrial accidents, the use and disposal of petroleum products, and natural oil seeps. Their toxicity, environmental persistence, and lipid solubility increase as the number of benzene rings increases.

Dehalogenation, the process of removing halogens, converts most halogenated aromatic chemicals to chemicals that are not toxic. Dehalogenation occurs by an enzymatic reaction, using dioxygenase, that replaces the halogen on the benzene ring with a hydroxyl group. The same enzymes that transform halogenated aromatic compounds also convert the polycyclic aromatic hydrocarbons to other nontoxic chemicals such as catechol or protocatechuate.

A wide variety of bacteria, algae, and fungi can metabolize PAHs. For example, cyanobacteria in the genera *Nostoc, Anabaena, Aphanocapsa,* and *Oscillatoria* oxidize naphthalene. Bacterial degradative pathways may be engineered to expand the range of chemicals that can be degraded. Because wastes generally are mixtures of toxic chemicals—oil spills and those containing heavy metal wastes are examples—bacterial strains that can degrade a wide array of chemicals are highly desirable.

As mentioned earlier, degradative pathways often are plasmid encoded; for example, one plasmid might encode enzymes that degrade toluene and xylene, while a second plasmid encodes a gene that degrades the herbicide 2,4-D (2,4-dichlorophenoxyacetic acid). By transferring plasmids encoding enzymes for specific degradative pathways into a recipient microorganism, a variety of chemicals can be degraded.

The first engineered microorganisms with advanced degradative properties were generated by Ananda Chakrabarty and his colleagues in the 1970s. They transferred plasmids into a bacterial strain that could degrade several compounds in petroleum (Figure 5.12). Chakrabarty obtained the first U.S. **patent** for a genetically engineered microorganism. Although the bacterial strain never was commercialized or used to

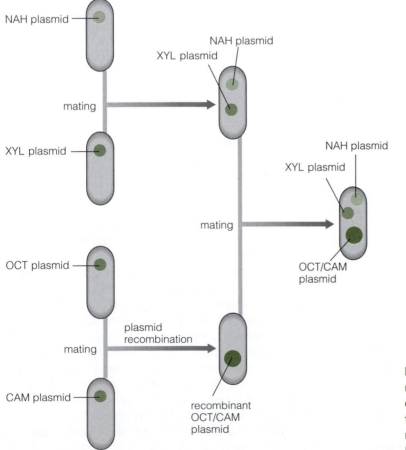

FIGURE 5.12 The production of a camphor-, naphthalene-, xylene-, and octane-degrading strain of bacteria (CAM, NAH, XYL, and OCT, respectively) through the recombination of plasmids during matings of several bacterial strains.

clean up oil spills, the development of an oil-eating microbe was a significant achievement. Chakrabarty's work revolutionized industrial biotechnology in the sense that biotechnology companies now could use patents to protect their living inventions exactly as other companies protect, for example, pharmaceuticals and organically synthesized chemicals.

Genetic engineering allows us also to produce new toxin-degrading bacterial strains by manipulating the genes within an organism's biodegradative pathways; to enhance degradative capabilities by increasing the levels at which enzymes are transcribed; to create mutants with altered enzyme activities that further degrade products of existing pathways into still less toxic compounds; to combine gene transfer and mutation to increase the number of compounds that are degraded; and to exploit new hosts that enhance degradation of specific chemicals. For example, after the addition of *Pseudomonas putida* genes encoding toluene dioxygenase, *E. coli* can oxidize carcinogenic chemicals such as trichloroethylene, benzene, toluene, xylene, naphthalene, and phenol. Transformed *E. coli* cells seem less susceptible than *Pseudomonas* to damage by some chemicals and can maintain degradation rates longer. Thus, bacterial species, such as *E. coli*, with an increased tolerance to toxic compounds may serve as efficient hosts for the biodegradative enzymes from other bacteria.

Future developments will provide new solutions for degrading and detoxifying wastes. Research on cell-free enzyme digestion of aqueous toxic wastes—particularly pesticides—in soils and streams seems very promising. The use of these enzymes *in situ* or in high-volume fermenters would facilitate large-scale cleanup. Microbes and their compounds can be used to chelate (bind) insoluble metals and produce soluble complexes for their removal (see the following discussion). Bacteria release compounds that bind metals in a soluble form and transport them into the cell for metabolism. This process may be exploited in the future to concentrate absorbed or insoluble metals in polluted soils, sludges, groundwater, and aquifers.

Heavy Metals

Metals, important elements in the earth's crust, are essential to living organisms. Several are required in trace amounts (examples are magnesium, manganese, copper, selenium); they are necessary for the activities of many enzymes in metabolic reactions (for example, prosthetic groups) and for maintaining the integrity of membranes and other cellular structures. Many metals are toxic only above naturally occurring concentrations. By contrast, heavy metals, such as arsenic, cadmium, lead, and mercury, do not play a role in metabolic processes and are highly toxic even in low concentrations (for example, they can disrupt meta-

bolic reactions, bind to DNA, and increase mutations). The main sources of heavy metals in the environment are smelters, power plants, waste incinerators, and vehicle exhaust. These metals remain in the atmosphere for long periods (up to 10 days) and accumulate in the food web. Methods of removal rely on adsorption of metals to bacterial cells or to nonliving substrates, precipitation of contaminating metals, and transformation and degradation of toxic metal wastes by microorganisms.

Many bacteria have evolved effective resistance mechanisms, most of them plasmid encoded, ensuring survival in the presence of high metal concentrations. The main mechanism for heavy-metal resistance involves transport of the metal ions outside the cell. Other modes of resistance include (1) the intracellular bioaccumulation of metals in an inaccessible form so that they do not cause harm to the microorganism (the compartmentalization or chelation of metal ions) and (2) the chemical transformation of a toxic compound to a less toxic one. The latter process shows much promise for bioremediation.

The complete microbe-mediated transformation of hazardous metals to such nontoxic compounds as CO_2 and H_2O is not possible. However, processes that contribute to the conversion of toxic to less toxic substances or the sequestering of toxic metals by bacteria are being explored. Bacterial genes are being identified that function in bacterial resistance to mercury, cadmium, cobalt, zinc, chromate, copper, lead, tin, arsenic, and tellurium. Microorganisms that have been genetically modified to constitutively express operons for heavy-metal resistance are being studied to determine their effectiveness in detoxifying metal-containing compounds.

OIL AND MINERAL RECOVERY

The potential exists for the enhanced recovery of oil and minerals in the future. Oil and minerals are nonrenewable resources that must be extracted at great expense and often by compromising the environment. As oil reserves are reduced, prices increase and the reliance on foreign oil has serious international consequences. According to a 1985 estimate by the Department of Energy, the known U.S. oil reserve is estimated to have originally totaled 488 billion barrels. Of that, 133 billion barrels have been produced through 1983. Existing technologies can remove approximately 28 billion additional barrels. However, approximately 327 billion barrels most likely cannot be recovered by conventional methods. Only 5% or less of the total U.S. production has been recovered using alternative technologies, such as microbial-enhanced oil recovery (MEOR).

As the worldwide demand for minerals increases, new methods of extraction must be developed. Traditional methods of mineral extraction include chemical treatment and smelting of metal-rich ores. These methods are costly and generate airborne pollutants that become acid rain. Biotechnology offers opportunities to develop economical, energy-efficient, and environmentally sound methods of extracting metals such as copper, antimony, gold, silver, tin, cobalt, molybdenum, nickel, zinc, and lead. Biomining also can be used to recover metals (for example, copper) from low-grade ores and dumps left by mining operations.

Oil Recovery

MEOR may become an economically attractive alternative for recovering oil. Bacteria can be used for oil recovery in several ways. Bacteria are injected into the oil reservoir along with nutrients for growth and colonization. Alternatively, the growth and metabolism of indigenous microbes are stimulated by the injection of appropriate nutrients. Sometimes microbial products such as extracellular polysaccharides and biosurfactants from fermenters are injected into the oil reservoir to enhance oil recovery. A polysaccharide called xanthan gum from bacteria is often pumped with water into the ground near oil wells to help loosen oil that adheres to rock. Oil is then flushed underground to the well as a dense paste with the water–gum mixture. Bacterial cells that grow and produce extracellular polysaccharides in pores, crevices, and highly permeable areas also help increase oil recovery. These bacteria and their products improve oil recovery by bringing the recovery fluid (the water–gum mixture) into contact with a larger quantity of oil within the reservoir.

Anaerobic bacteria that produce a variety of solvents, such as acetone, and a variety of alcohols reduce interfacial and surface tension forces, thereby increasing oil mobility and releasing oil from rock surfaces and small cracks or pores. The solubilization of gases produced by microbes results in the precipitation of certain compounds that allows the oil to flow better. Because reservoirs lack oxygen, anaerobic bacteria generally prevail in these environments and are the microorganisms of choice for MEOR.

Technical difficulties must be solved before MEOR can be used on a large scale. Some of the most serious problems include the natural biodegradation of crude oil and microbial corrosion and souring (an increase in hydrogen sulfide content caused by sulfate-reducing bacteria) of the oil. Aerobic microorganisms are undesirable in oil recovery. If oxygen is introduced, hydrocarbon-degrading aerobes can grow and metabolize the lighter and more easily recovered oil fractions, including the low molecular weight aromatic and heterocyclic components of oil. Anaerobes will grow on the by-products of aerobic bacteria and produce undesirable reduced sulfur compounds. These processes reduce the quantity and quality of recovered oil.

Metal Extraction

Microorganisms are used to extract metals from sulfide and iron-containing ores and other minerals (called *biomining*). Using oxidation, they convert insoluble metal compounds (for example, nickel, copper, and zinc) in ores to soluble compounds (metal sulfates) that can be recovered. Bacteria and fungi can be used in the recovery of metals from solution because the negatively charged cell surfaces of polysaccharides trap and concentrate positively charged metal ions. Thus, valuable metals can be recovered or hazardous industrial wastes containing toxic heavy metals can be removed by adsorption.

Chemoautotrophic bacteria (a broad group of prokaryotes that include sulfur, iron, hydrogen, and nitrifying bacteria) obtain energy by the oxidation of inorganic materials such as minerals and require only carbon dioxide as a carbon source; they use, for example, substrates containing iron, such as ferrous ions (Fe^{2+}), and reduced sulfur compounds such as sulfides (for example, hydrogen sulfide, H_2S), elemental sulfur (S^0), and thiosulfate ($S_2O_3^{2-}$). The oxidation of minerals by chemoautotrophs can be exploited for the extraction of metals. In addition to a supply of oxygen, the oxidation of a sulfidic mineral requires a nutrient solution containing ammonium, magnesium, potassium, sulfate, nitrate, calcium, chloride, and phosphate ions. Also required are the trace elements zinc and manganese, which usually are present in the minerals being extracted. Thus, to catalyze mineral oxidation reactions under natural conditions, the bacteria must have access to these nutrients in addition to oxygen.

Thiobacillus ferrooxidans seems to be the dominant oxidizing organism of most acidic environments at temperatures below 40°C. This microorganism was isolated in 1947 from ore deposits in a coal mine in West Virginia and now is used routinely to leach copper and uranium from rock. *T. ferrooxidans* also may be useful for the mining of cobalt, nickel, zinc, and lead. It oxidizes iron sulfide to iron sulfate for energy; one of the by-products is sulfuric acid, which in conjunction with sulfate, leaches out metallic minerals while eroding rock. *T. ferrooxidans* also can concentrate metals intracellularly; among these metals are cobalt, zinc, copper, nickel, arsenic, antimony, cadmium, and even strontium and rubidium.

The final products of bacterial oxidation of an ore or mineral concentrate is the treated solid and a liquid. The valuable metal may be found in the solid or the liquid or both. Each type of oxidation process requires the use of different metal recovery methods for liberated gold in solids, for solubilized metals, and for the separation of metals in the form of insoluble sulfates from solid residues. The by-products in effluents of all extractions contain sulfuric acid, iron, and sometimes arsenic (cyanide for gold extraction), which is toxic.

Modern biotechnology may be used to enhance the efficiency of metal extraction by engineering microorganisms with increased resistance to heavy-metal ions such as cadmium and mercury, to acidic conditions, and to high temperatures. Bacterial cell adsorption of metals may become a common method of recovery in the future. Bacterial cell surfaces, including the many extracellular polysaccharide polymers that make up capsules and slime, are able to bind metals in the environment because of their anionic properties. Bacterial adhesion to mineral surfaces seems to occur in two phases. Electrostatic interactions (due to the electrostatic charge on the bacterial cell surface) probably play a role in initial adsorption and a second, stronger bonding occurs through the formation of chemical bonds (for example, sulfur–sulfur bonds) between the cell and the mineral surface. Different bacterial surface properties appear to bind different metals. Thus, specific bacteria could be used to obtain various metals or minerals. Metal selectivity might even be enhanced if the anionic properties of the surface could be modified for selective metal binding. Localized high concentrations of metals on surfaces and subsequent mineralization or scavenging by bacterial exopolymers may one day have important industrial applications and be used to concentrate and retrieve valuable metals.

Future of Bioremediation

Although bioremediation shows much promise, major breakthroughs have not been forthcoming and there are no new innovative technologies on the horizon at this time. One of the primary limiting factors has been the microorganisms and plants that have shown promise in bioremediation processes. These organisms are limited by their metabolism and home range, and therefore, any improvements in their capacity for cleaning up wastes in inhospitable environmental conditions will most likely have to come from genetic manipulation.

Most of the technologies developed will be in pollution prevention. At this time, most progress has been in biomining and the processing of industrial pollutants. Other new developments will be the reduced use by industry of hazardous chemicals that are detrimental to the environment and the increased use of bioproducts (substances produced by living organisms); the reduction of the amount of toxic wastes produced; and the amount of fossil fuels used for energy consumption (replaced with alternative fuels).

MICROORGANISMS AND THE FUTURE

Microorganisms are fascinating, flexible, living machines that are capable of producing a vast array of useful compounds. Research is focusing on the identification and characterization of new strains. We can expect to see microorganisms play a more prominent role in bioremediation and in the production of newly identified metabolites, novel compounds from bioconversions, and the expression of foreign proteins. The future is very bright for microbial biotechnology.

General Readings

K.H. Baker. 1994. *Bioremediation*. McGraw-Hill, New York.

J. Barrett, M.N. Hughes, G.I. Karavaiko, and P.A. Spencer. 1993. *Metal Extraction by Bacterial Oxidation of Minerals*. Ellis Horwood, London.

A.L. Demain. 2000. Microbial biotechnology. *Trends Biotechnol*. 18:26–31.

Y. Murooka and T. Imanaka, eds. 1994. *Recombinant Microbes for Industrial and Agricultural Applications*. Marcel Dekker, Inc., New York.

D.E. Rawlings, D. Dew, and C. du Plessis. 2003. Biomineralization of metal-containing ores and concentrates. *Trends Biotechnol*. 21:38–44.

P.F. Stanbury and A. Whitaker. 1984. *Principles of Fermentation Technology*. Pergamon Press Ltd., Oxford.

B. Zechendorf. 1999. Sustainable development: How can biotechnology contribute? *Trends Biotechnol*. 17:219–225.

Additional Readings

R.M. Atlas. 1995. Bioremediation of petroleum pollutants. *BioScience* 45:332–338.

T. Bachmann. 2003. Transforming cyanobacteria into bioreporters of biological relevance. *Trends Biotechnol*. 21:247–249.

A. Banerjee, R. Sharma, Y. Chisti, and U.C. Banergee. 2002. *Botryococcus braunii*: A renewable source of hydrocarbons and other chemicals. *Crit. Rev. Biotechnol*. 22:245–279.

C.A. Batt and A.J. Sinskey. 1984. Use of biotechnology in the production of single cell protein. *Food Technol*. 38:108–111.

J.W. Blackburn and W.R. Hafker. 1993. The impact of biochemistry, bioavailability and bioactivity on the selection of bioremediation techniques. *Trends Biotechnol*. 11:328–333.

J.R. Bragg, R.C. Prince, E.J. Harner, and R.M. Atlas. 1994. Effectiveness of bioreme-diation for the Exxon Valdez oil spill. *Nature* 268:413–418.

M.G. Bramucci and V. Nagarajan. 2000. Industrial wastewater bioreactors: Sources of novel microorganisms for biotechnology. *Trends Biotechnol*. 18:501–505.

B.C. Carlton and C. Gawron-Burke. 1993. Genetic improvement of *Bacillus thuringiensis* for bioinsecticide development. In L.

Kim, ed. *Advanced Engineered Pesticides.* Marcel Dekker, Inc., New York, pp. 43–61.

D.E.A. Catcheside and J.P. Ralph. 1999. Biological processing of coal. *Appl. Microbiol. Biotechnol.* 52:16–24.

A.M. Chakrabarty. March 1981. Microorganisms having multiple compatible degradative energy-generating plasmids and preparation thereof. U.S. patent no. 4,259,444.

M. Charles. 1985. Fermentation scale-up: Problems and possibilities. *Trends Biotechnol.* 3:134–139.

Y. Chisti. 2003. Sonobioreactors: using ultrasound for enhanced microbial productivity. *Trends Biotechnol.* 21:89–93.

J.D. Coates and R.T. Anderson. 2000. Emerging techniques for anaerobic bioremediation of contaminated environments. *Trends Biotechnol.* 18:408–412.

N. Cochet and P. Widehem. 2000. Ice crystallization by *Pseudomonas syringae*. *Appl. Microbiol. Biotechnol.* 54:153–161.

R.P. Elander. 2003. Industrial production of β-lactam antibiotics. *Appl. Microbiol. Biotechnol.* 61:385–392.

R.J. Frederick and M. Egan. 1994. Environmentally compatible applications of biotechnology. *BioScience* 44:529–535.

K. Furukawa. 2003. "Super bugs" for bioremediation. *Trends Biotechnol.* 21:187–189.

M.L. Ghirardi, L. Zhang, J.W. Lee, T. Flynn, M. Seibert, E. Greenbaum, and A. Melis. 2000. Microalgae: A green source of renewable H$_2$. *Trends Biotechnol.* 18:506–511.

D. Ghosal, I.S. You, D.K. Chatterjee, and A.M. Chakrabarty. 1985. Microbial degradation of halogenated compounds. *Science* 228:135–142.

R.J. Giorgio and J.J. Wu. 1986. Design of large-scale containment facilities for recombinant DNA fermentations. *Trends Biotechnol.* 4:60–65.

B. Guieysse and B. Mattiasson. 1999. Fast remediation of coal-tar related compounds in biofilm bioreactors. *Appl. Microbiol. Biotechnol.* 52:600–607.

M. Gupte, P. Kulkarni, and B.N. Ganguli. 2002. Antifungal antibiotics. *Appl. Microbiol. Biotechnol.* 58:46–57.

L.-A. Jaykus. 2003. Challenges to developing real-time methods to detect pathogens in food. *ASM News* 69:341–347.

V.C. Kalia, S. Lal, R. Ghai, M. Mandal, and A. Chauhan. 2003. Mining genomic databases to identify novel hydrogen producers. *Trends Biotechnol.* 21:152–156.

A. Kiessling. 1993. Nutritive value of two bacterial strains of single-cell protein for rainbow trout (*Oncorhynchus mykiss*). *Aquaculture* 109:119–130.

T. Klaus-Joerger, R. Joerger, E. Olsson, and C.-G. Granqvist. 2001. Bacteria as workers in the living factory: Metal-accumulating bacteria and their potential for materials science. *Trends Biotechnol.* 19:15–20.

S.Y. Lee, J.H. Choi, and Z. Xu. 2003. Microbial cell-surface display. *Trends Biotechnol.* 21:45–52.

S.E. Lindow. 1993. Biological control of plant frost injury: The ice- story. In L. Kim, ed. *Advanced Engineered Pesticides.* Marcel Dekker, Inc., New York, pp. 113–128.

S.E. Lindow. 1994. Control of epiphytic ice nucleation active bacteria for management of plant frost injury. In L. Gusta, G. Warren, and R. Lee, eds. *Biological Ice Nucleation and Its Application.* Amer. Phytopath. Soc. Press, St. Paul, Minnesota, pp. 239–256.

A. Lomoascolo, C. Stentelaire, M. Asther, and L. Lesage-Meessen. 1999. Basidiomycetes as new biotechnological tools to generate natural aromatic flavours for the food industry. *Trends Biotechnol.* 17:282–289.

G.J. Lye, P. Ayazi-Shamlou, F. Baganz, P.A. Dalby, and J.M. Woodley. 2003. Accelerated design of bioconversion processes using automated microscale processing techniques. *Trends Biotechnol.* 21:29–37.

M. Mejare and L. Bulow. 2001. Metal-binding proteins and peptides in bioremediation and phytoremediation of heavy metals. *Trends Biotechnol.* 19:67–73.

M.W. Mohn, C.Z. Radziminski, M.-C. Fortin, and K.J. Reimer. 2001. On site bioremediation of hydrocarbon-contaminated Arctic tundra soils in inoculated biopiles. *Appl. Microbiol. Biotechnol.* 57:242–247.

H. Mooibroek and K. Cornish. 2000. Alternative sources of natural rubber. *Appl. Microbiol. Biotechnol.* 53:355–365.

Y. Motarjemi. 2002. Impact of small scale fermentation technology on food safety in developing countries. *Int. J. Food Microbiol.* 75:213–219.

R.V. Nonato, P.E. Mantelatto, and C.E.V. Rossell. Integrated production of biodegradable plastic, sugar, and ethanol. 2001. *Appl. Microbiol. Biotechnol.* 57:1–5.

Y. Poirier, C. Nawrath, and C. Somerville. 1995. Production of polyhydroxyalkanoates, a family of biodegradable plastics and elastomers, in bacteria and plants. *Bio/Technology* 13:142–150.

A. Pollio and G. Pinto. 1994. Progesterone bioconversion by microalgal cultures. *Phytochem.* 37:1269–1272.

R.C. Prince. 1997. Bioremediation of marine oil spills. *Trends Biotechnol.* 15:158–160.

P.J. Punt, N. van Biezen, A. Conesa, A. Albers, J. Mangus, and C. van den Hondel. 2002. Filamentous fungi as cell factories for heterologous protein production. *Trends Biotechnol.* 20:200–206.

B. Qi, W.M. Moe, and K.A. Kinney. 2002. Biodegradation of volatile organic compounds by five fungal species. *Appl. Microbiol. Biotechnol.* 58:684–689.

J.L. Ramos, A. Wasserfallen, K. Rose, and K.N. Timmis. 1987. Redesigning metabolic routes: Manipulation of TOL plasmid pathway for catabolism of alkylbenzoates. *Science* 235:593–596.

T. Robinson, D. Singh, and P. Nigam. 2001. Solid-state fermentation: A promising microbial technology for secondary metabolite production. *Appl. Microbiol. Biotechnol.* 55:284–289.

R. Rolle and M. Satin. 2002. Basic requirements for the transfer of fermentation technologies to developing countries. *Int. J. Food Microbiol.* 75:181–187.

M.C. Ronchel. 1995. Construction and behavior of biologically contained bacteria for environmental applications in bioremediation. *Appl. Envir. Microbiol.* 61:2990–2994.

S.K. Samanta, O.V. Singh, and R.K. Jain. 2002. Polycyclic aromatic hydrocarbons: Environmental pollution and bioremediation. *Trends Biotechnol.* 20:243–248.

S. Sasaki and I. Karube. 1999. The development of microfabricated biocatalytic fuel cells. *Trends Biotechnol.* 17:50–52.

R.P.J. Swannell and I.M. Head. 1994. Bioremediation comes of age. *Nature* 368:396–397.

K.N. Timmis and D.H. Pieper. 1999. Bacteria designed for bioremediation. *Trends Biotechnol.* 17:201–204.

J. Van Rie, W.H. McGaughey, D.E. Johnson, B.D. Barnett, and H. Van Mellaert. 1990. Mechanism of insect resistance to the microbial insecticide *Bacillus thuringiensis*. *Science* 247:72–74.

L.P. Wackett. 2000. Environmental biotechnology. *Trends Biotechnol.* 18:19–21.

A.E. Wheals, L.C. Basso, D.M.G. Alves, and H.V. Amorim. 1999. Fuel ethanol after 25 years. *Trends Biotechnol.* 17:482–487.

M.D. White, B.R. Glick, and C.W. Robinson. 1994. Bacterial, yeast and fungal cultures: effect of microorganism type and culture characteristics on bioreactor design and operation. In J.A. Asenjo and J. Merchuk, eds. *Bioreactor System Design*. Marcel Dekker, Inc., New York, pp. 1–34.

K.D. Wittrup. 2001. Protein engineering by cell-surface display. *Curr. Opin. Biotechnol.* 12:395–399.

H.A. Wood and R.R. Granados. 1991. Genetically engineered baculoviruses as agents for pest control. *Annu. Rev. Microbiol.* 45:69–87.

T. Yamamoto and G.K. Powell. 1993. *Bacillus thuringiensis* crystal proteins: Recent advances in understanding its insecticidal activity. In L. Kim, ed. *Advanced Engineered Pesticides*. Marcel Dekker, Inc., New York, pp. 3–42.

M. Yoshida and H. Minato. 1987. Assessment of the pathogenicity of bacteria used in the production of single cell protein. *Agri. and Biol. Chem.* 51:241–242.

G.J. Zylstra. 1994. Molecular analysis of aromatic hydrocarbon degradation. In S.J. Garte, ed. *Molecular Environmental Biology*. CRC Press, Inc., Boca Raton, Florida, pp. 83–115.

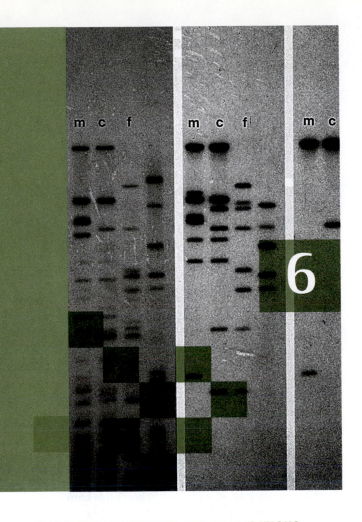

PLANT BIOTECHNOLOGY

Plants are important for so many reasons. They are major energy transducers—they transform energy from the sun into chemical energy that living organisms can use. They are the base of our food supply. Not only do humans use plants for food, but plants provide us with fiber, pharmaceuticals, industrial compounds, and fuels. Biotechnological developments and genetic engineering methods are revolutionizing agriculture and the production of plants. Some major benefits of this revolution are crops that resist pests and disease and new agricultural products such as human proteins for medical purposes.

The focus of this chapter is the propagation of plants and the genetic manipulation of plants using the tools of biotechnology. Transgenic plants are created by permanently integrating new DNA into a plant's DNA. Seeds, tissues, or cells are transformed and then regenerated into a whole plant. Transgenic plants can be used to study how a plant controls the transcription and translation of its genes. Although these plants are important to basic research, the knowledge acquired often leads to new applications. Natural characteristics are being altered or new characteristics are being added to make plants more desirable. Many plants are being genetically manipulated: tomato, soybean, carrot, canola, potato, corn, rice, sunflower, rye, beet, alfalfa, pear, apple, and cabbage are just a few. Sophisticated plant breeding methods are also generating many new varieties of agronomically important plants. Genetically engineered crops are rapidly becoming more prominent in farmers' fields and in grocery stores throughout the United States. In 1994, the first genetically engineered plants were grown commercially in the United States. Within 2 years, 6 million acres of genetically engineered plants were grown in fields. By 1998, 58 million acres were devoted to U.S. genetically engineered crops, 70 million acres worldwide. Estimates suggest that approximately 70% of foods in the United States contain at least some genetically engineered crop plant.

PLANT TISSUE CULTURE AND APPLICATIONS

Plant Tissue Culture

A plant is composed of a variety of cell types that make up different organs such as stems, leaves, and roots. Plant tissue culture refers to the sterile, *in vitro* cultivation of plant parts such as organs, embryos, and seeds, as well as single cells on either solidified or liquid nutrient media. Unique to plants is the ability of nondividing, differentiated cells to revert to an undifferentiated state that has meristematic (capable of cell division) activity. These cells have the ability to divide and differentiate into a whole plant.

The first experiments, though unsuccessful, were initiated in 1902 to culture plants through *in vitro* propagation. The first true plant tissue cultures were successfully conducted in 1939 when undifferentiated tissues of carrot were continuously cultured for prolonged periods.

Plant tissue culture offers many practical benefits. Scientists are able to promote the mass propagation of desirable plants through *in vitro* cultivation. Plants usually propagated by seed can now be cultured *in vitro* to yield thousands of identical plants; a number of agronomically important plants, such as the oil palm in the tropics, have been successfully propagated by large-scale cloning. Specific characteristics such as disease and herbicide resistance also can be selected for while plants are in culture. Regenerating flowering plants in culture offers an additional advantage: there are no viruses in cells of meristematic regions—regions of plant tissue made up of cells that divide to give rise to the tissues and organs of flowering plants. Thus, plants regenerated from **meristematic tissue** will be virus free. Each growing season, commercial potato and strawberry producers start with virus-free plants from tissue culture. Plant tissue culture is invaluable when traditional plant breeding cannot generate plants with desired traits. Often, tissue culture shows faster results than traditional plant breeding.

In plant tissue culture, plant growth is usually initiated from a piece of tissue or an organ that is removed from a plant (called an explant). When grown on a specific nutrient medium in the presence of a specific ratio of growth hormones, the nondividing cells revert to an undifferentiated, meristematic state whereby they form callus tissue (Figure 6.1). This is called dedifferentiation. Callus tissue has the ability to differentiate into a

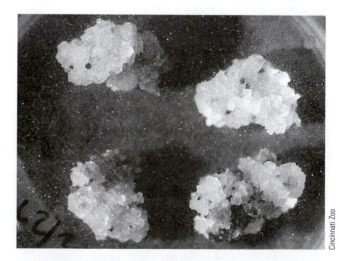

FIGURE 6.1 Callus is an undifferentiated mass of cells from which plants can be regenerated.

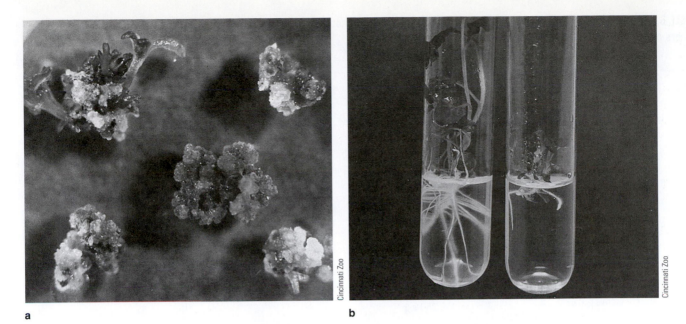

a   b

FIGURE 6.2 Shoots (a) and roots (b) develop from callus tissue in the presence of specific plant growth hormones.

whole plant or plant organ in a process called redifferentiation. The ability of a plant cell to give rise to a whole plant through the process of dedifferentiation and redifferentiation is called totipotency. This ability is unique to plants. In general, a callus must be initiated before cells can be regenerated into a whole plant.

Plant tissue culture is a broad term used to define different types of *in vitro* plant culture. Six different types of *in vitro* cultures are recognized. Each type can result in a whole plant.

1. Callus culture—culture of differentiated tissue from an explant that dedifferentiates (Figure 6.2)

2. Cell culture—culture of cells or cell aggregates (small clumps of cells) in liquid medium

3. Protoplast culture—culture of plant cells with their cell walls removed

4. Embryo culture—culture of isolated embryos

5. Seed culture—culture of seeds to generate plants

6. Organ culture—culture of isolated plant organs such as anthers, roots, buds, and shoots

Micropropagation

Tissue culture is an excellent alternative way of propagating plants asexually. Although it has been used most often for crop improvement, the horticulture industry also uses tissue culture as a way to propagate esthetically desirable plants that produce beautiful flowers.

Desirable plants are cloned through tissue culture to produce genetically identical plants in a process called *in vitro* clonal propagation (also called micropropagation). The plant that supplies the material, cells or tissues, that is cultured is called the *parent* plant.

Researchers and horticulturists have exploited plant regeneration to propagate large numbers of plants originating from a single plant. Table 6.1 compares traditional propagation of yams using tubers (fleshy underground stems) with micropropagation. Micropropagation is practiced in both laboratories and plant nurseries. There is tremendous economic value in the propagation of plants using tissue culture. Examples include the large variety of ornamental plants, such as those sold in supermarkets; agricultural crops such as strawberry, banana, potato, and tomato; and a variety of medicinal plants. Micropropagation forms the basis of a multimillion dollar industry.

Today's high yield oil palms on plantations in places such as Malaysia were generated beginning in the 1960s by tissue culture methods. These palms not only produce more oil—approximately 30% more than normally cultured palms—but are significantly shorter, making harvesting less expensive. High yield micropropagated palms are replacing old plantation trees, because palm oil is used extensively in foods. Thus, micropropagation provides the ability to produce thousands of genetically identical plants by asexual reproduction.

There are four stages in micropropagation (Figure 6.3).

TABLE 6.1 Comparison of Micropropagation and Traditional Culture Methods for Tubers over 2 Years

	Traditional	Micropropagation	
Year 1	100 g tuber	100 g tuber	
	↓	↓	
	One mature plant	One preconditioned plant	
		↓	
		10 nodal cultures	
		↓	
	1600 g tuber	Shoot multiplication	
	↓	↓	
Year 2	--------------------------	650,000 minitubers	Dormant
	16 mature plants	617,500 plants	Season
	↓	↓	
	16 × 600 g tubers	617,500 × 500 tubers	
	= 25.6 kg	= 308,750 kg	

Note: The reduced number of plants from minitubers between year one and two reflects an assumed loss of 5% on establishment in soil.

S.H. Mantell et al., *Principles of Plant Biotechnology,* Table 6.5. Blackwell Scientific Publications, London, 1985. Used by permission of Blackwell Science Ltd.

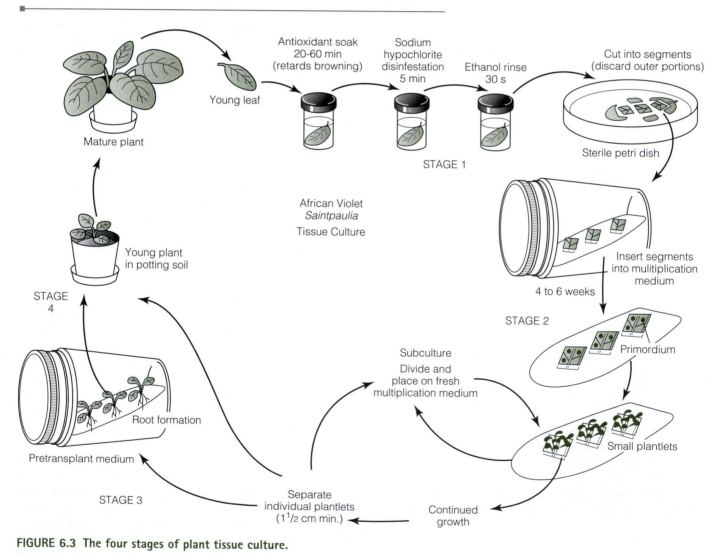

FIGURE 6.3 The four stages of plant tissue culture.

Stage 1—Initiation of sterile explant culture: the selection of explants, sterilization of tissue surface to prevent contamination, and transfer of explants to nutrient media

Stage 2—Shoot initiation: multiplication of shoot tissue from explants on nutrient media

Stage 3—Root initiation: multiplication of root tissue from explants on nutrient media

Stage 4—Transfer of plants to sterile soil or other substrate under controlled conditions

Cultured plant tissue must be supplied with various ratios and concentrations of mineral nutrients, certain vitamins, sucrose, and plant growth hormones such as **auxin** and **cytokinin** (Figure 6.4). Adding very small amounts of auxin and cytokinin to the culture induces multiple shoots to appear from a single-shoot apex within several weeks of culture. These shoots can be separated and each grown into a complete plant with desirable traits from a single individual in a small space relatively quickly (Figure 6.5).

Somatic Embryos One method of plant tissue culture produces embryo-like structures, called embryoids, from cultured somatic plant tissues. Somatic embryo-

FIGURE 6.5 Plant tissue culture room where plants are regenerated from cells and tissues.

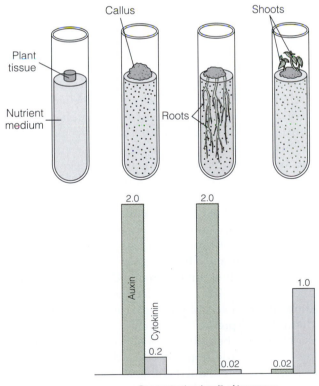

Concentration (mg/l) of hormones placed into the nutrient medium

FIGURE 6.4 Callus can form from an excised piece of plant tissue. The development of shoots and roots depends on the concentrations of the plant growth hormones auxin and cytokinin.

genesis induces cultured somatic, or nonsex, cells (gametes are the sex cells) to form embryoids from vegetative tissues. Embryoids form directly from tissue by using hormones such as auxins to disrupt normal tissue development and to initiate **embryoid** formation. Embryoids also can be formed from callus tissue; a change in hormones induces these cells to form embryoids instead of shoots and roots. During callus formation, numerous embryoids form on the surface, and each will develop into a normal plant. Thousands of little plants called plantlets can be produced by this method.

Somatic embryoids also can be produced in a liquid suspension culture. The source material can be single cells or tissue such as from leaf or callus. If tissue is used, the cells must be separated; the cell walls are digested with enzymes called pectinase and cellulase to yield protoplasts. In the liquid media cell walls regenerate around the protoplasts, cell division begins, and the new cells aggregate to form small clumps. These cell aggregates can be plated onto solid culture media to induce the formation of shoots and roots. Important crop plants have been produced using protoplast regeneration and include rice, sweet potato, potato, tomato, flax, sunflower, and cabbage.

Liquid propagation is often the preferred method for large-scale commercial production. Propagation in liquid cultures using large-volume bioreactors yields millions of embryoids. Food crops such as asparagus, celery, tomato, potato microtubers and numerous ornamentals such as begonias, African violet, Boston fern, and lilies have been successfully propagated. Automation allows 8000 or more plantlets per hour to be transferred to sterile soil.

Somatic embryoids may be packaged as functional seeds. These artificial seeds are encapsulated for distribution and are produced to mimic naturally formed embryos (called *zygotic embryos* based on fertilization of eggs). Somatic embryoids are protected within a capsule made of a complex of agar and other gel-forming compounds. Each embryoid is enveloped in a protective, hydrated gel containing nutrients, growth regulators, fungicides, and other supplements.

Chemicals from Plants In addition to the primary metabolites such as amino acids and nucleotides, plants produce a complex array of secondary metabolites or compounds not directly involved in the organism's growth and development—there are probably at least 50,000 such chemicals in the plant kingdom. Although these compounds are not required for photosynthesis and respiration or for biosynthetic processes, many protect plants against predation by mammals, insects, and pathogens. Other compounds are involved in attracting pollinators and fruit dispersers. Many plant-derived compounds, extracted primarily from field-grown plants, are used as flavorings, colorings or dyes, fragrances, pharmaceuticals, and insecticides (Table 6.2). Although many of these compounds can be synthesized in the laboratory, the synthesis is often sufficiently complex that yields are quite low. Many compounds cannot be or have not been reproduced in the laboratory.

Indigenous peoples have been using compounds extracted from plants as medicines for thousands of years. In fact, 75% of the world's population still relies on herbal medicines. Although in the United States herbal medicines are not widely accepted, more than 25% of our pharmaceuticals come directly from plants. Many of these plants were identified because of their use in traditional medicine. Some pharmaceuticals once came from plants but now are chemically synthesized. Pharmaceuticals that were originally isolated from plants include acetylsalicylic acid, or aspirin; digitoxin, a cardiac glycoside that increases heart muscle contractions; and anticancer drugs such as vinblastine, vincristine, and taxol. Typically, the development of a drug begins with the identification of an herbal medicine that is widely used, often by indigenous people. Herbal extracts are collected and the active compound is identified. Sometimes the compound can be chemically synthesized and modified to increase its effectiveness. Clinical trials follow.

TABLE 6.2 Selected Plant Products and Estimated World Market Value

Product	Approximate Retail Market Value (millions, U.S.$)
Medicinal	
Ajmalicine	5
Codeine	90–100
Corticosteroids	300
Ephedrine, pseudoephedrine	100
Quinine	50
Vinblastine, vincristine	50–75
Flavor and fragrance	
Cardamon	25
Cinnamon	4–5
Spearmint	85–90
Vanilla	15
Agriculture insecticide	
Pyrethrins	20

Chrispeels and Sadava, *Plants, Genes, and Agriculture.* Copyright © 1994 Boston: Jones and Bartlett Publishers, Sudbury, MA. www.jbpub.com. Reprinted with permission.

Usually plants of interest are grown on tropical and subtropical plantations and harvested for extraction. This method is very costly and energy intensive, poses a potential cost to the native biodiversity, and often is practiced in politically unstable regions; moreover, fertile land often is unavailable. The synthesis of desirable compounds by plant cell culture would alleviate these and many other associated problems. Suspension cultures of plant cells cultured in bioreactors may one day replace plantations in the same way that microorganisms have been used for years in fermentation processes to produce a wide variety of important compounds.

One benefit of using cell cultures is that they may offer much higher metabolite concentrations per cell and synthesis rates than whole plants. Usually in large-scale productions, a bioreactor provides cells with a highly controlled environment. Cells are harvested after an appropriate time so that the metabolites can be extracted from vacuoles in the cells. When cells are induced to overproduce a particular product, the cost of isolation may be reduced. Also, cells in culture may provide an endless supply of product if dividing cell cultures are maintained and only a subset of cells removed and processed.

An example of an important industrial plant metabolite produced by cell culture is shikonin, a red pigment found in the purple roots of a perennial herb, *Lithospermum erythrorhizon*, that grows in Japan, China, and Korea. Shikonin was first commercially marketed in 1985; it is used in many creams, cosmetics, and dyes.

The production process was developed by two Japanese scientists, M. Tabata and Y. Fujita, from Kyoto University. They demonstrated that single cells or protoplasts were required to start clones. Critical variables in the production of shikonin were the amounts and types of hormones, nutrients, and nitrogen sources as well as the temperature, light quality, and amount of oxygen supplied to the cultures. Even increasing the amount of exogenous calcium present was beneficial. Shikonin production was significantly increased after much experimentation. In the first culturing step, cells were cultured in a 750-liter bioreactor. After 2 weeks, 1.2 kg of shikonin was produced, equivalent to 1.6 grams per liter of cell culture. Today the commercial output of shikonin is 1.5–4.0 grams per liter of cell culture.

Although a number of important secondary metabolites have been produced commercially from cell and root cultures, the costs of research and development keep culturing from being cost-effective in most cases. It is not unusual to invest 10 years in research and development for one product. Sales must recover costs and usually can do so only if the product's price and market demand are high. Sometimes the costs of the compound may be high but the total market is low; for example, jasmine sells at $5000/kg, but annual sales total only $500,000. On the other hand, there may be a very low unit price but a huge market; for example, spearmint oil sells at $30/kg, but annual sales total $100 million.

Plants can also be manipulated to produce compounds not normally found in plants: The gene (or genes) for a specific enzyme can be inserted into a plant to enable its cells to synthesize a particular metabolite if provided with a readily available precursor compound (see the discussion on molecular farming later in the chapter). Large-scale commercial use of both whole plants and plant cells will become an effective way to produce compounds and chemicals for which industrial screening programs have identified important uses, such as in pharmaceuticals and pesticides. *In vivo* processes may replace some laboratory synthetic reactions in the future.

The production of commercially valuable plant-derived compounds through new biotechnologies may have a negative impact on the economies of many developing countries. Tropical and subtropical farms produce the plants that make many of the world's commercial secondary metabolites. Because many metabolites must be directly extracted from plants, cultivation has been the only way to obtain them. However, the development of biotechnologies in industrialized nations to produce metabolites from cell cultures might eliminate the need to purchase them from developing countries. The routine use of plants and plant cells as living bioreactors (factories) to produce commercial products may adversely affect economies that currently rely on the production and export of plant extracts to industrialized nations.

Other Uses of Tissue Culture

Protoplast Fusion Protoplasts can be transformed with foreign genes (to introduce new characteristics) using such methods as microinjection and electroporation (described in Chapter 3, Basic Principles of Recombinant DNA Technology). The genetic characteristics of cells from two unrelated plant species or genera or from plants of the same species with different characteristics can be combined by fusing two protoplasts from different genetic backgrounds. In protoplast fusion, two cells are fused using chemicals or electroporation, their genetic material is mixed, and the fusion produces a hybrid cell with new characteristics (Figure 6.6). Hybrid plants, genetically different from the donor plants, result from this fusion (Figure 6.7). Hybrids are screened for desirable traits. Protoplast fusion is particularly useful if species are sexually incompatible or cross with difficulty.

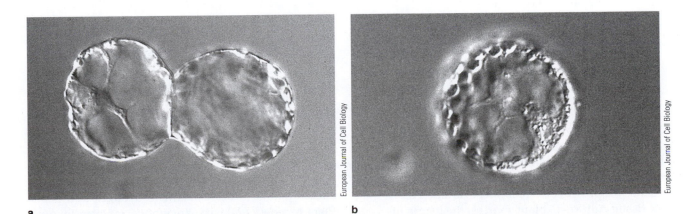

a   b

European Journal of Cell Biology

FIGURE 6.6 Two tobacco plant protoplasts, obtained by digesting away the cell wall, are fused (a and b) to produce a cell that acquires some of the characteristics of both genetic backgrounds and can be regenerated into a plant with some traits from both parental plants.

a   b

FIGURE 6.7 (a) A wild diploid non–tuber-bearing species (*Solanum brevidens*) that is resistant to potato leaf roll virus (PLRV) and frost (left). A diploid tuber-bearing potato clone from *Solanum tuberosum* (right). Protoplasts fused from the two parental lines of potato generated a tetraploid somatic hybrid that was resistant to PLRV and taller (center). (b) Flower and calyx (leaves not shown) of *S. brevidens* (left) and *S. tuberosum* (right) are smaller than the fusion hybrid (center). (b) Plant Science. 39:75–82 (1985); Austin, Baer, and Helgeson. Used by permission from Elsevier Science Ireland Ltd. and the authors.

Somaclonal Variation   Plant tissue culture also can generate useful genetic variation. The genetic variability produced by plant tissue culture is called somaclonal variation. Variability can be exploited to improve characteristics of crop and ornamental plants. Examples of somaclonal variants and their improved characteristics obtained in different crop species include

- Corn—herbicide resistance
- Wheat—grain color and height
- Barley—grain yield
- Soybean—height, maturity, and seed protein and oil content
- Tomato—dwarf growth habit, early flowering, and orange fruit color
- Carrot—higher carotene content
- Oats—increased seed weight, seed number, and grain yield
- Potato—yield
- Sugarcane—sugar content, yield, and disease resistance

Additional examples of beneficial variation include stress resistance such as salt tolerance, heavy-metal tolerance, drought tolerance, insect resistance, and improved seed quality. A number of different factors influence the presence of somaclonal variation during plant tissue culture. Explant sources, the length of time cells are in culture, the culture conditions such as growth hormone types and concentrations, and selective agents such as herbicides or other toxins that are introduced in low doses influence the degree of success of finding plant variants in culture.

This variability is caused by changes in the chromosome number and structure in cells of plant explants being cultured. Examples of changes that influence genetic variability include chromosome rearrangements, single-gene mutations, gene amplification (increase in gene copy number), and the activation of transposable elements. The resulting plants differ from the original parent plant. These mutations occur randomly so that the regenerated mature plant is not genetically identical to the parent tissue, rather the plant is a mosaic with some tissue like the parent and other tissue that differs genetically. Although the process of somaclonal variation is not entirely understood, it is one way that new, potentially desirable traits are obtained if they can be stably maintained and passed on to progeny.

Germplasm Storage   Germplasm is the genetic material of an organism. Plant breeders rely on the germplasm of wild relatives and old varieties of modern crop plants. Ancient crops and wild relatives, most often found in developing countries, are often tolerant to a variety of insect pests, bacterial and fungal infection, harsh environmental conditions such as drought, and other stresses due to thousands of years of genetic selection. This germplasm contains genes that confer important characteristics and increase hardiness in plants. For example, some genes enable plants to be resistant to drought and pests. Plant breeders are using ancient germplasm to introduce desirable traits such as insect resistance into modern crops by crossing susceptible plants with more resistant varieties.

Unfortunately, environmental degradation, urbanization, and changes in farming practices are among the major reasons that valuable germplasm is becoming extinct. The loss of traditional farming practices, the clearing of old fields, and the displacement of old crops by modern varieties increase the rate at which valuable germplasm of wild relatives and ancient landraces (different varieties of old crops) is being lost. Saving ancient crop varieties and their plant relatives for future use is very important to modern agriculture.

Gene banks, storage facilities for valuable germplasm, are located in many countries throughout the world and contain thousands of plant accessions (see Table 1.2). The plant material housed within these facilities is so valuable that there is often a sophisticated security system to prevent theft, and the material is protected from destruction by fire, earthquake, or flood. *Ex situ* conservation involves collecting and storing plants in a central location, usually in the country that is a center of origin or diversity for that particular variety. Unfortunately, most seeds and vegetative tissues (for example, stems) have a limited shelf life and cannot be stored for long periods of time. Usually seeds are stored if possible; however, they must be maintained in an environment of controlled temperature and humidity and must be tested periodically to determine viability. Seeds sometimes must be germinated and plants grown so that seeds of the next generation can be harvested. Many seeds cannot be stored. Some plants can be propagated by tubers or bulbs; however, these also cannot be stored for long periods.

Plant tissue culture is important to the conservation effort. If seeds are not produced or the crop does not reproduce vegetatively, cultured tissue must be used. The germplasm of many wild relatives and ancient varieties of important crop plants can be cultured for later use or research. Researchers are identifying new ways to store plant tissue for future use and when the need for the germplasm arises; cryopreservation—storing tissues at ultracold temperatures in liquid nitrogen—is an example of how germplasm is preserved. Cell and tissue culture in conjunction with cryopreservation provides a way for important germplasm to be maintained.

GENETICALLY ENGINEERED PLANTS

One of the most rapidly expanding areas of biotechnology is the genetic engineering of crop plants. Genetic engineering has been used to add a variety of new traits to important crops. Currently, there are thousands of field trials being conducted worldwide (although mostly in the United States). At this time, six different types of traits have been introduced into plants. Twelve genetically engineered plant species

have been approved for commercialization in the United States. The gene source for these traits comes from a variety of sources.

Traits	Genetically Modified Plants
Insect resistance	Corn, cotton, potato, tomato
Herbicide resistance	Corn, soybeans, cotton, flax, rice, sugarbeets, canola
Virus resistance	Squash, papaya, potato
Delayed fruit ripening	Tomato
Altered oil content	Canola, soybeans
Pollen control	Corn, chicory

Plant Transformation and *Agrobacterium tumefaciens*

As discussed earlier in the chapter, it is generally quite straightforward to grow a complete plant from a piece of tissue or single cells grown in culture. Scientists have learned how to insert a foreign gene (that is, a new trait) into plant cells in culture and then induce them to divide and differentiate, thus generating a whole plant. Using this approach, new traits can be added to plants. Plant transformation is the process of inserting a foreign gene into a plant and having the gene be expressed and inherited as a part of the plant genome. The result is a transgenic plant, and the added gene is called a transgene. Scientists use several different methods for transforming plants that are described in Chapter 3, Basic Principles of Recombinant DNA Technology.

One of the most frequently used transformation methods exploits the ability of a common pathogenic soil bacterium called *Agrobacterium tumefaciens* to facilitate gene transfer to host plants. *Agrobacterium* causes crown gall disease, a disease that affects some agronomically important plant species. In nature, *Agrobacterium* cells divide rapidly after infecting a site of injury on a plant. A large plasmid called the Ti (tumor-inducing) plasmid harbors a region of DNA called T-DNA (transferred DNA). Genes within the Ti plasmid, called *vir* (virulence) genes, cause tumorous growths—crown galls—to form on the plant surface near the infection site. *Vir* genes encode proteins that transfer the T-DNA to cells at the site of the wound (composed of undifferentiated callus tissue) and insert the DNA into a region of the plant chromosome. Because the T-DNA genes (of eukaryotic origin) have plant promoters, the genes are not expressed in the bacterial cells. However, upon infection of plant cells, plant hormones are produced to cause tumor cells to grow. For plant transformation, plant molecular biologists replace the tumor-inducing genes in the T-DNA region with a foreign gene (Figure 6.8). *Agrobacterium* can no longer induce crown gall disease after infection, and instead, the foreign gene (with the remnant T-DNA) is transferred to the plant genome (Figure 6.9).

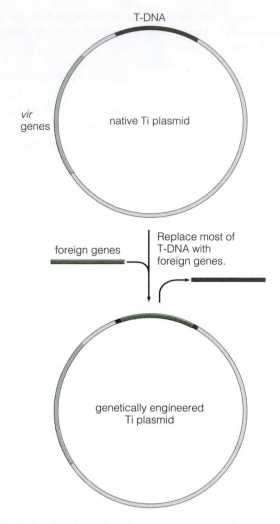

FIGURE 6.8 Ti plasmid and the replacement of much of the T-DNA with foreign genes.

Examples of plant cells and tissue to be transformed include explants using leaf discs, *in vivo* inoculation of seedlings or buds of plants, and incubation of protoplasts in a suspension of *Agrobacterium*. Because only some of the cells receive the new gene, media containing either an antibiotic or chemical is used to select for transformants. After selection, explants and dividing cells are cultured on solid media with nutrients and plant growth hormones to support the callus tissue. Shoot and root growth are promoted by different combinations of hormones as discussed earlier in this chapter. Plants are examined to establish whether the foreign gene has been inserted and expressed in the desired tissue. In this manner, many genetically engineered plants will express new, desirable traits.

Challenges of Foreign Gene Expression

Regions of DNA are involved in the transcriptional and translational control of a gene. As discussed in Chapter 3, Basic Principles of Recombinant DNA Technology, at the 5′ end there is a promoter region that is involved in the initiation of transcription. Other elements may include enhancer or silencer regions that are involved in regulation of expression. Two factors determine temporal and spatial regulation (the expression of genes at the right time and in the appropriate tissue). First, a gene must be delivered to all cells and stably maintained for transmission to progeny. The promoter region must be recognized by the host cell so that RNA polymerase binds. The promoter can be regulated or constitutively expressed, depending on the type of gene and the desired outcome. Termination and polyadenylation signals also must be provided. Thus, consideration of the appropriate regulatory re-

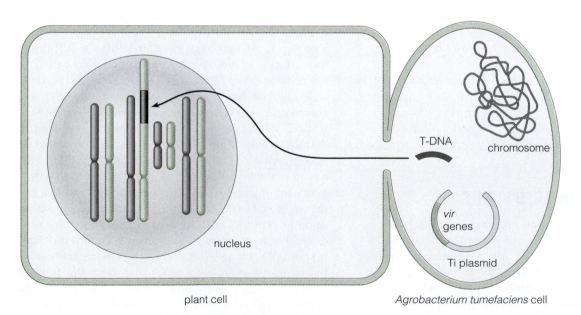

FIGURE 6.9 T–DNA is transferred during plant cell infection by *Agrobacterium tumefaciens*. *Agrobacterium* inserts its T-DNA into the chromosomes of plant cells.

gions of a foreign gene introduced into plant cells is important. Regions include enhancer and promoter, as well as a terminator of transcription and 3' polyadenylation signals. Organelle targeting sequences may also be required.

Different promoter sequences are used for different outcomes. For strong constitutive expression, the cauliflower mosaic virus 35S (CaMV 35S) promoter often is used. A highly regulated promoter responds to a certain signal—heat, light, nutrients, and so on—by turning the gene on. For example, the promoter for the gene encoding ribulose 1,5 bisphosphate carboxylase small subunit is light regulated, that is, activated when the plant is in the presence of light. If a gene product is to be produced primarily in a specific region of the plant, tissue-specific regulatory regions (such as for leaves) must be included in the construct.

Also to be considered is codon usage—the codons and amino acids used by an organism; some codons are used more than others depending on the species of plant or other organism. Genes from nonplant sources especially may specify amino acids that do not match the plant's tRNA and amino acid pools. Recently, researchers have been able to use **codon engineering**: They resynthesize the donor genes if the codons or amino acids do not match the host system. Codon engineering helps ensure that the host plant will have the appropriate tRNAs and amino acids to synthesize the protein. For example, to increase the production of *Bacillus thuringiensis* toxin by potato, tomato, rice, and cotton, the Bt toxin gene has been engineered to match the host plant's translation machinery and to remove incompatible sequences.

APPLICATIONS OF PLANT GENETIC ENGINEERING

Crop Improvement

A major goal of plant genetic engineering is to improve agronomically important crops by transforming them with foreign genes. The first plant biotechnology products are emerging. Much research has focused on agronomic traits for controlling insects, viruses, bacterial and fungal plant diseases, and weeds. Other modifications include developing more nutritious food products, manipulating petal color of flowers, and synthesizing biodegradable, organic polymers. A very promising area of research is the production of specialty oils for detergents, cosmetics, and lubricants and the modification of the lipid composition of seed crops to reduce levels of saturated fats. Examples of modified oils include those with increased levels of laurate and myristate for shampoos and soaps and those that solidify at room temperature without hydrogenation. Partially hydrogenated oils (such as margarine), which are added to many foods, have been shown to be unhealthy. Other types of spe-

cialty oils include cooking oil with reduced saturated fat, an oil substitute for cocoa butter in chocolate, oils for cosmetics, and liquid wax for lubricants. Because many oils are obtained from tropical plants, the production of oil substitutes from genetically engineered oilseed crops grown in the United States, primarily rapeseed, soybean, and sunflower, would decrease the dependency on tropical oils. Genetically engineered crop plants are beginning to be used for the production of industrially important chemicals and pharmaceutically active compounds.

The loss of prime farmland may one day require that crops be cultivated in areas that are less suitable or marginal for agriculture. Crops that can be grown in these regions are being developed. For example, stress-tolerant plants are being engineered that are able to live in colder, drier regions. A 2002 study compiled by the nonprofit research organization National Center for Food and Agricultural Policy (www.ncfap.org), documented through 40 case studies of 27 crops that the production of hardier crops through biotechnology will produce an additional 14 billion pounds of food, improve farm income by $2.5 billion, and at the same reduce the use of pesticides by 163 million pounds. Biotechnology has produced six crops currently in the marketplace—soybeans, corn, cotton, papaya, squash, and canola—resulting in the production of 4 billion pounds of additional food on the same area of land, the improvement of farm income by $1.5 billion, and the decreased use of pesticides by 46 million pounds.

Genetically Engineered Traits: The Big Six

Although numerous field trials are being conducted on genetically engineered plants, most are experimental and may never enter the commercial market. Currently, six different traits have been engineered into crop plant species that have been approved for commercial production: herbicide, insect, and virus resistance; altered oil content; delayed fruit ripening; and pollen control.

Herbicide Resistance   Because weeds growing with crop plants can significantly reduce yields by competing for nutrients, sunlight, and water, farmers must apply herbicides to control weeds, often mixtures of different herbicides. Herbicides are used annually in agricultural areas to decrease the impact of weeds on crops. Between 1966 and 1991, the use of herbicides in agriculture in the United States more than quadrupled, reaching an estimated 495 million pounds and a cost of $10 billion. Although farmers routinely applied more than 100 chemical herbicides, weeds still reduced crop productivity by approximately 12%. The sensitivity of traditional crops to herbicides limits the types that can be used. Because many herbicides do not discriminate between crop and weed, they must sometimes be applied early, before crop emergence. Often the chemical persists in fields when the crops are germinating and can kill them.

By modifying crop plants so they are resistant to a broad-spectrum (kills all or most plants) herbicide, a single chemical is effective without killing the crop plants and the number of applications may be reduced. A larger selection of biodegradable or less-toxic herbicides can be made available to the farmer. Companies that develop herbicides also are engineering herbicide-tolerant plants to accompany their chemicals (Monsanto is an example). Critics, however, claim that if a chemical approach to weed control continues, chemical use will most likely increase. In this case, instead of creating a safer and cleaner environment, biotechnology will have perpetuated our dependence on toxic chemicals. This dependence may generate more herbicide-resistant weeds through increased abundance of chemical applications or perhaps by accidental crossing of herbicide-resistant crops with neighboring related wild plants.

There are several different types of genetically engineered herbicide resistance. One way that herbicides kill plants is by binding to a specific target in a plant's biochemistry. Scientists study how specific herbicides affect a plant and then can modify the target (usually a protein) so that it no longer binds the herbicide. These plants are herbicide resistant and tolerate exposure to that particular herbicide. Another way that plants become resistant is to produce a new protein that inactivates or detoxifies the herbicide. Crop plants have been engineered to be resistant to four types of herbicides.

Herbicide	Resistance-Modified Crops
Glyphosate	Soybeans, corn, canola, cotton, and sugarbeets
Glufosinate	Soybeans, corn canola, cotton, sugarbeets, and rice
Bromoxynil	Cotton
Sulfonylurea	Cotton and flax

Glyphosate Resistance One of the most commonly used broad-spectrum herbicides is glyphosate (marketed as Roundup, Rodeo, and others). Most genetic engineering uses a single bacterial gene that confers resistance to an herbicide. For example, glyphosate-resistant plants have been engineered to readily degrade the herbicide Roundup (made by Monsanto) into nontoxic compounds (Figure 6.10). The herbicide normally inhibits an important enzyme, EPSPS (5-enolpyruvyl-shikimate-3-phosphate synthase), in the aromatic amino acid synthesis pathway (the shikimate pathway) in both plants and bacteria. The gene encoding EPSPS in a glyphosate-resistant *E. coli* strain was isolated, placed under control of a plant promoter, and transferred into plant cells. Monsanto has marketed Roundup Ready crops that are resistant to glyphosate such as soybeans, corn, rapeseed (canola), and cotton. Monsanto also has developed Roundup Ready sugarbeets not yet on the market.

Glufosinate Resistance The active ingredient (phosphinothricin) in glufosinate herbicide (for example, market names Basta and Liberty) mimics the structure of the amino acid glutamine. This compound binds to the plant enzyme glutamine synthase, which is required for nitrogen metabolism and blocks this metabolic pathway (by inactivating glutamine synthase). Herbicide-resistant plants have been generated by inserting a gene from the bacterium *Streptomyces*, whose protein product inactivates phosphinothricin in the herbicide. Genetically engineered glufosinate-resistant varieties of soybean, corn, and rapeseed have been approved and are on the market. Sugarbeets and rice have been approved but are not marketed at this time. Roundup Ready soybeans, first grown in 1996, are the most widely cultivated genetically engineered plant on the market.

a b

FIGURE 6.10 (a) Glyphosate–resistant callus after transfer of a bacterial EPSPS gene to cells. (b) Glyphostate-resistant soybeans before (left) and after (right) application of glyphosate (the product Roundup®). Note the absence of weeds in the row of soybeans treated with the herbicide.

FIGURE 6.11 Transgenic tobacco plant leaves with (right) and without (controls on left) a protease inhibitor gene were exposed to larvae. Extensive damage to controls occurred, while transgenic leaves were protected.

Clarence A. Ryan

Bromoxynil Resistance Bromoxynil (marketed as Buctril) inactivates photosynthesis in plants. Bromoxynil-resistant plants are produced when a gene encoding the enzyme bromoxynil nitrilase (BXN) is transferred into plants from the soil bacterium *Klebsiella pneumoniae*. Nitrilase inactivates bromoxynil before this herbicide can kill the plant. Only cotton has been engineered for bromoxynil resistance (Monsanto's BXN cotton). The EPA allows only approximately 10% of the cotton crop to be sprayed with bromoxynil because of exposure limits considered to be safe.

Sulfonlurea Sulfonylurea kills plants by blocking an enzyme required for the synthesis of three amino acids, valine, leucine, and isoleucine. Herbicide resistance was obtained by first modifying the enzyme by gene mutation in tobacco, and then transferring the mutated gene into crop plants. Monsanto's cotton is the only crop currently on the market.

Insect Resistance Biopesticides, pesticides produced by living organisms, have had a bright beginning in agriculture. One type of biopesticide is produced by the Bt toxin gene found in the bacterium *Bacillus thuringiensis*. Chapter 5, Microbial Biotechnology, discusses the use of the Bt gene to generate insect-resistant plants. Bt-based insect resistance has been engineered in varieties of corn and cotton that are on the market. Most Bt-corn varieties produced by Monsanto are also glyphosate resistant. Aventis marketed a variety of Bt corn called *StarLink* that also included glyphosate resistance (see Cause for Concern?). Potato varieties have been produced, but they have not been marketed since 2001.

In the 1990s many experiments were conducted to explore the use of plant **protease** inhibitors as biopesticides. Insect pests attack stored cereal grains, peas, and

beans and result in huge losses, and evidence showed that protease inhibitors might be able to protect plant products, especially in storage. Protease inhibitors, naturally produced by plants, are produced in response to wounding. There are several classes of protease inhibitors that correspond to insect gut proteases that aid in digestion. After protease inhibitors are ingested by insect larvae, their digestive enzymes are inhibited, and starvation results (Figure 6.11). Examples of successful laboratory experiments include the transfer of the cowpea trypsin inhibitor gene into tobacco and the potato protease inhibitor into rice. Garden pea seeds also have been engineered that resist attack by two species of weevils that damage crops in storage (Figure 6.12). A protein, α-amylase inhibitor, blocks the action of the enzyme α-amylase, which the insects use to digest the starch in seeds. In studies, most of the weevils that fed on pea seeds either died or suffered inhibited development. The α-amylase inhibitor is expressed only in the seeds of peas. Other experiments have shown that a cowpea protease inhibitor gene effectively protects oil palms against attack by bagworm larvae.

Although Bt toxins have been very successful in combating insect pests, results with protease inhibitors have been mixed. Because plant pests have co-evolved with plants, they have long been exposed to plant protease inhibitors, and it is thought that many insects have developed mechanisms to tolerate these enzymes. Therefore, not all crops or even varieties of the same crop are protected by protease inhibitor genes.

Virus Resistance Many crops are lost to viral diseases, resulting in losses of millions of dollars each year. Chemicals are widely used to help control the insect vectors (for example, aphids) that spread viral diseases from one plant to another. However, controlling viral

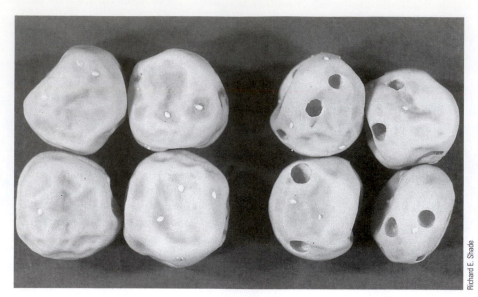

Richard E. Shade

FIGURE 6.12 Pea seeds genetically modified with the α–amylase inhibitor gene resist attack by weevil larvae (left). Larvae have burrowed into the pea seeds that do not harbor the gene (right).

diseases is very difficult. Viral diseases cause symptoms such as yellowing and mottling of leaves, deformed fruits, and stunted growth.

When a virus infects a plant, its DNA enters a plant cell (the protein coat is removed). The virus reproduces in the plant cell by making copies of the virus and packaging each DNA molecule into a protein coat. Thousands of virus copies are made and leave the cell to infect other cells of the plant.

Genetic engineering may provide a more desirable alternative to chemicals. Research has focused on isolating genes involved in resistance to diseases caused by viruses, bacteria, and fungi. Genetically engineered virus resistant plants harbor a viral coat protein gene whose protein product is overproduced. In this way, plants are immunized against the virus because the virus is unable to reproduce in the plant host cell. The plant shuts off the gene of the overproduced coat protein, and at the same time, the coat protein gene of the invading virus is shut off. This prevents the virus from reproducing. This type of resistance is called coat protein–mediated viral resistance.

Numerous plant viruses have been identified, and many of them are host specific. The coat protein genes of a number of viruses have been used to genetically engineer resistance in crops. Among the viruses are cucumber mosaic virus, alfalfa mosaic virus, tobacco streak virus, tobacco etch virus, tobacco rattle virus, potato virus Y and potato virus X, and potato leaf roll virus. Resistance genes for a variety of diseases—not just viral infections—have been isolated from plants: a gene isolated from *Arabidopsis thaliana* (which serves as a model for the study of plant molecular genetics) and tomato confers resistance to the bacterial pathogen *Pseudomonas syringae*; a tomato gene is involved in resistance to the fungal pathogen *Cladosporium fulvum*; a

flax gene has been isolated that confers resistance to fungal rust disease; and a tobacco gene protects against tobacco mosaic virus. One day these genes may be transferred to other agronomically important crops.

To date, yellow squash, potatoes, and papaya have been genetically engineered to be virus resistant. Virus resistance was transferred from yellow squash to zucchini by traditional breeding.

Yellow Squash and Zucchini Asgrow Seed markets several different squash varieties that are resistant to three devastating viruses that cause huge losses in squash-related crops such as zucchini, pumpkins, squash, watermelon, and cucumber: watermelon mottle virus 2, zucchini yellow mosaic virus, and cucumber mosaic virus.

Potato Although no longer marketed (as of 2001) because of poor sales, Monsanto developed two varieties of potatoes resistant to the potato leaf roll virus and potato virus Y, called NewLeafPlus and NewLeafY, respectively. These varieties also contain a Bt-resistance gene that was first marketed as NewLeaf to resist Colorado potato beetle larvae. Two factors caused the demise of NewLeaf varieties (those with both the Bt gene and virus resistance): large chain restaurants such as McDonald's and Burger King and manufacturers of potato products such as Procter and Gamble and Frito-Lay do not use genetically engineered potatoes because of public pressure, and a newly developed insecticide controls insect pests.

Papaya Two varieties of genetically engineered virus-resistant papaya were developed by Cornell University and researchers in Hawaii after an outbreak of papaya ring spot virus that reduced crop productivity by 40% from 1992 to 1997. Since 1998 these varieties, called

Rainbow and SunUp, have been provided to papaya farmers at no cost.

Altered Oil Content   Oils from plants (extracted from the seeds) serve many functions, from use as food additives such as salad and cooking oil and margarine, as well as additions to processed foods, to commercial applications such as in detergents, soaps, cosmetics, lubricants, and paints. Plants produce different types of oils, the properties of which are determined by the types of fatty acids (long carbon–hydrogen chains) the plant makes. Each type of plant oil can be used for a specific industrial purpose. For example, peanut and canola oils are used for cooking, while other types of oils are good for cosmetics (for example, jojoba oil).

Several varieties of soybean and canola have been genetically engineered to produce oils with better cooking and nutritional properties. Imported tropical coconut oil and palm oil is generally used to make soap and detergent, but a genetically engineered canola oil (called high laurate canola) may eventually substitute. The fatty acid type is changed by genetically modifying an enzyme in the fatty acid synthesis pathway. Changes vary from changing carbon length of the fatty acid chain to varying the degree of saturation (number of carbons on the carbon chain).

Delayed Fruit Ripening   The only genetically engineered tomatoes that have been marketed are those with delayed-ripening genes. This desirable trait allows more time to pass between removal from the vine to the market and, thus, increases the shelf life of the tomato. Although a typical ripe tomato tastes better, it becomes too soft (during ripening) and rots before or during shipment. Typically, tomato varieties are picked before the onset of ripening to avoid overripening and softening by the time they reach the market. Although the green, unripe tomatoes are gassed with ethylene (a plant growth hormone that induces ripening) to turn them red, this process does not allow them to develop any flavor. Some genetically engineered tomatoes produce a reduced amount of ethylene so that these tomatoes can fully develop on the vine but stop just before ripening and turning red. In this way, the tomato can develop more flavor on the vine before shipment to markets. An example of this type of genetically engineered delayed ripening tomato is Endless Summer by DNA Plant Technologies. At this time, there are no genetically engineered tomatoes on the market.

What Happened to the Flavr Savr Tomato?   The first genetically modified food product was the Flavr Savr tomato developed by Calgene, Inc. (now part of Monsanto), a biotechnology company that was in California. The bioengineered tomato generated much discussion and controversy but offered many benefits, including a garden-fresh taste year round. Flavr Savr tomatoes were produced by blocking the expression of the polygalacturonase (PG) gene. Fruit softens because the PG enzyme degrades pectin, a major component of plant cell walls. When the amount of PG is decreased, the fruit can remain on the plant longer without becoming soft and is redder and riper by the time it is picked for distribution. In addition, shelf life is increased and tomato products are thicker because of a higher pectin-to-water ratio.

Tomato plants were transformed with the antisense PG gene, which encodes mRNA that is complementary to PG mRNA. The two messenger RNAs (anti-PG mRNA and PG mRNA) may bind to one another so that they are degraded and translation of PG protein is prevented. Antisense technology is being developed to block the production of many other proteins that will have application in agriculture and medicine.

The Flavr Savr tomato was the first genetically engineered food to be reviewed by the U.S. regulatory network. The Food and Drug Administration (FDA) approved the Flavr Savr tomato in May 1994, making it the first genetically engineered whole food to gain government approval in any country. The FDA ruled that the Calgene tomato was as safe as a conventional tomato and did not require labeling in markets to indicate which gene was added or that it was genetically modified. Calgene identified their tomatoes as MacGregor's brand. In the summer of 1994, the tomatoes were first sold in Chicago-area markets. Labels identified them as genetically modified. They were a success; however, the tomatoes were delicate and many were damaged during shipping. In addition, the cost of developing the Flavr Savr had been high, the tomatoes did not grow well in Florida, and tomato prices were low. To add to Calgene's financial problems, Monsanto filed a patent-infringement lawsuit against Calgene. To settle the suit, Calgene sold 55% of its shares to Monsanto. The Flavr Savr tomato was sold through 1996, including in a small market in Canada. By 1997, Monsanto purchased the remaining shares, and the Flavr Savr tomato was no longer marketed.

In the mid-1990s, Monsanto, DNA Plant Technologies (DNAP), and Agritope developed and received regulatory approval for genetically engineered delayed-ripening tomatoes using a different approach from Calgene's Flavr Savr. None of these varieties have been marketed. Although genetically engineered tomatoes are not on the market, numerous companies (for example, Calgene, Agritope, Seminis, Aventis, and DNA Plant Technologies) are still developing genetically engineered varieties of delayed-ripening tomatoes.

Pollen Control   Hybrid crops are primarily used in agriculture today because they often have improved traits over the parent plants. A hybrid is made by crossing two distantly related varieties of the same crop plant. For example, two varieties of soybean are crossed. The

Cold and Drought Tolerance and WeatherGard Genes

Low temperature is an important environmental factor that limits where crops can grow and reduces yields in a growing season. Such stress causes significant reductions in crop yield. Even when low temperatures do not result in losses in yield, it often results in a significant reduction in crop quality. Crop loss from frost damage is significant and has at times reached $2 billion worldwide in a single year. In Florida, because of occasional freezes, citrus groves are now planted farther south. Some commercially useful subtropical and tropical plants cannot grow in the southeastern United States because of occasional freezing. For example, eucalyptus, which grows rapidly and has commercially valuable wood, is susceptible to intermittent freezes and cannot be easily grown in the United States. The amount of rainfall also affects crop productivity and quality. Between 1978 and 1995, average crop losses from drought in the United States was more than $1 million each year. The particularly devastating droughts in 1996, 1998, and 1999 cost farmers approximately $5 billion, $7 billion, and $1 billion, respectively. Thus, drought and low temperatures significantly affect agricultural productivity.

Plants vary greatly in their ability to withstand low-temperature stress (called cold hardiness). Crops such as corn, rice, and cassava that originated in tropical regions of the world can be killed or damaged at above freezing temperatures.

Crops with a temperate origin can tolerate temperatures below freezing from −5° C to as low as −30° C. Many biochemical and physiological changes are known to occur during cold acclimation. With the onset of low temperature, it is thought that temperature sensors (similar to a thermometer) at the cell membrane transmit a stress signal that is amplified and sent on to other parts of the cell. The message eventually reaches the nucleus and controls the activation of specific genes that code for transcription factors (proteins) that act as master switches to turn on groups of genes, which code for proteins that protect the cell from freezing damage. One example is the gene that codes for antifreeze protein. The genes that are induced during cold treatment are called cold-regulated (COR) genes.

In 1997, Michael Thomashow and his colleagues at Michigan State University identified a transcription factor called CBF, a master switch that controls the activation (expression) of a group of COR genes. They patented it as WeatherGard. Genetically engineered plants containing the CBF gene showed a significant improvement in tolerance to freezing compared with nonengineered plants. Interestingly, WeatherGard genes also increased drought tolerance and tolerance to high salt soils. All major crop species examined so far, including corn, soybean, rice, wheat, rice, tomato, barley, canola, alfalfa, and many vegetable and tree species, contain CBF genes.

The researchers from Michigan State University and a San Francisco–based company called Mendel Biotechnology, which has licensed WeatherGard technology, engineered canola plants using CBF genes. The engineered plants are able to survive freezing temperatures as much as 4–5° C lower than the nontransgenic controls, a very significant increase in cold hardiness. WeatherGard crops are plants that contain regulatory genes that allow them to be more tolerant to drought, low temperatures, and high salt soils. These traits increase crop yields and quality, and also increase the land available to agriculture. In addition, crops may not require extensive irrigation, an important characteristic in regions where water is scarce.

Mendel Biotechnology estimated that a 1% increase in grain production due to an increase in drought and cold tolerance will generate $3 to $4 billion per year.

Because many of the most important crops in developing countries, such as cassava, corn, and rice, are of tropical origin, they can be very sensitive to low temperatures. Biotechnology has the potential to increase food production and decrease temperature-related and possibly drought-related losses in the developing countries. A collaborative effort will be necessary to transfer cold tolerance technology to developing countries. Toward this end, Mendel Biotechnology has donated WeatherGard technology for the development of temperature- and drought-tolerant crops in Africa and Asia.

resulting hybrid may be taller, produce more seeds and higher yields, and may be more resistant to environmental pressures. The mixing of pollen (male) and ovule (female) for fertilization must occur in a controlled manner to ensure that the correct parental plants are used in the cross. Because many types of crop plants have both male and female reproductive structures on the same plant, methods must be applied to prevent the plant from fertilizing itself.

Plant breeders use different methods to control pollen transfer. One way is to remove the male flower parts by hand before pollen is released. An example is removing the tassel (male part) on a corn stalk (female part is the ear). To simplify this process, scientists have engineered male-sterile plants by inserting in one variety of a particular crop plant a gene that comes from the bacterium *Bacillus amyloliquefaciens* that produces a protein (barnase) in pollen-producing tissues. Barnase blocks pollen production and renders the male reproductive structure sterile. In a different plant variety of the same crop plant, a gene encoding a protein that blocks barnase (called *barstar*) is inserted. When pollen from the barstar plant variety is used to fertilize the male-sterile barnase plant variety, a hybrid plant is produced that is fertile because the barstar and barnase genes will mix. In nature *Bacillus amyloliquefaciens* produces barnase to degrade the RNA of a foreign invader, while barstar is produced to inactivate barnase by binding to it. Two crops, corn and chicory, have been genetically modified with the barnase-barstar genes to make hybrids.

Genetically Engineered Foods

More than 60% of processed foods in the United States contain ingredients that come from genetically engineered plants. Most of these plant ingredients are genetically engineered corn and soybeans, with the rest of the ingredients coming from canola or cottonseed oil. Although 12 different genetically engineered plants have been approved in the United States, with more than one variety of each type of plant, not all the approved varieties are on the market. Many approved genetically modified plant varieties were marketed only briefly and removed from stores, while some have never been marketed or are available in only some regions or food products.

Soybean In 2002, approximately 74% of the U.S. soybean crop was genetically engineered. Soybean products include vegetable oil, soy flour, and soy protein. Several soybean varieties have been modified to be resistant to broad-spectrum herbicides (see the previous discussion on soybeans and herbicides).

In 2003, scientists at the U.S. Department of Agriculture (USDA), Pioneer Hi-Bred International, and Dupont removed a primary antigen from soybean. This protein, called P34, is a major soybean allergen and can cause a severe allergic response in sensitive individuals. The increased usage of soybean in processed foods poses a risk to people allergic to soybean; however, the elimination of a major allergen should allow people to ingest these foods without a major immune response. Although this research was experimental, it should have important implications for products that cause allergies such as peanut-containing processed foods.

Corn In 2002 approximately 32% of field corn in the United States was genetically engineered. Products include corn oil; corn syrup; corn flour; baking powder; corn starch; gluten; sweeteners such as fructose, dextrose, and sorbital; alcohol; and nutritional supplements such as vitamin C. Sweet corn makes up less than 3% of genetically engineered corn, and there is not a genetically modified popcorn. Bt insect resistance is the most commonly genetically engineered trait in corn and provides resistance to insect pests such as the European corn borer, a moth larva that burrows into corn stalks and damages the plants. Other genetically engineered traits include resistance to different herbicides. By 2000 approximately 25% of the U.S. corn crop was genetically engineered, with 72% being Bt corn, 24% herbicide-resistant varieties, and 4% a combination of both (Bt and herbicide resistant).

Canola Canola, extracted from the rapeseed plant, is grown primarily in Canada. More than 60% of the crop in 2002 was genetically engineered. Canola is found in many processed foods and is a common cooking oil.

Cotton More than 71% of the 2002 cotton crop was genetically engineered. Genetically modified cottonseed oil is common in processed foods such as pastries, snack foods, fried foods such as potato chips, peanut butter, and candies, to name a few.

Other Crops Genetically engineered papaya (only in Hawaii) and squash (a small number of farmers plant these varieties) are grown only rarely in the United States. Other plants approved for commercialization include tomato, rice, sugar beet, flax, and red heart chicory, although these are not marketed in the United States and are not present in foods.

Nutritionally Enhanced Plants—Golden Rice: An International Effort

Much of the world's population suffers from malnutrition. It is the hope of many that biotechnology can provide the means to help increase the production of food, as well as to provide nutritionally enriched foods to those who need it. Because more than one-third of the

The Case of StarLink Corn

In September 2000, Americans learned that a genetically engineered corn variety that was approved only for animal consumption, not human, had been detected in Taco Bell taco shells. This genetically modified corn, called StarLink, was later detected in a wide variety of corn-containing products in many countries. It has been estimated that there may have been several hundred different corn products that were recalled by manufacturers of processed foods. Kraft Foods, Inc., the distributor of Taco Bell taco shells, immediately recalled any products that might have contained StarLink corn.

StarLink was the only genetically modified variety of corn approved for animal consumption, but not for human use. Aventis CropScience, the developers of StarLink, were confident that the appropriate precautions would be taken to prevent farmers from mixing the genetically engineered corn and keep it from entering the human food supply. Unfortunately, the precautions failed and StarLink was detected in processed foods. Regulators later determined

that many farmers growing StarLink were not aware that it was not approved for human use. It is believed that farmers sold the corn directly to mills where genetically engineered and nongenetically engineered varieties were mixed. Although only 1% of the total corn harvest in 1999 and 2000 was StarLink, because it was mixed with other corn varieties at different mills, the contamination of corn-containing processed foods was extensive. Some have estimated that 50% of the corn harvest of 2000 was contaminated with StarLink. Aventis purchased the remaining StarLink corn harvests of 1999 and 2000, but approximately several million bushels most likely entered the human food supply.

StarLink contained two new genes: (1) one gene provided resistance to butterfly and moth caterpillar pests such as the European corn borer, and (2) the other gene enabled StarLink corn to resist certain herbicides such as Basta and Liberty. StarLink was novel in that the gene for pest resistance was a modified form of the *Bacillus thuringiensis* (Bt) toxin protein Cry9C (Aventis proprietary technology).

The FDA and EPA approved StarLink corn for only animal consumption because of a concern that modified Cry9C may be an allergen to humans. Allergenic proteins are typically small and resistant to stomach enzyme digestion and heat. This means that a protein would remain intact for a longer period in the digestive tract and have the potential to cause an allergic response. Although most Bt toxin proteins break down quickly in the stomach, the modified Cry9C produced by StarLink corn was more heat stable and resistant to digestion. The EPA concluded that StarLink posed a moderate allergy risk to humans based on testing. Although many thought the risk of an allergic reaction was very small because of the low concentration of protein in foods, the product was approved only for animal use.

To date, no confirmed cases of allergic reaction to StarLink have been documented. Even so, in 2001, the EPA ruled that even traces of StarLink would not be allowed in foods for human consumption. Thus, Aventis withdrew its regulatory approval to sell StarLink corn in the United States.

world's population relies on rice as a food staple, one of the first projects was to enhance the quality of rice. Unfortunately, natural varieties of rice do not supply vitamin A. Vitamin A deficiency causes 500,000 cases of irreversible blindness in children each year and other vitamin A–deficiency related diseases, especially in the heavily populated areas of Latin America, Asia, and Africa.

Genetic engineering technology has been used to create a variety of rice, Golden Rice, with high levels of beta-carotene and other carotenoids. These carotenoids are precursors to vitamin A. Golden Rice was developed by two scientists, Dr. Ingo Potrykus of the Insti-

tute for Plant Sciences at the Swiss Federal Institute of Technology in Zurich, Switzerland, and Dr. Peter Beyer of the Center for Applied Biosciences at the University of Freiburg in Germany. This international effort was funded in part by the Rockefeller Foundation and the European Union in the hopes that Golden Rice can be distributed free of charge to farmers in developing countries. A number of agencies are now working to distribute Golden Rice worldwide. The inventors of this technology have sold the rights to Syngenta, one of the world's leading agricultural biotechnology companies (www.syngenta.com), with sales in 1999 of approximately $7 billion. The recent surge in the healthy

Genetically Engineered Foods and Public Concerns

Because of their newness, genetically engineered foods have raised concerns about safety. However, as future research demonstrates their safety and still more genetically engineered plant products are introduced to the public, many concerns and fears will most likely diminish, as they did in the 1970s when cloning methods were first developed. The release of the first genetically engineered food, the Flavr Savr tomato, generated much discussion about the potential risks of genetically engineered food. The primary public fear was that genetically engineering a plant might produce unexpected results. Plants produce secondary metabolites that may be toxic to humans or livestock or may alter the quality of foods. Many people have food allergies that vary from a mild reaction such as a rash to a more severe reaction such as anaphylactic shock. There has been concern that the allergenic properties of food from a donor plant might be conferred on the host plant, and that people may not be aware that the genetically engineered food contains the new protein. Others have questioned whether the nutritional quality of engineered foods could change unexpectedly (a disruption in the balance of nutrients). Perhaps a particular nutrient might change to a form that cannot be metabolized or absorbed properly. A very different type of concern has focused on food choice. Some religious and ethnic groups avoid eating certain foods (for example, Muslims and Orthodox Jews do not eat pork). Vegetarians might also avoid some genetically engineered foods. How will engineered produce affect food selection when, for example, an apple contains an animal gene? What if a cucumber contains a human gene? Does the cucumber possess a human trait? Of course, these questions address hypothetical situations, but they represent the types of questions posed by concerned individuals and critics of genetically engineered foods.

Another source of debate has encompassed the use of antibiotic resistance genes as selectable markers to indicate when an organism has been transformed. These enzymes have the potential to inactivate clinically invaluable antibiotics. Although evidence has not supported this concern, opponents of genetically engineered foods still consider antibiotic resistance an issue. Scientists are developing ways to remove selectable markers from genetically engineered crop plants to assuage public concerns.

Some fear that deleting genes also may introduce harmful side effects when the product is ingested. For example, plants produce secondary compounds that may protect them from fungal and bacterial infection. If secondary compounds are removed, people may be exposed to cancer-causing compounds produced by the fungus. An example often cited is decaffeinated coffee. Research has demonstrated that caffeine may inhibit fungal synthesis of aflatoxins; if a gene involved in caffeine biosynthesis is removed, fungal aflatoxins may then contaminate the coffee bean.

Other fears include the potential that uncharacterized DNA included with the gene and selectable markers will code for additional, unknown proteins or produce unexpected, harmful side effects or undesirable traits in the engineered plant. Fear also has been expressed that the introduced trait in genetically engineered crops may be spread through the pollen to wild plants by wind or insect pollinators. This dispersal may irreversibly alter the ecosystem and even affect wildlife that feed on these wild plants. The development of male-sterile crop plants (nonpollen producers) reduces the chance that pollen will be transferred to wild plant populations.

---

foods market in the industrialized world may pave the way for nutritionally enriched genetically modified plants. Syngenta is exploring commercial opportunities for Golden Rice in the United States and Japan, as well as providing expertise in making this product available at no cost to developing countries. Several other biotechnology companies also are helping in the endeavor by providing assistance to developing countries. For example, Monsanto will provide royalty-free licenses for its technology, which will aid in the development of Golden Rice and other vitamin A–enriched rice varieties. Although Golden Rice is not yet available for planting and consumption, plans are to have it ready for distribution in the near future after additional testing.

Other nutritionally enhanced crops that are being produced by genetic engineering include iron-enriched rice and a tomato variety with three times the normal amount of beta-carotene. Conventional breeding methods are also being used to increase the vitamin and mineral content of rice, wheat, corn, beans, and cassava.

Molecular Farming

A new field is emerging whereby plants (and animals) are being genetically engineered to produce important pharmaceuticals, vaccines, and other compounds of value. This process is called molecular farming. In the not-too-distant future, plants will be used as living bioreactors (just as microbes already are being used) by inserting a foreign gene to produce proteins of therapeutic and industrial value. An inexpensive, readily available complex metabolite could be used as a nutrient source, and cells could then convert this compound into a valuable product, which accumulates until the cells are harvested. Whole plants or seeds would then be harvested and the product extracted, allowing producers to avoid complex *in vitro* biosynthetic reactions in the laboratory.

Soybean plants have been used as bioreactors to produce a variety of monoclonal antibodies with therapeutic value (such as for treatment of colon cancer). Other potential medical products from plants include corn, rice, and tobacco that produce clot-busting drugs and human blood products such as albumin and hemoglobin. In research designed to help slow the pace of HIV infection, Epicyte Pharmaceuticals has produced the first plant lines to produce human antibodies to HIV. Antibody production is a first step in the prevention of HIV, which currently infects 900,000 people in the United States and is predicted to infect as many as 25 million people in India and 15 million in China by the end of this decade. Epicyte is the first U.S. company to begin Phase I clinical trials with a human herpes antibody produced in plants. This company also has planted a corn crop that produces an antibody to treat respiratory syncytial virus (RSV).

The cost of producing foreign compounds in plants may be much lower than producing them in bacteria and mammalian cells for a number of reasons. (1) Commercial scale-up involves simply planting seeds rather than using costlier fermenters. (2) Proteins are produced in high quantity (for example, soybeans have a protein content of 40–45%). (3) Foreign proteins produced are active, unlike those produced by bacteria, which are denatured and must be renatured after isolation. (4) Foreign proteins in stored seeds are very stable (for example, in experiments, they have remained active after 5 years). (5) Pathogens, toxins, viruses, and other contaminants that are harmful to humans or animals are not likely to be present. In addition, because plants are eukaryotes, posttranslational processes (for example, glycosylation, phosphorylation, proteolytic cleavage) are present.

Edible Vaccines   Vaccines are the major tools for keeping people and animals healthy and free from disease. In industrialized countries infants (and even livestock) are routinely vaccinated against many diseases such as polio, measles, and hepatitis B. Vaccines are expensive to produce and require proper handling and refrigerated storage. Most vaccines are administered by injection, although the polio vaccine can be given orally.

Unfortunately, people, especially children, in developing countries do not have access to most vaccines. Some diseases have devastating consequences in developing countries, while few people are affected in the industrialized countries. For example, diarrhea causes up to 10 million deaths each year among children in developing countries. The cholera vaccine could help alleviate this problem; however, this vaccine is not available. How are vaccines made available to people in countries where refrigeration is not readily available and the vaccine's cost is prohibitive? One idea, suggested by Charles Arntzen at Texas A&M University, was to engineer edible plants that produce vaccines. When the plant food is consumed, the person becomes vaccinated by producing antibodies to the protein in the plant.

In a landmark experiment, potatoes were transformed with a gene encoding a protein that makes up part of the *E. coli* enterotoxin (gut toxin that causes diarrhea and illness). Mice that consumed transgenic potato produced antibodies specific for the toxin. This was the first demonstration that plants have the ability to produce vaccines and important pharmaceuticals. Later, the potato triggered an immune response to the same toxin protein in humans. Currently, research on edible vaccines is focusing on human diseases, although livestock and other animals will benefit. Although potatoes must be consumed raw, other candidates for vaccine production include banana and tomato. Livestock vaccines can be distributed in alfalfa, corn, and wheat. Other diseases are being investigated, such as hepatitis B virus, which kills more than 1 million people each year and is a leading cause of liver cancer. Although vaccines are available for many diseases such as hepatitis B, an edible vaccine would reach people in developing countries. Global eradication may be possible for diseases like polio, hepatitis B, and others.

Biopolymers and Plants   Recently, the advances made in plant genetic engineering methods have enabled multiple genes to be moved into plants. This provides opportunities for plants to acquire new multistep biochemical pathways. A promising area of plant biotechnology is in the area of biopolymers. As discussed in Chapter 5, Microbial Biotechnology, a class of biodegradable thermoplastics called polyhydroxyalkanoate (PHA) polymers has much commercial value, and is produced by microbial fermentation technology.

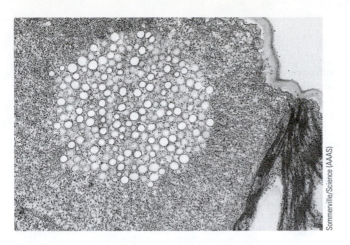

FIGURE 6.13 Transmission electron micrograph of an *Arabidopsis thaliana* leaf mesophyll cell with an accumulation of polyhydroxybutyrate (PHB) granules in the nucleus.

The production of biopolymers in plant seeds would be advantageous in that large amounts of product could be produced and readily extracted. Previous research demonstrated that a type of PHA polymer called *poly(β-hydroxybutyrate)* (PHB) is produced in the experimental plant model *Arabidopsis* (Figure 6.13). PHB is synthesized in a multistep pathway requiring three bacterial genes. Researchers have moved these genes from the bacterium *Alcaligenes eutrophus* into canola (rapeseed) for synthesis of PHB in seeds. They were successful in producing the polymer as 1% of canola seed weight. Other biopolymers also are being investigated. For example, the PHA polymer PHBV (poly-3-hydroxybutyrate-co-3-hydroxyvalerate) is currently manufactured from fermentation of *Alcaligenes* and sold under the name *Biopol* by Monsanto. Plants are able to produce PHBV using genes from *Alcaligenes*.

A BRIGHT FUTURE

Exciting new agricultural biotechnologies are providing solutions to important problems in agriculture. The wise use of modern biotechnology should embrace safer, less toxic agricultural practices as well as the conservation and use of germplasm. Genetically engineered plants can contribute in positive ways toward sustainable agriculture and should provide an alternative to poor agronomic practices that ultimately contribute to environmental degradation. Questions to be asked include how we can become less dependent on chemicals, hybrid seed, and other costly amendments, rather than encouraging high-input solutions that promote soil loss, increased use of chemicals, and the loss of germplasm diversity. Plant genetic engineering will offer sound alternatives that will help decrease our dependence on agronomic practices that promote environmental degradation. Nutrient-enriched plant products and heartier and higher-yield crops will help impoverished areas where malnutrition is high. Important characters such as drought tolerance and increased tolerance to other environmental stresses will enhance crop productivity. Plant-generated vaccines and pharmaceuticals will one day provide low-cost, readily available, and safe medical products.

New technologies such as microarrays and DNA chips (Chapter 3, Basic Principles of Recombinant DNA Technology) and the genome sequencing of *Arabidopsis*, rice, corn, and other important crop plants (see Chapter 9, Genomics and Beyond) will have far-reaching implications for future agriculture. These new developments will revolutionize plant biotechnology and human health care. The future is bright for plant biotechnology.

General Readings

S.S. Bhojwani, ed. 1990. *Plant Tissue Culture: Applications and Limitations.* Elsevier, Amsterdam.

H.S. Chawla. 2002. *Introduction to Plant Biotechnology*, Science Publishers, Inc., Enfield, New Hampshire.

V.L. Chopra, V.S. Malik, and S.R. Bhat, eds. 1999. *Applied Plant Technology*, Science Publishers, Inc., Enfield, New Hampshire.

M.J. Chrispeels and D.E. Sadava. 2001. *Plants, Genes, and Agriculture.* Jones and Bartlett Publishers, Boston, Massachusetts.

H. Daniell and A. Dhingra. 2002. Multiple gene engineering: Dawn of an exciting new era in biotechnology. *Curr. Opin. Biotech.* 13:136–141.

H. Daniell and S.E. Harding. 2000. Genetically modified food crops: Current concerns and solutions for next generation crops. *Biotech. Genet. Eng. Rev.* 17:327–352.

H. Daniell, K. Wycoff, and S.J. Streatfield. 2001. Medical molecular farming: Production of antibodies, biopharmaceuticals and edible vaccines in plants. *Trends Plant Sci.* 6:219–226.

R. Fischer, S. Schillberg, and N. Emans. 2001. Molecular farming of medicines: A field of growing promise. *Outlook on Agriculture* 30:31–36.

K. Rajasekaran, T.J. Jacks, and J.W. Finley, eds. 2002. *Crop Biotechnology*. Oxford University Press.

I.K. Vasil. 2003. *Plant Biotechnology: 2002 and Beyond*, Kluwer Academic Publishers, Boston, Massachusetts.

J.D. Williamson. 2002. Plant biotechnology: Past, present, and future. *J. Amer. Soc. Hort. Sci.* 127:462–466.

Additional Readings

A.F. Bent and I.-C Yu. 1999. Applications of molecular biology to plant disease and insect resistance. *Adv. Agron.* 66:251–298.

B.B. Buchanan. 2001. Genetic engineering and the allergy issue. *Plant Physiol.* 126:5–7.

U. Commandeur, R.M. Twyman, and Rainer Fischer. 2003. The biosafety of molecular farming in plants. *AgBiotechNet* 5:ABN 110.

H. Daniell. 2002. Molecular strategies for gene containment in transgenic crops. *Nature Biotechnol.* 20:581–586.

G. Giddings, G. Allison., D. Brooks, and A. Carter. 2000. Transgenic plants as factories for pharmaceuticals. *Nature Biotechnol.* 18:1151–1155.

R.D. Hall, ed. 1999. Plant Cell Culture Protocols. In *Methods in Molecular Biology.* Vol. 111. Humana Press, Totowa, New Jersey.

K.K. Hatzios. 1994. Herbicides and Herbicide Resistance. In *Encyclopedia of Agricultural Science.* Vol. 2. Academic Press, pp. 501–512.

E.M. Herman, R.M. Helm, R. Jung, and A.J. Kinney. 2003. Genetic modification removes an immunodominant allergen from soybean. *Plant Physiol.* 132:36–43.

F. Kempken, K. Esser, U. Luettge, J.W. Kadereit, and W. Beyschlag. 2001. Transgenic crops for the Third Millennium. *Progress in Botany* 62:114–139.

P. Lucca, I. Potrykus, M. Laimer, W. Reucker, eds. 2003. Genetic Engineering Technology against Malnutrition. In *Plant Tissue Culture: 100 Years since Gottlieb Haberlandt.* Springer-Verlag, New York, pp. 167–174.

S.R. Padgette et al. 1996. New weed control opportunities: Development of Soybeans with a Roundup Ready™ Gene. In S.O. Duke, ed. *Herbicide-Resistant Crops: Agricultural, Environmental, Economic, Regulatory, and Technical Aspects.* CRC Lewis Publisher, Boca Raton, Florida, pp. 53–84.

Y. Poirier, D.E. Dennis, K. Klomparens, and C. Somerville. 1992. Polyhydroxybutyrate, a biodegradable thermoplastic, produced in transgenic plants. *Science* 256:520–523.

I. Potrykus. 2001. Golden Rice and beyond. *Plant Physiol.* 125:1157–1161.

A.S. Rishi, N.D. Nelson, and A. Goyal. 2001. Molecular farming in plants: A current perspective. *J. Plant Biochem. Biotechnol.* 10:1–12.

W. Schuch. 1994. Improving tomato quality through biotechnology. *Food Technol.* 48:78–83.

G.H. Toenniessen, J.C. O'Toole, and J. DeVries. 2003. Advances in plant biotechnology and its adoption in developing countries. *Curr. Opin. Plant Biol.* 6:191–198.

J. Winterer. 2002. The mixed success of protease inhibitors to combat insect pests in transgenic crops. *AgBiotechNet* 4:ABN 082.

J. Yu, J.E. Carter, W.H.R. Langridge, G. Dietrich, and W. Goebel, eds. 2002. Edible vaccines. In *Vaccine Delivery Strategies.* Horizon Scientific Press, Norfolk, United Kingdom, pp. 369–397.

J. Yu and H.R. Langridge. 2001. A plant-based multicomponent vaccine protects mice from enteric disease. *Nature Biotechnol.* 19:548–552.

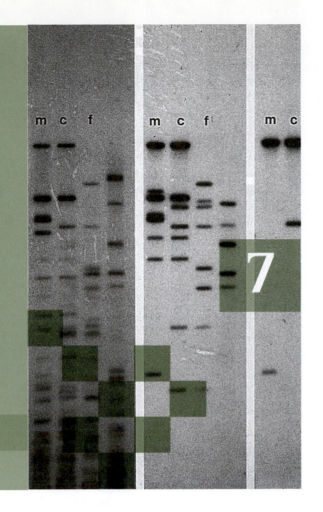

ANIMAL BIOTECHNOLOGY

GENE TRANSFER METHODS IN ANIMALS

Microinjection
Embryonic Stem Cell Gene Transfer
    Gene Targeting in Mice
Retrovirus and Gene Transfer

TRANSGENIC ANIMALS AND THEIR APPLICATION

Mice
Cows
Pigs, Sheep, and Goats
    Xenotransplantation
■ *Biotech Revolution*
    Humanizing Pig Tissues and Organs
Birds

ANIMAL HEALTH

Foot-and-Mouth Disease
Mad Cow Disease
Coccidiosis
Trypanosomiasis
Theileriosis

ANIMAL PROPAGATION

Artificial Insemination
Animal Clones

CONSERVATION BIOLOGY

Embryo Transfer
■ *Cause for Concern?*
    Animal Clones: Promise and Controversy

REGULATION OF TRANSGENIC ANIMALS

PATENTING GENETICALLY ENGINEERED ANIMALS

Selective animal breeding has been practiced for nearly 10,000 years to produce desirable traits in livestock, as we saw in Chapter 1, Biotechnology: Old and New. Today, breeders are under increasing market pressure to produce livestock that grow faster and convert animal feed to lean tissue, with less total fat. Breeders have focused not only on manipulating the composition of animal products but also on increasing production, such as by enhancing reproduction rates. Increased milk production, decreased fat content, better wool quality, faster maturation rate, disease resistance, and the increase in frequency of egg laying are but a few of the changes brought about by selective breeding. However, such methods of producing desirable traits take time; many generations of animals must be born and bred before an appropriate trait is finally expressed. Hundreds of gene pairs are involved in the modification of measurable traits such as egg and milk production, fat composition, and disease resistance. Consider the numbers the breeder is up against: If 20 heterozygous gene pairs are involved, the number of possible gene combinations is approximately 3.5 billion! Simply maintaining the rate of genetic improvements made in the past, let alone increasing the rate, requires intensive screening and selection programs.

As we saw in Chapter 6, Plant Biotechnology, the availability of tools for genetically manipulating plants is revolutionizing agriculture. Recombinant DNA technology now allows us to introduce foreign genes into organisms for the expression of specific new traits. Animals also are engineered for a variety of purposes, ranging from use as human disease models to the introduction of desirable traits into a variety of agronomically important animals—including fish, as we see in Chapter 8, Marine Biotechnology. Approaches that combine breeding with molecular genetics and recombinant DNA technology are being used to produce animals that express a variety of desirable traits. Powerful genetic methods such as marker-assisted-selection (MAS) technology help identify chromosomal loci associated with physical traits. MAS identifies relationships between gene markers mapped to specific chromosomal regions or loci and measurable traits such as disease resistance, milk production, and fat content. MAS has identified more than 75 genes that influence growth. Twenty-nine of these encode either growth factors or growth-factor receptors. Genome projects are identifying new genes and their functions (Chapter 9, Genomics and Beyond). New biotechnologies are affecting animals in many significant ways. More efficient transformation methods are available to generate transgenic animals, and animal cloning provides many opportunities in the experimental sciences (for example, molecular genetics), as well as for commercial biotechnology (new products, improved livestock characteristics).

Genetic engineering has enabled scientists to transfer genes across species, families, and even kingdoms. These evolutionary boundaries, which were obstacles in the past, are now readily crossed with modern gene transfer methods that allow the transfer of DNA into living cells and fertilized eggs of animals. Transgenic animals not only provide invaluable research tools for studying gene regulation and disease, but they may be genetically modified and cloned for the production of pharmaceuticals, **vaccines**, and rare chemicals as well as for food production. This is an exciting yet often controversial new frontier in animal biotechnology.

Major goals of biotechnology include the generation of livestock that are more economically produced and products that are more nutritious to the consumer. Livestock research has several primary focuses: increasing growth rate and muscle development with growth-hormone genes from various sources, producing antibodies and recombinant vaccines to increase disease resistance, and cloning superior animals (that is, animals that express highly desirable traits). Animals are being engineered as bioreactors for producing rare pharmaceuticals and other medical compounds. Genetically engineered livestock are producing important products in milk for treating a variety of human diseases and health needs. Examples of important products that have potential include

- Human hemoglobin, which could be used during trauma when much blood is lost. Hemoglobin is more desirable than whole blood or red blood cells for transfusions, because it requires no refrigeration and is compatible with all blood types, eliminating the need for blood typing

- Human protein C, which helps prevent blood clotting

- Human tissue plasminogen activator, which is used to treat patients after a heart attack

- Human α_1-antitrypsin (ha1AT), which may be useful in treating the more than 20,000 people in the United States who have ha1AT deficiency, which predisposes them to a life-threatening type of emphysema.

GENE TRANSFER METHODS IN ANIMALS

One challenge in creating transgenic animals is to ensure that the **transgene** turns on at the right time and in the right tissue. To be functional, the integrated gene must be expressed and regulated appropriately (for example, in a tissue-specific manner). Thus, the gene to be transferred must be accompanied by the appropriate promoters and regulatory sequences, as described in Chapter 2, From DNA to Proteins. Some

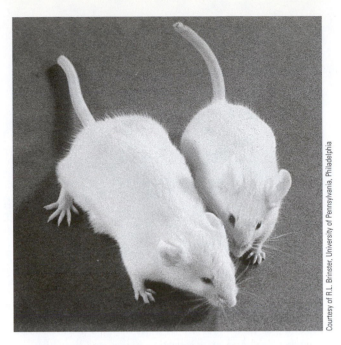

Courtesy of R.L. Brinster, University of Pennsylvania, Philadelphia

FIGURE 7.1 Two 10-week-old sibling male mice, one of them the product of genetic engineering experiments. The mouse on the left harbors a new gene comprising the mouse metallothionein promoter fused to the rat growth hormone structural gene. This mouse weighs 44 g, whereas its control sibling weighs 29 g. The gene is passed on to offspring that also grow larger than controls. In general, mice that express the gene grow two to three times faster than controls and up to twice normal size.

genes require an enhancer that may be located far from the promoter. Gene regulation can be extremely complex during embryonic development, when many genes are regulated temporally and tissue specifically. The engineering of organisms requires fusion of the correct promoter and enhancer and gene coding sequence. The construct must then be incorporated into the chromosome where gene expression is regulated. Figure 7.1 shows the results of a gene-transfer experiment.

Since foreign genes were first introduced into mice by **pronuclear microinjection** of DNA in 1980, a tremendous effort has been expended to develop efficient methods of producing transgenic animals. In addition to microinjection, methods have included retrovirus-mediated and **embryonic stem cell**–mediated gene transfer into the germ line, the reproductive cell line. Other methods, described in Chapter 3, Basic Principles of Recombinant DNA Technology, have included electroporation, virus infection, and DNA precipitation onto the cell surface using calcium phosphate.

Microinjection

Microinjection, first developed in the mouse, has been the method of choice for most transgenic research. Once a gene has been characterized and appropriately expressed in eukaryotic cells, a transgenic animal can be made by microinjection of the cloned gene into the fertilized eggs (ova) of a donor animal. The foreign gene must be injected before the first cell division, or cleavage, occurs so that all cells of the organism harbor the gene.

Donor females are superovulated (induced to produce exceptional numbers of ova) and mated, and the fertilized eggs are removed and placed in a sterile dish with buffer. Before fusion occurs to make a diploid **zygote** with one nucleus, male and female pronuclei are separate. A very thin pipette or needle injects DNA into the large male pronucleus (refer to Figure 3.14). Microinjected eggs are implanted into the oviduct of a surrogate female made pseudopregnant with hormones. After birth, to identify which of the young are transgenic, tissue samples are subjected to DNA analysis by Southern blot hybridizations with the new gene as a probe or by PCR amplification of the new gene (Figure 7.2). Founder animals—those animals with the new gene integrated into their germ cell line and somatic cells—are bred to establish new genetic lines with the desired characteristics.

The steps involved in microinjection are

1. Identification (and sometimes modification such as mutation) of a foreign gene of interest

2. Insertion of the foreign gene into an appropriate vector

3. Microinjection of DNA directly into the pronucleus of a single fertilized egg

4. Implantation of the microinjected egg into a surrogate mother

5. Allowing the embryo to develop to term

6. Demonstrating that the foreign gene has been stably incorporated into the host genome and that it is heritable in at least one of the offspring

7. Demonstrating that the gene is expressed and regulated correctly in the host organism

Some microinjected fertilized eggs (for example, the morula or blastocyst in cows) must be cultured *in vitro* until they reach a certain stage, and others (such as those of pigs) are transferred to hosts almost immediately after microinjection. Other species have presented the problem that the opacity of the cytoplasm makes it difficult to observe the pronuclei. A special type of microscopy—**differential interference contrast microscopy**—has been used for rabbit, sheep, and goat

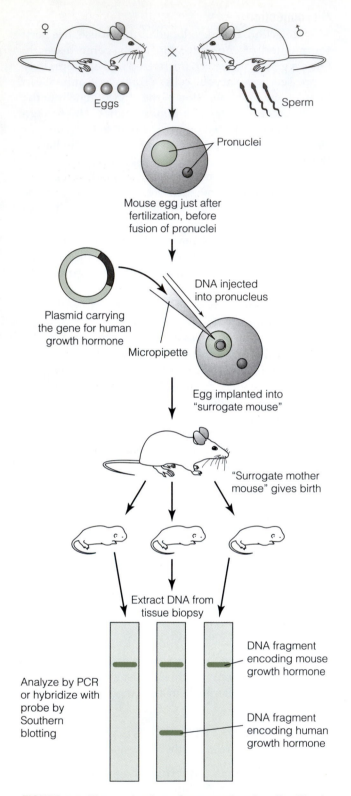

FIGURE 7.2 The production of transgenic mice. Fertilized eggs are collected before the male and female pronuclei fuse. DNA is injected into the male pronucleus, cultured, and then implanted into a surrogate female mouse. DNA taken from progeny is used for Southern blot hybridization or PCR to confirm the presence or absence of foreign DNA.

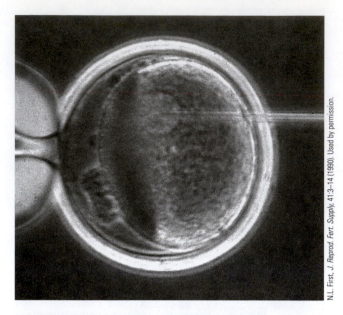

N.L. First, *J. Reprod. Fert. Supply*, 41:3–14 (1990). Used by permission.

FIGURE 7.3 Microinjection of DNA into the pronucleus of a bovine egg after centrifugation has moved the cytoplasm to make the pronuclei more visible.

eggs. However, cow and pig eggs must be centrifuged to stratify the cytoplasm so that the pronuclei are visible (Figure 7.3).

Microinjection is not without problems. Few injected eggs survive, and not all of those retain the new DNA. The new DNA is integrated at random if it is not targeted to a specific chromosomal locus (Unfortunately, **homologous recombination**—the direct replacement of a gene with its counterpart or homologue—and gene targeting efficiency are low in animals). This can lead to unexpected changes in gene expression, especially when the inserted DNA disrupts other genes. Microinjected genes often integrate into a single chromosomal locus as head-to-tail concatamers. In addition, transgenic animals are often mosaics because the gene has integrated into only some of the cells. This mixture of transformed and untransformed cells is especially problematic if the gene has not incorporated into the germ cells, or is in only a portion of the germ cells. All germ cells must be transformed to ensure that the gene is transmitted to progeny. There are also instances in which a gene is incorporated but expression is low or nonexistent.

Embryonic Stem Cell Gene Transfer

Stem cells are undifferentiated cells that divide to produce differentiated progeny cells, while maintaining an undifferentiated stem cell population. Some stem cells are able to differentiate into all the different tissues of the body. These cells are called **pluripotent** (capable of differentiating into other cell types) stem cells. Other stem cells appear to be more specialized and only develop into a few tissue types. For example, stem cells in bone marrow develop into the different blood cells; neural stem cells differentiate into the various types of stem cells of the nervous tissue. Research has recently demonstrated that the more specialized stem cells might have the ability to develop into other tissues in the laboratory—neural stem cells can differentiate into blood cells when transferred into bone marrow, and bone marrow stem cells can develop into muscle cells. Thus, even specialized stem cells have the ability to differentiate into a variety of cell types. Stem cells have a variety of important applications—from medical (see Chapter 10, Medical Biotechnology) to the production of transgenic animals.

One method of producing transgenic animals by targeted gene replacement (homologous recombination) is through the use of cultured embryonic stem (ES) cells (ES cell lines are available in culture). ES cells are isolated from the inner cell mass of donor blastocysts of early embryos and can be cultured *in vitro* before **transfection** with a specific gene. A gene is targeted to specific areas by homologous recombination and the use of selectable markers; treated cells are then screened by either PCR or a selection method to determine whether the gene has been integrated. Transformed ES cells are microinjected into animal blastocysts so that they can become established in the somatic and germ line tissues. They are then passed on to successive generations by breeding founder animals (Figure 7.4).

At present, the ES method has been most successful with mice because mouse ES cells are pluripotent and, when integrated into blastocysts, can divide and differentiate in the mouse embryo (compare this with plant regeneration, described in Chapter 6, Plant Biotechnology).

The mouse is used as a model in the study of gene function and is useful in many areas of medicine (see the section on lung disease in Chapter 10, Medical Biotechnology). More than 5000 human diseases are caused by genetic defects; thus, by using gene targeting to introduce mutations in experimental animals to produce human disease models, researchers may identify effective treatments and gene therapies. Specific genes are inactivated in mouse models to simulate human diseases such as Alzheimer's disease, cancer, atherosclerosis, amyotrophic lateral sclerosis (Lou Gehrig's disease), and cystic fibrosis. Methods are used to inactivate gene function and look for changes in the phenotype. Gene targeting also can be used to add genes encoding desirable characteristics. ES cell transformation technology has been used to inactivate specific genes to determine their effect on growth and development or to develop effective treatments.

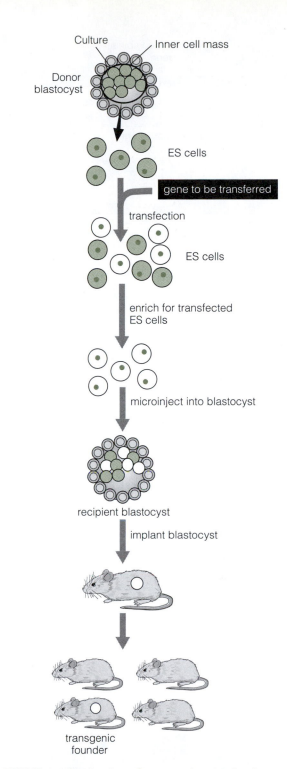

FIGURE 7.4 Production of transgenic mice by the transfer of embryonic stem cells containing a gene of interest. Stable transgenic lines are obtained by crossing founder animals that have the gene in their germ cells.

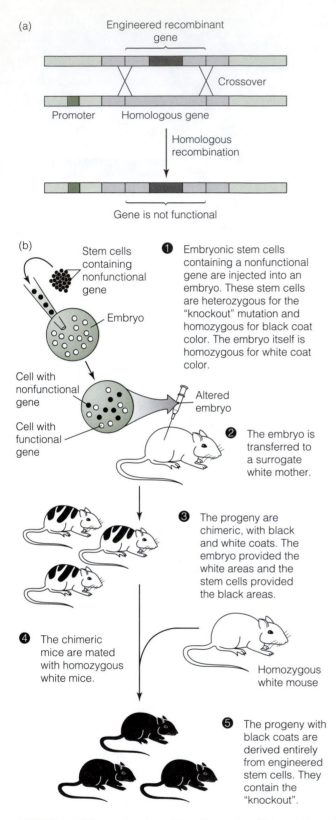

(a)

Engineered recombinant gene

Crossover

Promoter   Homologous gene

Homologous recombination

Gene is not functional

(b)

Stem cells containing nonfunctional gene

Embryo

Cell with nonfunctional gene

Cell with functional gene

Altered embryo

❶ Embryonic stem cells containing a nonfunctional gene are injected into an embryo. These stem cells are heterozygous for the "knockout" mutation and homozygous for black coat color. The embryo itself is homozygous for white coat color.

❷ The embryo is transferred to a surrogate white mother.

❸ The progeny are chimeric, with black and white coats. The embryo provided the white areas and the stem cells provided the black areas.

❹ The chimeric mice are mated with homozygous white mice.

Homozygous white mouse

❺ The progeny with black coats are derived entirely from engineered stem cells. They contain the "knockout".

Gene Targeting in Mice Gene targeting is the insertion of DNA into a specific chromosomal location; it is often used to inactivate a specific gene (knockout) in the genome. In mice (called knockout mice), for example, gene inactivation is a common means of elucidating how a specific gene functions. In gene targeting, a cloning vector (called a targeting vector), which harbors the gene to be inserted, recombines with a region of the target chromosome that is homologous with a DNA region on the targeting vector (Figure 7.5). A selectable marker gene (for example, bacterial neomycin phosphotransferase, which confers resistance to the antibiotics neomycin and kanamycin) ensures that the desired recombination event is selected. Antibiotics included in the growth medium of cells transformed with the targeting vector and inserted gene permit only those cells that have the correct insertion into the chromosome to survive.

A gene is first isolated, altered *in vitro* according to the requirements of the experiment (for example, inactivated), and then targeted to its counterpart on a specific chromosome in cells. A portion of the gene is disrupted by inserting an antibiotic resistance gene, such as neomycin, into it, with regions of the gene flanking the marker remaining on either side. This marker enables cells that contain targeted DNA to be selected (positive selection).

ES cells from early mouse embryos are used in gene targeting experiments. ES cells containing the inactivated gene are injected into early mouse embryos and contribute to the developing mouse. A **chimera** is produced where some cells are normal and others have the mutated gene. If germ cells, the precursors to eggs and sperm, come from stem cells harboring the mutant gene, progeny of the chimeric mouse will have one copy of an inactivated gene in every cell. Because there are two copies of the gene —one normal and one inactivated—the mice will be heterozygous (for this particular gene). By mating the heterozygous mice with each other, homozygous knockout mice are produced that do not express a functional protein. (Figure 7.6). These genetically manipulated embryos are placed in surrogate mothers, where they develop.

FIGURE 7.5 The production of knockout mice. Through homologous recombination (a), a selected gene is inactivated and (b) transferred into embryonic stem cells. Each stem cell is transferred into an early mouse embryo, where it contributes to the developing mouse. Embryos are implanted into surrogate mothers. If the inactivated gene is present in the germ cells that give rise to eggs and sperm, the chimera will have an inactivated copy of the gene. Breeding these heterozygous mice will produce homozygous knockout progeny with both genes inactivated.

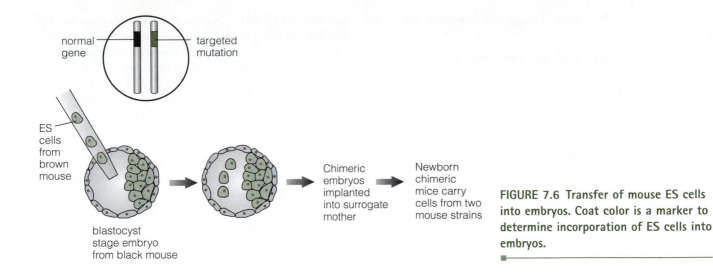

Normal gene / targeted mutation

ES cells from brown mouse

blastocyst stage embryo from black mouse

Chimeric embryos implanted into surrogate mother

Newborn chimeric mice carry cells from two mouse strains

FIGURE 7.6 Transfer of mouse ES cells into embryos. Coat color is a marker to determine incorporation of ES cells into embryos.

Coat color serves as a marker to track the incorporation of introduced ES cells in mouse embryos. If a mouse has both black and *agouti* coat genes, it will be a chimera with a readily detectable black and brown coat. ES cells from a brown mouse, for example, would contain the *agouti* gene, expressed as a light coat even if in single copy, as well as the gene to be targeted. The ES cells to be transferred will contain a gene for a different coat color than the host embryo's.

When newborn mice are screened for the presence of a targeted gene, coat color indicates whether the mice contain transformed ES cells. For example, male chimeric mice with a black and *agouti* coat inherit ES cells from both the host embryo (black coat gene) and the introduced ES cells (*agouti*). These chimeras are selected for mating to a female with a black coat (non-*agouti*). Only progeny with brown coats are selected for further screening of the DNA to determine which individuals have inherited the targeted gene. Those male and female mice with a targeted gene copy are mated to obtain mice with two copies of the targeted gene. Confirmation of mice **homozygous** for the targeted gene is obtained by a DNA test such as PCR.

Retrovirus and Gene Transfer

Retroviruses can efficiently infect animal cells and integrate into their genomes. They have been used to transfer DNA into a variety of animals during very early development. Retroviruses use RNA as their genome instead of DNA and can infect and replicate in host cells without killing them. During infection, two copies of RNA and the enzyme reverse transcriptase enter the host cell. After infection, reverse transcriptase catalyzes the synthesis of a double-strand DNA using the RNA as a template. The DNA then inserts into the host genome (provirus). The formation of new virus parti-

cles requires viral entry and integration of the DNA copy of the RNA genome and expression of viral genes.

Retroviruses can infect a wide variety of cell types and a large number of those cells, each with a single DNA copy. However, the type of foreign DNA that can be used is severely restricted because only small DNA fragments of approximately 8 kb can be transferred into an organism. To ensure safety during transfection, the virus used as a vector is disabled by the deletion of genes that encode reverse transcription and viral particle packaging. A retrovirus **helper virus** is required for packaging. As an extra safety factor, the helper virus cannot be packaged because of genome modifications. One potential risk, however, is that both disabled vectors may recombine to generate a fully functional retrovirus that can integrate into the host genome.

Precautions such as those described do not completely alleviate the risks of transferring DNA into organisms (for example, the death of a patient undergoing an experimental gene therapy occurred recently; see Chapter 10, Medical Biotechnology for gene therapy). Because of the special risks of retroviruses, other gene transfer methods are usually used when genetically engineered products (the animals themselves or products from them, for example, proteins) are commercially produced.

TRANSGENIC ANIMALS AND THEIR APPLICATION

Transgenic animals are providing many benefits. Transgenic animals provide research scientists with important animal models in which to study human diseases. Gene knockout (inactivation of a specific gene) experiments (and knockin—the addition of a new gene) provide information on how specific proteins function in the body, and animal disease models allow new drugs and therapies to be tested. Transgenic livestock are being developed to

Increase production (for example, eggs, meat, milk), while decreasing costs to farmers

Provide healthier, more robust animals

Provide more nutritious and healthier foods (for example, lower cholesterol beef and pork, eggs with increased vitamin E)

Produce disease-resistant animals

Increase wool quality and quantity

The most promising developments are transgenic animals that harbor human genes. These living biofactories produce important human proteins in their milk. They show promise in providing low cost, high quality pharmaceuticals and biologicals.

Mice

As discussed earlier, mice have been used as a model for many human diseases; examples include arthritis, hypertension, Alzheimer's disease, coronary heart disease, certain cancers, and various neuromuscular disorders such as the family of muscular dystrophy diseases. If mice lack a particular human disease counterpart, they often can be genetically engineered, by knockout technology or by replacement of the normal gene with a mutated counterpart, to acquire it; numerous examples exist such as cystic fibrosis, β-thalassemia, atherosclerosis, retinoblastoma, and Duchenne muscular dystrophy. In this manner, therapeutic compounds can be tested and the molecular biology of the disease can be studied in a convenient animal model (see Chapter 10, Medical Biotechnology).

The potential for using animals as living factories has been explored in mice, and the technology that has developed is being transferred to other, larger animals. Transgenes introduced into mice or other animals (for example, cows and goats) are expressed and their products (for example, interleukin-2, α₁-antitrypsin, and clotting factor IX) secreted in the milk for isolation. The gene construct must include mammary-specific promoters (for example, a promoter from the β-casein gene) and appropriate enhancers. One important protein that is being studied in this way is the cystic fibrosis transmembrane regulator (CFTR). CFTR normally acts as a chloride channel, allowing chloride ions to move in and out of cells. When CFTR is mutated, the channel is disrupted. The resulting disruption of ion flow causes mucus to accumulate in organs, especially in the lungs and pancreas. The mucus attracts bacteria that later die and release their DNA, further inhibiting normal organ function. Because the CFTR protein is associated with membranes of transformed cells, it has not been successfully expressed *in vitro*. Secretion into milk facilitates isolation because the mammary glands produce fat globules in milk that are encapsulated by plasma membrane. In transgenic animals, a heterologous transmembrane protein could be associated with the plasma membrane of the globule and isolated readily from milk during lactation. The CFTR protein secreted in milk can be used to study protein function so that effective therapies can be developed.

Cows

Transgenic cattle with specific traits have been produced by microinjection of eggs in the following steps:

Egg collection from slaughter houses

Maturation of eggs *in vitro*

Fertilization of eggs *in vitro*

Centrifugation of eggs to clear the yolk to see the pronuclei

DNA microinjection into male pronucleus

Develop embryos to blastula stage

Screen cells from blastula-stage embryos for the foreign gene (called a transgene) using PCR

Implant embryos into surrogate females

Birth of calves

Genetic engineering technology is being used to alter the composition of milk and provide important human proteins. One goal is to increase two of the four proteins, κ-casein and β-casein (phosphoproteins in milk) in cow's milk by overexpressing the genes. The end result is the improved processing characteristics of milk and milk products such as cheese. Another desirable genetic modification would be to remove lactose from milk. A lactose-free product would be beneficial to the millions of people who are lactose intolerant and, therefore, cannot digest milk that contains lactose.

Other potential modifications include increased resistance of animals to bacterial, viral, and other pathogenic diseases. For example, mastitis, a bacterial infection of the mammary glands, is a common affliction of cows. Resistance in cows would save farmers money, reduce the use of antibiotics, and prevent much suffering. *In vivo* immunization, transferring the antibody genes for specific antigens to animals, would protect livestock from many diseases and eliminate the need for vaccinations. Genetic engineering could eventually reduce the number of vaccinations, drugs, and veterinarian visits and ultimately the cost of raising cattle.

The production of recombinant bovine somatotropin (rBST) for use in cows to increase milk production has created a recent controversy. Somatotropins occur naturally in the milk and meat of certain animals, and BST, a protein hormone produced in the cow's pituitary gland, is essential for milk production. When injected into cows, BST stimulates increased milk production and facilitates the efficient conversion of feed into milk rather than body fat. In the past, somatotropin has been obtained from cow pituitaries by extraction, an expensive and time-consuming method. Today, rBST can be produced in *E. coli* after transfer of the *bst* gene. After injection into cows, rBST increases milk production by approximately 25%. Although extensive testing has shown that rBST is not toxic to humans and that cows injected with rBST do not have elevated levels of the protein in their bodies, rBST has remained controversial. The FDA has approved rBST as an animal drug and has formally stated that the milk and meat of treated cows are safe for human consumption. Nevertheless, opponents of rBST express many concerns that have focused primarily on economics and human health.

The production of important human proteins in milk also shows promise. It has been estimated that 20 transgenic cows could produce approximately 100 kg of protein each year (the annual output of milk per cow is approximately 10,000 liters, recombinant protein per cow in approximately 60 kg each year). It would take only one cow to produce enough factor VIII (blood clotting factor) to treat the world's hemophiliacs. Other examples of proteins produced by animal bioreactors in milk include

Insulin

Erythropoietin

Fibrinogen

Hemoglobin

Monoclonal antibodies

Tissue plasminogen activator (TPA)

Nerve growth factor

Granulocyte colony-stimulating factor

Growth hormone

Interleukin-2

The future may bring cost-effective methods for generating transgenic cows that grow faster, require less feed, produce more milk, or are leaner.

Pigs, Sheep, and Goats

As with cows, sheep and goats also are being genetically engineered as bioreactors to produce important human proteins or pharmaceuticals. Proteins such as factor VIII and IX, interferon, interleukin, growth hormone, plasminogen activator, CFTR (to treat cystic fibrosis), TPA (to counter the effects of heart attack) and a variety of antibodies are being mass produced in this way. The secretion of proteins from transgenes in the milk of sheep and goats seems to have no ill effects on most host animals or their progeny. Pigs that harbor a foreign gene, on the other hand, seem to have a variety of problems, which include lameness, lethargy, thickened skin, kidney dysfunction, inflamed joints, peptic ulcers, pericarditis, severe osteoarthritis, and a propensity toward pneumonia.

In experiments similar to those conducted with cows, the porcine somatotropin gene has been cloned into *E. coli* and the product used to treat pigs. Recombinant porcine somatropin increases an animal's growth rate, as well as feed efficiency, and decreases fat deposition. This recombinant hormone may have potential for enhancing production of pigs in the future.

Xenotransplantation Xenotransplantation is another area where recombinant pigs will make important contributions. Xenotransplantation refers to the use of animal organs in human patients. The global shortage of human organs available for transplant is fueling research in this area. The first attempt to transfer an animal organ into a human occurred in 1964 when a chimpanzee's heart was transplanted into a patient. Other unsuccessful attempts to transfer hearts from primates followed. The most famous patient was Baby Fae who survived 20 days with a baboon heart in 1984. Two transplants using pig hearts are known to have occurred: one in 1991 in Poland and another in 1996 in India. Transgenic pigs were not used and the hearts failed within 24 hours.

Pigs have been the animal of choice in recent xenotransplantation research. They are inexpensive to raise, easy to breed, and have organs that are similar in size to human organs. They also can be genetically engineered and cloned. Apart from the ethical concerns (for example, humane considerations, the production of animals with some physical abnormalities, the significant degree of loss of litter mates due to either cloning or genetic engineering, the use of animal organs in humans—that is, humanizing pig organs), one of the primary fears is that porcine endogenous retrovirus (PERV), a virus similar to HIV, may be transferred to humans by way of organ transplantation. Pig clones will provide a way to produce disease-free lines of pigs for medical purposes.

The greatest challenge of xenotransplantation is how to avoid the massive immune response by the body. Scientists are focusing on producing genetically engineered disease-free pigs that have modified cell surface antigens to reduce the problems of rejection by the immune system. Recent developments include the production of cloned pigs lacking one copy of a key functional gene involved in immune rejection (see Biotech Revolution). To do this, one gene copy was inactivated by knockout technology. The nucleus of a knockout cell was transferred into an enucleated pig egg (see Animal Clones).

The first application of xenotransplantation technology will most likely be the transplantation of human insulin-producing porcine islet cells from

Humanizing Pig Tissues and Organs

The biotechnology company PPL Therapeutics Ltd. produced the first cloned pigs in March 2000. By April 2001, the same company introduced a foreign gene into cloned pigs, and by December 2001 cloned knockout pigs were produced. These pigs harbor an inactivated gene that would normally code for an enzyme, 1,3 galactosyl transferase (GT), that catalyzes the transfer of the sugar galactose to the surface of pig cells. The human immune system recognizes this sugar as foreign and aggressively attacks the organ with this sugar (see Chapter 4, Basic Principles of Immunology). The result is rejection of the transplanted organ within minutes. Before

xenotransplants can be used, both genes must be inactivated before organs will be accepted by the human immune system. This will involve the breeding of cloned knockout pigs to produce offspring with both GT genes inactivated.

Cloned miniature pigs are also being genetically engineered to lack cell surface antigens that trigger the human immune response. Miniature pigs do not carry PERV and, therefore, are highly desirable for xenotransplantation technology.

In 2003, researchers from BioTransplant Inc. and Novartis Pharma AG, in a joint venture, announced that they had identified the receptors used by PERV to enter and infect cells. This will allow scientists to develop strategies to prevent

virus transmission, although the risk of transfer is thought to be extremely small. PERV has been shown to infect some human cells in culture; however, PERV has never been detected in patients who receive kidney dialysis using pig cells. Recently, a screening method has been developed that can identify the presence or expression of PERV in cells. This assay will be useful as diagnostic tool to help ensure the safety of xenotransplantation. A patent (no. 6190861), Molecular Sequence of Swine Retrovirus and Methods of Use, has been granted to three U.S. companies—BioTransplant Inc., Immerge BioTherapeutics Inc., and Massachusetts General Hospital—for the development of the assay.

---

knockout pigs (that do not produce pig insulin) for the treatment of diabetes. Heart and kidney transplants will be the first organs used. Because the liver produces so many different proteins, pig proteins are likely to be somewhat different from those of humans; therefore, liver transplants may not be available for quite a while. Xenotransplantation clinical trials may begin within the next several years. It has been estimated that the market could be more than $5 billion for solid organs and $6 billion for cellular therapies for diabetes, Parkinson's disease, and Alzheimer's disease.

Birds

Genetic engineering will allow the production of healthier birds used for food in the near future. Chickens, ducks, geese, and small game birds could be made resistant to viral and bacterial diseases. Poultry products could be improved by decreasing the fat content of meat and the cholesterol in chicken eggs. Chicken or duck eggs also could serve as biofactories for producing valuable proteins. Examples of proteins include growth hormone, insulin, monoclonal antibodies, and interferon. Typically, birds secrete a large amount of ovalbumin in their reproductive tracts that becomes part of the egg. If transgenic birds are engineered to

also secrete other types of proteins into the egg, these products could readily be isolated within the egg. As an alternative, eggs could be eaten so that a therapeutic protein is also consumed.

Retrovirus vectors have been used to generate transgenic birds. After being disabled to prevent adverse side effects, the virus is used to infect cells at the blastoderm stage of development. As discussed earlier, problems associated with retrovirus vectors include the instability of the introduced gene and restrictions on the size of DNA inserts. (There is also, among some, the fear of using retrovirus DNA when the transgenic organism may be a source of food.)

In avian species, the mode of egg fertilization and the structure of the egg membranes prevent the microinjection of foreign genes. After fertilization, the egg becomes encased in a tough inner membrane as well as shell membranes that are difficult to penetrate. The male pronucleus that will fuse with the female pronucleus cannot be identified because more than one sperm enters the ovum. DNA microinjected into the cytoplasm will not be incorporated into the genome.

Alternative methods of transformation are being investigated. A promising technique involves introducing blastoderm cells, transformed with lipsomes, into

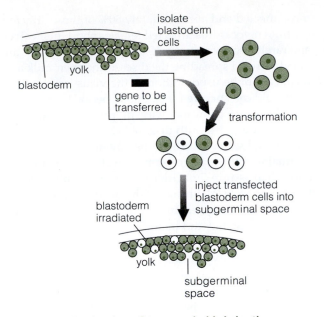

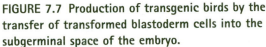

FIGURE 7.7 Production of transgenic birds by the transfer of transformed blastoderm cells into the subgerminal space of the embryo.

the subgerminal space of host embryos (Figure 7.7). A chimeric organism would be produced having some cells from the host and others derived from the transformed donor cells. If the introduced transformed cells become a part of the germ cell line, stable transgenic lines can be produced through matings of founder animals.

ANIMAL HEALTH

Disease prevention and detection is another area where modern biotechnology can make important contributions. Many animal diseases are very contagious and can be economically devastating, especially in developing tropical countries. Livestock, in particular, are susceptible to many types of dysentery and diseases such as African horse sickness, bovine leukosis, bovine infectious rhinotracheitis, brucellosis, and Rift Valley fever. Developing countries often have no way of coping with the widespread outbreak of disease. Even in the United States, diseases account for significant livestock losses.

Recombinant DNA technology may be the only way of preventing some of the more widespread and devastating animal diseases found in many developing countries. Monoclonal antibodies and recombinant vaccines produced by this technology are effective against deadly bacterial and viral diseases, many of which will be controlled or eradicated once low-cost vaccines are developed and in widespread use. Genetic engineering enables immunogenic proteins produced in bacterial and yeast cells to be used as vaccines that offer several advantages over traditionally produced vaccines (that is, pure preparations of viable pathogenic organisms that are attenuated—weakened—so they cannot cause disease but can stimulate the immune response against a virulent pathogen). The mass production of antigens from bacteria or yeast cells can dramatically reduce production costs. A pure protein is often produced, eliminating the need for extensive and costly testing, because the disease-causing pathogen is not present. Recombinant DNA technology also eliminates the need for inactivated vaccine (attenuated live virus) by enabling the antigen to be synthesized separately from the pathogen. The genes of different proteins can be fused so that a variety of antigens are present in a single vaccine.

Traditional (attenuated virus) vaccines for livestock have serious drawbacks. First, many virus types and subtypes are specific to a particular region of the world. Thus, the most effective vaccines are polyvalent—specific for more than one virus. Some virus strains, however, cannot be grown in large quantities to provide antigens for vaccination. Second, after a period of time, vaccines may have to be modified to counter new virulent strains. Virus particles of traditional vaccines are unstable at less than pH 7 and must be refrigerated to remain potent. In the tropics, refrigeration may not be available where the vaccine is needed.

Rapid new diagnostic tools are being developed to diagnose common animal diseases. Examples of diseases that are detected by modern diagnostics include

Companion animals—canine and feline heartworm disease, feline leukemia, and various pathogenic fungal diseases

Large livestock—brucellosis, trypanosomiasis, swine fever, mastitis, trichinosis, foot-and-mouth virus, rinderpest, and equine infectious anemia virus

Poultry—avian reovirus, coccidiosin, salmonellosis, Newcastle disease virus, avian encephalomyelitis virus, avian leukosis virus, avian influenza virus, and hemorrhagic enteritis virus

Other new diagnostics are being developed to assess the state of an animal's health. For example, a very recent development has been an inexpensive, but sensitive antibody assay that can determine very early kidney disease in dogs—before symptoms begin and before the standard kidney-function blood test detects a problem. This early detection can potentially extend an animal's life if precautions are taken (for example, use of a special diet). A similar assay is expected for cats in the very near future. Kidney failure is a common ailment in older cats and is most often first detected when kidney function has been reduced by approximately 70%.

New diagnostic tests are being developed to increase the sensitivity with which parasites can be detected in animals (traditional serological tests have proven to be inadequate because the presence of antibodies does not always indicate current infections; antibodies can remain in an animal for long periods of time). Instead of detecting the antibodies produced in response to an infection, new diagnostic methods detect parasite proteins (antigens), as well as DNA. Parasite antigen detection kits are available for detecting trypanosome proteins and tick-borne protozoan parasites. Mouse monoclonal antibodies raised against specific parasite proteins are used to identify parasite antigens in animals. Such assays are highly sensitive and specific. Not only can these assays detect current infections they can also distinguish between different strains or species of parasite (for example, between the major tsetse-transmitted species of trypanosomes). Parasite-specific DNA assays are desirable because of their sensitivity and the stability of the DNA. Samples can be collected in the field and delivered to a laboratory for DNA analysis (using PCR). **Polymorphic** DNA sequences and their probes allow different parasites, such as tick-borne parasites and trypanosomes, to be detected and distinguished (using PCR or Southern hybridization). Other molecular-based diagnostics tests also are used such as ribotyping.

In many cases, biotechnology is only beginning to offer effective treatments and diagnostic tools. Developing countries in particular have special needs because of climate (often high temperatures), lack of modern facilities (for example, refrigeration), and a need for low-cost treatments. Edible vaccines may offer an effective way to vaccinate livestock in developing countries (see discussion on human vaccines in Chapter 4, Basic Principles of Immunology). Five economically devastating diseases are discussed in the following sections.

Foot-and-Mouth Disease

Foot-and-mouth disease is one of the most highly contagious diseases known (it takes only 10 virus particles to infect a cow). It is one of the most devastating to livestock and very costly to the industry. It is caused by a virus that infects cattle, swine, goats, and sheep (and other cloven-hoofed animals, as well as other animals such as elephants, rats, mice, and hedgehogs). Infections in humans are very rare and, if they occur, the disease is not fatal. The United States has not had an outbreak since 1929, although the disease has been found in Great Britain, France, Eastern Europe, Asia, Africa, the Middle East, and South America. A devastating recent example was the 2001 outbreak in the United Kingdom where millions of animals had to be destroyed. Worldwide, approximately 30% of cloven-hoofed animals are affected. Although only about 5% of adult animals die from the disease, young cattle and pigs are most affected and approximately 50% of those afflicted die from myocarditis. Infected females abort their calves and rarely produce young. Usually, infected animals are killed to avoid spreading the disease and their milk and meat cannot be used, although the virus is extremely hardy and can survive for extended periods.

The most common way of protecting animals against foot-and-mouth disease is by vaccination with inactivated virus vaccine (when available). However, thermal instability, short-term protection (4–6 months), and virus strain specificity present problems for treatment and prevention with vaccines. Experimental vaccines using recombinant DNA technology are being investigated and include a recombinant mixture of different strains of inactivated antigen for cross-protective activity to different viral strains.

Mad Cow Disease

Mad cow disease (also referred to as bovine spongiform encephalopathy, or BSE) is a chronic and, unfortunately, fatal neurodegenerative disease found primarily in older dairy cattle from the United Kingdom (first diagnosed in 1986). Other countries where it has been confirmed include Belgium, Austria, Finland, Greece, Luxembourg, The Netherlands, Portugal, Spain, Switzerland, Germany, Italy, Denmark, and Japan. It belongs to a general class of illnesses found in mammals (for example, sheep, goats, deer, elk) called transmissible spongiform encephalopathies (TSE). Humans cannot get BSE, however, they can acquire new variant Creutzfeldt-Jakob disease (nvCJD), a fatal neurodegenerative TSE. All these diseases are characterized by pathological changes in the brain that cause the nervous system tissue to appear spongelike. Because the incubation period can be as long as 12 years, the specific source of the disease may be difficult to identify. The cause of BSE and related diseases is thought to be a prion—a poorly understood infectious agent that is common to these diseases. It is not a virus or bacterium but a protein. Scientists do not thoroughly understand prions, although molecular biology research is ongoing. Prions are very resistant to sterilization and do not evoke any detectable immune response or inflammatory reaction in hosts; therefore, unfortunately, there are no treatments or preventative drugs for BSE at this time.

Coccidiosis

Coccidiosis is one of the most economically devastating diseases of the cattle industry, costing $100 million each year. Nine parasitic protozoan species in the genus *Eimeria* invade epithelial cells of the digestive tract and associated glands in livestock such as cattle, sheep, and poultry. Coccidiosis also can be extremely costly to the poultry industry, where overcrowded, stressful conditions promote the spread of disease (from fecal contamination) and make treatment very difficult. Developmental problems are found in the young, and 20% of those infected die. In industrialized countries, many feeds contain an anticoccidial drug to help prevent the disease. Several drugs are available both for treatment and prevention of coccidiosis. A gene for a particular protein produced by the oocysts of the protozoan has been cloned so that purified antigen can be prepared for an effective vaccine. *In vivo* trials have been ongoing to determine the appropriate mixture of purified antigens for a vaccine that will boost immunity to coccidiosis. However, an effective vaccine has not yet been marketed.

Trypanosomiasis

The trypanosome, due to its many unique characteristics, is one of the most studied parasites—more is known about the biochemistry and molecular biology of this parasite than any other nonmammalian cell type. However, there is not yet a drug to prevent the disease called *African sleeping sickness* or *trypanosomiasis*. More than 60 million cattle, as well as humans, pigs, domestic buffalo, camels, horses, and small ruminants, in sub-Saharan Africa are susceptible to trypanosomiasis, a disease transmitted by the tsetse fly. Trypanosomiasis is difficult to treat because the single-celled trypanosomes continually change their surface antigens during infection and can escape detection by the host's immune system. Some drug-discovery programs are involved in screening molecules from the chemical libraries of pharmaceutical companies; however, overall, funding has been inadequate. The most common method of preventing this disease in livestock is the routine use of chemical sprays or dips. Research is focusing on ways to prevent infection by preparing vaccines from trypanosome components involved in the pathological process, such as the proteolytic enzymes used to degrade host molecules or proteins responsible for the suppression of the cattle immune system. Researchers are trying to isolate trypanosome resistance genes in West African cattle so that transgenic trypanotolerant cattle can be produced. Host genes that inhibit parasite cell division may one day be identified and, through their increased expression, used to prevent the spread of the parasite within the host. Global parasite genome initiatives for in-depth analyses on African trypanosomes (and other parasites that cause devastating illnesses such as malaria and leishmaniasis) are ongoing with hopes that new preventatives and treatments will be developed.

Theileriosis

Theileriosis, or East Coast fever, is a deadly disease of cattle prevalent in 12 countries of east, central, and southern Africa, Several strains of tick-transmitted protozoans cause theileriosis. Losses to African farmers exceed $170 million U.S. dollars. Once ingested by the tick, the parasite develops and forms sporozoites in the

tick's salivary gland that are released into cattle bitten by infested ticks. Infected lymphocytes in the cattle become leukemic and eventually lympholysis occurs, usually causing death within 1 month of infection. Ticks feeding on the blood of cattle infected with the protozoan transmit the disease to other cattle. At least 24 million cattle are at risk of infection, and high yield exotic breeds are especially susceptible. Spraying or dipping cattle in a chemical insecticide has traditionally been used to control the disease. However, insect resistance can develop, the milk and meat may become contaminated with the pesticide, and chemicals can harm the environment. A crude preparation of live, infected ticks with an antibiotic has been used to stimulate cattle immune systems, but immunity is not always obtained and the antibiotic may not inhibit development of the parasite. Recombinant DNA technology offers a means of producing new drugs for treatment, synthesizing monoclonal antibodies for typing strains in geographical areas, and producing sporozoite antigen for vaccination in large quantities. If specific proteins expressed in the sporozoite stage of the parasite can be cloned and expressed in bacteria and insect cultures, these recombinant proteins in vaccine form may offer immunity in cattle.

ANIMAL PROPAGATION

Artificial Insemination

Artificial insemination allows genetically desirable animals to be bred more efficiently. Before this method was developed, very few females produced offspring from a single, selected male. Now, artificial insemination allows a diluted sperm sample from one bull to inseminate 500 to 1000 females. This process has been used in the beef and dairy industries for more than 40 years to increase the frequency of desirable characteristics. This method also has been used to increase the genetic diversity of endangered animals in zoos by artificially inseminating a female in one zoo with the sperm of a genetically unrelated male in another zoo.

Animal Clones

Cloning livestock such as sheep and cattle has been practiced commercially for more than 20 years. The two most common methods have been embryo splitting and nuclear transfer. Embryo splitting—also called twinning or cloning—to produce identical twins has been used in cattle breeding (Figure 7.8). Eggs are collected from a select female and fertilized *in vitro*. After the embryo reaches a specific stage of development (8- to 16-cell embryo), the cells are separated from one another. The cells are incubated further and the embryos are transplanted into surrogate mothers. This technique produces two superior animals instead of one and can icrease the frequency of specific traits in an animal population. Medical research has used

FIGURE 7.8 Two different sets of identical Holstein calf twins that were produced by embryo splitting. The two calves on the right came from one embryo, and the ones on the left derived from a different embryo.

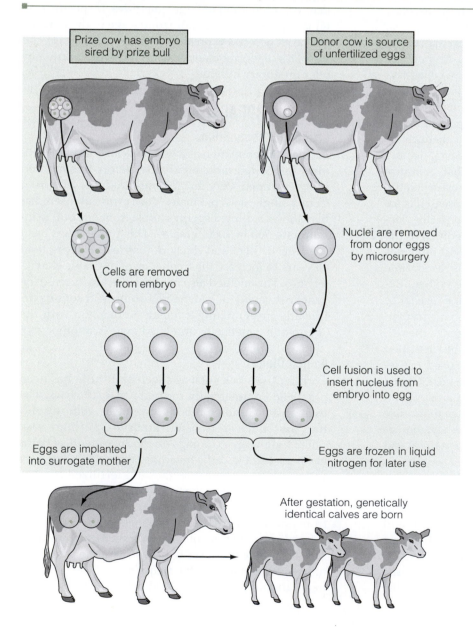

Prize cow has embryo sired by prize bull

Donor cow is source of unfertilized eggs

Cells are removed from embryo

Nuclei are removed from donor eggs by microsurgery

Cell fusion is used to insert nucleus from embryo into egg

Eggs are implanted into surrogate mother

Eggs are frozen in liquid nitrogen for later use

After gestation, genetically identical calves are born

FIGURE 7.9 The production of cloned cows by cell fusion of enucleated donor eggs with cells removed from an embryo. Each resulting egg contains a genetically identical nucleus. Eggs can be frozen for future use or implanted into surrogate mothers.

animal (specifically mice) embryo splitting to genetically manipulate one embryo while the other twin is left untouched as the control. This cloning method was extended to human embryos. In preliminary experiments, cells of very early human embryos were separated and cultured *in vitro*. Although these were developmentally defective embryos and were not transferred to a human host, the potential for increasing the number of eggs in human *in vitro* fertilization is very real. Embryo splitting may eventually increase the number of eggs available to women who produce few eggs during superovulation.

Nuclear transfer methods increase the number of offspring from a female animal to perhaps hundreds or even thousands (Figure 7.9). The first successful animal cloning by nuclear transplantation was conducted in 1986. Sheep eggs were collected, the nuclei removed from each egg, and each enucleated egg was fused with a cell from a 16- to 32-cell embryo. Fused cells were grown into embryos and transplanted into surrogate mothers. Thus, theoretically, a 32-cell embryo can give rise to 32 animal clones.

The first cloning experiments were actually conducted in the 1950s when developmental biologists Robert Briggs and Thomas King developed a method for nuclear transplantation in the leopard frog to study cell differentiation and totipotency. They transplanted a nucleus from different developmental stages—a blastula, a gastrula-stage embryo (a developing embryo beginning cell differentiation), and a neurula (a later embryo developing neural tissue)—into an enucleated egg (ultraviolet irradiation destroys the nucleus). Blastula cell nuclei were totipotent and could direct egg development into a complete frog (Figure 7.10). As cells differentiated in older embryos, the ability to develop into a frog was dramatically reduced. In the 1960s John Gordon extended the nuclear transplantation experiments using the South African clawed frog; however, nuclei from intestinal epithelial cells of tadpoles were used. Approximately 1% of the embryos developed into normal adult frogs. They were clones of the nuclei donor frogs.

Commercialization of cloning will allow desirable traits to be readily maintained and propagated. Cloning has applications in agriculture and medicine. Animal breeders would like to propagate animals with superior traits, such as cows with high milk production or a champion racehorse, and animals that harbor, for example, a gene for a human protein. A prize animal is not identified until it matures; however, to clone such an animal, differentiated cells must be used.

In a landmark experiment in 1996, Scottish scientists Ian Wilmut and Keith Campbell at the Roslin Institute in Edinburgh, Scotland, produced a clone of an adult Dorset sheep using a nucleus from a differentiated cell (somatic cell) rather than from an embryonic

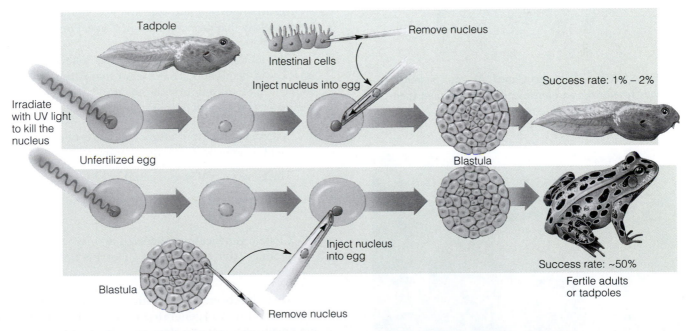

FIGURE 7.10 The demonstration of nuclear totipotency. Amphibian nuclear transfer experiments show that early embryonic cells have the ability to direct the development of a complete frog. However, the rate of success decreases dramatically when nuclei from later developmental stages are used. In the South African clawed frog, nuclei from tadpole epithelial intestine cells are able to direct the development of embryos—1–2% of the embryos develop into a normal frog (top panel). In the leopard frog, embryos produced from nuclei taken from late developing embryonic cells (after gastrula stage) do not develop.

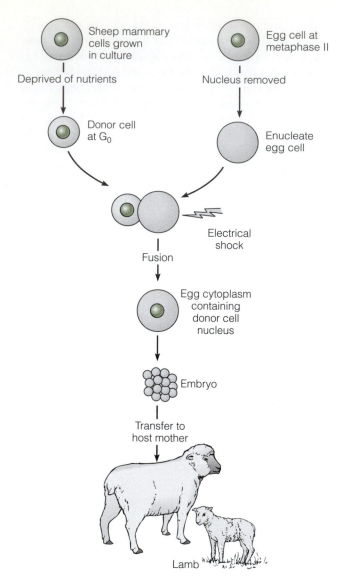

FIGURE 7.11 How Dolly was cloned. An adult mammary cell was fused with an enucleated sheep egg and then implanted into a surrogate mother.

FIGURE 7.12 Dolly the sheep with her surrogate mother. Dolly was cloned by nuclear fusion of an adult cell with an enucleated egg.

cell (Figure 7.11). They named the clone Dolly and in 1997 announced the first animal clone to the world. In this experiment, mature mammary epithelial cells with intact nuclei were taken from the udder of a 6-year-old sheep. These cells were cultured and then starved for certain nutrients to put them into a nondividing, dormant state in the cell cycle (G_o phase) (gene expression was turned off). Eggs were isolated from sheep through superovulation and the nuclei were removed. The nuclei from the cultured udder cells were transferred into the enucleated eggs (by electrofusion, in which an adult cell is fused to an enucleated egg cell), and the cells were cultured. Developing embryos were implanted into surrogate sheep (of a different breed than the donor) that were prepared hormonally for implantation. One sheep gave birth to Dolly (Figure 7.12). The result was a sheep clone of the host that provided the genetic material from its mammary cell. Although Dolly was the only result of 277 cloning attempts, the potential exists for the cloning of animals using the DNA from differentiated cells. The clone was confirmed by the following: the surrogate sheep were a different breed than the donor (from which the nuclei were collected) sheep and Dolly was a Dorset lamb like the donor breed; the DNA fingerprints (genotype characterization) of Dolly matched the donor sheep. Dolly died in 2003 of a virally induced lung cancer commonly found in older sheep kept inside. Being a clone was thought not to result in her death; however, postmortem examination of Dolly's chromosomes revealed the ends of the chromosomes (called telomeres) were shorter than normal and that she was perhaps aging more rapidly than her chronological age. Furthermore, Dolly suffered from arthritis.

This landmark experiment was the first time an animal could be cloned. That meant that differentiated cells could be coaxed into totipotency. It also generated excitement for several reasons

Scientists could now study the effects of environment on genetically identical animals.

The genetics of diseases could be studied.

The genetics of development could be more thoroughly studied.

Cloned, genetically engineered animals could produce large quantities of a therapeutic human protein in their milk.

There has been some controversy that centers on whether Dolly was generated from a fully differentiated mammary cell or a stem cell. Mammary cells contain a variety of cell types, some of which are stem cells, which have the ability develop into other types of cells. Stem cells divide and some remain as stem cells while others differentiate. It is not completely known whether the nucleus that gave rise to Dolly came from a completely differentiated udder cell or a less differentiated stem cell.

The cloning of Dolly was the beginning of animal cloning from mature, differentiated cells. In 1998, scientists cloned more than 24 mice using direct injection of nuclei into enucleated eggs (Figure 7.13). Cloned goats, cows, horses, pigs, mules, rabbits, Bantengs (an endangered Southeast Asian hoofed mammal), and cats, have closely followed. Scientists are working hard to clone nonhuman primates for research purposes, deer—larger bucks with bigger antlers that would appeal to hunters—and endangered animals such as cheetahs and pandas. Other animal clones will most certainly follow.

CONSERVATION BIOLOGY

With the world's human population surpassing six billion and expected to reach more than nine billion in the not-so-distant future, population pressures contribute to increasing pollution, loss and fragmentation of wildlife habitat, species extinction, and extensive environmental degradation. The most effective solutions to these problems include preservation of important habitat, increased control of the exotic animal trade, reestablishment of wild populations using captive animals, a shift in the distribution of wealth in developing countries, and human population control. However, these are not often practical or viable solutions, and other methods to conserve biological diversity must be investigated.

Preservation of genetic diversity presents a formidable challenge. Biotechnology is playing a more important role in conservation as biologists look toward new developments to maintain or even increase captive animal populations. In the past few years, significant advances have been achieved in *in vitro* fertilization; in techniques for embryo transfer, oocyte maturation, and control of the ovarian cycle; and in the production of transgenic laboratory and livestock animals. Controlled and assisted breeding, used in domestic livestock and in human reproduction, is being extended to conservation biology to help preserve endangered animals. Many zoos, for example, have instituted captive breeding programs for endangered species to ensure their *ex situ* survival.

Currently, frozen semen is in widespread use for only a small number of species, but in the future, cryopreservation of gametes and embryos will allow the creation of comprehensive frozen germplasm banks. Material could be collected and preserved to provide optimal genetic diversity in animal populations or to restore lost or severely reduced diversity in such populations. A coordinated international effort would be required to salvage some of the remaining genetic diversity of highly endangered species.

To sustain genetic diversity, careful genetic analysis of populations must be conducted. DNA finger-

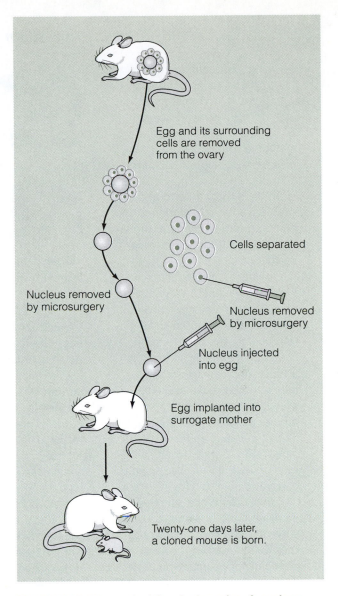

Egg and its surrounding cells are removed from the ovary

Cells separated

Nucleus removed by microsurgery

Nucleus removed by microsurgery

Nucleus injected into egg

Egg implanted into surrogate mother

Twenty-one days later, a cloned mouse is born.

FIGURE 7.13 The method for cloning mice. A nucleus from a differentiated cell is removed and directly injected into an enucleated egg. Developing embryos are transferred to surrogate mothers.

printing (described in Chapter 11, DNA Profiling, Forensics, Other Applications) enables individuals to be readily distinguished and is being used in the management of captive breeding programs. Fingerprinting is becoming an accepted way to resolve genetic differences between closely related individuals in breeding populations and can identify relationships of highly inbred animals.

Embryo Transfer

Embryo transfer has been used in the cattle industry for many years. Females that have been superovulated are artificially inseminated with select bull

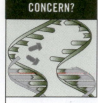
Animal Clones: Promise and Controversy

Cloning animals using cell fusion and microinjection is revolutionizing agriculture and medicine. Sheep, cattle, goats, pigs, horses, and other cloned animals will allow the production of animals with uniform and desirable traits. Animal clones, like plants, afford many opportunities in the agriculture and medical sectors.

Animal cloning is not completely rosy—only a very small percentage of embryos survive, newborn animals often succumb to problems shortly after birth, and cloned animals suffer from arthritis and respiratory and circulatory problems. Other animals are oversized (large-offspring syndrome, a common occurrence in cloned animals) and have physical deformities. Some cloned animals must be euthanized. Thus, with current technology, for every animal that is born and survives, many embryos and animals do not survive. Reports of cloning success rates for a New Zealand biotechnology company in New Zealand, AgResearch, have suggested that almost 25% of calves and lambs cloned from adult animals die within the first 3 months of life compared with 5% of normal calves, and others die after the first 3 months from abnormalities caused most likely by problems with the patterns of gene expression. AgResearch claims that it has achieved a 6% survival rate of cloned embryos through

FIGURE 7.14 Hypothetical human and horse clones. (top) Courtesy of Lisa Starr. (bottom) PhotoDisc/Getty Images.

the first 3 months of life—a success rate that is much higher than its competitors.

We must decide whether the current state of animal cloning is justified by the benefits. What is ethically acceptable? Cloning deer for large antlers? Cloning a prize bull so that the supply for his semen can meet the demand? How do we track the offspring of clones such as this? Most likely the products of cloned animals, such as milk, meat, and cheese, will not be identified as coming from cloned animals.

Cyagra, the cloning company that produced the bull (named Full Flush) clones collectively called the Dream Team, has created more than 100 cloned calves and plans to encourage producers to consider cloning elite animals. They sell biopsy kits for $950 for skin collection required for cloning (using fibroblast cells). Cells are grown in the laboratory until approximately 6 million cells

are obtained, and then they are frozen. The nuclei are removed from 60–70 cells for cloning. Clones cost $19,000 each. Do we want to move in a direction in which, in the future, domesticated animals such as cows and other livestock are identical to one another and animal production is carefully designed and engineered? Will we become so accustomed to cloning and what we call elite animals that it will be easy to move to the next step—cloning humans? In this brave new world of animal cloning, will the reverence for life and the appreciation of the uniqueness (and perhaps not-so-elite characteristics) of living organisms be compromised? Or is this no different from what we have done for thousands of years—select for desirable characteristics to achieve an anticipated result? Perhaps this time we have more powerful tools with which to seek perfection (Figure 7.14).

sperm. Fertilized eggs or embryos are retrieved from the female and frozen in liquid nitrogen at −196° C. These embryos can be implanted into surrogate mothers.

Embryo transfer has been practiced in some zoos in recent years. The embryos can easily be transported to various regions of the world for implantation. Genetic stocks can thus be transferred to regions lacking genetic diversity or requiring new genetic lines of livestock. Because it eliminates the use of natural reproductive processes, embryo transfer can increase the individual's annual production of progeny, and it allows surrogate mothers of a different, although closely related, species to be used. The ability to implant embryos of rare and endangered species in common surrogate hosts is important for the captive reproduction of these species.

The Center for Reproduction of Endangered Wildlife (CREW) at the Cincinnati Zoo in Ohio has used embryo-transfer technology since 1984, when Holstein cows were first used as surrogate mothers for the rare Malaysian ox, or Gaur (Figure 7.15). Holstein cows and elands also have served as hosts for embryos of the endangered bongo antelope (Figure 7.16).

Embryo transfer was extended to natural populations. In Kenya CREW used surrogate eland antelope to reintroduce bongo antelope in the hopes of increasing dwindling wild populations. The success rates of implantation in native animals versus those in captivity could be compared, as could the success rates of native-collected embryos versus captive-collected embryos. The CREW Kenya Project was the first binational conservation effort involving interspecies and intraspecies embryo transfers. In the initial phases of the project, cryopreserved bongo embryos (in liquid nitrogen at −196° C) were transferred from the Cincinnati Zoo to Kenya for implantation into eland surrogates. Cryopreserved embryos collected from wild bongo at the Mt. Kenya Game Ranch in Nanyuki were implanted into both eland and bongo surrogates at the Cincinnati Zoo. Successful transfers using embryo transfer technology under closely monitored conditions provided a knowledge base from which to increase bongos and their genetic diversity in the wild. This was a landmark experiment in which biotechnology complemented conservation biology.

REGULATION OF TRANSGENIC ANIMALS

The commercialization of transgenic animals has elicited more public concern than transgenic plants. The question whether genetically engineered animals should be regulated (and how) has stirred a hotbed of controversy. Several issues have elicited heated debate among consumer groups, concerned citizens, governmental officials, and even scientists. The major fears expressed have been the potential detrimental effect on the environment, the health risk of consuming genetically engineered foods, and the health risk to the engineered animal.

At present, there is little federal control of environmental impacts of transgenic animals. Because few transgenic animals existed in 1988, the federal framework published in that year said little about environmental risks of genetically engineered animals, focusing instead primarily on the risks of using engineered animals as food. In 2003 legislation—Minor Use and

FIGURE 7.15 Male Gaur calf born to a surrogate Holstein.

FIGURE 7.16 Bongo calf born to an eland surrogate.

Minor Species Animal Health Act—was introduced to improve the access of farmers to animal drugs. Because the FDA is planning to regulate genetically engineered animals using the same authority it has for regulating animal drugs, critics (health, consumer, and environmental groups) have expressed concerns that transgenic animals (for example, fast-growing salmon) would be marketed more quickly and without an extensive government review, an unintended, but potential consequence of the proposed bill.

PATENTING GENETICALLY ENGINEERED ANIMALS

Winning approval for the patenting of genetically engineered animals is a relatively recent victory for biotechnology companies in what has been a long struggle to obtain patents on living organisms. In 1930, the U.S. Congress passed the Plant Patent Act, which allowed patents to be granted for innovations in horticulture involving asexually propagated plant varieties; 40 years later sexually propagated plants became patentable. In 1980, after numerous appeals, the U.S. Supreme Court delivered a landmark decision stating that an "invention" was not unpatentable simply because it was a living organism (*Diamond v. Chakrabarty*). Finally, on April 7, 1987, Donald J. Quigg, the commissioner of patents and trademarks, announced that the Patent and Trademark Office considered nonnaturally occurring nonhuman multicellular organisms (which of course includes animals) to be patentable.

On April 13, 1988, Harvard University obtained patent number 4,736,866 for a genetically altered mouse used as a model for studying how genes contributed to breast cancer. The Harvard mouse set off an emotional debate. Although patenting living organisms is not new—plant varieties, lower animals, and microorganisms have been patented—the idea of patenting mammals has far-reaching implications and has caught the attention of animal-rights groups, farmers, and consumer groups.

The issue of patenting genetically engineered animals has been addressed by lawmakers seeking ways to protect **intellectual property**, which includes patents, copyrights, trade secrets, and trademarks. The patent debate has been fueled by dramatic developments in biotechnology in the past 15 years: recombinant DNA technology, cell fusion methods, monoclonal antibody technology, animal biofactories, and animal cloning. Although biotechnology, broadly defined, is very old (as we saw in Chapter 1, Biotechnology: Old and New), such recent innovations have ignited the controversy over patenting living organisms. Much of the debate has focused on patenting animals.

The debate has extended beyond the appropriateness of animal patents to encompass questions about the consequences of commercial uses of patented organisms and about the merits of the technology itself. Reluctance to embrace animal patents arises from myriad concerns. These include the environmental implications, the welfare of the engineered animal, and the ethics of creating a novel living organism and changing the course of evolution. Just a few of the many objections raised against animal patents are excessive interference in the natural world; the increased suffering of animals in the research, industrial, and agricultural sectors; a decline in genetic diversity in commercialized animal species; and the demise of traditional family farms. In the past, farmers have been especially concerned about the long-term implications of animal patents. New technologies and patented livestock may accelerate the trend toward larger and fewer farms: small farmers are not equipped to adopt new, patented biotechnologies, nor could most afford them. The high cost of using patented animals could erode the profit margins of many farmers. Unable to compete effectively with larger operations that can adapt to biotechnological developments, small farms will decline still further in number, and fewer, larger farms will remain. However, high-tech farming—for example, animals that produce specific human proteins—may provide new opportunities (for example, specialty products) for small farms.

We are witnessing a tremendous increase in the number of patented products, both living and nonliving, resulting from modern biotechnology. The open discussion of ethical, social, and legal implications is required to ensure that the end products of modern technology contribute in a positive way toward a better world.

General Readings

E.T. Bloom. 2001. Xenotransplantation: Regulatory challenges. *Curr. Opin. Biotechnol.* 12:312–316.

M.H. Hofker and J. van Deursen, eds. 2003. *Transgenic Mouse Methods and Protocols.* Humana Press, Totowa, New Jersey.

L.-M. Houdebine. 2003. *Animal Transgenesis and Cloning.* Wiley, Hoboken, New Jersey.

National Research Council. 2002. *Animal Biotechnology: Science Based Concerns.* The National Academies Press, Washington, D.C.

National Research Council. 2002. *Scientific and Medical Aspects of Human Reproductive Cloning.* The National Academies Press, Washington, D. C.

Additional Readings

M.R. Capecchi. 1994. Targeted gene replacement. *Sci. Am.* 270:52–59.

D.-Y. Chen et al. 2002. Interspecies implantation and mitochondria fate of panda-rabbit cloned embryos. *Bio. Reprod.* 67:637–642.

S.-H. Chen et al. 2002. Efficient production of transgenic cloned calves using preimplantation screening. *Bio. Reprod.* 67:1488–1492.

J.G. Cloud. 1990. Strategies for introducing foreign DNA into the germ line of fish. *J. Reprod. Fertil. Suppl.* 41:107–118.

L.V. Cundiff, M.D. Bishop, and R.K. Johnson. 1993. Challenges and opportunities for integrating genetically modified animals into traditional animal breeding plans. *J. Anim. Sci.* 71 (Suppl. 3):20–25.

S. Damak, H. Su, N.P. Jay, and D.W. Bullock. 1996. Improved wool production in transgenic sheep expressing insulin-like growth factor 1. *Bio/Technology* 14:185–188.

P.L. Davies, C.L. Hew, M.A. Shears, and G.L. Fletcher. 1990. Antifreeze Protein Expression in Transgenic Salmon. In R.B. Church, ed., *Transgenic Models in Medicine and Agriculture*. Wiley-Liss, New York, pp. 141–161.

R.H. Devlin et al. 1994. Extraordinary salmon growth. *Nature* 371:209–210.

P. DiTullio, S.H. Cheng, J. Marshall, R.J. Gregory, K. Ebert, H.M. Meade, and A.E. Smith. 1992. Production of cystic fibrosis transmembrane conductance regulator in the milk of transgenic mice. *Bio/Technology* 10:74–77.

K.M. Ebert. 1989. Gene transfer through embryo microinjection. In L.A. Babiuk, J.P. Phillips, and M. Moo-Young, eds. *Animal Biotechnology*. Pergamon Press, Oxford, pp. 233–250.

T.A. Ericsson et al. 2003. Identification of receptors for pig endogenous retrovirus. *Proc. Natl. Acad. Sci. USA* 100:6759–6764.

P. Giraldo and l. Montoliu. 2001. Size matters: Use of YACs, BACs, and PACs in transgenic animals. *Transgenic Res.* 10:83–103.

Z. Gong and C.L. Hew. 1995. Transgenic fish in aquaculture and developmental biology. *Curr. Top. Dev. Biol.* 30:177–214.

J. Guenet. 1995. Animal models of human genetic disease. In M.A. Vega, ed., *Gene Targeting*. CRC Press, Inc., Boca Raton, Florida, pp. 149–154.

L.M. Houdebine. 2000. Transgenic animal bioreactors. *Transgenic Res.* 9:305–320.

P.J. Hines, B.A. Purnell, and J. Max. 2000. Stem cells branch out. *Science* 287:1417.

J.T. Hune and R.W. Doms. 2000. Closing in on the amyloid cascade: Recent insights into the cell biology of Alzheimer's disease. *Mol. Neurobiol.* 22:81–98.

J. Kahana. 2001. A new BACE therapy for Alzheimer's disease. *Trends Cell Biol.* 11:192.

Y. Kato, T. Tani, Y. Sotomaru, K. Kuokawa, J. Kato, H. Doguchi, H. Yasue, and Y. Tsunoda. 1998. Eight calves cloned from somatic cells of a single adult. *Science* 292:2095–2098.

C. Kubota, H. Yamakuchi, J. Todoroki, K. Mizoshita, N Tabara, M. Barber, and X. Yang. 2000. Six cloned calves produced from adult fibroblast cells after long-term culture. *Proc. Natl. Acad. Sci. USA* 97:990–995.

N.G. Levinsky. 2001. Xenotransplantation: An Ethical Dilemma. In N.G. Levinsky, ed. *Ethics and the Kidney*. Oxford University Press, New York.

J.F. Loring. 1999. Designing animal models of Alzheimer's disease with amyloid precursor protein (APP) transgenes. *Meth. Mol. Med.* 32:249–270.

D. Malakoff. 2000. The rise of the mouse, biomedicine's model mammal. *Science* 288:248.

A. McLaren. 2000. Cloning: Pathways to a pluripotent future. *Science* 288:1775.

C. Michie. 2001. Xenotransplantation, endogenous pig retroviruses and the precautionary principle. *Trends Mol. Med.* 7:62–63.

H.D.M. Moore. 1992. *In Vitro* Fertilization and the Development of Gene Banks for Wild Animals. In H.D.M. Moore, W.V. Holt, and G.M. Mace, eds. *Biotechnology and the Conservation of Genetic Diversity*. Oxford University Press, Oxford, pp. 89–99.

M. Müller and G. Brem. 1994. Transgenic strategies to increased disease resistance in livestock. *Reprod. Fertil. Dev.* 6:605–613.

H. Niemann and W.A. Kues. 2000. Transgenic livestock: Premises and promises. *Anim. Reprod. Sci.* 60-61:277–293.

E. Notarianni and M.J. Evans. 1992. Transgenesis and Genetic Engineering in Domestic Animals. In J.A.H. Murray, ed. *Transgenesis*. Wiley, New York, pp. 251–281.

K.W. O'Connor. 1993. Patents for genetically modified animals. *J. Anim. Sci.* 71(Suppl. 3):24–40.

A. Onishi, M. Iwamoto, T. Akita, S. Mikawa, K. Takeda, T. Awata, H. Hamada, and A.C. Perry. 2000. Pig cloning by microinjection of fetal fibroblast nuclei. *Science* 289:1188–1190.

R.D. Palmiter and R.L. Brinster. 1986. Germ-line transformation of mice. *Annu. Rev. Genet.* 20:465–499.

J.L. Platt, ed. 2002. *Xenotransplantation: Basic Research and Clinical Applications*. Humana Press, Totowa, New Jersey.

D.P. Pollock, J.P. Kutzko, E. Birck-Wilson, J. L Williams, Y. Echelard, and H.M. Meade. 1999. Transgenic milk as a method for the production of recombinant antibodies. *J. Immunol. Meth.* 231:147–157.

D.J. Porteous and J.R. Dorin. 1993. How relevant are mouse models for human diseases to somatic gene therapy? *Trends Biotechnol.* 11:173–181.

D. Powell. 1995. Safety in the Contained Use and Release of Transgenic Animals and Recombinant Proteins. In G.T. Tzotzos, ed. *Genetically Modified Organisms: A Guide to Biosafety*, The University of Arizona Press, Tucson.

D.A. Powers, T.T. Chen, and R.A. Dunham. 1992. Transgenic Fish. In J.A.H. Murray, ed. *Transgenesis*. Wiley, Chichester, pp. 233–249.

R.S. Prather, F.L. Barnes, M.M. Sims, J.M. Robl, W.H. Eyestone, and N.L. First. 1987. Nuclear transplantation in the bovine embryo: Assessment of donor nuclei and recipient oocyte. *Biol. Reprod.* 37:859–866.

V.G. Pursel and C.E. Rexroad, Jr. 1993. Status of research with transgenic farm animals. *J. Anim. Sci.* 71 (Suppl. 3):10–19.

R. Renaville and A. Burny, eds. 2001. *Biotechnology in Animal Husbandry*. Kluwer Press, Boston, Massachusetts.

H. Sang. 1994. Transgenic chickens: Methods and potential applications. *Trends Biotechnol.* 12:415–420.

R.M. Shuman. 1991. Production of transgenic birds. *Experientia* 47:897–905.

C. Stewart. 1997. Nuclear transplantation. An udder way of making lambs. *Nature* 385:769–771.

A. Teale. 1993. Improving control of livestock diseases. *Bioscience* 43:475–483.

P.B. Thompson. 1993. Genetically modified animals: Ethical issues. *J. Anim. Sci.* 71 (Suppl. 3):51–56.

A.S. Waldman. 1995. Molecular Mechanisms of Homologous Recombination. In M.A. Vega, ed. *Gene Targeting,* CRC Press, Inc., Boca Raton, Florida, pp. 45–64.

I. Wilmut, A.L. Archibald, M. McClenghan, J.P. Simons, C.B.A. Whitelaw, and A.J. Clark. 1991. Production of pharmaceutical proteins in milk. *Experientia* 47:905–912.

I. Wilmut, A.E. Schnieke, J. McWhir, A.J. Kind, and K.H. Campbell. 1997. Viable offspring derived from fetal and adult mammalian cells. *Nature* 385:810–813.

I. Wilmut and C.B. Whitelaw. 1994. Strategies for production of pharmaceutical proteins in milk. *Reprod. Fertil. and Dev.* 6:625–630.

D.P. Wolf and M. Zelinski-Wooten, eds. 2001. *Assisted Fertilization and Nuclear Transfer in Mammals.* Humana Press, Totawa, New Jersey.

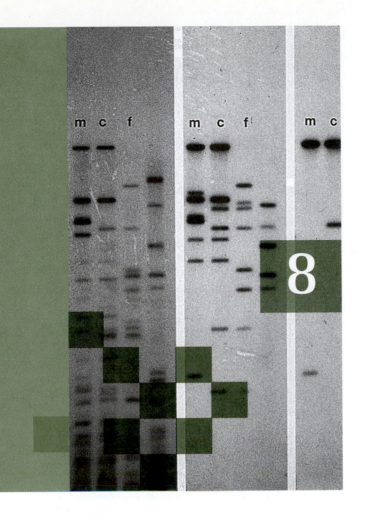

MARINE BIOTECHNOLOGY

8

The Office of Technology Assessment defined marine biotechnology in 1991 as "any technique that uses living marine organisms (or parts of these organisms) to make or modify products, to improve plants or animals, or to develop microorganisms for specific uses." The chief traditional biotechnologies have been **aquaculture** and the isolation of natural products from marine organisms for commercial applications. Few such products have been developed, perhaps in part because marine biological diversity is vast, and most of it remains unexplored. Modern marine biotechnology, although still in its infancy, encompasses such new technologies as protein engineering, recombinant DNA technology, and hybridoma and monoclonal production using marine organisms or their components.

The marine environment covers almost 71% of the earth's surface and contains 97% of the available water. The oceans play an important role in global ecology; they regulate global climate, moderate fluctuation of temperatures, produce a third of the global supply of oxygen, regulate levels of carbon dioxide, and recycle 90% of the world's carbon in the marine food web. A vast number of people depend on the biological diversity in the oceans. Approximately 100 million tons of seafood is harvested each year to feed a growing world population. The global marine harvest accounts for 16% of animal-protein consumption and is especially important in developing countries. In Asia alone, fish is a main protein staple for more than one billion people. The world population is expected to almost double by 2025. Eventually, there may not be enough arable land or energy available to grow enough grain and raise enough livestock and poultry to sustain so many people. Fish may play a larger role in feeding this population (although we are placing this role at risk by rapidly depleting and polluting the oceans).

The oceans are an ancient ecosystem where life, in the form of bacteria, originated about four billion years ago. Marine organisms provide us with a valuable gene pool that is beginning to be tapped. To ensure a healthy and stable marine food web, pollution must be controlled, threatened and endangered species must be protected (and perhaps organisms cultured that are in high commercial demand), and the link between the ocean's primary producers (that is, plankton) and other organisms must be understood.

The oceans offer a wealth of untapped resources for research and the development of products (including medical). The vast majority of marine organisms have not been identified (the majority are microorganisms) and little is known about many of the organisms already identified. Marine organisms are of great interest to scientists and industry for two important reasons: (1) They represent a major proportion of the Earth's organisms. Most major groups of organisms on Earth are either mostly or completely marine. There is much genetic and metabolic information to be obtained from these organisms. (2) Marine organisms have unique metabolic pathways and other adaptive functions, such as sensory and defense mechanisms, reproductive systems, and physiological processes, that allow them to live in extreme environments, from cold Arctic seas ($-2°$ C) to very hot hydrothermal vents to pressures at great depths.

In recent years increasing funds have been devoted to marine biotechnology research; the United States, for example, spent some $55 million in 1995. Unfortunately, only approximately 2% of federal funds devoted to biotechnology have been earmarked for marine biotechnology and aquaculture. Although funds are limited, research has been promising. Additional funding will allow for the more rapid development of new technologies and pharmaceuticals, increased cultivation of aquatic organisms, and improvement of bioremediation processes.

Five primary research areas have been the major focus in marine biotechnology:

1. Identifying bioactive compounds produced by marine organisms and elucidation of their function and mode of action. The results of this research will lead to the production of new drugs in the medical field and industrial chemicals.

2. Increasing our understanding of the environmental factors, nutritional requirements, and genetic factors that control the production of primary and secondary metabolites in marine organisms. This will aid in the isolation of compounds and the production of new bioproducts.

3. Understanding the genetics, biochemistry, physiology, and ecology of marine organisms. This knowledge will aid in the identification of new research projects and contribute to marine conservation strategies.

4. Developing tools and diagnostics for the improvement of health, reproduction, growth, and cultivation of marine and freshwater organisms. This area of biotechnology will enhance the aquaculture industry.

5. Developing bioremediation tools for better waste processing and disposal, the clean-up of coastal areas, and the remediation of oil spills. This will help improve the cultivation of economically important marine organisms and increase desirable fish populations.

AQUACULTURE

In many regions of the world overfishing has dangerously reduced fish and **shellfish** stocks, and some areas now are closed to commercial fishing. People are therefore beginning to look to aquaculture: the propagation of aquatic animals and plants at high densities in fresh, brackish, or salt water. Salt water, or marine, culture (called **mariculture**) and freshwater aquaculture use similar methods. It has been projected that the worldwide seafood demand will increase by 70% in the next 35 years, and that with the decline in fisheries, aquaculture must produce seven times as much seafood as is currently produced to meet the global demand.

Aquaculture has been practiced throughout the world for thousands of years. Ancient aquaculture, which was primarily freshwater, began in the Far East; Chinese aquaculture dates back probably at least 3000 years. The earliest known record, in 473 BC, reports monoculture of the common carp. As methods were refined, polyculture was practiced, combining species with different food preferences; ponds contained varieties of finfish, shellfish, or **crustaceans**.

Today, high density aquaculture is practiced in many countries. In the Philippines, for example, the giant tiger prawn, *Penaeus monodon,* is cultured in ponds at densities of 100,000 to 300,000 prawns per hectare. Significant advances have been made in the culturing of oysters, clams, abalone, scallops, brine shrimp, blue crab, and a variety of finfish.

Traditional aquaculture encompasses low input technologies to propagate, harvest, and market fish, algae, crustaceans, and **mollusks**. An aquaculturist must know the chemical composition of the soil where the pond will be situated, the amount and type of water to be used, and the type and quantity of animal feed necessary for maximum production. Unfortunately, aquaculture ponds have the potential of destroying delicate habitats through pollution by wastes or by decimating, for example, mangroves to make room for shrimp ponds. Biotechnology may be able to help solve some of the problems created by intensive aquaculture practices.

Marine and freshwater biotechnologies are used today to increase the yield and quality of finfish, crustaceans, algae, and various **bivalves** such as clams and oysters. Recently, transgenic organisms with desired characteristics have been generated for future commercial production. Modern Japanese mariculture is more productive than freshwater aquaculture and has recently accounted for more than 92% of Japan's total aquaculture yield. The major food products have been the Japanese oyster, *Crassostrea gigas,* and a red alga called nori (*Porphyra*). Nori provides the largest seaweed harvest. Other products include the fish yellowtail jack, the red sea bream, as well as a scallop and two types of brown algae (wakame and kombu).

Although the primary products of aquaculture have been food for human and animal consumption, it has also yielded food supplements, natural products, pharmaceuticals and medicines, ornamental jewelry, and ornamental fish for home aquariums. Production of cultured organisms has increased dramatically in re-

cent years (from almost 10 million tons in 1985 to 14 million tons in 1993, with projections of more than 25 million tons by the year 2005). The United Nations Food and Agricultural Organization (FAO) predicts that by the end of the century products from aquaculture will account for 20–25% of the world's fisheries production by weight. The world fisheries have virtually reached the maximum sustainable yield of 100 million tons per year, and in some regions the fisheries have collapsed—which means that aquaculture will become increasingly important. According to a 1991 FAO report, the wild marine and freshwater fish catch accounts for 82% and 6% of the world total, respectively, while freshwater and marine aquaculture make up 7% and 5%, respectively, of the total fish catch.

Improved hatchery methods would increase the production of threatened or endangered fish and shellfish (examples include striped bass, oysters, and scallops), thereby enhancing marine fisheries for both recreation and commercial use. Improving marine fisheries could benefit other species that are used extensively for human consumption or sport, such as snook, tarpon, haddock, halibut, cod, red snapper, and flounder (Table 8.1).

The by-products of fish culture and agriculture are often used complementarily: pond humus can be used to fertilize fields, and livestock waste can serve as pond fertilizer to stimulate the growth of plankton, which is food for the fish. Crop by-products also can be used as fish feed. Anaerobic digesters sometimes are used to produce methane gas as an energy source from aquacultural and agricultural wastes.

Gastropod, Bivalve, and Crustacean Production

The high worldwide demand for clams, oysters, mussels, abalone, crabs, shrimp, and lobsters favors the development of efficient culturing methods to increase their production. A variety of methods exist, from rearing in tanks to using floating platforms and substrates for the growth of organisms (Figure 8.1). Genetic manipulation in culture promotes faster growth and maturation, increased disease resistance, and triploidy. Normal **diploid** oysters spawn in the summer and lose their flavor because they form a massive amount of reproductive tissue. Instead of the two sets of chromosomes that diploids have, triploid Pacific oysters have three (two from the female and one from the male). **Triploid** oysters are obtained in culture by treating eggs with cytochalasin B. This inhibitor of normal cell division doubles the number of chromosomes so that, when the eggs are fertilized with normal sperm, zygotes have three sets of chromosomes. Because triploid oysters are sterile and do not form reproductive organs, they are more flavorful and meatier year round. They also grow larger and faster than diploid oysters and can be harvested sooner. Triploid oysters represent a significant portion of the total oyster production in the

TABLE 8.1 Examples of Aquacrops (all have potential for culture in the U.S.)

Common Name (scientific name)	Culture and Importance
Alligator (*Alligator mississippiensis*)	Grown in southern U.S., primarily Louisiana and Texas; flesh is used for food and the skin is used for leather.
Bullfrog (*Rana catesbiana*)	Cultured for food and laboratory use; mostly imported to U.S.
Carp (common) (*Cyprinus carpio*)	Many nations produce and consume; not particularly popular in the U.S.
Carp (grass or white amur) (*Ctenopharyngodon idellus*)	Consumes huge amounts of vegetation; produced in China.
Catfish (channel) (*Ictalurus punctatus*)	Widely grown in southern U.S.
Chinese waterchestnut (*Eleocharis dulcis*)	Plant grown for corms; production is experimental in the U.S.; popular and successful in China; imported food to U.S.
Crawfish (red swamp) (*Procambarus clarkii*)	Primarily grown in Louisiana and Texas for food.
Eel (Japanese) (*Anguilla japonica*)	Popular food in Italy, Japan, and other countries; cultured in Taiwan, Japan, and other countries.
Goldfish (*Carassius auratus*)	Popular ornamental fish in U.S.; grown in Arkansas and Missouri.
Freshwater prawn (*Macrobrachium rosenbergii*)	Experimental in U.S.; successful production in Asian and other countries.
Minnow (fathead) (*Pimephales promelas*)	Popular baitfish in U.S.; grown in Arkansas and other areas.
Oyster (American) (*Crassostrea virginica*)	Limited culture success in Chesapeake Bay area and Eastern Shore of U.S.; successful culture in Japan, Taiwan, and other areas.
Salmon (chinook) (*Oncorhynchus tshawytscha*)	Popular food fish; harvested in several locations.
Shrimp (Chinese white) (*Penaeus chinensis*)	China is world leader in shrimp production; other species produced elsewhere.
Striped bass (*Morone saxatilis*)	Thought to hold much potential for culture, particulary its hybrids, still experimental.
Tilapia (blue) (*Tilapia aurea*)	Grown in warm water or intensive systems; sometimes considered a pest.
Trout (rainbow) (*Salmo gairdneri*)	Popular; produced in northern U.S.
Watercress (*Nasturtium officinale*)	Grown in temperate and subtropical climates, such as in Hawaii; the only aquatic foliage consumed on a regular basis in U.S. in raw salads or cooked.

Compiled from J. E. Bardach, J. H. Ryther, and W. O. McLarney, *Aquaculture: The Farming and Husbandry of Freshwater and Marine Organisms*. New York: John Wiley & Sons, Inc., 1972; H. K. Dupree and J. V. Huner, *Third Report to the Fish Farmers*, Washington D.C.: U.S. Fish and Wildlife Service, 1984; B. Rosenberry, *World Shrimp Farming*, San Diego, *Aquaculture Digest*, 1990; and S. W. Waite, B. C. Kinnett, and A. J. Roberts, *The Illinois Aquaculture Industry: Its Status and Potential*, Springfield, State of Illinois, Department of Agriculture, 1988.

J. S. Lee and M. E. Newman, *Aquaculture: An introduction*, Table 1-1. Interstate Publishers, Inc., 1992. Used by permission. See this book for original references.

United States (making up almost 50% of the Pacific Northwest oyster hatchery). Concerns about the safety of using cytochalasin B may soon lead to the production of triploid oysters by mating tetraploids (having four sets of chromosomes) and normal diploid oysters.

Bivalves such as oysters and **gastropods** such as abalone (which has a commercial value of $20 to $30 per pound) are cultured by the manipulation of the reproductive cycle. Hydrogen peroxide added to the seawater induces synthesis of prostaglandin, a type of hormone that triggers spawning. Larvae are then induced to settle by the addition of the amino acid aminobutyric acid (GABA), an important neurotransmitter in animals. Larvae exposed to GABA settle onto the substrate and begin developmental metamorphosis and cellular differentiation. Growth-accelerating genes are being identified and cloned for the eventual production of compounds from these genes, which are important in the efficient, controlled culturing of abalone and other

shellfish. Productivity is also being increased by crossing specific genetic lines to increase, for example, the growth rate of oysters (by up to 40%). The use of recombinant fish growth hormone may also produce faster-growing shellfish.

MARINE ANIMAL HEALTH

Biotechnology provides opportunities for improvement of the health of cultivated aquatic organisms (and to prevent the transfer of diseases from cultivated and wild stocks). More than 50 diseases affect fish and shellfish and cause millions of dollars in losses each year in the United States. Like terrestrial animals, marine animals are susceptible to **protozoan**, bacterial, viral, and fungal diseases. Marine pollution, which is increasing dramatically worldwide, often promotes growth of **pathogens** (such as the fungus *Saprolegnia* and the common marine bacterial pathogen *Vibrio*) that infect ma-

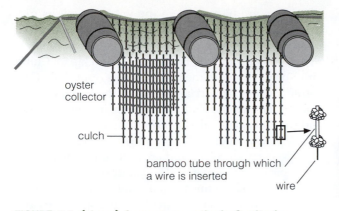

FIGURE 8.1 (above) A common method of culturing oysters: A raft floated by buoys holds strings of culches with oyster larvae attached. (right) Aquaculture facility where shrimp are cultured in fiberglass tanks. An enclosure similar to a greenhouse helps regulate the temperature of the aquatic environment.

rine organisms. Ninety percent of the marine catch comes from coastal waters, which are the areas most susceptible to pollution.

Potential benefits from biotechnology include the early detection of diseases, the understanding of organisms' susceptibility to disease and pathogens, the development and application of new antibiotics and vaccines, and the development of pathogen-free stocks.

High densities of animals in aquaculture also increase the chance of disease. In fact, disease contributes to a significant loss of production each year, with devastating economic consequences. Shrimp aquaculture provides an important source of revenue to many Asian and Latin American countries, primarily through export. For example, because of a virus that decimated black tiger shrimp, Taiwan's production of these shrimp dropped from 114,000 metric tons in 1987 to 50,000 in 1988 and 30,000 in 1991 (of a world total, in 1991, of 690,100 metric tons). At this time there are no vaccines to protect shrimp and other shellfish. If disease-resistant organisms could be cultivated, they would lower production costs and dramatically increase the amount of biomass available for human consumption.

Methods of bacterial control include the use of antibacterial agents such as disinfectants and antibiotics (tetracycline, chloramphenicol, penicillin, etc.) in ponds or culturing pens. However, the heavy use of antibiotics—Norway, for example, used 8000 kilograms in salmon aquaculture facilities in 1987—poses an ultimate risk to human health: Antibiotic resistance could be transferred from bacteria in culturing facilities to human bacterial pathogens that cause serious diseases such as typhoid, dysentery, and cholera, normally treated with antibiotics. Antibiotic residues also could

remain in the fish, crustacean, or bivalve consumed by humans.

Salmon in aquaculture facilities sometimes become infected with viruses. Two of the most deadly viral diseases are infectious hematopoietic necrosis (IHN) and infectious pancreatic necrosis (IPN). IHN was first found in 1953, when a tremendous number of salmon died in Washington State; it has since spread worldwide. Protozoans such as flagellates and ciliates may not kill the fish but can cause extensive damage by feeding on parts of the fish so that it can no longer be used for food. Viral diseases pose a particularly difficult problem for fish aquaculture facilities. Infected animals often become carriers and must be destroyed. Sometimes all the residents of a pond must be destroyed and the pond decontaminated. The lost revenue is recovered through higher prices to the consumer. Efficacious fish vaccines would greatly benefit the aquaculture industry, but very few are available. Recently, scientists have developed a recombinant vaccine against IHN that protects both commercial trout and salmon stocks from the deadly virus. The vaccine is based on a cloned subunit of the viral protein coat that is produced in bacteria. The vaccine is being tested in large-scale field trials.

Because few marine disease-causing organisms have been identified, additional research is needed before vaccines can be generated. Moreover, fish are not easily vaccinated. The usual method is injection or immersion in water containing the vaccine. However, the aquatic environment dilutes concentrations. A new method involves immersing fish and using ultrasound for approximately 10 minutes, which somehow facilitates entry of the vaccine into the fish.

Researchers are seeking a vaccine that is effective against IHN. Such a vaccine might also protect wild populations of salmon that are susceptible to this disease. Several types of vaccines have been tested that confer resistance. A recombinant vaccine, produced by isolating and expressing genes encoding viral proteins, induces the formation of antibodies in fish. The development of new vaccines and efficient delivery systems will enhance productivity from the egg through larval stages (a risky part of the life cycle). Marine metabolites and other constituents also may prove to be invaluable in the fight against disease. For example, a shellfish extract has been shown to increase immunity against infection in blue crabs and prawns as well as to protect eels from *Aeromonas* infection.

Biotechnology can also be used against the spread of disease in aquaculture through the development of diagnostics. If diseases can be detected early, before they spread, timely preventive measures can be taken and infected organisms can be removed.

ALGAL PRODUCTS

Algae are a diverse group of photosynthetic eukaryotes that are used throughout the world for various products, including food. Of the two types of algae, microalgae and macroalgae, the latter is the more economically important. The three major groups of eukaryotic macroalgae are green (Chlorophyta), red (Rhodophyta), and brown (Phaeophyta).

Algae are harvested in the wild as well as produced in culture. The world algae harvest is about four million tons a year, which is worth approximately $1 billion. Most of the production is from Japan, China, and Korea, although the United States (especially California, where brown algae, or kelp, is harvested), Canada, France, the United Kingdom, Indonesia, Chile, and the Philippines also contribute.

Wild seaweeds were collected for food and medicines as early as 900 BC and are still being collected in some parts of the world. The California kelp forests (Macrocystis, sometimes called giant kelp because they can grow to 30 m) have been harvested since the early 1900s (Figure 8.2). Kelp was used as fertilizer and as a source of potash and acetone for the production of explosives. The Kelco Company of San Diego, founded in 1929, was the world's first producer of algin products; the first product was kelp meal for livestock feed. Later, algin was extracted. Today approximately 70 algin products are manufactured for many applications (Table 8.2).

Kelp has been used for many years as a food supplement and is an important source of potassium, iodine, and other essential minerals; carbohydrates; and vitamins. Alginates are the main structural components of the cell wall and intercellular matrix of brown seaweeds. They are used as food thickeners and stabilizers (they retain moisture and assure smooth texture and uniform thawing of frozen foods). Algin is added to desserts, dairy products (just check your favorite ice cream), canned foods, salad dressings, cake mixes, and

Dale Glantz

Dale Glantz

a

b

FIGURE 8.2 (a) Sunlight filters through a giant kelp forest. One of the fastest-growing plants, giant kelp can grow as much as two feet a day (Coronado Islands, Baja California). (b) Large ships like this kelp harvester prune the kelp canopy, leaving enough submerged fronds to grow new canopy (Cedros Island, Baja California).

TABLE 8.2 Some Applications of Alginates in Food and Industrial Products

Property	Product	Performance
Food Applications		
Water-holding	Frozen foods	Maintains texture during freeze-thaw cycle.
	Pastry fillings	Produces smooth, soft texture and body.
	Syrups	Suspends solids, controls pouring consistency.
	Bakery icings	Counteracts stickiness and cracking.
	Dry mixes	Quickly absorbs water or milk in reconstitution.
	Meringues	Stabilizes meringue bodies.
	Frozen desserts	Provides heat-shock protection, improved flavor release, and superior meltdown.
	Relish	Stabilizes brine, allowing uniform filling.
Gelling	Instant puddings	Produces firm pudding with excellent body and texture; better flavor release.
	Cooked puddings	Stabilizes pudding system, firms body, and reduces weeping.
	Chiffons	Provides tender gel body that stabilizes instant (cold make-up) chiffons.
	Pie and pastry fillings	Cold-water gel base for instant bakery jellies and instant lemon pie fillings. Develops soft gel body with broad temperature tolerance; improved flavor release.
	Dessert gels	Produces clear, firm, quick-setting gels with hot or cold water.
	Fabricated foods	Provides a unique binding system that gels rapidly under a wide range of conditions.
Emulsifying	Salad dressings	Emulsifies and stabilizes various types.
	Meat and flavor sauces	Emulsifies oil and suspends solids.
Stabilizing	Beer	Maintains beer foam under adverse conditions.
	Fruit juice	Stabilizes pulp in concentrates and finished drinks.
	Foutain syrups and toppings	Suspends solids, produces uniform body.
	Whipped toppings	Aids in developing overrun, stabilizes fat dispersion, and prevents freeze-thaw breakdown.
	Sauces and gravies	Thickens and stabilizes for a broad range of applications.
	Milkshakes	Controls overrun and provides smooth, creamy body.
Industrial Applications		
Water-holding	Paper coating	Controls rheology of coatings; prevents dilatancy at high shear.
	Paper sizings	Improves surface.
	Adhesives	Controls penetration to improve adhesion and application.
	Textile printing	Produces very fine line prints with good definition and excellent washout.
	Textile dyeing	Prevents migration of dyestuffs in pad dyeing operations. (Algin is also compatible with most fiber-reactive dyes.)
Gelling	Air freshener gel	Firm, stable gels are produced from cold-water systems.
	Explosives	Rubbery, elastic gels are formed by reaction with borates.
	Toys	Safe, nontoxic materials are made for impressions or puttylike compounds.
	Hydro-mulching	Holds mulch to inclined surfaces, promotes seed germination.
	Boiler compounds	Produces soft, voluminous flocs easily separated from boiler water.
Emulsifying	Polishes	Emulsifies oils and suspends solids.
	Antifoams	Emulsifies and stabilizes various types.
	Lattices	Stabilizes latex emulsions, provides viscosity.
Stabilizing	Ceramics	Imparts plasticity and suspends solids.
	Welding rods	Improves extrusion characteristics and green strength.
	Cleaners	Suspends and stabilizes insoluble solids.

Courtesy of The NutraSweet Kelco Company, a unit of Monsanto Company.

FIGURE 8.3 Nori has been cultivated as a food source for hundreds of years.

FIGURE 8.5 Wakame is cultured in Japan and China and has a large market value.

even beer for foam stabilization. Their industrial applications include paper coatings and textile printing, and they are used in pharmaceuticals (for example, antacids, pill coatings, and capsules) and cosmetics.

The red alga Porphyra, or nori, has been cultured as a food source in Japan since 1570 (Figure 8.3). The culturing process relied on the natural dispersal of spores (actually, propagules called conchospores) from wild populations in the ocean. In the beginning, the substrate for algal spores was plant debris placed in the ocean; much later, horizontal nets were used (Figure 8.4). The spores germinate and grow into fronds that are harvested, dried, and processed. A significant amount of research has been devoted to this economically important alga, and much is known about spore formation, storage of inoculum, strain improvement, disease resistance, and nutrient composition. With the elucidation of the *Porphyra* life cycle, large sheets of nori are produced from controlled seeding of culture nets.

Other algae have been similarly cultured. The brown algae *Undaria* (wakame, shown in Figure 8.5) and *Laminaria* (kombu) are grown off the coasts of Japan and China. Ropes are seeded with spores, the germinated organisms are propagated in environmentally controlled culturing tanks, and the mature plants are grown in coastal waters on large floating rafts and then harvested and dried. Wakame and kombu have a variety of commercial applications; they are used in noodles, soups, and salads, or with meats. Annual production of wakame exceeds 20,000 tons, and the annual market value of both algae is $600 million.

Alginic acids (alginates) from brown algae and phycocolloid polysaccharides (agars, carrageenans) from red algae were commercially produced early in the twentieth century. In seventeenth-century France, soda ash was obtained from wild kelp (brown algae). By the mid-nineteenth century, iodine was extracted. Agar production originated in China and Japan, probably in the seventeenth century, while carrageenans were produced in the 1830s, first in Ireland and then in the United States.

Today, alginic acids and phycocolloid polysaccharides are used in food, industrial products, fertilizer, and energy production. Alginates are in large demand by several industries: of the more than 35,000 tons produced, the textile industry uses 50%; foods use 30%; pharmaceuticals use 5%; and paper uses 6%. Carrageenan is used extensively as an extender in foods such as evaporated milk and ice cream, in toothpaste, and in a variety of cosmetics. Agars are used primarily in foods but also in pharmaceuticals (for example, as a component in capsules holding medication) and in scientific laboratories for making gels (for example, for gel electrophoresis) and solidified culture media. The annual production of agar products is approximately 11,000 tons, with a value of approximately $160 million, according to 1990 figures.

Tom Mumford

FIGURE 8.4 Nori is cultured on horizontal nets. The blades are dried and processed for human consumption.

TABLE 8.3 Microalgal Products and their Approximate Market Value

Algal Products	Uses and Examples	Approx. Value, $/kg	Approx. Market*	Algae Genus or Type	Product Content	Production System	Current Status
Isotopic compounds	Medicine Research	>1000	Small	Many	>5%	Tubular, indoors	Commercial
Phycobili-proteins	Diagnostics	>10,000	Small	Red	1–5%	Tubular, indoors	Commercial
	Food colorings	>100	Medium	Cyanobacteria		Tubular, indoors	Commercial
Pharmaceuticals	Anticancer Antibiotics	Unknown (very high)	Large Large	Cyanobacteria Other	0.1–1%	Indoor ponds, fermenters	Research
β-Carotene	Food supplement Food colorings	>500 300	Small Medium	Dunaliella	5%	Lined ponds	Commercial
Xanthophylls	Chicken feeds Fish feeds	200–500 1000	Medium Medium	Greens, diatoms, etc.	0.5%	Unlined ponds Lined ponds	Research Commercial(?)
Vitamins C & E	Natural vitamins	10–50	Medium	Greens, unknown	<1%	Fermenters	Research
Health foods	Supplements	10–20	Medium to large	Chlorella, Spirulina	100%	Lined ponds	Commercial
Polysaccharides	Viscocifiers, gums	5–10	Medium to large	Porphyridium	50%	Lined ponds	Research
	Ion exchangers		Large	Greens, others			Commercial
Bivalve feeds	Seed production Aquaculture	20–100 1–10	Small Large	Diatoms, Chrysophytes	100%	Lined ponds	Commercial Research
Soil inocula	Conditioner, Fertilizers	>100	Unknown Unknown	Chlamydomonas N-fixing species	100%	Indoor ponds Lined ponds	Commercial Research
Amino acids	Proline Arginine, aspartate	5–50 5–50	Small Small	Chlorella Cyanobacteria	10% 10%	Lined ponds Lined ponds	Research Conceptual
Single cell protein	Animal feeds	0.3–0.5	Large	Greens, others	100%	Unlined ponds	Research
Vegetable oils	Foods, feeds	0.3–0.6	Large	Greens, others	30–50%	Unlined ponds	Research
Marine oils	Supplements	1–30	Small	Diatoms, others	15–30%	Lined ponds	Research
Waste treatment	Municipal, industrial	1 per kg algae	Large	Greens, others	n.a.	Unlined ponds	Commercial
Methane, H₂, liquid fuels	General uses	0.1–0.2	Large	Cyanobacteria, greens, diatoms	30–50%	Unlined ponds	Research

*Market sizes ($ million); small <$10; medium $10–100; large >$100.

J. R. Benemann in *Algal and Cyanobacterial Biotechnology*, Cresswell et al., eds., Table 11.1. Addison Wesley Longman, London.

The demand for products like carrageenans and agars far exceeds the red algae harvest. Although the process of isolating these compounds is straightforward, genetic manipulation of cultured algal strains might enhance growth and hydrocolloid production. Genetically improved strains might decrease the cost of products—agar and agarose can be expensive, ranging from $2 to $200 and from $250 to $40,000 per kilogram, respectively, depending on purity.

Experiments using algal cell culture are in progress to increase the amounts synthesized of such products as agar. Macroalgae, or seaweeds, can be cultured *in vitro* by producing protoplasts and callus tissue from which these algae can be regenerated. Cell and tissue culture allows new traits to be selected or genes for spe-

cific characteristics to be transferred. Protoplast fusion allows desirable characteristics from two different organisms to be mixed. *In vitro* propagation and selection may one day lead to the development of disease resistance; faster growth; tolerance to variations in light, temperature, and nutrients; and increased production of metabolites and nutrients.

The microalgae comprise a diverse group of both eukaryotic algae (for example, green algae) and prokaryotic photosynthetic bacteria (for example, cyanobacteria, sometimes called blue-green algae). Although used primarily as food, microalgae are also a source of pigments such as phycoerythrin, phycocyanin, β-carotene, and zeaxanthin (Table 8.3). Five types of microalgae have been most exploited: *Dunaliella*, *Scenedesmus*, *Spirulina* (a

TABLE 8.4 Comparison of the Gross Chemical Composition of Human Food Sources and Different Algae (percentage of dry matter)

Commodity	Protein	Carbohydrates	Lipids	Nucleic Acid
Baker's yeast	39	38	1	—
Rice	8	77	2	—
Egg	47	4	41	—
Milk	26	38	28	—
Meat muscle	43	1	34	—
Soya	37	30	20	—
Scenedesmus obliquus	50–56	10–17	12–14	3–6
Scenedesmus quadricauda	47	—	1.9	—
Scenedesmus dimorphus	8–18	21–52	16–40	—
Chlamydomonas rheinhardii	48	17	21	—
Chlorella vulgaris	51–58	12–17	14–22	4–5
Chlorella pyrenoidosa	57	26	2	—
Spirogyra sp.	6–20	33–64	11–21	—
Dunaliella bioculata	49	4	8	—
Dunaliella salina	57	32	6	—
Euglena gracilis	39–61	14–18	14–20	—
Prymnesium parvum	28–45	25–33	22–38	1–2
Tetraselmis maculata	52	15	3	—
Porphyridium cruentum	28–39	40–57	9–14	—
Spirulina platensis	46–63	8–14	4–9	2–5
Spirulina maxima	60–71	13–16	6–7	3–4.5
Synechococcus sp.	63	15	11	5
Anabaena cylindrica	43–56	25–30	4–7	—

E. W. Becker, *Microalgae: Biotechnology and Microbiology,* Table 12.1. Cambridge University Press, 1994. Used by permission of the publisher and the authors.

cyanobacterium), *Porphyridium*, and *Chlorella*. The last three in particular are sources of protein and vitamins, especially in developing countries and in health-food stores (Tables 8.4 and 8.5).

Mass culturing enables large quantities of microalgae to be grown and harvested in outdoor ponds. The potential exists for lowering costs so that microalgae products can compete commercially with other compounds. For example, arachidonic acid is an essential dietary fatty acid and precursor to prostaglandins and other important compounds. One of the few natural sources, the red unicellular algae *Porphyridium*, has one of the highest concentrations of arachidonic acid—36% of the total fatty acids. This organism also can be mass cultivated in open ponds.

Phycobiliproteins are pigments associated with the photosynthetic apparatus of red algae, cyanobacteria, and cryptomonads. *Poryphryidium* is an excellent source of phycobiliproteins, because compounds can easily be extracted and cells are readily cultured. *Por-*

phyridium phycobiliproteins (for example, phycoery-thrin in red algae) are a potential source of phycofluors, used to label or tag biologically active molecules such as immunoglobulin, protein A, and biotin.

Records indicate that indigenous people in Mexico were culturing the cyanobacterium *Spirulina* for food in 1524. People living around Lake Chad in West Africa also were consuming *Spirulina* centuries ago. This cyanobacterium is a desirable food source because it is readily harvested, its cell wall constituents are easily digested, and it is approximately 70% protein by dry weight. The largest culturing facility is near Mexico City, at Lake Texcoco, an ancient lake, now dry, that was an Aztec site. At this lake, the Indians harvested *Spirulina* and another cyanobacterium, *Oscillatoria*. Today, *Spirulina* is propagated in ponds at this lake by pumping underground water into a spiral evaporation system created by dikes (Figure 8.6). At a certain site in the spiral, the water has enough salinity so that *Spirulina* can grow. There are culturing facilities in Thailand, Is-

TABLE 8.5 Comparison of the Recommended Daily Intake of Vitamins with the Content of Human Food Sources and Different Algae

Vitamin	RDI (mg d^{-1})	Beef liver (fresh)	Spinach (fresh)	mg kg^{-1} dry matter					
				1	2	3	4	5	6
Vitamin A	1.7	360	130	840	225	—	230	554	480
Thiamine (Vit B$_1$)	1.5	3	0.9	44	14	55	8	11.5	10
Riboflavin (Vit. B$_2$)	2.0	29	1.8	37	28.5	40	36.6	27	36
Pyridoxine (Vit. B$_6$)	2.5	7	1.8	3	1.3	3	2.5	—	23
Cobalamine (Vit. B$_{12}$)	0.005	0.65	—	7	0.3	2	0.4	1.1	0.02
Vitamin C	50	310	470	80	103	—	20	396	—
Vitamin E	30	10	—	120	—	190	—	—	—
Nicotinate	18	136	5.5	—	—	118	120	108	240
Biotin	—	1	0.07	0.3	—	0.4	0.2	—	0.15
Folic acid	0.6	2.9	0.7	0.4	—	0.5	0.7	—	—
Pantothenate	8	73	2.8	13	—	11	16.5	46	20

Notes:
RDI, Recommended daily intake (adults).
1, *Spirulina platensis* (Faggi, 1980); 2, *Spirulina maxima* (Jaya, Scarsino & Spadoni, 1980); 3, *Spirulina sp.* (Durand-Chastel, 1980); 4, *Scenedesmus obliquus* (Becker, 1984); 5, *Scenedesmus quadricauda* (Cook, 1962); 6, *Chlorella pyrenoidosa* (Fisher & Burlew, 1953).

E. W. Becker, *Microalgae: Biotechnology and Microbiology,* Table 12.5. Cambridge University Press, 1994.
Used by permission of the publisher and the author. See this book for original references.

rael, Japan, Taiwan, and the United States; total world-wide production is approximately 850 tons a year. This cyanobacterium is marketed as dried flakes, and is found primarily in health-food stores, in fish food, and in Japanese cuisine. Current cost is about $10 per kilogram, but it has been as high as $150 per kilogram—most likely because of an earlier interest in its potential therapeutic value.

The cost of culturing algae as a protein source is much higher than the cost of soybean production ($2 to $10 per kilogram versus $0.20 per kilogram). However, in developing countries protein is desperately needed, fertile farmland is in short supply, and high-technology amendments to the soil such as chemical fertilizers and herbicides are not affordable. Thus, the low-technology cultivation of high protein algae may be attractive. (Single-cell protein is discussed in Chapter 5, Microbial Biotechnology.)

Modern algal biotechnology involves genetic manipulation, chiefly through mutation and selection, to produce algae that grow faster in culture, are more disease resistant, synthesize more of a particular metabolite, and even produce new, unique products. Recombinant DNA technology has not been used extensively, because gene transfer methods must be established for each type of organism, and specific genes and promot-

FIGURE 8.6 An aerial view of the spiral algal pond at Lake Texcoco, in Mexico.

Cambridge University Press

ers must be isolated that would allow the expression of a desired trait or product.

Over the past 20 years, algae have received much attention for their pharmacological potential, their use as agricultural fertilizers and as energy biomass, and their use in cell cultures for producing unusual and rare chemicals. The variety of potential products is extremely broad—encompassing polysaccharides, proteins, lipids, pigments, carotenoids, sterols, vitamins, antibiotics, enzymes, pharmaceuticals, and fine chemicals, as well as biofuels such as hydrocarbons, methane, and alcohol. Moreover, commercially available marine polysaccharides have antiviral, anti-ulcer, antitumor, anticoagulant, and cholesterol-lowering activities. Algae are an efficient, renewable, environmentally friendly source of chemicals, pigments, and energy.

Many unusual compounds with unique structures have been isolated from algae. Unfortunately, some marine pharmacology research programs have declined in recent years because of the tremendous (and costly) screening effort required to identify bioactive compounds, difficulties in collecting and culturing, and the small amounts in which many algal compounds are isolated. Most algal products do not seem to show much economic potential, especially in competition with chemical syntheses. Two exceptions have been β-carotene and glycerol. The pigment β-carotene is found in green plants, but some algae such as the halotolerant, unicellular green alga *Dunaliella salina* produce very high amounts and can be mass cultured in shallow brackish ponds. Glycerol, an osmoregulatory product of *Dunaliella*, has been commercially produced. However, because glycerol can also be produced during petroleum extraction, the future of algal glycerol is uncertain.

Currently, algal biotechnology represents only a small fraction of modern biotechnology. Commercial interests and the bulk of investments have focused on the development of genetically engineered plants and animals, diagnostics, new pharmaceuticals from plants and synthetic compounds, and industrial fermentations using animal, plant, and bacterial cells. In the future, however, increased awareness of the importance of algae should stimulate research activities, and algal biotechnology will most likely represent a larger portion of modern biotechnology.

FUELS FROM ALGAE

Nonrenewable fossil fuels provide most of the world's energy (78%). In 1992 oil provided 26% of domestic energy production in the United States and 41% of energy consumption; coal has been the largest source of domestic energy (33%) and is the least expensive fossil fuel per energy unit ($1.42 per million BTU). Although

fossil fuel stores are not inexhaustible, no one knows exactly how much remains. Readily accessible reserves are being depleted, and new sources must be identified. Continued extraction and combustion of fossil fuels also pose environmental problems. For example, coal has been a major source of air pollution, contributing 66% to total U.S. sulfur dioxide emissions and 36% to greenhouse gases (human-generated carbon emissions). Projections of the worldwide market for clean power technologies range from $270 billion to $750 billion during the next 20 years (1996 figures). In 1996 renewable energy (such as wind and solar) and nuclear power represented only 18% and 4%, respectively, of global energy production.

A potential alternative to fossil fuels is photosynthetically generated biomass, a resource that is renewable and will not damage the environment. Approximately 40% of all photosynthesis (primary energy production) occurs in the oceans. In photosynthesis, seaweeds, phytoplankton (small algae and cyanobacteria), and seagrasses take up carbon dioxide and use energy from the sun to produce sugars (the carbon atoms come from the carbon in carbon dioxide) and oxygen. It has been estimated that the oceans contain 50 times as much carbon dioxide as the atmosphere and that photosynthesis converts 35 gigatons of carbon into marine biomass each year. Therefore, this is a tremendous source of potential fuel for the production of energy. The biomass can be converted to, for example, methane. From 47% to 65% of the available carbon is converted, and the gas produced varies from 60% to 72% methane, with carbon dioxide and trace amounts of nitrogen making up the rest. Unfortunately, marine biomass is not competitive with other more traditional sources of biomass such as soybean. In addition, biomass is not economically competitive with other types of fuels at this time. Therefore, marine biomass has not been utilized commercially.

Bacteria used in the bioconversion are obtained from agricultural wastes, high saline environments, and estuaries. Biomass conversion to methane can use a variety of algae, which provide all the nutrients that the bacterial culture requires. The amount of methane produced depends in part on the composition of the algal culture, because the amounts of polysaccharides, lignin, protein, and other components that influence the efficiency of the bioconversion vary greatly among species. At the present time, biological gasification is not competitive with other sources of natural gas. Algal harvesting and production account for 70% of the total cost of the methane.

Although *Dunaliella* produces up to 85% of its dry weight in glycerol, this alcohol is not a good fuel because its high viscosity and oxygen content give it a very low energy value. However, bacterial fermentation and catalytic conversion can convert glycerol into more useful products. Several bacteria, such as *Klebsiella* sp., *Clostridium pasteurianum*, and *Bacillus* sp., can ferment glycerol from *Dunaliella* biomass to higher-energy, nonviscous compounds. *Klebsiella* produces 86% 1,3-propanediol, 9% acetate, and 3% ethanol; *Clostridium* produces 72% *n*-butanol, 21% ethanol, 1% acetate, and trace amounts of 1,3-propanediol; and *Bacillus* produces 92% ethanol and 8% acetate. Ethanol, butanol, and 1,3-propanediol can be used as liquid fuels and have higher energy contents than glycerol. Gasohol, a gasoline with an ethanol additive produced from the bacterial fermentation of corn, is available in the Midwest of the United States. A simple chemical conversion can further convert ethanol to alkenes, alkanes, and aromatics similar to those found in gasoline.

Algae can potentially produce fuel directly. Algae—primarily microalgae—can produce large amounts of hydrocarbons derived from either fatty acid or isoprenoid biosynthesis. Resulting hydrocarbons are high in energy, having two times the energy of any other biological compound per comparable mass. Small isoprenoids with fewer than 16 carbons can be used directly as liquid fuels. Larger molecules can be chemically converted to liquid fuels, although economical conversion methods still must be developed. Naturally occurring, high-yield microalgae that redirect a large portion of fixed carbon into lipid production must be identified and mass cultured in saltwater ponds for use as fuel. The lack of technologies may be due to the fact that some hydrocarbons such as triglycerides are more valuable to the food industry than as a source of energy.

In the future algae may be genetically modified so as to synthesize gasoline-type fuels such as acyclic hydrocarbons (among them, alkanes and alkenes). Some brown algae (*Macrocystis*) and cyanobacteria (for example, *Anacystis nidulans*) already synthesize small amounts of fuels from fatty acids. Genetic engineering may increase production.

For the time being, algal fuels will not replace fossil fuels. Research first must find ways to increase culturing efficiency and increase biofuel yields. However, even if biofuels account for only 2–3% of the total global energy use, this will represent an enormous amount of energy and would decrease the impact of fossil fuel use on the environment.

Biotechnology can be used to make biomass energy production more economically feasible in a variety of ways. Examples include the enhancement of photosynthesis by genetically engineering the enzyme responsible for capturing carbon dioxide (ribulose-1,5-bisphosphate carboxylase/oxygenase), making it a more efficient enzyme. The chemical composition of biomass can be modified to make alternative fuels for specific applications (for example, increase lipid content of microalgae).

ALGAL CELL CULTURE

Algal culture is used chiefly to generate biomass from which cells and metabolites can be isolated. Culturing often takes place in large ponds or raceways (Figure 8.7 shows several facilities). Cultured algal cells also are used to maximize the production of high cost or rare compounds. This type of culturing usually takes place in a fermenter or bioreactor rather than in ponds. Culture conditions are highly controlled, and products undergo significant downstream processing of product collection, purification, and packaging. Currently, microalgae are the main organisms used in culture, although macroalgae yield valuable agars and agarose (polysaccharide-containing polymers) that are used in research and diagnostic laboratories.

The world demand for agars far exceeds production because of seasonal variations in yield and quality, inadequate production methods, and the scarcity of agar-producing algae. Cell culture technology may be able to contribute to agar production. International collaborators are spending many millions of dollars to develop adequate cell culture production technologies. Red algal cells are being isolated and selected for efficient growth in culture, as well as being genetically engineered for high rates of agar and agarose production.

Specialty chemicals are produced from microalgal cultures. For example, a particular strain of *Chlorella* (green algae that produce a large quantity of amino acids) has been developed that produces 30% more proline than other strains produce. The proline can be readily extracted from the culture for commercial use (such as for food supplements and cell culture media). A variety of hydrocarbons (including important lipids), polysaccharides, and other important compounds also can be produced. Gene transfer from other organisms can yield new algal products; for example, after genes are transferred from bacteria, the green alga *Chlamydomonas* can produce the rare amino acid octopamine.

Although algal cell culture is a high cost operation, the products that are isolated can be very profitable. For example, amino acids fetch $5 to $100 per kilogram; food coloring phycobiliproteins, $100 per kilogram; β-carotenes and other pigments, $300 to $500 per kilogram; and medical phycobiliproteins, $10,000 per kilogram. Other industrial compounds, such as dihydroxyacetone, gluconic acid, hydrogen, and acetic acid, are produced by immobilized microalgal cells. One algal culture in a large fermenter potentially can generate thousands of dollars of specialty chemicals each month.

FIGURE 8.7 A variety of algal culture facilities: (a) Circular algal ponds with rotating agitator in Taiwan. (b) Concentric algal pond at Auroville, Indiana. (c) A sloping algal pond in Peru has floating injectors. (d) A primitive algal pond with a polyethylene lining in Madras, India.

MEDICAL APPLICATIONS

Marine organisms also make important contributions to medical research. Sharks may not form cancerous cells and so are being used to study the immune system; compounds in shark cell extracts might hold the key to cancer suppression. Other marine organisms serve as models for the study of human disease. For example, because the sea cucumber's abdominal cavity is filled with bacteria, the creature is being used to study resistance to peritonitis, and because the icefish lacks red blood and hemoglobin, it is a model for anemia.

Marine Natural Products and Their Medical Potential

Marine organisms produce many metabolites, which are usually referred to as natural products. Chemists and pharmacologists are studying these natural prod-

ucts because of their unusual and sometimes unique chemical structures. Unlike primary and intermediate metabolites and cofactors, secondary metabolites are not essential for growth and reproduction, but most probably confer some selective advantage on the organism (or did so at some time during the organism's evolution). The many compounds provide much diversity of chemical structure and function.

Researchers are evaluating these marine natural products for their usefulness. In the past, studies were limited to cataloging the metabolites and other chemicals isolated from marine organisms. Today, investigators focus on identifying applications in medicine and agriculture. Compounds that are pharmaceutically active or that can help control insect pests destructive to agriculture are being studied. Chemists also use these compounds to synthesize new derivatives or analogs to generate more compounds that have potential value. Secondary metabolites exhibit several types of activities: antitumor, antiviral, anti-inflammatory, enzymatic, insecticidal and herbicidal (a highly neglected area), antibacterial (antibiotic), and toxic. Some secondary metabolites may one day be developed commercially.

For the past 100 years, much energy has been devoted to screening the world's organisms for useful chemicals. Approximately 20,000 chemicals have been characterized, and many of these come from marine organisms such as bacteria, algae, **sponges**, corals, jellyfish, bryozoans, mollusks, ascidians (**tunicates**), and **echinoderms**. Most of the new natural products identified each year have come from terrestrial organisms: annual sales of pharmaceuticals from plants exceed $10 billion in the United States. However, plants were virtually untapped until recently, so natural products from marine organisms may hold similar promise. Indeed, marine organisms have been shown to be an excellent source of new compounds that exhibit a variety of biological activities. From 1977 to 1987, approximately 2500 new metabolites from a variety of marine organisms were reported. In 1992 alone, some 200 published articles characterized more than 500 marine natural products. Most of these compounds came, in descending order of abundance, from sponges, ascidians, algae, mollusks, and soft and gorgonian corals. Sponges have recently become a primary object of investigation because of their wide range of biosynthetic capabilities. Metabolites from echinoderms (such as starfish and sea stars) also have been isolated for natural products. Microbial organisms such as bacteria, cyanobacteria, and actinomycetes also are important sources of pharmaceutical compounds.

Anticancer and Antiviral Compounds

Many compounds from marine organisms are extremely toxic to humans; however, controlling toxicity by diluting concentrations exploits their therapeutic value. The National Cancer Institute tests marine extracts and compounds against many *in vitro* cell lines that represent various cancers (such as kidney, brain, colon, lung, skin, blood cells, and ovary, as well as a lymphoblastic cell line infected with the AIDS virus).

A variety of compounds isolated from marine organisms show promise as anticancer drugs. Didemnin B (Figure 8.8a), a cyclic peptide from a Caribbean tunicate (*Trididemnum solidum*), was shown experimentally to be very effective against leukemia and melanoma cells in mice. Although in human clinical trials the compound did not exhibit significant anticancer properties, didemnin derivatives may nevertheless provide cancer-fighting properties. Didemnin B is also an effective immunosuppressive agent—it is ap-

proximately 1000 times more effective than the standard cyclosporin A. Further study may prove it useful in repressing rejections of transplanted organs.

The Romans knew of the toxic effects of crude extracts from sea hares (a type of mollusk) containing dolastatins and by the year 150 were using such extracts in the treatment of various diseases. However, not until the 1970s were dolastatins known to be effective against lymphocytic leukemia and melanoma. Different structural forms of dolastatins that show antitumor properties have since been isolated. Dolastatin 10 (see Figure 8.8b) is a linear peptide from the sea hare *Dolatella auricularia*, found in the Indian Ocean. This compound is the most active of the isolates and is a potent antimitotic (inhibitor of cell division) because it inhibits the polymerization of tubulins into microtubules, which are involved in cell division. This compound is undergoing clinical trials and is being compared with other antitubulin polymerization drugs, such as vinblastine, which is isolated from the Madagascar periwinkle.

Marine sponges seem to be an excellent source of natural products. Alkaloids, terpenoids, and sterols are but a few of the hundreds of compounds isolated from them. Some compounds may actually be produced by symbiotic organisms living within the sponge, such as cyanobacteria, dinoflagellates, microalgae, and an assortment of bacteria. The South Pacific marine sponge *Luffariella variabilis* is the source of manoalide, a member of the terpene group of chemicals. Pharmacological studies have demonstrated that it has anti-inflammatory, analgesic, and antifungal properties, in addition to being effective in treating leukemia. Scientists have synthesized more than 300 analogs of manoalide, many of which are in clinical trials as anti-inflammatory compounds. Luffariellin A (see Figure 8.8c) and B, also isolated from *Luffariella*, may be anti-inflammatory agents. A variety of potentially important chemicals from several marine sponges have exhibited antitumor activities: halichondrins (see Figure 8.8d), mycalamides, onnamide A, calyculins, swinholides (Figure 8.8e), and misakinolides (Figure 8.8f). Halichondrins in particular have generated special interest by demonstrating potent activity against leukemia and melanoma cell cultures.

Other organisms such as nudibranchs (mollusks that are essentially snails lacking a shell) extract potent compounds from invertebrate prey and use them for protection. The ulapualides, compounds demonstrating activity against both leukemia and fungi, can be transmitted to the nudibranch's eggs to protect them from predators seeking a readily available meal.

Useful antiviral drugs (such as aciclovir and AZT) are extremely rare. However, marine organisms may be a source of antiviral agents. Researchers worldwide are screening compounds from marine organisms for an-

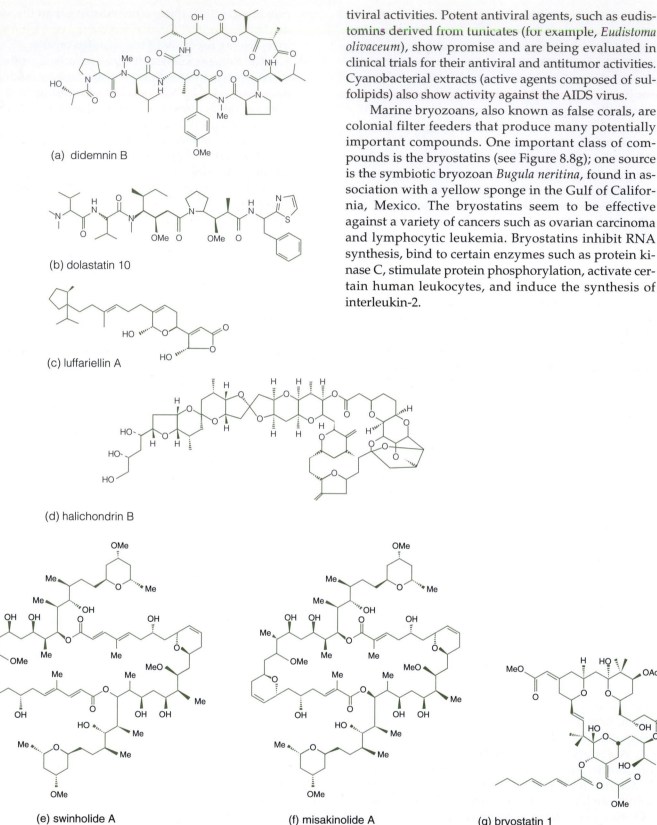

(a) didemnin B

(b) dolastatin 10

(c) luffariellin A

(d) halichondrin B

(e) swinholide A

(f) misakinolide A

(g) bryostatin 1

tiviral activities. Potent antiviral agents, such as eudistomins derived from tunicates (for example, *Eudistoma olivaceum*), show promise and are being evaluated in clinical trials for their antiviral and antitumor activities. Cyanobacterial extracts (active agents composed of sulfolipids) also show activity against the AIDS virus.

Marine bryozoans, also known as false corals, are colonial filter feeders that produce many potentially important compounds. One important class of compounds is the bryostatins (see Figure 8.8g); one source is the symbiotic bryozoan *Bugula neritina*, found in association with a yellow sponge in the Gulf of California, Mexico. The bryostatins seem to be effective against a variety of cancers such as ovarian carcinoma and lymphocytic leukemia. Bryostatins inhibit RNA synthesis, bind to certain enzymes such as protein kinase C, stimulate protein phosphorylation, activate certain human leukocytes, and induce the synthesis of interleukin-2.

FIGURE 8.8 Chemical structures of (a) didemnin B, (b) dolastatin 10, (c) luffariellin A, and (d) halichondrin B. Chemical structures of (e) swinholide A, (f) misakinolide A, and (g) bryostatin 1. Me = methyl group, CH_3.

Unfortunately, the amount of any marine natural product that is available is limited by how many organisms are harvested that produce it. Studies that focus on the chemical synthesis of pharmaceutically active marine compounds will be indispensable not only for developing new medical treatments but also for preventing the severe depletion of the marine sources of the compounds. Recent advances in biochemical pharmacology and marine chemistry, and the availability of sophisticated screening methods, will lead to new therapeutics from marine metabolites in the near future.

Antibacterial Agents

Many marine organisms—including cyanobacteria; green, brown, and red algae; sponges; dinoflagellates; jellyfish; and sea anemones—produce secondary metabolites with antibiotic properties. One promising class of broad-spectrum antibiotics, squalamine, has recently been isolated from the stomach of the dogfish (*Squalus acanthias*); it has activity against a wide variety of bacteria, fungi, and protozoa. Unfortunately, antibiotic isolation and characterization has not been extensive, and few compounds from marine organisms are being studied, most likely because of the costs of identifying and testing commercially important antibiotics.

Marine Toxins

Toxins are thought to be produced by marine organisms for a variety of purposes, including predation, defense from predators and pathogenic organisms, and signal transduction in the autonomic and central nervous system. Many of these marine natural products are extremely toxic; their effects can range from dermatitis to paralysis, kidney failure, convulsions, and hemolysis. Livestock and humans have become ill and died when specific toxin-producing algae or cyanobacteria have contaminated water. Dinoflagellates produce saxitoxins that are 50 times more toxic than curare. When bivalves that have incorporated saxitoxins are ingested by humans, paralytic shellfish poisoning can lead to severe illness or death. Dinoflagellates also produce the potent ciguatoxin. When small grazing fish that have ingested dinoflagellates while feeding on algae are then eaten by larger fish, ciguatoxin is transferred and can poison humans who consume the fish.

Marine toxins are being studied for their antitumor, anticancer, and antiviral compounds, tumor promoters, and antiinflammatory properties. When concentrations are controlled, marine toxins show promise as antitumor agents, analgesics, and muscle relaxants. They are useful neurophysiology and neuropharmacology research tools for studying such processes as signal transduction in the nervous system. Even compounds that are too toxic to be used therapeutically are being used as models for new, chemically synthesized drugs.

Many neurotoxins act only on specific receptors and ion channels. Each toxin has its own mode of action and so can be used to probe a different step in signal transmission. Ion channels and membrane excitation are being studied by means of specific sodium channel blockers; an example is tetrodotoxin, from the puffer fish, which blocks peripheral nerve transmission. Tetrodotoxin and saxitoxin (Figure 8.9a and b)

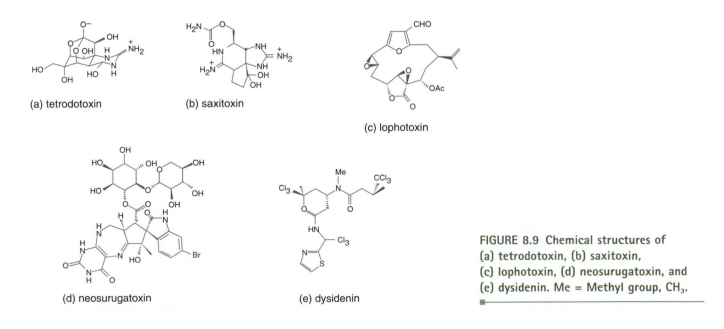

(a) tetrodotoxin

(b) saxitoxin

(c) lophotoxin

(d) neosurugatoxin

(e) dysidenin

FIGURE 8.9 Chemical structures of (a) tetrodotoxin, (b) saxitoxin, (c) lophotoxin, (d) neosurugatoxin, and (e) dysidenin. Me = Methyl group, CH_3.

have been used to study the assembly and function of sodium channels, although the mode of toxin action is not completely understood. Potent conotoxins from the predatory cone snail in the genus *Conus* are used to study calcium and sodium channels, as well as to examine binding to vasopressin and neuromuscular receptors. Other compounds inhibit nerve-induced muscle contraction by blocking postsynaptic signals at neuromuscular junctions; the paralytic lophotoxin (Figure 8.9c) from the gorgonia *Lophogorgia* is an example. Still others act to block ganglia; two such compounds are neosurugatoxin (Figure 8.9d), from the Japanese ivory mollusk, which is an acetylcholine receptor antagonist, and prosurugatoxin, from the shellfish *Babylonia japonica*. To study the role of adenosine as a second messenger in signal transduction, researchers are using a toxin isolated from a red alga as an adenosine analog for binding to adenosine receptors. Other researchers are using sponge dysidenin (Figure 8.9e), an inhibitor of iodide transfer in thyroid cells, to study iodide transport.

PROBING THE MARINE ENVIRONMENT

The study of the interaction of marine organisms with each other and with their environment is an exciting area of marine biology. Modern biotechnology methods have made important contributions in many areas of marine research, including the rapid identification and quantification of new species and populations of microscopic marine organisms, such as bacteria, protozoans, microalgae, and viruses; the tracking of commercially important organisms; the bioremediation of marine niches; the development of diagnostics to detect contaminants, pollutants, and pathogens in seafood and the environment; and the use of marine organisms in biomedical research.

The marine environment harbors a wealth of new organisms, especially new bacteria. Biotechnology methods are increasing our understanding of the abundance, distribution, and rates of growth of various marine organisms. Specific DNA probes and the polymerase chain reaction (PCR, described in Chapter 3, Basic Principles of Recombinant DNA Technology) allow different organisms to be detected and identified in marine environments without their having to be isolated and cultured. Larval invertebrates and microbes, which are often difficult to isolate, can readily be tracked in the environment. The country of origin of important fish stock populations can be established as well. Molecular techniques can determine the level of genetic variation in organisms. Genetic diversity in a group of organisms is an indicator of a healthy population and one that is more likely than an inbred one to tolerate environmental fluctuations and resist disease.

DNA probes or PCR can identify currently unknown intermediate life stages of important marine organisms such as shellfish. Once these intermediates are identified for study, researchers may be able to isolate and characterize the chemical signals that induce settling and metamorphosis. Often such research findings can be applied to the culture and propagation of economically important crustaceans, bivalves, and gastropods.

Microplankton have been shown to be the base of the food chain that supports most of the other marine animal species in the oceans. Microplankton can be identified by fluorescent DNA probes (probes labeled by fluorescent tags rather than radioisotopes; detection is by fluorescence). Microbial primary producers are being identified and quantified by PCR and flow cytometry (which distinguishes different types of cells by defined characteristics). Thus, ecological processes as critical as primary production in the oceans can be studied more easily with the help of biotechnology. In polluted areas, such as areas of industrial waste and raw sewage outfalls, the numbers and types of microbes that cause disease can be readily identified. For example, hepatitis A and E and cholera affect large numbers of people exposed to polluted water through swimming, washing, or harvesting food. Mussels, oysters, and clams, often consumed by humans, reside in offshore areas where ocean dumping occurs and can harbor pathogens.

Many of the pathogens released into the marine environment remain unidentified. Diagnostic kits using DNA probes and monoclonal antibodies will help researchers track pathogens along polluted and contaminated shores and enable health officials to determine what areas should be closed to the public. These detection kits will also be invaluable for detecting pathogens and toxins in commercial seafood.

CONSERVATION

Biotechnology has begun to play an important role in the conservation of threatened and endangered marine fish and mammals. By using the DNA from animals, scientists can determine whether protected animals are illegally being used for food, jewelry, or traditional medicines and potions. Makeshift DNA laboratories are being used in countries worldwide to determine whether markets and restaurants are selling endangered species in violation of international law. Most recently, DNA methods identified unlabeled meat from humpback (protected since 1966) and fin (protected since 1989) whales in Japan; other samples included dolphin, northern minke whale, and a beaked whale about the size of a killer whale. PCR also detects illegal animal meat and parts, and has been used to test samples from markets and shops. PCR is performed in a thermal cycler and can be conducted wherever electricity is available to run the instrument. DNA extracted from food and animal samples is amplified by PCR and then is sent to a laboratory for analysis. For example, examination of a region of PCR-amplified mitochondrial DNA allows both the whale species and its ocean of origin to be identified. This fingerprinting (described further in Chapter 11, DNA Profiling, Forensics, and Other Applications) is a relatively inexpensive and easy method to detect poaching.

TERRESTRIAL AGRICULTURE

Marine organisms have much to offer terrestrial agriculture but until now have largely been untapped. Soil-dwelling and freshwater cyanobacteria and other aquatic microalgae have been used as natural fertilizers, especially in developing countries. In Asia and the Middle East, mass culturing has produced cyanobacteria for fertilizer (usually distributed as dried flakes) and for use in rejuvenating nutrient-exhausted soils. Marine plants, macroalgae and microalgae, and animals have characteristics that, through the application of modern biotechnology, can benefit agriculture. Marine organisms are salt-tolerant (the most tolerant aquatic plant is *Dunaliella*, from the Dead Sea), and many, like flounder, survive at very low temperatures. Genes encoding adaptive traits are being experimentally transferred to agronomically important crop plants to determine whether they can express these traits. Salt tolerance might extend the range of important crops or sustain ranges that would otherwise be lost. Intensive agricultural practices increase the salinity of soil, often to such a level that crops can no longer grow. Salt-tolerant crops could grow in these regions and would increase agricultural productivity substantially, which would be especially important to very poor, malnourished populations. Many highly populated regions of the world lack adequate fresh water but have plenty of water high in saline. Such water might be used for irrigation—reserving scarce fresh water for drinking—if salt-tolerant crops were introduced.

A terrestrial halophyte (an organism requiring salt water) was recently discovered in the Sonora desert in Mexico, where extremes of both temperature and dryness are found. The halophyte, *Salicornia bigelovii*, is a potential high yield oilseed crop from which locals can obtain an unsaturated vegetable oil for commercial production. Cold- or freeze-tolerant plants can grow at high altitudes, extending the range of important crops. These may be especially beneficial to developing countries with undeveloped land at high altitudes, such as

Nepal, Peru, Chile, Bolivia, Afghanistan, and Ecuador. Future research should result in many exciting discoveries in the next decade, ranging from new knowledge about marine organisms and their environment to new medical and agricultural applications.

TRANSGENIC FISH

Some of the first recombinant DNA applications focused on increasing the growth and weight of cultured finfish. In early experiments, the rainbow trout growth-hormone (GH) gene was transferred to the bacterium *Escherichia coli*. The recombinant growth hormone was then isolated from *E. coli* cells and injected into young trout. Treated fish grew in size and weight faster than untreated fish. In other experiments, recombinant salmon GH and tuna GH were injected into rainbow trout and Japanese snapper, respectively, with promising results. The process of injecting fish is labor-intensive and poses technical problems that the use of transgenic fish alleviates. Moreover, transgenic fish have much commercial value.

Direct gene transfer allows fish to harbor the desired gene and express the appropriate product (such as GH). Fish are excellent organisms for gene transfer because their eggs are large and transparent, making the developmental process readily visible. Gene transfer by microinjection into the nucleus is straightforward, and has been the method of choice for fish transgenics. In addition to transgenic fish with genes for GH, fish with genes for antifreeze protein are being field tested, in the hope that economically important fish can extend their range of survival into colder waters.

Transgenic techniques are being developed to introduce desirable traits into many species of fish. For fish transgenics to be successful, appropriate promoters, enhancers, and tissue-specific DNA sequences must be identified. To regulate transgenes (for example, to turn them on and off and to ensure expression in specific tissues), light, temperature, and diet may be used as inducers of transcription. For example, a light-regulated promoter would be activated if fish were exposed to light of a certain color or intensity.

Methods of gene transfer include electroporation, direct pronuclear injection of the egg, and injection through the micropyle (the opening in the egg where sperm enter). Fish pronuclei are often too small to be seen clearly under a microscope and are obscured by an opaque chorion and large volume of cytoplasm. Thus, the gene is often transferred by microinjection into the cytoplasm of fertilized eggs (already released from the female). This method is easier than pronuclear injection of unfertilized eggs (which must be isolated, cultured, and fertilized *in vitro*). However, because the

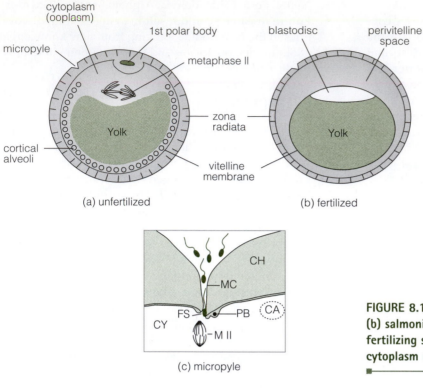

cytoplasm (ooplasm)

microcyle

1st polar body

metaphase II

cortical alveoli

Yolk

zona radiata

vitelline membrane

(a) unfertilized

blastodisc

perivitelline space

Yolk

(b) fertilized

CH

MC

FS

CY

PB

CA

M II

(c) micropyle

FIGURE 8.10 Schematic of unfertilized (a) and fertilized (b) salmonid eggs. Sperm enter the micropyle (MC) (c); fertilizing sperm (FS), chorion (CH), first polar body (PBI), cytoplasm (CY), cortical alveoli (CA), metaphase II (M II).

first cell division occurs only 60 minutes after the egg is fertilized, not much time is available to inject eggs. Staggered fish matings enable more freshly fertilized eggs to be injected.

Salmonid eggs (from, for example, Atlantic salmon, rainbow trout, Arctic char, or brown trout) have a hard chorion (Figure 8.10). For these, injection through the micropyle with very fine glass needles (2–3 μm outside diameter) inserts the transgene very near the male pronucleus, which is just under the micropyle. Another procedure uses sturdy glass needles (15 μm or less in diameter) to penetrate the chorion and insert the DNA into the **blastodisk** just after fertilization but before the chorion completely hardens. Another successful method is to cut a minute hole in the egg after the chorion hardens for insertion of a glass needle (approximately 10 μm in diameter) into the blastodisk just before the first cell division.

Embryos can be left to develop in temperature-controlled tanks. The survival rates of microinjected fish embryos are much higher (from 35% to 80%) than for mammals. From 10% to 70% of the surviving embryos are transgenic. To determine which fish have received the foreign gene, DNA is isolated from fish scales and the nucleated red blood cells are analyzed by PCR. The founder fish are mated to produce stable transgenic lines.

Many important fish species cannot be cultured because of low tolerance to certain environmental conditions or diseases. A variety of environmental conditions

may limit the home range. For example, rainbow trout and salmon cannot be raised in warm water; on the other hand, Atlantic salmon, a highly desirable commercial fish, is susceptible to freezing at temperatures below −0.7° C. (The body fluids of most fish freeze between −0.5° C and −0.8° C; seawater, with a much higher salt concentration than fish, freezes between −1.7° C and −2.0° C.) The waters of the Atlantic off the coast of Canada and the polar oceans reach subzero temperatures, and floating ice is abundant during the winter months. Fish such as winter flounder that survive these freezing temperatures do so by synthesizing antifreeze proteins in the liver that circulate in the blood. These fish are protected from freezing at temperatures as low as the freezing temperature of seawater. Research is underway to produce freeze-resistant salmon by introducing the winter flounder antifreeze gene. Stable transgenic lines of Atlantic salmon have been produced that may eventually be cultured in sea pens along the Atlantic coast.

Fish are being genetically modified to make them grow faster and larger and to allow them to flourish through aquaculture. Most studies have focused on the effect of GH genes. With the success of recent research, fast-growing transgenic fish strains may become an important future aquaculture commodity. In one study, engineered catfish containing a trout GH are being studied in experimental ponds. Researchers have achieved mixed results, because only three of five engineered catfish families grew 40% larger than

Fluorescent Fish Pets

Genetically engineered glow-in-the-dark fish are one of the latest developments in biotechnology. Taikong Corporation in Taiwan is marketing a fluorescent aquarium fish (rice fish approximately 1 inch long) that glows bright green when a black light illuminates the tank. This fish is the first genetically engineered pet novelty designed to glow in the dark, and one biotechnology product produced for enjoyment only. The genetically engineered fish were produced in the research laboratory of Dr. H.J. Tsai at National Taiwan University in 2001 as a research tool. The gene, from jellyfish, encodes a fluorescent protein that is used as a genetic marker to study the development of embryonic fish. Dr. Tsai produced fish with fluorescent hearts that researchers used to study heart development; however, unexpectedly, fish were produced that fluoresced brightly in every cell. The fish were discovered by Taikong, a company that sells aquarium equipment and fish food globally. Marketed in Taiwan in 2003 as Night Pearls, these fish come with accessories such as black-light tanks, fluorescent plastic coral, and fluorescent fish food. Before the fish are sold, they are sterilized in the chance that they are released into wild populations (the sterilization procedure has an approximately 90% success rate). Several other color varieties have been produced—a fluorescent red fish from a gene from coral and a half green–half red fish. Aquarium enthusiasts in the United States have been eager to acquire these fish; however, the reaction has been different in Europe where there has been much resistance to genetic engineering.

normal. Nevertheless, this area of research looks promising.

Recently, researchers produced transgenic Pacific salmon that harbor a metallothionein-B promoter (regulated by metals and glucocorticoid) and the sockeye salmon GH gene and regulatory elements. The construct was microinjected into the blastodisc of salmon eggs. Out of 3000 injected eggs, approximately 6.2% transgenic fish survived for 1 year. As Figure 8.11 shows, these fish were more than 11 times larger than their nontransgenic counterparts (one fish was 37 times larger!).

Because male fish grow faster than female fish, methods for producing only male fish are being identi-

Robert H. Devlin, Fisheries & Oceans, Canada

FIGURE 8.11 Comparison of 1-month-old coho salmon siblings; nonengineered fish are at left, transgenic fish are at right. The largest fish (top right) is 41.8 cm in length.

The Commercialization and Release of Transgenic Fish

The development of transgenic fish with desirable characteristics such as faster growth, cold tolerance, and disease resistance offers many benefits to the aquaculture industry. Research has shown that new technologies, particularly those relating to genetically engineered organisms take time to achieve acceptance within society. Furthermore, the more extreme the modifications, the more vital it is to have programs and policies in place to help educate the public and ensure the safety of the product. Along these lines, environmental impact studies are necessary to help alleviate fears, assess risks,

and ensure the safety of potential introductions. Scientists have an obligation to transmit scientific information to the pubic in a manner that can be understood. If the positive outcomes (applications, safety, and ethical nature) of their research cannot be clearly understood, the technologies and new products will not be embraced.

A special problem also exists with transgenic fish developed for commercial distribution. There are many examples of nongenetically engineered fish ending up in bodies of water where they are detrimental to the environment. Often these fish do well in their new habitat and outcompete the native species (by overreproducing or

even consuming them), resulting in the disruption (and often local extinction or depletion of native species) of the ecological balance. Once an undesirable fish enters a lake or stream, it is almost impossible to remove it. Extreme care must be taken to ensure that fish remain in aquaculture facilities and undergo only highly controlled breeding. This is a daunting task and one that must be taken very seriously. Before embarking on an extensive global transgenic fish aquaculture program, there must be technologies and policies in place to keep these fish from mixing with wild fish populations (see **Transgenic Fish**).

---

fied. Recent experiments using classical breeding methods have produced genetically homogeneous all-male tilapia (a very important fish in Asian aquaculture), channel catfish, and salmon. To produce these fish, homogeneous male and female breeding stocks must first be generated.

Another area of intense interest is the production of tropical fish and temperate-zone food fish that are resistant to the viral, fungal, and bacterial diseases that can decimate aquaculture stocks.

Concerns about environmental safety have been raised that focus on the containment of transgenic fish. Case studies demonstrate that engineered fish might not remain isolated in commercial facilities, but could accidentally end up in ponds, lakes, streams, and even oceans, where they could decimate already threatened wild fish populations. They could disrupt trophic food webs, increase competition for food and spawning sites, and disturb natural mating systems. Examples of exotic fish introductions that have decimated native fish populations and their ecosystems abound. To give just two, the sea lamprey entered Lake Erie through the Welland Canal and spread to all the Great Lakes, where it destroyed 97% of the trout population; the Nile perch was

intentionally introduced into Africa's Lake Victoria and is eating indigenous fish species—by the 1980s 300 native species had become extinct.

Fish that have been engineered to occupy an extended home range incorporating previously inhospitable habitats might displace their wild counterparts in those habitats; extending home range can be equivalent to introducing an exotic species into a particular habitat (for example, lake trout introduced into Yellowstone Lake, in Yellowstone National Park, are displacing the native cutthroat trout). Even small genetic modifications that dramatically alter an animal's home range could be introduced into wild populations through breeding.

Both physical and biological controls are needed to prevent release of transgenic fish into the environment. Because physical containment is always subject to compromise by such factors as weather and human error, a biological barrier must ensure that mating between cultured and wild fish does not occur. The elimination of reproductive capabilities can be engineered into transgenic stock; methods include triploid females, sterilization by high doses of **androgens**, and genetically engineered sterility.

General Readings

A.R. Beaumont and K. Hoare. 2003. *Biotechnology and Genetics in Fisheries and Aquaculture*. Blackwell Science, Oxford.

E.W. Becker. 1994. *Microalgae: Biotechnology and Microbiology*. Cambridge University Press, Cambridge.

R.R. Colwell. 2002. Fulfilling the promise of biotechnology. *Biotechnol. Adv.* 20:215–228.

J. de la Fuente and F.O. Castro, eds. 1998. *Gene Transfer in Aquatic Organisms*. Springer-Verlag, New York.

I. Karunasagar, I.Karunasagar, and A. Reilly, eds. 1999. *Aquaculture and Biotechnology*. Science Publishers, Enfield, New Hampshire.

Y. Le Gal and H.O. Halvorson. 1998. *New Developments in Marine Biotechnology*. Plenum Press, New York.

J.S. Lee and M.E. Newman. 1992. *Aquaculture—An Introduction*. Interstate Publishers, Inc., Danville, Illinois.

Additional Readings

B. Austin. 1998. Biotechnology and diagnosis and control of fish pathogens. *J. Marine Biotechnol.* 6:1–2.

B. Baker. 1996. Building a better oyster. *Bioscience* 46:240–244.

V.S. Bernan. 1997. Marine microorganisms as a source of new natural products. *Adv. Appl. Microbiol.* 43:57–90.

B.K Carté. 1993. Marine natural products as a source of novel pharmacological agents. *Curr. Opin. Biotechnol.* 4:275–279.

B.K. Carté. 1996. Biomedical potential of marine natural products. *Bioscience* 46:271–286.

D.J. Chapman and K.W. Gellenbeck. 1989. An Historical Perspective of Algal Biotechnology. In R.C. Cresswell, T.A.V. Rees, eds. *Algal and Cyanobacterial Biotechnology*. Longman Scientific and Technical, Essex, United Kingdom, pp. 1–27.

Y. Chisti. 2000. Marine biotechnology: A neglected resource. *Biotechnol. Adv.* 18:547–548.

G. Cimino, S. De Rosa, and S. De Stefano. 1984. Antiviral agents from a gorgonian, *Eunicella cavolini. Experientia* 40:339–340.

R.R. Colwell. 1984. The industrial potential of marine biotechnology. *Oceanus* 27:3–12.

D.J. Faulkner. 1992. Biomedical uses for natural marine chemicals. *Oceanus* 35:29–35.

D.J. Faulkner. 1995. Marine natural products. *Nat. Prod. Rep.* 12:223–269.

W. Fenical. 1997. New pharmaceuticals from marine organisms. *Trends Biotechnol.* 15:339–341.

J.A. Frederick, D. Jacobs, and W.R. Jones. 2000. Biofilms and biodiversity: An interactive exploration of aquatic microbial biotechnology and ecology. *J. Indust. Microbiol. Biotechnol.* 24:334–338.

A. Fuji. 1987. Aquaculture and mariculture. *Oceanus* 30:19–23.

K.-W. Glombitza and M. Koch. 1989. Secondary metabolites of pharmaceutical potential. In R.C. Cresswell, T.A.V. Rees, eds. *Algal and Cyanobacterial Biotechnology*. Longman Scientific and Technical, Essex, United Kingdom, pp. 161–238.

M.E. Gonzalez, B. Alarcón, and L. Carasco. 1987. Polysaccharides as antiviral agents: Antiviral activity of carrageenan. *Antimicrob. Agents Chemother.* 31:1388–1393.

M.G. Haygood, E.W. Davidson, K. Seana, and D. Faulkner. 1999. Microbial symbionts of marine invertebrates: Opportunities for microbial biotechnology. *J. Mol. Microbiol. Biotechnol.* 1:33–43.

Y. Hirata and D. Uemura. 1986. Halichondrins—Antitumor polyether macrolides from a marine sponge. *Pure Appl. Chem.* 58:701–710.

I.L. Hook, S. Ryan, and H. Sheridan. 1999. Biotransformation of aromatic aldehydes by five species of marine microalgae. *Phtytochem.* 51:621–627.

A. Ianora. 2001. Biotechnology and plankton: Ecological aspects of marine biotechnology. *Biol. Internat.* 41:65–70.

H. Ikenoue and T. Kafuku. 1992. *Modern Methods of Aquaculture in Japan*. Kodansha Ltd., Tokyo.

M. Indergaard and K. Østgaard. 1991. Polysaccharides for food and pharmaceutical uses. In M.D. Guiry and G. Blunden, eds. *Seaweed Resources in Europe: Uses and Potential*. Wiley, Chichester, pp. 169–184.

C.M. Ireland, B.R. Copp, M.P. Foster, L.A. McDonald, D.C. Radisky, and J.C. Swersey. 1993. Biomedical potential of marine natural products. In D.H. Attaway and O.R. Zaborsky, eds. *Marine Biotechnology: Pharmaceutical and Bioactive Natural Products.* Vol. I, Plenum Press, New York, pp. 1–43.

J. Kahle, G. Gerdes, and G. Liebezeit. 2003. Culture of *Flustra floliacea* (bryozoa) for secondary metabolite production. *Biomol. Eng.* 20:80–81.

I. Kitagawa and M. Kobayashi. 1992. Anticancer drugs from marine organisms. In S. Baba, O. Akerele, and Y. Kawaguchi, eds. *Natural Resources and Human Health—Plants of Medicinal and Nutritional Value*. Elsevier, Amsterdam, pp. 123–132.

R. Leon, M. Martin, J.M. Vega, J. Vigara, and C. Vilchez. 2003. Microalgae mediated biotransformations in non-conventional media. *Biomol. Eng.* 20:44–45.

C.I. Liao and N.-H. Chao. 1997. Developments in aquaculture biotechnology in Taiwan. *J. Marine Biotechnol.* 5:16–23.

G. Liles. 1996. Gambling on marine biotechnology. 1996. *Bioscience* 46:250–253.

R.E. Lyons and Y. Li. 2002. Crustacean genomic: Implications for the future in aquaculture. *AgBiotechNet* 4:1–6.

R.H. McPeak and D.A. Glantz. 1984. Harvesting California's kelp forests. *Oceanus* 27:19–26.

M. Nickel, S. Leininger, G. Proll, and F. Bruemmer. 2001. Comparative studies on two potential methods for the biotechnological production of sponge biomass. *J. Biotechnol.* 92:169–178.

W. Pennell and B.A. Barton, eds. 1996. *Principles of Salmonid Culture*. Elsevier, New York.

E. Pennisi. 1996. Sorcerers of the sea. *Bioscience* 46:236–239.

R.J. Radmer. 1996. Algal diversity and commercial algal products. *Bioscience* 46:263–270.

R.J. Radmer and B.C. Parker. 1994. Commercial applications of algae: Opportunities and constraints. *J. Appl. Phycol.* 6:93–98.

D. Sipkema, R. Osinga, J. Tramper, and R.H. Wijffels. 2003. Production of new pharmaceuticals with functional marine sponges in a bioreactor. *Biomol. Eng.* 20:51–52.

G. Skjåk-Bræk and A. Martinsen. 1991. Applications of some algal polysaccharides in biotechnology. In M.D. Guiry and G. Blunden, eds. *Seaweed Resources in Europe: Uses and Potential.* Wiley, Chichester, pp. 219–258.

L. Tangley. 1996. Ground rules emerge for marine bioprospectors. *Bioscience* 46:245–249.

J.H. Tidwell and G.L. Allan. 2001. Fish as food: Aquaculture's contribution. *EMBO Reports* 2:958–963.

H. Yamamoto. 1995. Marine adhesive proteins and some biotechnological applications. *Biotechnol. Genet. Eng. Rev.* 13:133–165.

T. Yasumoto and M. Murata. 1993. Marine toxins. *Chem. Rev.* 93:1897–1909.

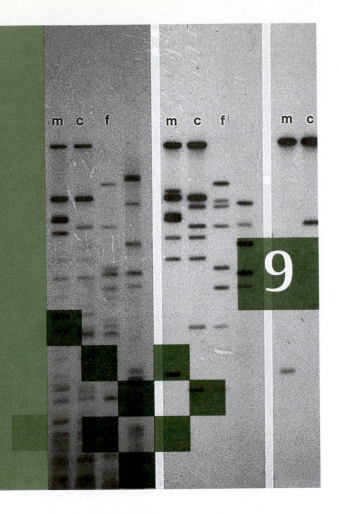

9

GENOMICS AND BEYOND

Genomics is a field of science that involves the elucidation of the content, organization, and evolution of the genomes of various organisms—different microorganisms, plants, and animals. Today the genomes of several hundred different organisms are being analyzed. This includes DNA sequencing, as well as the expression and function of genes and the proteins they encode. Closely related to genomics is the area of bioinformatics, which encompasses the use of computers, computational tools, and databases to organize and analyze DNA and protein data.

The goals of genomics include

1. The assembly of physical and genetic maps of the genomes of different organisms

2. The compilation and organization of both expressed gene sequences and other nongene regions

3. The generation of gene expression (that is, transcription, or the production of transcripts) profiles under different conditions

4. Finding the location of all genes in a given genome and providing annotation for each gene

5. The elucidation of gene function and regulation (functional genomics)

6. The identification of all proteins in a genome and their functions, including the detection of protein–protein interactions (called proteomics)

7. The characterization of DNA polymorphisms within the genomes of a given species

8. The comparison of genes and proteins in the genome of one species with the genomes of other species (comparative genomics)

9. The implementation and management of databases and genome-based research tools that are accessed through the Internet (that is, web-based resources)

GENETIC, CYTOLOGICAL, AND PHYSICAL MAPS

Maps aid in determining the location of genes within a genome. Two types of maps are generated: **Genetic linkage maps** determine the relative arrangement and approximate distances between genes and markers on the chromosomes; physical maps specify the physical location (in base pairs) and distance between genes or DNA fragments with unknown functions that are mapped to specific regions of the chromosomes (Figure 9.1).

Maps have different levels of resolution, ranging from low to high, and the degree of resolution that is appropriate depends on whether, for example, a large fragment of DNA is to be studied or a more detailed picture of a small DNA region is needed. A genomic library consists of random DNA fragments (such libraries are described in Chapter 3, Basic Principles of Recombinant DNA Technology) and is used to establish sets of ordered, overlapping cloned DNA fragments, or contigs, for each chromosome of the genome—in other words, high-resolution maps.

After mapping is complete, the DNA must be sequenced to determine the order of all the nucleotide bases of the chromosomes and the genes in the DNA sequence must be identified. In all phases of the project, a major focus has been on developing instrumentation to increase the speed of data collection and analysis. New, automated technologies are significantly increasing the speed and accuracy of DNA sequencing while decreasing the cost. Software and database systems manage the data generated from mapping and sequencing projects, and database management systems store and aid in distributing genomic information.

Genetic linkage mapping is analogous to assigning city locations and other landmarks along a highway on a road map. The more information—detail—that the map presents, the more useful it is. A map of the United States showing only Boston and San Francisco as end points on a coast-to-coast highway would be of little value. Adding other cities—Pittsburgh, Chicago, Denver, Salt Lake City—would increase its utility, as would adding national and state parks, landmarks, monuments of interest, and the intersections of other highways.

Genetic linkage maps show the order and genetic distance between pairs of **linked** genes—that is, genes on the same chromosome—that determine variable **phenotypic traits**. (The difference between genetic distance and physical distance is explained below.) Genetic linkage maps enable geneticists to follow the inheritance of specific traits (that is, genes) as they are passed from generation to generation within families.

Linkage maps also determine the arrangement of genes or markers with unknown functions on the chromosomes; they show the order of linked genes and pairwise distances between their loci. During **meiosis**, as the haploid egg and sperm cells form, homologous chromosomes (maternally and paternally derived) line up, and DNA segments can be exchanged between the homologs. Figure 9.2 shows that new combinations of alleles result from this process of homologous recombination. During meiosis, each human chromosome pair is involved, on average, in 1.5 crossover events. The likelihood of **crossing over** increases as the distance between the two loci increases. Crossing over between two genes or markers on the same chromosome sometimes can occur if there is enough distance between them. If two genes are very close, they are linked, and recombination is unlikely to occur between them. Thus, the frequency of recombination is a quantitative index of the linear distance between two genes on a genetic linkage map. Distances are measured in **centiMorgans** (cM), named for the famous geneticist Thomas Hunt Morgan. If genes

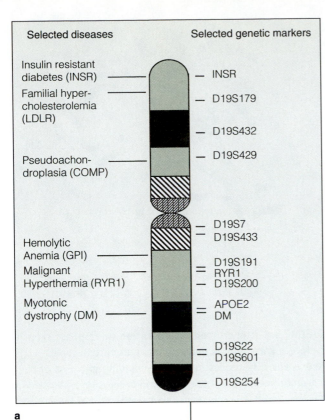

a

Selected diseases | **Selected genetic markers**

Insulin resistant diabetes (INSR) — INSR

Familial hyper-cholesterolemia (LDLR) — D19S179

— D19S432

— D19S429

Pseudoachon-droplasia (COMP)

— D19S7
— D19S433

Hemolytic Anemia (GPI)

Malignant Hyperthermia (RYR1) — D19S191
— RYR1
— D19S200

Myotonic dystrophy (DM) — APOE2
— DM

— D19S22
— D19S601

— D19S254

FIGURE 9.1 (a) Diagram of human chromosome 19 with the locations of selected disease genes and genetic markers. (b) Diseases found on chromosome 19 as of 2003.

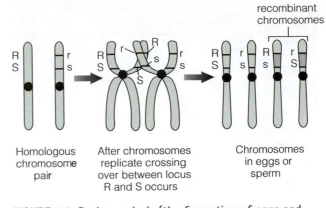

Homologous chromosome pair | After chromosomes replicate crossing over between locus R and S occurs | Chromosomes in eggs or sperm

recombinant chromosomes

FIGURE 9.2 During meiosis (the formation of eggs and sperm), homologous chromosomes sometimes cross over and exchange DNA. Crossing over produces genetic diversity, since new combinations of alleles are established within the gametes.

67 million bases

19

Coxsackie virus sensitivity
Cyclic hematopoiesis
Fucosyltransferase-6 deficiency
Hypocalciuric hypercalcemia, type II
Leukemia, myeloid/lymphoid or mixed-lineage
Wegener granulomatosis autoantigen
Bleeding disorder
Persistent Mullerian duct syndrome, type I
Mucolipidosis
Glutaricaciduria, type I
Leprechaunism
Rabson-Mendenhall syndrome
Diabetes mellitus, insulin-resistant
Ichthyosis
Leukemia, T-cell acute lymphoblastoid
Liposarcoma
Mycobacterial and salmonella infections, susceptibility to
Eye color, green/blue
Hemiplegic migraine, familial
Episodic ataxia, type 2
Ataxia, spinocerebellar and cerebellar
Leukemia, acute myeloid
Mannosidosis, alpha, types I and II
Alzheimer disease, late onset
Glomerulosclerosis, focal segmental
Deafness, autosomal dominant
Hypercalcemia, familial benign, Oklahoma type, type III
Orofacial cleft
Charcot-Leyden crystal protein
Hemolytic anemia
Hydrops fetalis
Malignant hyperthermia susceptibility
Central core disease
Osteodysplasia, polycystic lipomembranous
Maple syrup urine disease, type Ia
Camurati-Engelmann disease
Myotonic dystrophy
Heart block, progressive familial, type I
Optic atrophy
3-methylglutaconicaciduria, type III
Cystic fibrosis modifier
Meconium ileus in cystic fibrosis, susceptibility to
Cone dystrophy
Leber congenital amaurosis
Retinitis pigmentosa, late-onset dominant
Diabetes mellitus, noninsulin-dependent
Hyperferritinemia-cataract syndrome
Hypogonadism, hypergonadotropic
Retinitis pigmentosa, autosomal dominant
Ectrodactyly, ectodermal dysplasia, cleft lip/palate

Ataxia, cerebellar, Cayman type
Convulsions, familial febrile
Guanidinoacetate methyltransferase deficiency
Muscular dystrophy
Hirschsprung disease
Peutz-Jeghers syndrome
Leukemia, acute lymphoblastic
Atherosclerosis, susceptibility to
Malaria, cerebral, susceptibility to
Sicca syndrome
Glioblastoma
Thyroid carcinoma, nonmedullary
Low density lipoprotein receptor
Hypercholesterolemia, familial
Arteriopathy, cerebral
Pseudoachondroplasia
Epiphyseal dysplasia, multiple
Severe combined immunodeficiency disease
Hair color, brown
Leigh syndrome
MHC class II deficiency
Exostoses, multiple, type 3
Benign familial infantile convulsions
Leukemia/lymphoma, B-cell
Spondylocostal dysostosis, autosomal recessive
Prostate-specific antigen
Spastic paraplegia, autosomal dominant
Cystinuria, types II and III
Nephrosis, congenital, Finnish type
Generalized epilepsy with febrile seizures plus
Ovarian carcinoma
Microcephaly, autosomal recessive
Hyperlipoprotinemia, types Ib and III
Myocardial infarction susceptibility
Cytochrome P450 (coumarin resistance)
Nicotine addiction, protection from
X-ray repair
Excision repair
Xeroderma pigmentosum, group D
Trichothiodystrophy
DNA ligase I deficiency
Polio virus receptor
Herpes virus entry mediator B
Glutaricaciduria, type IIB
Colorectal cancer
Leukemia, T-cell acute lymphoblastic
Shaw-related subfamily genes
Melanoma inhibitory activity
Cardiomyopathy, familial hypertrophic

b

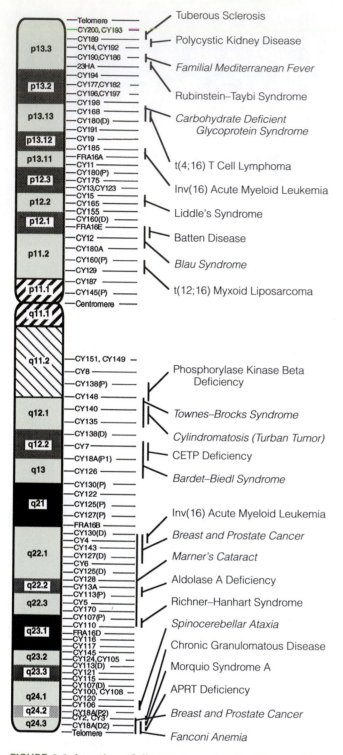

FIGURE 9.3 Location of disease genes on chromosome 16. Italicized genes have not yet been cloned.

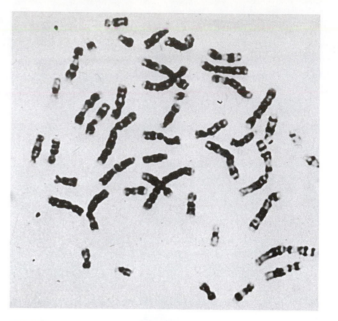

FIGURE 9.4 A microscopic preparation (metaphase squash before a karyotype is made) of human chromosomes shows the differences in size and banding patterns of the chromosomes. Dept. of Energy, Human Genome 1991–1992 Program Report, Primer on Molecular Genetics, pp 198 and 216. Prepared by the Human Genome Management Information System, Oak Ridge National Laboratory.

marker is also inherited in individuals with the disease but is not found in individuals who do not have that disease. Exact chromosomal locations have been found for many disease genes, including fragile X syndrome, cystic fibrosis, and Huntington's disease. Figure 9.3 shows the locations of numerous disease genes and markers on chromosome 16.

Another type of map that has been used is the cytological map, which sometimes helps align the genetic linkage map with the physical map. In the past, this map has encompassed the banding pattern of the different chromosomes. Chromosomes are distinguished from one another by banding patterns and size differences (see Figure 9.4). Chromosomes can be detected with a light microscope during metaphase of the cell cycle, and special stains are used to detect the pattern of banding. A pattern of alternating light and dark bands is observed in **metaphase chromosomes**. The pattern reflects the differences in the ratio of the bases adenine and thymine to guanine and cytosine. A picture of all the chromosomes grouped by size and type (1 through 22 with X and/or Y) is called a **karyotype** (an example is in Figure 9.5). Karyotyping is still used today to diagnose a variety of human genetic disorders. Some chromosomal abnormalities, such as broken or missing chromosomes or extra copies (for example, people with Down syndrome have three occurrences of chromosome 21, as Figure 9.5[c] shows), are readily visualized. However, karyotyping does not detect most

(for example, A and B) are separated by recombination 1% of the time—that is, if 1 out of 100 products of meiosis is recombinant—they are 1 cM apart. A genetic distance of 1 cM represents a physical distance of approximately 1 million base pairs, or 1 megabase (Mb). Genetic maps are very powerful: An inherited disease gene can be located on the map if a second gene or DNA reference

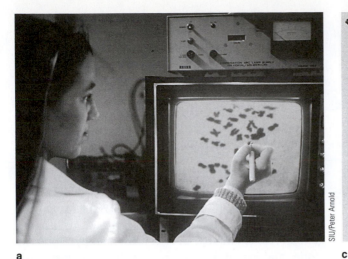

a

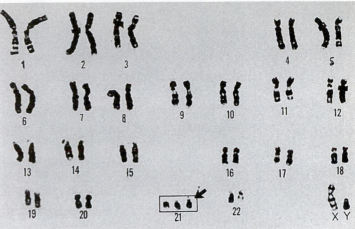

c

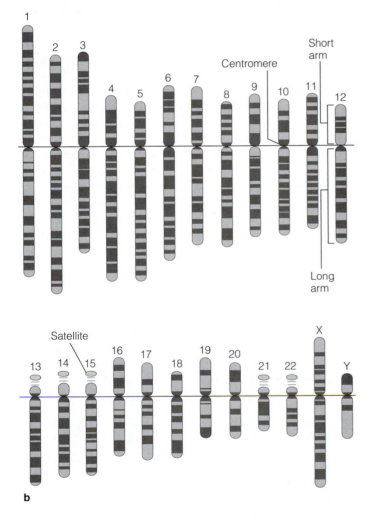

b

FIGURE 9.5 Karyotyping. (a) The use of an image analysis computer to prepare a karyotype. Homologous chromosomes are matched and organized by size. (b) Diagram of a human karyotype. Chromosomes are numbered in order of size, although chromosome 21 is smaller than chromosome 22. Banding patterns are shown. (c) An individual's 46 chromosomes (22 pairs of autosomes and one pair of sex chromosomes, in this case, X and Y) can be arranged by size and banding pattern into a karyotype for the diagnosis of genetic disease. In this example, an extra copy of chromosome 21 identifies the individual as having Down's syndrome. (c) from Dept. of Energy, Human Genome 1991–1992 Program Report, Primer on Molecular Genetics, pp 198 and 216. Prepared by the Human Genome Management Information System, Oak Ridge National Laboratory.

genetic diseases such as cystic fibrosis, Huntington's disease, or certain cancers. Although karyotyping detects large changes in the complement of chromosomes, such as chromosome breakage, duplications, and translocations, it does not detect small mutations, such as base changes and small insertions or deletions.

A more recent method of constructing cytological maps is the use of fluorescence *in situ* hybridization (FISH) of probes to metaphase chromosomes. Meta-

phase chromosomes are spread out on a microscope slide, and a solution containing a fluorescent-tagged DNA probe is added. Under appropriate conditions, the probe hybridizes to its DNA complement on the chromosome and is detected with a fluorescence microscope (Figure 9.6). The relative orientation of genes and DNA fragments can be assigned to specific chromosomes, and the gaps between mapped cosmids can be bridged. Cytological maps are sometimes used to locate genetic markers that are associated with observable traits.

Physical maps involve the assembly of contiguous stretches of DNA, called contigs, where the distances between landmark DNA sequences are expressed in kilobases (1 kb = 1000 nucleotides) (Figure 9.7). Physical maps provide information about the physical organization of the DNA; examples are the location of restriction enzyme sites and the order of chromosomal restriction fragments. Entire genomes can be studied with a library of genomic DNA; however, library clones are uncharacterized, random fragments and are not placed in order as they are on the chromosome.

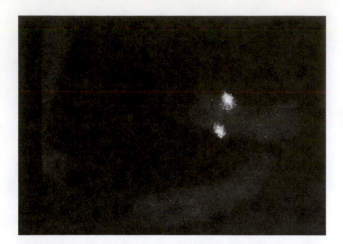

FIGURE 9.6 Fluorescence *in situ* hybridization (FISH) of a cosmid probe to chromosome 16. The light spots show where the probe hybridized to the chromosome. FISH helps assign DNA to specific chromosomes and fill in gaps between mapped cosmids. Metaphase chromosomes comprise two sister chromatids; therefore, a probe binds in two locations on each chromosome. Courtesy of Norman Doggett, Center for Human Genome Studies, Los Alamos National Laboratory, Los Alamos, New Mexico

The ultimate physical map is the complete DNA sequence of the genome. Markers of various types (for example, restriction fragment length polymorphisms [RFLPs]) can then be placed on the sequenced physical map to provide (1) high-resolution genetic and linkage maps and (2) detailed comparisons of collinear regions of, for example, two genomes.

Because many genomes are very large, large DNA fragments must be cloned into vectors to maintain a manageable number of clones in the library. Yeast artificial chromosomes (YACs) and bacterial artificial chromosomes (BACs) have been used as cloning vectors to clone thousands of kilobases and up to 300 kb of DNA, respectively. For YACs, genomic DNA is attached to the yeast DNA and transferred into yeast host cells for replication. Only a small portion of the yeast's total DNA—origin of replication, telomere, and centromere—is required for replication, so most of the YAC is the foreign DNA insert. BACs are similar in organization to typical plasmid cloning vectors; however, the copy number within a cell is close to one to prevent possible recombination with other DNA. This helps maintain the structural integrity of the cloned DNA insert.

The average insert used in YAC libraries is 200,000 to 400,000 base pairs long. This range is 10 times larger than inserts used in other libraries, such as for bacteriophage and cosmids, where up to 20,000 to 40,000 base pairs, respectively, can be cloned. The human genome, for example, can be represented by 7500 YAC clones and is maintained and amplified in yeast host cells. YACs and their inserts are cut into smaller fragments and recloned or subcloned (for example, into cosmids) so that a detailed map of a YAC clone is obtained. DNA inserts also are analyzed by obtaining restriction maps, identifying polymorphic markers, or DNA sequencing. However, without an ordered physical map—one that refers to actual physical distances in base pairs between landmarks—the location of particular clones cannot be identified.

Another type of physical map is the complementary DNA (cDNA) map, which localizes coding regions (exons, for example) to specific chromosome regions or bands. The cDNA molecules are synthesized from an mRNA template. A cDNA map identifies the chromosomal location of specific genes, whether their functions are known or not. Researchers searching for a specific disease-causing gene, for example, can use cDNA maps to help locate it after having established a general location by genetic linkage methods.

High-resolution physical maps can be generated using what is called a shotgun approach. The genome is cut into small (many overlapping) fragments, each is cloned, and then sequenced. A computer assembles the fragments into contigs. Contigs are sets of DNA frag-

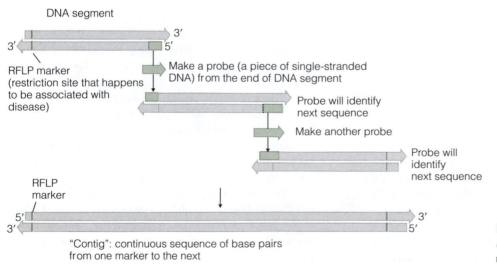

FIGURE 9.7 The assembly of a contig.

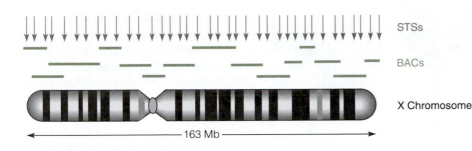

STSs

BACs

X Chromosome

← 163 Mb →

FIGURE 9.8 The X chromosome and sequence-tagged sites (STSs) scattered throughout the chromosome and on overlapping bacterial artificial chromosome (BAC) clones. From Discovering Geonomics, Proteonics, and Bioinformatics, by Campbell and Heyer. Copyright 2003, page 7, fig. 1.3. Published by Benjamin Cummings.

ments that are placed in order to form continuous, overlapping, linear stretches of DNA. This method generates a detailed map called a contig map.

Contig maps require several steps to produce. First, a library must be made that represents the particular genome—either the entire genome or a segment—in cloned DNA fragments. The DNA fragments within each clone must overlap other fragments. Overlap is accomplished by cutting the DNA with a specific restriction enzyme (or by random sharing of the DNA). If every restriction site on the DNA were cut, no fragments would overlap. Therefore, enzyme digestion is conducted in such a way that not every DNA restriction site is cut. This partial digestion randomly leaves many sites uncut so that overlapping DNA fragments are produced and the order along the chromosomes can be determined.

The order of the clones or contigs is determined by identifying the overlaps in the DNA fragments. Overlaps are detected when some of the DNA bands are the same—that is, two clones have bands in common. A universal reference system of DNA sequence identification has been developed. These regions serve as landmarks along a genome. Unique regions of 200 to 500 base pairs of partially sequenced DNA are used to identify clones, contigs, and long stretches of DNA. These **sequence-tagged sites** (STSs) are standard markers that are used for physical mapping (Figure 9.8). An STS also can be a region of cDNA—that is, an expressed sequence, called an expressed-sequence tag (EST). ESTs are used to represent landmarks along the map, thus helping to identify where pairs of clones overlap. These special sequences constitute a universal mapping language, enabling everyone to refer to a specific region of the genome by the same name, and enabling investigators to share information and construct compatible maps.

To identify an STS, a short DNA fragment is isolated from a chromosome-specific library of clones or from a small fragment within a clone from a contig. A small region 200 to 400 base pairs long is sequenced and compared with all known repeated sequences. Sophisticated computer analysis programs help identify unique sequences. Polymerase chain reaction (PCR) can be used to screen a library for clones containing STSs. PCR amplifies the unique DNA region from the total human genome. Gel electrophoresis will yield one

DNA band if the STS is unique and more bands on the gel if additional regions are amplified, indicating that the STS is not unique.

An STS also can include a repeated sequence if unique sequences flank the repeat on either side, for example, unique sequence GA … GA unique sequence. STSs then would be polymorphic in the region of the repeat (that is, repeated *n* times) and have many alleles in a population. Each individual carries two copies of the STS marker—one on each homologous chromosome. Thus, the inheritance of the STS can be determined. Once located on the physical map, polymorphic STSs aid in aligning the genetic linkage map with the physical map.

The use of STSs as probes for contig mapping allows clone overlap to be established; this method is called STS-content mapping. Clones that share an STS overlap belong in the same contig. To be most useful, STSs should be spaced every few thousand base pairs. STS information is stored in computer databases analogous to, for example, the large DNA sequence database called GenBank. STS sequences are identified from the library and are placed on the physical map as landmarks in the genome. They can be used as hybridization probes to fill in gaps or to identify an adjacent contig. Because assembling contigs is difficult and time-consuming, automation and sophisticated computer algorithms are used to greatly increase efficiency.

Several different approaches have been used to fill in the gaps that are likely to be present even after researchers generate detailed physical maps. For example, as Figure 9.9 shows, microdissection has been used to physically cut a piece of DNA from a specific region of a chromosome. This chromosomal piece can be cut into smaller fragments by restriction enzymes, then cloned, mapped, and sequenced by standard methods.

Alternatively, chromosome walking is often used (see previous discussion of STS). A small region at the end of the DNA fragment is used as a probe to screen the library for the adjacent clone. A DNA piece at the end of this second cloned fragment is used as the next probe. This process continues until a complete physical map has been obtained.

Chromosome-specific libraries can be constructed so that each chromosome has a contig map. Mapping is simplified if each chromosome is separated from the others before being cut by restriction enzymes and

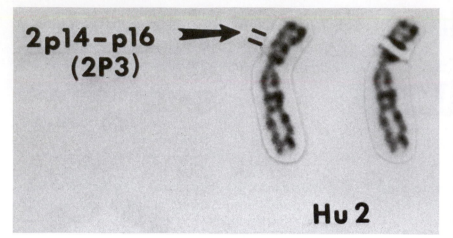

FIGURE 9.9 Microdissection of a DNA fragment (2p14–p16) from the short arm of human chromosome 2. A library of the DNA is constructed by cloning into a suitable vector. Clones from the library can be used to isolate larger regions of DNA from cosmid and YAC clones for the establishment of contigs and high-resolution physical maps. Left is before dissection, right is after dissection.

J. Yu, S. Tong, J. Qi, and F–T Kao, Somatic Cell and Mol. GE. 20:353–357 (1994). Reprinted by permission of Plenum Publishing Corp. and the authors.

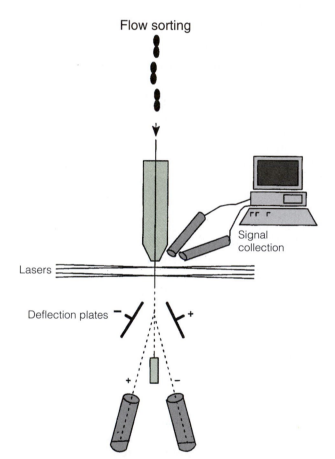

Flow sorting

Signal collection

Lasers

Deflection plates

FIGURE 9.10 A simple diagram of a dual laser flow cytometer and sorter that rapidly sorts chromosomes. Chromosomes fluoresce after staining with AT- and GC-specific dyes (each emitting light at a different wavelength). Laser beams excite the fluorescence of the dyes bound to the chromosomes so that each chromosome type emits different light intensities at each wavelength (which depends on the GC:AT content). This enables discrete chromosomes to be identified (with the exception of chromosomes 9 through 12 because of their similar fluorescence patterns). A particular chromosome can be separated by giving it an electric charge so that charged deflection plates deflect it into a collection tube.

cloned to make libraries. For the human genome, 24 libraries were required: 22 autosomal libraries and 1 each for the X and Y chromosomes.

Chromosomes can also be separated by size using a process called **flow cytometry**. Cell-sorting machines called flow cytometers (shown schematically in Figure 9.10) can analyze more than 2000 chromosomes per second and sort more than 50 chromosomes per second. The separation method uses the fluorescence of chromosomes that have been stained with two dyes, each one preferentially binding to DNA that is rich in either adenine and thymine (AT) or guanine and cytosine (GC). Two laser beams excite the fluorescence of each dye bound to the chromosomes. Because the AT and GC content of chromosome regions differ, each chromosome will emit different light intensities for each of the two dyes. A computer analyzes the fluorescence characteristics for each chromosome and uses this information to sort the chromosomes. Some chromosomes, such as chromosomes 9 through 12, cannot be separated because they give very similar fluorescence patterns. A specific chromosome can be separated from the rest by placing an electric charge on it and using charged deflection plates to channel it into a collection tube.

To summarize, the several types of maps range from coarse to fine resolution. The map with the lowest resolution is the genetic map, which measures the frequency of recombination between linked markers (which can be genes or noncoding DNA). The next level of resolution is the restriction map, on which DNA restriction fragments of 1–2 Mb are separated and mapped. The next higher level of resolution is achieved by placing in order 400,000 to 1,000,000 base pair fragments of overlapping clones from libraries of YAC clones. These clones then are further subcloned (with insert sizes of 20,000 to 40,000 base pairs) into other vectors to produce contig maps. Finally, the map with the finest resolution is the DNA sequence. This is the linear order of nucleotides or bases of the DNA for each chromosome within the genome.

HUMAN GENOME PROJECT

The human genome harbors approximately 30,000-40,000 genes (revised downward from the estimation of 80,000-100,000 genes before the generation of the DNA sequence): 23 pairs of chromosomes, one set from each parent, made up of one sex chromosome pair (XX or XY) and 22 autosomal chromosome pairs. The 3 billion base pairs in the haploid genome contain an amount of information equivalent to 200 telephone books of 1000 pages each (see Figure 9.11). The smallest chromosome (Y) is about 50 million base pairs, the largest (chromosome 1) is 250 million. If all the chromosomes in one human cell were removed, unwound, and placed end to end, the DNA would stretch more than 5 feet (although it would be only fifty-trillionths of an inch wide). The establishment of gene order within the genome, the determination of the nucleotide sequence of the genome, and the elucidation of gene function and regulation would be a huge boon to the identification and treatment of disease.

In 1988 a committee organized by the National Institutes of Health (NIH) and the Department of Energy (DOE) developed an action plan for the Human Genome Project. In 1990 a 5-year joint research proposal was submitted to Congress and in October 1990 the Human Genome Project officially began. The project was organized and supported primarily by the DOE and the NIH, which established working groups to address genome mapping; computational analysis to handle databases; and the social, legal, and ethical implications of human genome research. Congress provided funds for the project to the National Center for Human Genome Research (NCHGR, directed first by James Watson and after 1993 by Francis Collins) at the NIH, which in turn awarded grants and contracts to U.S. investigators. Congress provided additional funds to the DOE, which conducts research on human genetics at three national laboratories and funds independent investigators. About 50% of the funding supported study of the human genome; 25% supported study of other model organisms; another 20% supported conferences, training, and program administration; and 5% supported consideration of ethical issues. The Human Genome Organization (HUGO), established in 1988, facilitated the international scientific effort; Canada, Japan, France, the United Kingdom, and Italy established genome research programs, and the United Nations Educational, Scientific, and Cultural Organization (UNESCO) facilitated and promoted the inclusion of developing countries in international human genome initiatives.

Goals of the Human Genome Project

The goals of the Human Genome Project, listed in the following, originally were divided into two 5-year plans. The Human Genome Project has expanded as old objectives are met and new goals are added.

1. Generation of high-resolution genetic and physical maps that will help in the location of genes implicated in disease

2. Establishment of DNA sequence landmarks that will help to complete the entire genome sequence by 2005. Interestingly, two drafts of the sequence funded by both private and public initiatives were completed in 2000.

3. The identification of all the genes in the genome using bioinformatics (see later discussion), ESTs, and comparative genomics (see later discussion)

4. The establishment of polymorphism databases, in particular single-nucleotide polymorphisms (SNPs) (see Chapter 11, DNA Profiling, Forensics, and Other Applications) that will aid in disease diagnostics, the variation in human populations on a global scale, and human evolution.

Other goals include the development of new methods and tools for

1. DNA sequencing

2. The study of gene expression

human genome 200 telephone books
(1000 pages each)

model organism genomes		
Drosophila (fruit fly)	10 books	
yeast	1 book	
E. coli (bacterium)	300 pages	
yeast chromosome 3	14 pages	

FIGURE 9.11 Compiling the DNA sequence from the human genome into books would require 200 volumes, each the size of the 1000-page Manhattan telephone book.

3. The field of bioinformatics

4. The establishment and management of informational databases

The goals of the first 5 years (1993–1998) and the second 5 years (1998–2003) are outlined in Table 9.1. The scope of the Human Genome Project is enormous. Mapping and sequencing the genome was projected to take 15 years and be completed by 2005, at a cost of $3 billion—although these goals were completed by 2000 at a lower cost than originally predicted. Now that mapping and draft sequences are completed, many years will be needed to completely identify all the genes and determine the mechanisms of gene expression and the genetic basis of diseases. Other genome initiatives have arisen from the Human Genome Project.

Some of the historical achievements are listed here (Figure 9.12):

1994—The first high-resolution genetic map of the complete human genome was published. Twenty-three linkage groups (one on each chromosome) were identified with 3000 markers approximately 1 cM apart.

1995—The publication of a physical map constructed from 52,000 STSs approximately 60 kb apart. This formed the foundation for the public sequencing project of the International Genome Sequencing Consortium (IHGSC). Also in 1995, a database of 30,000 ESTs was released, as well as larger databases constructed by the commercial sector.

1998—A physical map was generated with 52,000 STSs.

2000—More than 1 million SNPs were mapped. This yielded a polymorphic marker at approximately every 2 kb in the genome. Approximately 85% of all exons were now within 5 kb of an SNP. Also in 2000, the first draft of the sequence for chromosome 21 was published.

2001—The public human genome sequence draft was published.

2002—Sequencing rate exceeds more than 1400 Mb per year at a cost of less than nine cents per base.

2003—Up to 99.9% of the gene-containing portion of the human genome sequenced to 99.9% accuracy, and 15,000 full-length cDNAs (DNAs obtained from mRNA, see Chapter 2, From DNA to Proteins) have been obtained. Other genome sequences have been completed—*E. coli*; the yeast *Saccharomyces cerevisiae*; the worm *Caenorhabditis elegans*; and the fruitfly *Drosophila melanogaster*. Whole-genome drafts of other genomes completed—several microbes, mouse, and rat.

TABLE 9.1 Goals of the Human Genome Project

First 5-year Plan: 1993–1998

1. THE GENETIC MAP

   Complete 2 to 5 cM map by 1995

   Develop new technology for rapid and efficient genotyping

2. THE PHYSICAL MAP

   Complete STS map to 100 kb resolution

3. DNA SEQUENCING

   Develop approaches to sequence highly interesting regions on Mb scale

   Develop technology for automated high throughput sequencing

   Attain sequencing capacity of 50 Mb per year; sequence 80 Mb by 1998

4. GENE IDENTIFICATION

   Develop efficient methods for gene identification and placement on maps

5. TECHNOLOGY DEVELOPMENT

   Substantially expand support for innovative genome technology research

6. MODEL ORGANISMS

   Finish STS map of mouse genome to 300 kb resolution

   Obtain complete sequence of biologically interesting regions of mouse genome

   Finish sequences of *E. Coli* and *S. cerevisiae* genomes

   Substantial progress on complete sequencing of *C. elegans* and *D. melanogaster*

7. INFORMATICS

   Continue to create, develop, and operate databases and database tools

   Consolidate, distribute, and develop software for genome projects

   Continue to develop tools for comparison and interpretation of genome information

8. ETHICAL, LEGAL, AND SOCIAL IMPLICATIONS (ELSI)

   Continue to identify and define issues and develop policy options

   Develop and disseminate policy regarding genetic testing

   Foster greater acceptance of human genetic variation

   Enhance public and professional education programs on sensitive issues

9. OTHER

   Training of interdisciplinary genome researchers

   Technology transfer into and out of genome centers

   Outreach

TABLE 9.1 Goals of the Human Genome Project *(continued)*

Second 5-Year Plan: 1998–2003

1. THE HUMAN DNA SEQUENCE

   Complete sequence by end of 2003, 2 years earlier than initially planned

   Complete one-third of sequence by end of 2001

   Finish working draft of 90% of genome in mapped clones by end of 2001

2. SEQUENCING TECHNOLOGY

   Continue to increase throughput and reduce cost

   Support research on new technologies and integration with genome projects

3. HUMAN GENOME SEQUENCE VARIATION

   Develop technologies for large-scale SNP identification and scoring

   Identify common variants in coding regions of most identified genes

   Create SNP map of at least 100,000 markers

   Develop intellectual foundations for studies of sequence variation

   Create public resources of DNA samples and cell lines

4. TECHNOLOGY FOR FUNCTIONAL GENOMICS

   Develop cDNA resources

   Support methods for study of function of non-protein coding sequences

   Develop technology for comprehensive analysis of gene expression

   Improve methods for genome-wide mutagenesis

   Develop technology for global protein analysis

5. COMPARATIVE GENOMICS

   Complete sequences of *C. elegans* and *D. melanogaster* genomes by 2002

   Develop mouse physical and genetic maps and cDNA resources

   Aim to complete mouse genome sequence by 2005

   Identify and start work on other important model organisms

6. ETHICAL, LEGAL, AND SOCIAL IMPLICATIONS (ELSI)

   Examine issues surrounding completion of human sequence and variation

   Examine issues relating to genetic technologies and public health activities

   Examine issues relating to genotype × environment interactions

   Explore interaction of genomics with philosophy, theology, and ethics

   Explore socioeconomic, racial, and ethnic factors relating to the HGP

7. BIOINFORMATICS AND COMPUTATIONAL BIOLOGY

   Improve content and utility of databases

   Develop better tools for data generation, capture, and annotation

   Develop and improve tools for comprehensive functional studies

   Improve tools for defining and analyzing sequence similarity and variation

   Create mechanisms for production of robust, exportable shared software

8. OTHER

   Train and nurture genome science academic career path

   Produce more scholars with genetic and ethical/legal/sociological training

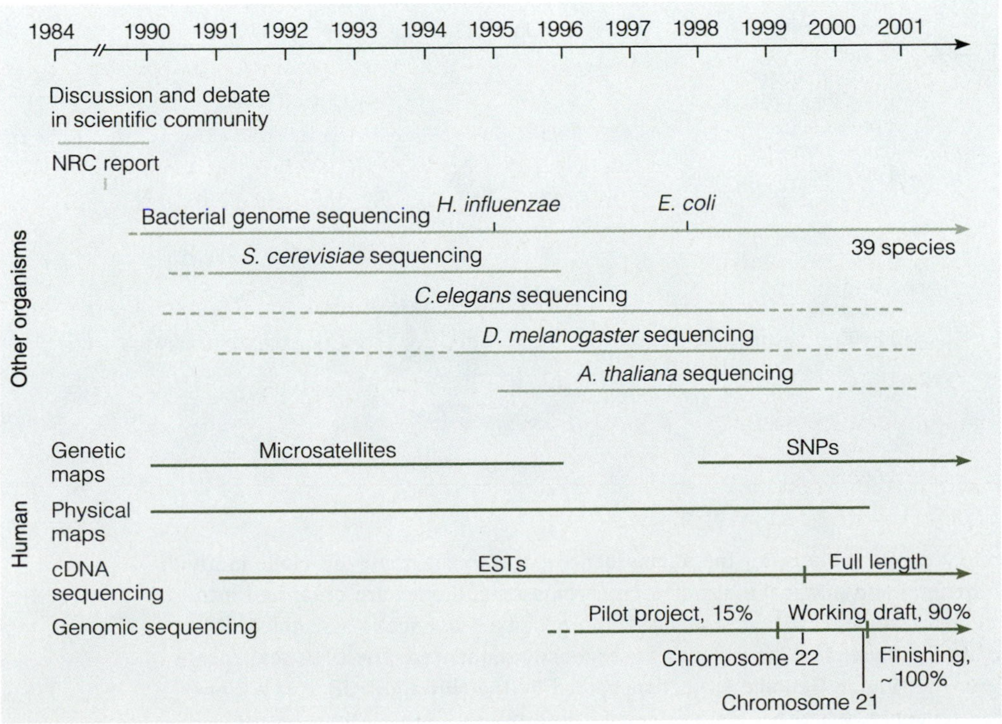

FIGURE 9.12 A history and timeline of the Human Genome Project and the genomes of other organisms. Reprinted with permission from *Nature,* 409: 860–921, fig. 1, p. 862. Copyright Macmillan Magazines Limited.

Detailed genetic and physical maps, the complete nucleotide sequence, and functional studies will ultimately be used to provide the location and identification of all the genes and other components—introns, repetitive sequences, and so on—of the genome. This has the potential to dramatically increase our understanding of human evolution and variation, gene regulation, human development, and human disease. Intricate knowledge of the human genome and gene function will revolutionize the medical field, leading to new areas of research, diagnostics, and treatments for genetic disorders. Areas such as genetic counseling and human genetics will be revolutionized. The Human Genome Project has stimulated the development of new technologies for analyzing and organizing the vast quantities of data being generated. These technologies are a vital component of the Human Genome Project.

Public and Private First Drafts of the Human Genome Sequence

Two completed drafts, one privately funded by Craig Venter of Celera Genomics and one public (led by Francis Collins, director of NCHGR) were published in February 2001, although completion of the DNA se-

quence was announced at a press conference in 2000. During the 9 months between announcement and publication, the DNA sequence was refined and improved (for example, closing gaps and verifying ambiguous regions).

The DNA used to sequence the human genome was obtained from anonymous donors (each donor signed a consent form). The publicly funded project through the IHGSC and the privately funded project by Celera Genomics used ethnically diverse genome libraries. However, only one person's DNA contributed 75% and 66% to the IHGSC and Celera genomic projects, respectively.

The public project by IHGSC used eight libraries for the assembly of the genome sequence (that is, physical map). They used only male DNA donors for construction of the first genome sequence draft, although one or more female DNA donors were included in later sequencing refinements. The ethnic identities of the donors are unknown and were chosen randomly from more than 50 ethnically diverse donors.

Celera Genomics obtained 21 ethnically diverse DNA donors from four different ethnic groups. Five libraries were used: one African, one Asian, one His-

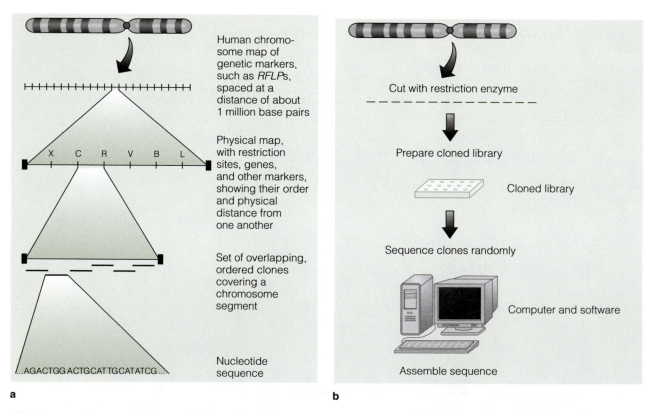

a

b

FIGURE 9.13 Two approaches to cloning and sequencing the human genome. (a) In the clone–by–clone method, a genomic library is used to make genetic and physical maps of each chromosome. Clones are organized into overlapping fragments and each cloned fragment is then sequenced. The genome sequence is assembled. (b) In the shotgun approach, a genomic library is prepared, and clones are randomly sequenced. The DNA sequence is assembled using computer software. The Human Genome Project sponsored by the NIH and DOE used a clone–by–clone method of assembling the sequence, whereas Celera Genomics used the shotgun method.

panic, and two Caucasian. Two were from male donors and three were from females.

The results of the human genome sequence were similar for both sequences (Figures 9.13 and 9.14). Some of the findings include the following:

1. The human genome contains a little more than 3 billion base pairs (3164.7 million nucleotides).

2. The number of genes found was close to one-third of the expected number of genes in the genome (30,000–40,000 protein-encoding genes rather than the predicted 80,000–100,000).

3. Only approximately 1.5% of the human genome encodes for the production of proteins. Repeated sequences that do not encode proteins make up at least 50% of the genome—they are thought to affect chromosome structure and dynamics. During the past 50 million years, there has been a striking decrease in the rate of accumulation of repeats in the genome.

4. The average length of a gene is 3000 nucleotides, although sizes vary greatly. The longest gene is the dystrophin gene, which is 2.4 million nucleotides.

5. Genes are not evenly distributed across a chromosome, rather genes concentrate in random regions, with large expanses of noncoding DNA between them. Gene-dense areas are composed primarily of G- and C-containing nucleotides. Gene-poor regions are richer in A- and T-containing nucleotides. GC- and AT-rich regions are seen as light and dark, respectively, bands on chromosomes using a microscope. There are repeated stretches of up to 30,000 G- and C-containing nucleotides (CpG islands) that are adjacent to gene-rich areas. These are thought to provide a barrier between genes and noncoding regions and may help regulate gene expression.

6. The Y chromosome has the fewest genes (231), while chromosome 1 has the most genes of all the chromosomes (2968).

7. Approximately 200 genes originated directly from bacteria.

8. Human proteins are more complex than the proteins that have similar functions in lower organisms.

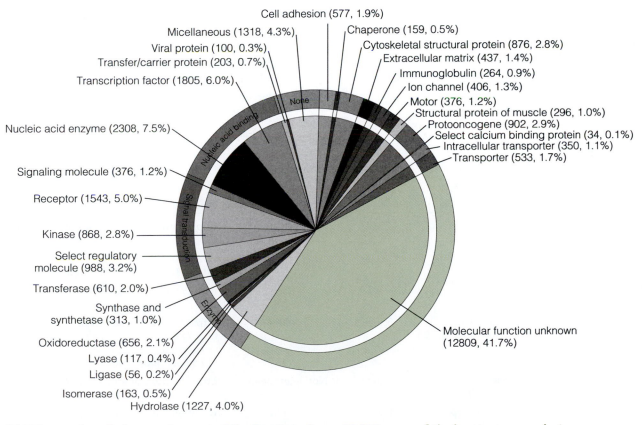

FIGURE 9.14 A preliminary assignment of the function of over 26,000 genes of the human genome. Just over 12,000 have no known function. Some classes of genes were assigned because they were identified from experimentation or were similar to other proteins of known function from other organisms. The most common genes are those that code for proteins involved in the metabolism of RNA and DNA—7.5% of all identified genes. Reprinted with permission from Venter, C. et al. (2001). The Sequence of the Human Genome. Science 291, 1304–1351, fig. 15, page 1335. Copyright 2001 AAAS.

TABLE 9.2 Organisms Associated with the Human Genome Project

Organism	Genome Size Million base pairs (Mb)	Estimated No. of Genes
Escherichia coli (bacteria)	4.64	4,300
Saccharomyces cerevisiae (yeast)	12	6,500
Caenorhabditis elegans (roundworm)	97	20,000
Arabidopsis thaliana (plant)	120	20,000
Drosophila melanogaster (fruitfly)	170	16,000
Homo sapiens (human)	3,300	35,000

9. The germline (sex cells) mutation rate in males is two times higher than in females.

10. The nucleotide sequence (order of the nucleotides) in all humans is 99.9% identical. There is no genome support for defined racial groups. The various ethnic groups share most of the sequence differences within the genome.

11. The functions for more than 50% of newly discovered genes is unknown.

Comparative Genomics

The Human Genome Project also has supported research on nonhuman organisms such as mouse, the fruitfly *Drosophila melanogaster*, and the nematode worm *Caenorhabditis elegans* (Table 9.2) because smaller, less complex genomes are easier to study, and correlations can be made to the human genome (referred to as comparative genomics). Sequencing information for nonhuman genomes generates invaluable information about gene organization, function, and evolution. During the evolution of organisms, many DNA regions, including genes, have remained almost unchanged. Comparisons of DNA sequences from different organisms (for example, mouse, yeast, nematode, fruitfly) identify linear regions of conservation (referred to as synteny—the conservation of gene order in divergent organisms) that can help determine function. One surprising outcome of comparative genomics is the large number of genes we share with other organisms. (Table 9.3 and Figure 9.15).

Human Genome Project and Disease

Many inherited diseases are caused by single genes and thus can be studied by genetic linkage analysis. Almost 5000 genetic disorders have been studied in this way. These maps, however, do not relate directly to the physical structure of DNA, and the gene of interest cannot be isolated on the basis of information from genetic linkage maps alone.

For human genome mapping, linkage analysis may involve the study of family members carrying a

TABLE 9.3 Comparison of Selected Genomes

Organism	Approximate Size of Genome (Date Completed)	Number of Genes	Approximate Percentage of Genes Shared with Humans	Web Access to Genome Databases
Bacterium (Escherichia coli)	4.6 million bp (1997)	4,403	Not determined	http://www.genome.wisc.edu/
Fruitfly (Drosophila melanogaster)	165 million bp (2000)	~13,600	50%	www.fruitfly.org/sequence.html http://Flybase.bio.indiana.edu/
Humans (Homo sapiens)	3 billion bp (February 2001)	30,000–40,000	100%	http://www.ornl.gov/hgmis/
Mouse (Mus musculus)	~3 billion bp (to be completed in 2001)	~35,000	~90%	http://www.informatics.jax.org/ http://www-genome.wi.mit.edu /genome_data/mouse /mouse_index.html
Plant (Arabidopsis thaliana)	125 million bp (2000)	~25,000	Not determined	http://www.arabidopsis.org/
Roundworm (Caenorhabditis elegans)	97 million bp (1998)	19,099	40%	www.genome.wustl.edu/gsc.C_elegans
Yeast (Saccharomyces cerevisiae)	12 million bp (1996)	~6,000	31%	http://genome-www.stanford.edu /Saccaromyces/

Source: Howard Hughes Medical Institute (2001), *The Genes We Share with Yeast, Flies, Worms, and Mice: New Clues to Human Health and Disease.*

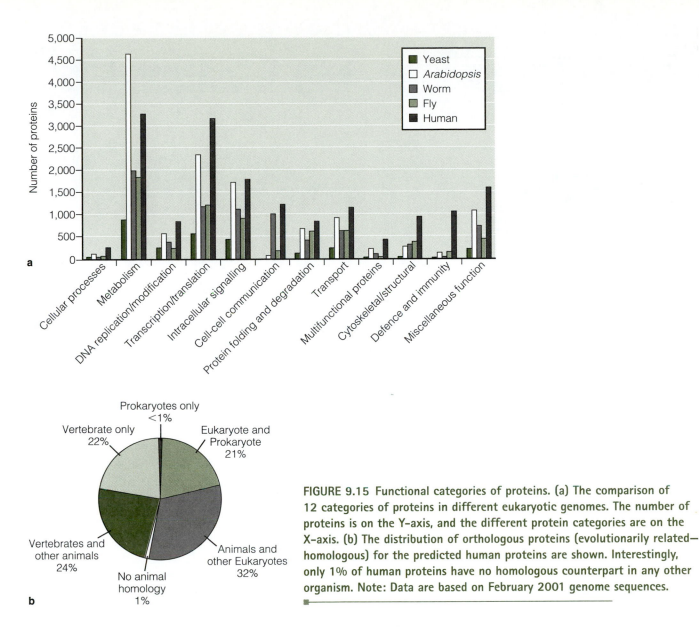

FIGURE 9.15 Functional categories of proteins. (a) The comparison of 12 categories of proteins in different eukaryotic genomes. The number of proteins is on the Y–axis, and the different protein categories are on the X–axis. **(b)** The distribution of orthologous proteins (evolutionarily related—homologous) for the predicted human proteins are shown. Interestingly, only 1% of human proteins have no homologous counterpart in any other organism. Note: Data are based on February 2001 genome sequences.

particular trait for an inherited disorder. Often, several generations of one family are studied to obtain enough information with which to infer linkage. Some family members must express the trait (gene) or genetic disorder, and the trait must vary among individuals (that is, there must be different alleles or forms of the gene). Analysis also requires that there be individuals who are **heterozygous** for DNA reference markers or who have a second gene linked to the gene in question. Heterozygous family members (members carrying two different forms of the trait or gene—one dominant and one recessive allele) enable geneticists to determine which chromosome of the homologous pair carries the allele for the genetic disorder and whether it is passed on to the offspring.

Specific physical regions of the DNA can be identified by noting actual variations in DNA sequence within a specific region for different individuals (also see Chapter 11, DNA Profiling, Forensics, and Other Applications). These sequence variations are called DNA polymorphisms. DNA polymorphisms that occur within a coding region or gene lead to changes in phenotypic or observable characteristics (required for genetic linkage mapping), such as eye or hair color or susceptibility to disease. Variations that occur in an intron or other noncoding region lead to little or no observable change in physical traits. Because these polymorphisms are detectable at the DNA level, they can serve as markers. Polymorphic markers also can be short, tandem, repeated sequences varying in the number of

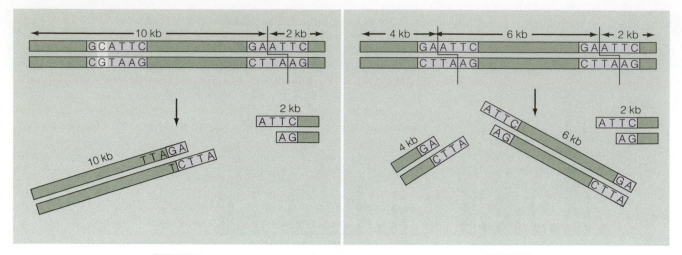

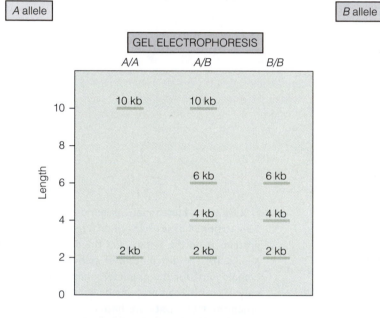

GENOTYPES		*RFLP* FRAGMENT SIZES
Homozygous	*A/A*	10 kb, 2 kb
Heterozygous	*A/B*	10 kb, 6 kb, 4 kb, 2 kb
Homozygous	*B/B*	6 kb, 4 kb, 2 kb

FIGURE 9.16 Restriction fragment length polymorphisms (RFLPs). A and B alleles represent alleles from homologous chromosomes. Arrows point to the site where a restriction enzyme cuts. Variation in the nucleotide sequence (highlighted) changes the enzyme recognition site. Thus, in allele A, the restriction site is missing. The variation in the site between allele A and B causes variation in the number of DNA fragments produced on a gel. In allele A, two fragments are produced (10 kb, 2 kb), whereas in allele B, three fragments are produced (6 kb, 4 kb, 2 kb). These variations are inherited, resulting in three possible genotypes: AA, BB, AB. The allele combination inherited by an individual can be determined by restriction enzyme digestion and gel electrophoresis followed by Southern blot hybridization. The banding pattern is shown for each genotype.

repeat units (that is, length). This type of linkage mapping is revolutionizing human genetics.

RFLPs reflect actual DNA sequence variations, which can be detected with restriction endonucleases (see Chapter 11, DNA Profiling, Forensics, and Other Applications). Restriction enzymes recognize specific nucleotide sequences and cut DNA at those sites. Fragments differ in length and number according to the location and number of restriction sites. A single base change within the recognition sequence can result in the gain or loss of a restriction endonuclease cut site. RFLPs are detected by gel electrophoresis and Southern blot hybridization (Figure 9.16).

Differences among individual family members can be detected when variations in DNA sequences exist. These sequence polymorphisms are passed from parent to offspring (Figure 9.17). These reference marker alleles are useful in studying the human genome. Because they are variable, they can be used just like variable phenotypic traits. Co-inheritance of DNA marker pairs can be used to determine the genetic distances between them, and thus a linkage map of DNA markers can be constructed. Co-inheritance of a DNA marker and a variable phenotypic trait can be used to determine the genetic distance between the marker and the gene. Polymorphic markers provide a way to relate loci on genetic linkage maps with the physical regions of human chromosomes. In addition, a mutant gene can be located by examining the co-inheritance of a reference marker and a disease in families—that is, a linkage of marker and mutant gene is established, as Figures 9.18 and 9.19 show. The physical location of the DNA marker on a chromosome can then be found by using the marker sequence as a DNA probe (as described in the next section, in the discussion of *in situ* hybridization). **Polymorphic DNA markers** serve as reference points, or landmarks, to help find a region of DNA that contains the gene of interest. If a gene is found between two DNA markers, that DNA region can be isolated for further study.

When a linkage map contains approximately 3300 polymorphic DNA markers, each separated by only 1 cM, gene hunting is much easier. Thus, for polymorphic DNA markers to be valuable, their linkage with a gene must be established, and their physical locations can be identified through the use of probes.

The use of polymorphic DNA markers for genetic linkage analysis is a powerful method for finding the approximate location of genes causing inherited diseases. If the polymorphic DNA markers are co-inherited or linked closely to the gene of interest, the markers and gene must be physically close to one another. To be amenable to further analysis, genetic linkage markers must be within 2 Mb (preferably no more

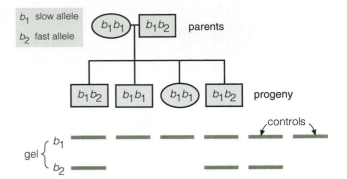

FIGURE 9.17 Schematic of Southern blot hybridization of DNA from parents and progeny cut with *Eco*RI and probed with DNA from locus b. b_1 is the "slow" allele, b_2 the "fast" allele. The inheritance of the restriction fragment length polymorphism (RFLP) at locus b in individuals can be demonstrated by the banding pattern of the autoradiogram.

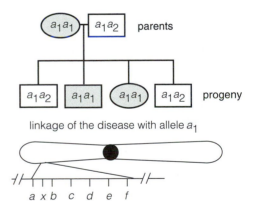

FIGURE 9.18 Demonstration of the co-inheritance or linkage of a marker (a) and a disease–causing allele (x) is a powerful way to determine the location of a disease gene. Shading identifies the individual with the disease x allele. When a and x are co-inherited, they are located so close together that a recombination event between them is rare. This linkage indicates that they are physically located close to one another. Southern blot hybridization with a probe for a is used to identify individuals (parents and progeny) with and without the disease and the a alleles they carry. Hybridization with a DNA probe for b also is necessary to establish linkage of b and the disease–causing allele x. A linkage map confirms the location of the disease-causing allele x.

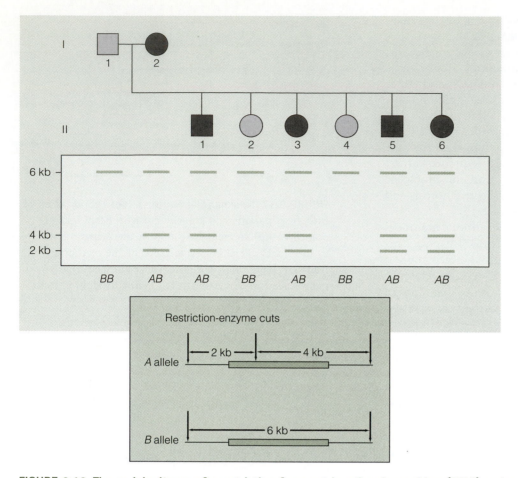

FIGURE 9.19 The co-inheritance of a restriction fragment length polymorphism (RFLP) and a dominant trait. A family pedigree is shown that illustrates the distribution of alleles A and B (refer to Figure 9.16) in family members. The individuals affected by a genetic disease are shown by solid symbols (males = squares; females = circles). The unaffected father is homozygous (has two copies) for allele B (6 kb), and the affected mother is heterozygous—AB. The affected offspring of the parents are shown with their genotypes. The example shows that these offspring have inherited the maternal allele A, whereas unaffected offspring have inherited allele B from both parents. This suggests that the mutant allele for the disease is on the same chromosome as (and close to) allele A.

than 1 cM) of the disease gene. To isolate the disease gene, a method called chromosome walking has been used (Figure 9.20). DNA markers are used to link the genetic linkage map to the physical map. Flanking DNA markers are used as a starting point in hybridizations to clones comprising contigs in the disease gene region. Once genetic linkage and physical maps are integrated, genes for inherited disorders can be identified, isolated, and sequenced.

The complete sequence of the human genome provides a physical map with the greatest resolution. The DNA sequence is invaluable in determining gene function, gene regulation, and the cause of genetic diseases. Genome diversity studies and SNP maps will be invaluable in the identification of disease-linked genes.

As of February 2003, approximately 3.7 million sites where SNPs can occur in the human genome have been identified.

OTHER GENOME PROJECTS

Today hundreds of genome projects are being undertaken that represent a range of organismal diversity—from viruses and microorganisms to plants and animals. Some of the genomes being studied are used in comparative genomics (for example, to help elucidate gene structure and function in the human genome), whereas other genome projects are initiated for biotechnological applications (microbial applications such as the development of renewable fuels).

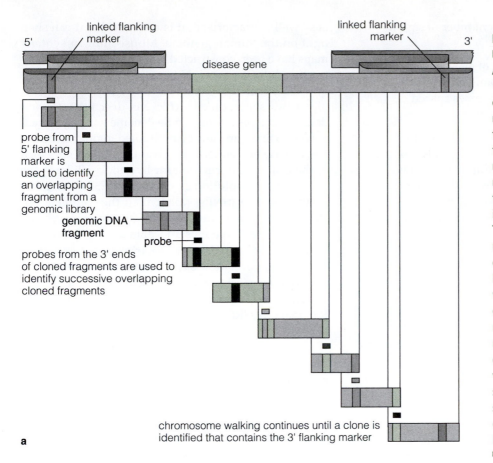

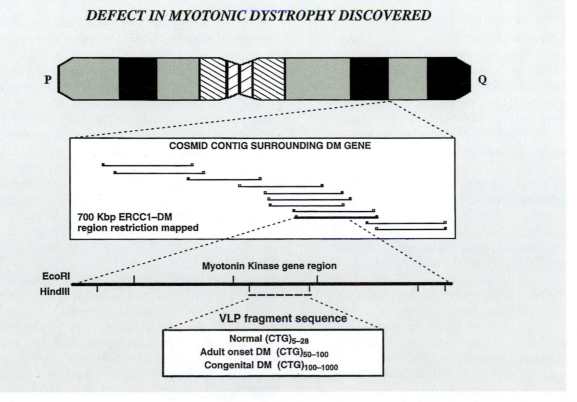

FIGURE 9.20 (a) Disease genes can be cloned by chromosome walking. After a marker is linked to within 1 cM of a disease-causing gene, chromosome walking from the marker to the gene is used to clone the gene. A probe from the 5′ flanking marker region identifies the next overlapping library clone. A restriction fragment from the end of that clone is used as a probe to isolate the next overlapping clone. This process of "walking" is repeated until the 3′ flanking marker on the other side of the disease–gene locus has been reached. The disease gene will be contained within one or more clones. (b) A section of the contig map of chromosome 19 spanning the region that is defective in the disease myotonic dystrophy. A variable-length polymorphism (VLP) sequence has been identified by sequencing DNA from one of the cosmid clones within the myotonin kinase gene region.

The human genome differs in a number of significant ways with other genomes.

1. Unlike the random distribution of gene-rich areas in the human genome, the genomes of other organisms are more uniform, with genes spaced more evenly throughout the genome.

2. Humans have, on average, three times as many kinds of proteins as the fruitfly or nematode worm because of alternative splicing (different combinations of intron or exon splicing or removal) of pre-mRNA and chemical modifications to proteins after translation. These processes can yield different protein products (and protein functions) from the same gene.

3. Humans share most of the same protein families with worms, flies, and plants, but the number of members (individual genes) within gene families has increased in humans, especially for those proteins involved in development and immunity.

4. The human genome has a much greater proportion of repeat sequences—50%—than the plant *Arabidopsis* (11%), nematode worm *Caenorhabditis* (7%), and fruitfly *Drosophila* (3%).

5. Although humans appear to have stopped accumulating repeated DNA, there does not seem to be a decrease in mice. This may be one of the fundamental differences between humans and rodents, although gene estimates are very similar.

Mouse and Other Animals

The use of rodents (mice and rats) has been a mainstay in medical research. As a result, mice and rats have been extensively studied. To address the needs of medical research, almost 100 well-characterized laboratory strains of mice have been produced. In addition, many mutant mouse lines exist to serve as human disease models (see Chapter 10, Medical Biotechnology), to study particular genes, and to extensively characterize the genome. Interestingly, although rodents and humans diverged long ago, regions of DNA sequence conservation remain, indicative of functional constraints placed upon molecules. In addition, many features of development, metabolism, and genetics (in particular, genetic disease) are also conserved. This is also why mouse models have been used to emulate human diseases. This also is one reason why the Mouse Genome Initiative (MGI) was formed (http://www.informatics.jax.org). At the MGI website, physical and genetic maps are available, as well as gene location, annotation, and other genomic information. Celera Genomics also provides the complete genome sequence for a fee through their Celera Discovery System (http://www.celera.com).

Researchers are very interested in identifying regions of conservation (regions of synteny) between the mouse and human genome. Because the mouse genome is so well characterized, it is thought that it can shed light on the human genome. Mouse–human synteny maps have been constructed and are available to scientists (see http://www.ncbi.nlm.nih.gov/Homology).

The genomes of a wide variety of animals —horses, pigs, cows, sheep, turkeys, chickens, deer, fish, cats, and dogs (most recently, the poodle genome was sequenced) to name a few—are now being studied for animal breeding, disease prevention, and evolution modeling. Much of the genome information for livestock will be in the form of high-resolution genetic maps, ESTs, and polymorphism data (SNPs) rather than the complete nucleotide sequence due to prohibitive costs. Livestock genome project activities and data are maintained on websites. The United States Department of Agriculture (USDA) supports these sites, sometimes in conjunction with veterinary schools where much of the research may be conducted. The National Animal Genome Research Program (NAGRP) manages and supports some ArkDB databases (http://www.thearkdb.org) (ArkDB database website is maintained by the Roslin Institute of Edinburgh, Scotland and links to each animal-specific genome project)—examples include PigBase at Iowa State University, the Bovine and Sheep Genome Project site at Texas A&M University, and the University of Kentucky's Horse Genome site. Other genome databases also are available such as the Fred Hutchinson Cancer Research Center Dog Genome Project (http://www.fhcrc.org/science/dog_genome/dog.html).

Invertebrates

The first two multicellular eukaryotic genomes to be sequenced were the fruitfly *Drosophila melanogaster* (FlyBase, http://flybase.bio.indiana.edu, and the Berkeley *Drosophila* Genome Project, http://www.fruitfly.org) and the nematode worm *Caenorhabditis elegans* (WormBase, http://www.wormbase.org). Both are widely used as model organisms for the study of the molecular biology of development. Because the organisms are readily mutated, the functions of all the genes can potentially be identified. This will also have important implications for the human genome because gene counterparts are being identified. Finally, nematode genome projects will have agricultural importance because some of these organisms are involved in crop damage.

Plants

Between 1999 and 2000, the genome sequence of the model plant *Arabidopsis thaliana* was published, one chromosome at a time (Arabidopsis Genome Initiative). Other genome projects, in particular those for agronomically important crops such as rice, wheat, alfalfa, sorghum, barley, soybean, and corn are providing valuable information for crop-breeding projects and the genetic improvement of plants. Researchers also are interested in plant genomes because they harbor plant-specific genes that encode enzymes for unique metabolic pathways that may have important biotechnological applications. Metabolic pathways unique to plants include

1. The production of secondary metabolites

2. Plant cell wall synthesis

3. Photosynthesis

4. Pathogen resistance

Microorganisms

Microorganisms have an ancient evolutionary history, dating back almost 4 billion years. They compose a significant portion of the Earth's biomass and are found in almost every habitat, with some microorganisms capable of surviving in extreme environments (high and low temperatures, salt, acid, and even in radioactive environments). Thus, microorganisms have adapted to a wide range of habitats by developing complex metabolic activities that enable them to survive. Scientists are interested in elucidating these mechanisms of survival and harnessing the vast capabilities of these organisms for use in a variety of applications—environmental, medical, industrial, agricultural, and energy (see Chapter 5, Microbial Biotechnology). Microbial genomics will provide information for the development of new biofuels, new bioremediation methods, and effective treatment for and protective treatment against biological warfare. Information from microbial genomes also will increase our understanding of disease and identify drug targets for treatment. Because we have identified only a fraction of the microbial diversity, much research is being spent on the search for new microorganisms, the elucidation of new metabolic pathways, and on microbial genome projects.

Examples of three of the first microbial genomes to be sequenced are *Mycoplasma genitalium* (589 kb, one of the smallest genomes of any free-living organism), *Haemophilus influenzae* (1830 kb, the first microbial genome sequenced, in 1995); and *Methanococcus jannaschii* (1660 kb). Interestingly, 56% of *M. jannaschii* genes (a total of 1738 genes) were new to science. Also, most genes for energy production, metabolism, and cell division were similar to those in bacteria, while genes for replication, transcription, and translation were more like those in eukaryotes. Currently, more than 45 microbial genome sequences have been published and more than 200 genomes are being sequenced.

The DOE initiated the Microbial Genome Project (MGP) in 1994 as an offshoot of the Human Genome Project. The overall goals of the MGP (http://www.ornl.gov/microbialgenomes) are to sequence and ultimately exploit entire microbial genomes. Numerous genomes have already been sequenced that will yield potential applications in the areas of cellulose degradation and energy production for the conversion of biomass into ethanol, methane, and hydrogen; carbon sequestration for the manage-

ment of global carbon to help stabilize climate; and bioremediation for the clean up of toxic wastes. The MGP is closely affiliated with two programs of the DOE's Biological and Environmental Research (BER) Program), Genomes to Life (DOEGenomestoLife.org) and Natural and Accelerated Bioremediation Research (NABIR) (http://www.lblgov/NABIR). Genomes to Life merges completed DNA sequence data with various advanced technologies to gain a fundamental understanding of living processes. This information will provide insights into the role of microorganisms in the processing of metals, carbon, nitrogen, and radioactive materials, as well as the biological foundation of climate change. By applying what is known about microbial metabolism, NABIR seeks to harness and improve natural microbial processes for the bioremediation of contaminated groundwater, soils, and sediments.

Also included in the realm of microbial genomics are the genome projects of yeast (the completed genome sequence published in 1997 for *Saccharomyces cerevisiae*, followed by sequences for *Schizosaccharomyces pombe* and *Candida albicans*, involved in human infections) and parasitic organisms. Parasitic organisms include trypanosomes, ticks, insect-borne viruses, malaria parasite, filarial nematodes, giardia, and mosquitoes. Parasitic genome project goals include

1. A better understanding of the genomes of important parasite vectors (for example, the mosquito *Anopheles* spp., which transmits the parasite *Plasmodium*)—to determine what portions of the genome are involved in transmission of the parasite

2. The accumulation of polymorphism data to gain insight into the genome diversity of specific parasites

3. The identification of species-specific genes (that code for antigens) that may lead to effective vaccines

4. A better understanding of parasite life cycles and points in the cycle that may be effective targets for drugs

FUNCTIONAL GENOMICS

After the complete sequence is obtained for a particular genome, the next step is to find all the genes and how they function. The DNA sequence is merely a meaningless linear nucleotide sequence without knowledge of what a genome encodes and how the genes function. Many genomes are being sequenced in record time (more than 100 genome projects) with the development of very rapid DNA sequencing technology, thus the limiting factor in terms of biotechnological applications will be the identification and function of all the genes in a particular genome. The study of the function of all the genes (and nongene regions) encoded by a particular genome is referred to as functional genomics. A variety of methods are used (molecular biology, genetics, biochemistry, recombinant DNA technology, and computational biology) to fully characterize the biochemical and metabolic activity, cell function, and physiological function of genes.

So much remains to be discovered about the human genome and how it functions. Here are examples we have yet to understand:

Exact locations and functions of all the genes in the genome

How genes are regulated

How chromosomes are organized

The roles of noncoding DNA

How proteins interact

The total content of proteins (proteome) and their functions

How SNP information correlates with disease and the overall health of an individual

What genes play a role in different multigene diseases

What genes are involved in the expression of different phenotypic characteristics (physical characteristics)

The level of evolutionary conservation among different organisms (for example, microbes, plants, humans, other animals, etc.)

PROTEOME AND PROTEOMICS

The next step after identifying all the genes in a genome is to understand the proteome. The proteome consists of all the proteins that are found in a specific cell under a defined set of conditions. Proteomics is the study of proteins encoded by the genome. This area encompasses several different aspects.

1. The expression and identification of proteins in a given cell

2. The structure of proteins (for example, tertiary folding)

3. Protein–protein interactions (for example, how proteins interact with each other in the cell)

4. Annotation of proteins in computer databases

It is thought that a major aspect of the complexity of an organism is due to the many different protein interactions within cells. Although the DNA sequence is an important first step in the elucidation of genomic function, the identification of proteins and their interrelationships are vital to this goal. In the post-genome era, data from genomics and proteomics provide the tools to understand complex protein pathways and interactions in cells. This gives a more sophisticated and meaningful interpretation of metabolic responses to mutation or changes that might be obtained from more traditional molecular biology methodology (for example, gene expression profiles using microarrays or Northern blot hybridizations). The addition of metabolomics data (see the following discussion) will provide even further information. Thus, we can go beyond transcriptional and translational regulatory networks to complex interactions between molecules and even pathways.

Proteomics presents a formidable challenge. Although each cell type of a multicellular organism such as a human harbors the same DNA, each cell type (for example, epidermal versus nerve cell) has a different proteome because a different array of proteins is synthesized (and characterize that cell type). In addition, the proteome of a given cell type can change depending on the physiological state of the organism, the age of the cell or organism, and the health of the individual. Other factors also can affect the proteome of cells such as genetic variation, the presence of medications, and diet.

OTHER ——OMES

The broad area of genomics now includes other types of ——omes in addition to the genome. The transcriptome is the full complement of expressed genes called transcripts that are made in a specific cell type under a specific set of conditions. Comparing a transcriptome with a reference transcriptome reveals changes in transcription for every gene in a genome. DNA microarray technology (see Chapter 3, Basic Principles of Recombinant DNA Technology) can potentially be used to measure the transcription level of every gene in a cell simultaneously.

The metabolome is the entire metabolic state of a cell, including the array of substrates, metabolites, and other small molecules that are synthesized in different cells and tissues. The study of the metabolome is referred to as metabolomics (the study of the structure, organization, and interaction of these components). From recent research on organisms (for example, the experimental model of plants, *Arabidopsis*) it appears that the metabolome may be even more complex than the proteome. For example, researchers compared the metabolomes of four different varieties of *Arabidopsis* and discovered that the metabolomes varied greatly. They were able to identify several hundred metabolites. Interestingly, in mutant varieties where only one protein was not expressed, the metabolomes were modified in very complex ways.

Therefore, to completely understand cells, different cell types at all levels must be studied—transcriptome, proteome, and metabolome—under various conditions.

Pharmacogenomics

Pharmacogenomics is a new field that studies how the genome is affected by and responds to different drugs. It combines pharmacology with genetics and genomics in the hope of offering patients safer and more effective treatments than the current one-drug-fits-all therapeutic approach. Personalized medicine will be an important outcome of this aspect of genome science. Tailored drug development involves linking genetic variation to specific diseases so that drugs are prescribed based on each patient's genetic profile. Scientists working in the area of pharmacogenomics are trying to identify all the genes that determine the overall effectiveness of drug treatments. These genes are involved in a variety of metabolic processes such as drug metabolism and toxicity, transporters, receptors, and signaling pathways. One complicating factor in this research is that many genes are involved in the response to a medication, not just a single gene.

The identification of individual human genome variation (including small polymorphisms such as SNPs and larger DNA sequence deletions or insertions) and gene function will provide vital information for the development of genome-tailored drugs that could reduce side effects, drug interactions, and even death. Interestingly, each individual responds to drug therapies differently, most likely due to subtle differences in the genome. Each individual has a unique collection of genomic variations that affects the predisposition to diseases, lifestyle behaviors such as drug and alcohol abuse, and even the response to medications. It is thought that safer, more effective drugs can be prescribed if they are tailored to an individual's genetic make-up. An example is the treatment of cancer, where more specific, tailored therapies can be developed based on the type of cancer cells and the genetics of the patient. Another patient may require a different drug treatment. Pharmacogenomics will have applications in the treatment of many different diseases. Other anticipated benefits include

Better vaccines

Advanced screening for diseases

More powerful drugs

Improvement in drug-discovery methods and more efficient approval processes

Decrease in health care costs

Most pharmaceutical and biotechnology companies are using pharmacogenomics during drug discovery (for example, to identify and eliminate toxic compounds) rather than develop high-tech personalized drugs linked to genetic markers. In July 2003 the commissioner of the Food and Drug Administration (FDA), Marck Mclellan, told the Joint Economic Committee of Congress that the FDA would evaluate how pharmacogenomics could serve as a predictor of drug safety and efficacy. In August 2003 a company called Gene Logic (of Gaithersburg, Maryland) announced that it will provide gene expression data (from microarray analysis) to the Center for Drug Evaluation and Research (CDER) for the evaluation of RNA expression standards for toxicogenomic (the study of toxic effects on the genome) data. The FDA is working with Gene Logic because the company has a gene chip with samples (for microarray analysis) necessary for toxicity testing. The goal of the CDER–Gene Logic agreement is to establish a reference database on toxicity-related genes that can be used in conjunction with genomic data and used as quality assurance in new drug applications.

Thus, the Human Genome Project has provided scientists with the DNA sequence, functional genomics is beginning to relate genes to diseases, and software companies are providing the analytical tools to sort through the immense quantity of raw sequence data with the overall goal to integrate genetic and clinical data. This information will provide pharmaceutical companies with valuable information for drug development.

BIOINFORMATICS

Bioinformatics is a new discipline of science that incorporates biology, computer science, and information technology. With the initiation of a variety of genome projects and the acquisition of large quantities of DNA sequence data, there has been a need for computerized databases to organize, catalog, and store sequence data, as well as tools to retrieve and analyze the data. Computational biology is a field within bioinformatics that involves the process of analyzing and interpreting collected data. Basically, bioinformatics is a discipline that provides tools, primarily computational, to help make sense of nucleic acid and protein sequences.

The overall goals of bioinformatics are

1. To develop tools that allow the efficient access and management of databases

2. To analyze and make sense of the large amount of DNA and protein sequences being generated. Molecular biology and computer science tools are used for gene identification, to help predict protein structure and function, to identify functional domains of proteins, and to conduct evolutionary analyses of genes and proteins.

3. To develop new algorithms (mathematical computations using computer programs) and other tools for the utilization and manipulation of data

Databases

A database is a computerized body of data that allows someone to retrieve specific pieces of information. DNA and protein databases contain nucleotide and protein sequences, respectively. A DNA database, for example, contains the nucleotide sequence, information about the DNA sequence (for example, coding sequence and intron locations), relevant publications, and sequence source or scientific name of the source organism. Some databases are updated daily. Specific data stored in a database may be retrieved using the Internet.

An example of a DNA database is GenBank, maintained at the National Center for Biotechnology Information (NCBI, http://www.ncbi.nlm.nih.gov). DNA sequence information can be sent to GenBank for storage (DNA sequences usually must be submitted to a DNA database such as GenBank before publication in a journal) and specific sequences can be searched and retrieved. The many databases at NCBI are linked by a search and retrieval system called Entrez. This allows specific information to be accessed and integrated from many databases. For example, a researcher can obtain information about the protein sequence (obtained from the protein database) as well as about the species from which the sequence was isolated (through the taxonomy database; taxonomy deals with the classification of organisms). A sequence entry includes a filename, the DNA or protein sequence, mutations, function, regulatory regions, encoded proteins if the sequence is DNA, and relevant references. Thus, a database can be searched for many different types of information.

Other databases include the European Molecular Biology Laboratory (EMBL, http://www.ebi.ac.uk) founded in 1980 and the DNA DataBank of Japan (DDBJ, http://www.ddbj.nig.ac.jp). GenBank, EMBL, and DDBJ have formed the International Nucleotide Sequence Database Collaboration (http://www.ncbi.nlm.nih.gov/collab), which enables the daily exchange of data. Another database, the Protein Information Resource (PIR), is located at the National Biomedical Research Foundation in Washington, D.C. The number of daily entries added have increased tremendously. All sequences must be annotated.

Databases can be used for data mining, a process by which the function and structure of a gene or protein sequence can be predicted by finding similar sequences in better-characterized organisms. Testable hypotheses regarding function can be generated by extracting well-characterized sequences from databases.

Tools of Bioinformatics

A variety of computational tools have been developed to analyze, for example, gene structure and function, gene and genome relationships, and the evolution of genes and proteins and organisms. Examples of bioinformatic tools include

1. Retrieval of DNA and protein sequences from databases

2. Database searches for similar DNA and protein sequences

3. Sequence alignment for comparison—both global and local regions (two or more sequences)

4. Prediction of RNA secondary structure—RNA folding (for example, ribosomal RNAs)

5. Protein classification (amino acid patterns, motifs, functional domains, structural features, prediction of function) and prediction of protein secondary structure—protein folding

6. Evolutionary relationships using DNA and protein sequences (that is, phylogenetic prediction or production of phylogenetic trees)

7. Gene prediction using DNA sequences—finding open reading frames (ORFs), promoters, searching for special sequence motifs

8. Genome analysis tools:

   Genome and sequence databases

   Gene nomenclature and annotation

   Identification of repetitive DNA sequences

   Identification of genes and gene families

   Gene location and gene order

   Identification of ESTs

   Microarray analysis and gene expression and regulation

   Proteome analysis and identification of protein families—those proteins of related sequence

   Metabolic pathway analysis

   Comparative genomics tools

   Evolutionary modeling

Ethical, Legal, and Social Implications

Although the success of the Human Genome Project offers a multitude of opportunities for medical advancements (new medical treatments and diagnostics) and studies of human evolution, it also poses potential societal and ethical problems. For example, some genetic disorders are now detectable long before they manifest themselves or can be cured. How is this information used? Controversy could also arise over reproductive issues if, for example, parents consider terminating pregnancies for reasons of genetic makeup or if there is social pressure to limit reproductive rights on genetic grounds. Issues of genetic discrimination and confidentiality in the insurance industry and in employment also must be addressed. How is personal genetic history protected? Early in the twentieth century, compulsory sterilization programs were implemented in some states for people viewed as abnormal.

In 1990 a working group of the NIH and DOE recommended that research and educational programs be established to address the Human Genome Project's ethical, legal, and social implications (collectively called ELSI). The objectives of ELSI (http://www.nhgri.gov/ELSI and http://www.ornl.v\gov/hgmis/elsi/elsi.html) were to

1. Examine the ethical, legal, and social implications of the Human Genome Project

2. Facilitate public discussion about the Human Genome Project

3. Develop policies on the beneficial use of the infor-

mation generated from the Human Genome Project

The National Human Genome Research Institute (NHGRI) currently commits 5% of its annual budget to ELSI. Funds support grants for research and education that focus on four primary areas.

1. The education of the public and professionals in the medical field

2. The integration of genetic technologies with clinical work

3. Issues that pertain to the implementation of clinical research such as consent, participation, and reporting

4. Privacy and fairness issues that pertain to the use of genetic information by, for example, employers, insurers, schools, court of law, and the military

The overall goal of ELSI is to identify the social effects of the project's discoveries and aid in genetics-related policymaking.

The Human Genome Project makes available genetic information that until only recently had not existed. Although much of this information will have a positive effect on society, other uses could potentially increase discrimination, restrict reproduction, affect employment and insurance coverage decisions, and cause personal distress for those who receive genetic information for which they are unprepared. Legislation may one day be necessary to protect individuals with known genetic predispositions to disease and with late-onset genetic disorders such as Huntington's disease.

How will everyone benefit from the knowledge generated by the Human Genome Project? That vast

amount of information must somehow be assimilated and disseminated so that it can be applied. This knowledge will most likely influence how healthcare workers, physicians, and genetic counselors are trained in the future. Future genetic breakthroughs will pose questions that must be answered, such as whom to screen, how to protect privacy, how to counsel clients, and how to train healthcare personnel. Should genetic screening be conducted when there are no treatments for the genetic disorders discovered? Genetic analysis, counseling, and treatments will become much more complex as we learn more about the human genome. Perhaps current training programs for genetic counselors, physicians, and social scientists will no longer be adequate. We must also anticipate how new genetic information will affect the medical field. For example, once genes for genetic diseases are identified, we will have to determine which diseases receive priority funding for developing treatments.

The information from the Human Genome Project has the potential to influence our lives more than any previous medical breakthrough. A major goal of ELSI is to help the general public understand issues related to the Human Genome Project such as the right to privacy, the potential for discrimination, and ethical considerations. Scientists should be trained in the areas of ethics, the legal system, and sociology to address public concerns. At the same time, social scientists, legislators, and lawyers must become familiar with the science to more effectively evaluate the long-term impact of the Human Genome Project.

Education is perhaps the most important way to address potential problems that the Human Genome

Continued

Project might present. ELSI projects at two sites (The University of Kansas Medical Center, http://www.kumc.edu/gec/prof/geneelsi.html and The Lawrence Berkeley Laboratory, http://www.lbl.gov/Education/ELSI/ELSI.html) address a wide variety of issues such as fetal genetic screening, cancer testing, eugenics, personal privacy, and medical databases.

In 1998, the ELSI Research and Planning Evaluation Group devised new goals in response to a NHGRI review of the Human Genome Project and future needs.

Some of the newly defined goals are

1. To assess the outcome of genetic testing and the impacts of public and individual knowledge

2. To determine how genetic information will be used in a variety of settings (for example, military, schools, courtroom)

3. To examine issues directly affected by the completion of the human genome se-quence such as the implications of our knowledge of human genetic variation

4. To examine how our knowledge of human genetics will affect our society, theology, philosophy of life, and our societal behaviors

5. To determine how socioeconomic factors, ethnicity, and race affect the use of genetic information

General Readings

A.M. Campbell and L.J. Heyer. 2003. *Discovering Genomics, Proteomics, and Bioinformatics*. Benjamin Cummings, San Francisco, California.

W.Y. Gibbs. 2003. The unseen genome: Gems among the junk. *Sci. Am.* 289:47–53.

G. Gibson and S.V. Muse. 2002. *A Primer of Genome Science*. Sinauer Associates, Inc., Sunderland, Massachusetts.

J. Grinwood and J. Schmutz. 2003. Six is seventh. *Nature* 425:775–776.

D.E. Kane and M.L. Raymer. 2003. *Fundamental Concepts of Bioinformatics*. Benjamin Cummings, San Francisco, California.

D.J. Kevles and L. Hood, eds. 1992. *The Code of Codes: Scientific and Social Issues in the Human Genome Project*. Harvard University Press, Cambridge, Massachusetts.

D.W. Mount. 2001. *Bioinformatics: Sequence and Genome Analysis*. Cold Spring Harbor Laboratory Press, Cold Spring Harbor, New York.

A.J. Mungall et al. 2003. The DNA sequence and analysis of chromosome 6. *Nature* 425:805–811.

M.A. Palladino. 2002. *Understanding the Human Genome Project*. Benjamin Cummings, San Francisco, California.

Additional Readings

J. Aach, W. Rindone, and G. Church. 2000. Systematic management and analysis of yeast gene expression data. *Genome Res.* 10:431–445.

M. Adams et al. 2000. The genome sequence of *Drosophila melanogaster*. *Science* 287:2185–2195.

Arabidopsis Genome Initiative. 2000. Analysis of the genome sequence of the flowering plant *Arabidopsis*. *Nature* 408:796–815.

D. Baker and A. Sali. 2001. Protein structure prediction and structural genomics. *Science* 294:93–96.

W.B. Barbazuk et al. 2000. The syntenic relationship of the zebrafish and human genomes. *Genome Res.* 10:1351–1358.

R. Bhalerao, O. Nilsson, and G. Sandberg. 2003. Out of the woods: Forest biotechnology enters the genomic era. *Curr. Opin. Biotechnol.* 14:206–213.

P. Brazhnik, A. de la Fuente, and P. Mendes. 2002. Gene networks: How to put the function in genomics. *Trends Biotechnol.* 20:467–472.

P.O. Brown and D. Botstein. 1999. Exploring the new world of the genome with DNA microarrays. *Nat. Genetics* 21 (Suppl):33–37.

S. Buckingham. 2003. Bioinformatics: Programmed for success. *Nature* 425:209–215.

C.J. Bult et al. 1996. Complete genome sequence of the methanogenic Archaeon, *Methanococcus jannaschii. Science* 273:1058–1073.

D.T. Burke. 1991. The role of yeast artificial chromosomes in generating genome maps. *Curr. Opin. Genet. Dev.* 1:69–74.

S.K. Burley. 2000. An overview of structural genomics. *Nat. Struct. Biol.* 7 (Suppl): 932–934.

J. Burris, R. Cook-Deegan, and B. Alberts. 1998. The Human Genome Project after a decade: Policy issues. *Nat. Genetics* 20:333–335.

H. Caron et al. 2001. The human transcriptome map: Clustering of highly expressed genes in chromosomal domains. *Science* 291:1289–1292.

J.M. Claverie. 2001. What if there are only 30,000 human genes? *Science* 252:1255–1237.

S.A. Chervitz et al. 1998. Comparison of the complete protein sets of worm and yeast: Orthology and divergence. *Science* 282:2022–2028.

B. Cohen, R. Mitra, J. Hughes, and G. Church. 2000. A computational analysis of whole-genome expression data reveals chromosomal domains of gene expression. *Nat. Genetics* 26:183–186.

D. Cohen, I. Chumakov, and J. Weissenbach. 1993. A first-generation physical map of the human genome. *Nature* 366:698–701.

F.S. Collins et al. 1998. New goals for the U.S. Human Genome Project: 1998–2003. *Science* 282:682–689.

P. Deloukas et al. 1998. A physical map of 30,000 human genes. *Science* 282:744–746.

K.M. Devine. 1995. The *Bacillus subtilis* genome project: Aims and progress. *Trends Biotechnol.* 13:210–216.

J. Dicks. 2002. Plant bioinformatics: Current status and future trends. *AgBiotechNet* 4:1–5.

W.F. Dietrich et al. 1994. A genetic map of the mouse with 4,006 simple sequence length polymorphisms. *Nat. Genet.* 7:220–225.

N.A. Doggett. 1994. The Polymerase Chain Reaction and Sequence-Tagged Sites. In N.G. Cooper, ed. *The Human Genome Project: Deciphering the Blueprint of Heredity.* University Science Books, Mill Valley, California, pp. 270–272.

S. Donadio, M. Sosio, and G. Lancini. 2002. Impact of the first *Streptomyces* genome sequence on the discovery and production of bioactive substances. *Appl. Microbiol. Biotechnol.* 60:377–380.

D. Duggan, M. Bittner, Y. Chen, P. Meltzer, and J. Trent. 1999. Expression profiling using cDNA microarrays. *Nat. Genetics* 21 (Suppl):10–14.

J.A. Eisen. 1998. Phylogenomics: Improving functional predictions for uncharacterized genes by evolutionary analysis. *Genome Res.* 8:163–167.

D.A. Fell. 2001. Beyond genomics. *Trends Genet.* 17: 680–682.

D. Fenyo. 2000. Identifying the proteome. Software tools. *Curr. Op. Biotechnol.* 11:391–395.

R.D. Fleischmann et al. 1994. Whole-genome random sequencing and assembly of *Haemophilus influenzae* Rd. *Science* 269:496–512.

A. Fraser, R. Kamath, P. Zipperlen, M. Martinez-Campos, M. Sohrmann, and J. Ahringer. 2000. Functional genomic analysis of *C. elegans* chromosome 1 by systematic RNA interference. *Nature* 408:325–330.

C.M. Fraser et al. 1995. The minimal gene complement of *Mycoplasma genitalium*. *Science* 270:397–403.

J. Fuhrman. 2003. Genome sequences from the sea. *Nature* 424:1001–1002.

K.L. Garver and B. Garver. 1994. The human genome project and eugenic concerns. *Am. J. Human Genet.* 54:148–158.

D. Gershon. 2003. Probing the proteome. *Nature* 424:581–587.

N. Goodman. 2002. Biological data becomes computer literate: New advances in Bioinformatics. *Curr. Opin. Biotechnol.* 13:68–71.

J. Hodgkin, R.H.A. Plasterk, and R.H. Waterston. 1995. The nematode *Caenorhabditis elegans* and its genome. *Science* 270:410–414.

G. Hofmann, M. Mcintyre, and J. Nielsen. 2003. Fungal genomics beyond Saccharomyces cerevisiae? *Curr. Op. Biotechnol.* 14:226–231.

H. Holtorf, M.-C. Guitton, and R. Reski. 2002. Plant functional genomics. *Naturwissenschaften* 89:235–249.

J. Hu et al. 2001. The Arkdb: Genome databases for farmed and other animals. *Nucleic Acids Res.* 29:106–110.

K.L. Hudson, K.H. Rothenberg, L.B. Andrews, M.J.E. Kahn, and F.S. Collins. 1995. Genetic discrimination and health insurance: An urgent need for reform. *Science* 270:391–393.

T.R. Hughes et al. 2000. Functional discovery via a compendium of expression profiles. *Cell* 102:109–126.

T. Ideker et al. 2001. Integrated genomic and proteomic analyses of a systematically perturbed metabolic network. *Science* 292:929–933.

International Human Genome Sequencing Consortium (F. Collins et al.). 2001. Initial sequencing and analysis of the human genome. *Nature* 409:860–921.

International SNP Map Working Group. 2001. A map of human genome sequence variation containing 1.42 million single-nucleotide polymorphisms. *Nature* 409:928–933.

L.M. Jat, L. Cao, K. Cheah, and D. Smith. 2001. Assessing clusters and motifs from gene expression data. *Genome Res.* 11:112–123.

J.M. Jeffords and T. Daschle. 2001. Political issues in the genome era. *Science* 252:1249–1251.

H. Jeong, B, Tombor, R. Albert, Z. Oltvai, and A. Barabasi. 2000. The large-scale organization of metabolic networks. *Nature* 407:651–654.

D.T. Jones. 2000. Protein structure prediction in the postgenomic era. *Curr. Op. Struct. Biol.* 10:371–379.

M. Keilbart. 2000. Societal and medical consequences of the Human Genome Project. *Funct. Integer Genomics* 1:146–149.

M. Kellis, N. Patterson, M. Endrizzi, B. Birren, and E.S. Lander. 2003. Sequencing and comparison of yeast species to identify genes and regulatory elements. *Nature* 423:241–254.

S.K. Kim, J. Lund, M. Kiraly, K. Duke, M. Jiang, J. Stuart, A. Eizinger, B. Wylie, and G. Davidson. 2001. A gene expression map for *Caenorhabditis elegans*. *Science* 293:2087–2092.

A. Koller et al. 2002. Proteomic survey of metabolic pathways in rice. *Proc. Nat. Acad. Sci. U.S.A.* 99:11969–11974.

B.M. Knoppers and R. Chadwick. 1994. The Human Genome Project: Under an international ethical microscope. *Science* 265:2035–2036.

A.M. McGuire, J. Hughes, and G.M. Church. 2000. Conservation of DNA regulatory motifs and discovery of new motifs in microbial genomes. *Nat. Genetics* 10:744–757.

M.H. Meisler. 1996. The role of the laboratory mouse in the human genome project. *Am. J. Hum. Genet.* 59:764–771.

A.P. Monaco and Z. Larin. 1994. YACs, BACs, PACs, and MACs: Artificial chromosomes as research tools. *Trends Biotechnol.* 12:280–286.

M. Morgante and F. Salammini. 2003. From plant genomics to breeding practice. *Curr. Opin. Biotechnol.* 14:214–219.

J.C. Murray et al. 1994. A comprehensive human linkage with centimorgan density. *Science* 265:2049–2054.

J. Nielsen. 1998. Metabolic engineering: Techniques for analysis of targets for genetic manipulations. *Biotechnol. Bioeng.* 58:125–132.

S. Paabo. 2001. The human genome and our view of ourselves. *Science* 252:1219–1220.

B. Palenik et al. 2003. The genome of a motile marine *Synechococcus*. *Nature* 424:1037–1042.

S.D. Pena. 1996. Third World participation in genome projects. *Trends Biotechnol.* 14:74–77.

S. Penn, D. Rank, D. Hanzel, and D. Barker. 2000. Mining the human genome using microarrays of open reading frames. *Nat. Genetics* 26:315–318.

F.B. Pichler, S. Laurenson, L.C. Williams, A. Dodd, B.R. Brent, and D.R. Love. 2003. Chemical discovery and global gene expression analysis in zebrafish. *Nature Biotechnol.* 21:879–883.

T. Read et al. 2003. The genome sequence of *Bacillus antrhacis* Ames and comparison to closely related bacteria. *Nature* 423:81–86.

J. Reboul et al. 2001. Open-reading-frame sequence tags (OSTs) support the existence of at least 17,300 genes in *C. elegans*. *Nat. Genetics* 27:232–336.

S. Rogic, A. Mackworth, and F. Ouellette. 2001. Evaluation of gene finding programs on mammalian sequences. *Genome Res.* 11:817–832.

P.R. Rosteck, Jr. 1994. The human genome project: Genetic and physical mapping. *Trends Endocrin. Met* 5:359–364.

G.M. Ruben et al. 2000. Comparative genomics of the eukaryotes. *Science* 287:2204–2215.

A.G. Rust, E. Mongin, and E. Birney. 2002. Genome annotation techniques: New approaches and challenges. *Drug Discovery Today* 7:S70-S76.

C. Schilling, J. Edwards, and B. Palsson. 1999. Toward metabolic phenomics: Analysis of genomic data using flux balances. *Biotechnol. Prog.* 15:288–295.

R.C. Stevens, S. Yokoyama, and I.A. Wilson. 2001. Global efforts in structural genomics. *Science* 294:89–92.

M. Stoneking. 1997. The human genome project and molecular anthropology. *Genome Res.* 7:87–91.

T. Strachan, M. Abitbol, D. Davidson, and J.S. Beckmann. 1997. A new dimension for the human genome project: towards comprehensive expression maps. *Nat. Genet.* 16:126–132.

J.C. Venter et al. 2001. The sequence of the human genome. *Science* 291:1304–1351.

J. Weissenbach. 1993. A second generation linkage map of the human genome based on highly informative microsatellite loci. *Gene* 135:275–278.

D.L. Wheeler et al. 2000. Database resources of the National Center for Biotechnology Information. *Nucleic Acids Res.* 28:10–14.

T. Wolfsberg, J. McEntyre, and G. Schuler. 2001. Guide to the draft of the human genome. *Nature* 409:824–826.

J.R. Yates III. 2001. Mass spectrometry: From genomics to proteomics. *Trends Genet.* 16:5–8.

10

MEDICAL BIOTECHNOLOGY

As our knowledge accumulates in the areas of genomics, proteomics, and pharmacogenomics, and research on stem cells and animal cloning progresses, a new generation of **therapeutics** is being developed that targets both genetic and acquired diseases, such as cystic fibrosis (CF), Parkinson's, hemophilia, familial hypercholesterolemia, heart disease, and cancer. Engineered vaccines and antibodies, designer drugs such as synthetic drugs and DNA drugs, and new methods of drug delivery are being developed. Revolutionary therapeutics are in clinical trials and many more are on the way. Although these new therapeutics will not completely replace traditional treatments, they will significantly enlarge the weapons arsenal against disease.

GENE THERAPY

Gene therapy—the use of genes in the treatment of patients—would have been a science fiction story only 20 years ago. New technologies and a more friendly regulatory environment are encouraging investigations into human gene therapy. The transfer of a normal gene into cells to correct a specific disorder (that is, permanent repair) is a desirable alternative to treating the problem with a drug that may not be a cure. If safety and efficacy can be established, genes may one day be regularly prescribed as "drugs" in the treatment, and perhaps in the prevention, of diseases.

Gene therapy has important implications in the treatment of acquired and genetic diseases, cancer, and even AIDS. Although the ethical implications of gene therapy continue to be debated, gene transfer trials continue to be approved by regulatory agencies. Much effort and funds have been devoted to the development of safe and efficient gene transfer methods. Although certainly much progress has been made, the road to success has not been smooth, and there have been some devastating results along the way (see later discussion). The development of successful gene therapy treatments has been a formidable challenge.

Two types of gene therapy are possible: somatic and germline. Somatic gene therapy is the use of only cells of the body so that the individual's disorder is corrected. This also includes embryonic gene therapy. Germline gene therapy refers to the permanent modification of a gene in the sex cells (eggs or sperm). This changes the progeny of the individual. Currently, only gene therapy of somatic cells in adults and children is allowed. At this time, there is a moratorium on germline gene therapy (which would require the same methods used for producing transgenic mice), although the issue has been hotly debated. The idea of manipulating the germline cell to pass the genetic correction or modification on to succeeding generations has significant ethical and safety implications.

Genetic Disorders

Genetic disorders are medical problems caused by a mutation in one or more genes on the chromosomes. Mutations are changes in the nucleotide sequence of a gene or nongene region of the DNA. A person can be born with a mutation or acquire a mutation anytime during his or her lifetime. Mutations cause problems when they affect the normal function of a gene and its products or regulation. Thus, when a mutation causes a specific medical disease, this is referred to as a *genetic disorder*. Genetic disorders can be grouped into four categories.

1. Single-gene changes—a mutation in one gene can result in a change in the protein product or possibly the elimination of the protein entirely. Sickle cell anemia is an example of a genetic disorder that is caused by the mutation of one gene (actually, only a change in one amino acid of the protein).

2. Multigene disorders—these disorders (sometimes called *multifactor disorders*) result from mutations in more than one gene, and sometimes in conjunction with environmental influences. These are difficult disorders to treat. Examples of these include heart disease, diabetes, and cancer.

3. Mitochondrial disorders—these diseases, affecting many organ systems, are caused by mutations in mitochondrial DNA.

4. Chromosome abnormalities—sometimes complete chromosomes or large regions of a chromosome are missing, duplicated, or modified in some way. Examples include Down syndrome.

Gene Target Selection

To be a candidate for gene therapy, several important factors must be considered. Only a single gene causing a genetic disorder can be a candidate for gene therapy (Table 10.1). Unfortunately, multiple genes or changes in a chromosome cause many diseases but are not candidates for gene therapy at this time due to the complexity of the disorders.

First, the normal and mutated genes must be identified and well studied. The disease that the mutation causes must be well understood. Generally, animal disease models are developed to study the disease and test possible treatments. Finally, an approved protocol, including a gene delivery method for the particular gene therapy, must be available. Another important aspect is the consideration of potential toxic effects of the gene or gene delivery vehicle (for example, virus—see later discussion) or whether the therapy produces an immune response. Additional aspects to be considered include

1. Gene targeting—the delivery of the gene to the right cells

TABLE 10.1 Examples of Genetic Disorders That Are Candidates for Gene Therapy

Disease	Frequency	Gene Product	Target for Therapy
Cystic fibrosis	1/2,500 (Caucasians)	Cystic fibrosis transmembrane regulator (CFTR)	Lung tissue
Duchenne's muscular dystrophy	1/10,000 (males)	Dystrophin (muscle)	Muscle tissue
Familial hypercholesterolemia	1/500	Liver receptor for low-density lipoprotein (LDL)	Hepatocytes
Hemoglobin defects (e.g., thalassemias)	1/600 for specific ethnic groups	Hemoglobin components	Bone marrow cells
Hemophilia	A 1/10,000 (males)	Blood clotting factor VIII	Fibroblasts or hepatocytes
	B 1/20,000 (males)	Blood clotting factor IX	Fibroblasts or hepatocytes
Severe combined immunodeficiency (SCID)	Very rare	Adenosine deaminase (ADA) in 25% of patients	Bone marrow or T lymphocytes

Adapted from I.M. Verma, *Sci. Am.* 263:68–84 (1990).

2. Genome integration—the integration of the gene into the host chromosome (if this is the intention of the therapy) so that the gene is not destroyed in the host cells

3. Gene activation—making sure that the gene is turned on (that is, transcription is initiated) at the right time, and that it is regulated appropriately

Gene Delivery Methods

The first step in gene therapy is to transform specific cells with a gene of interest. Major research has focused on finding methods for efficiently transferring genes to such cells as muscle cells, lymphocytes, hepatocytes, hematopoietic stem cells, and fibroblasts. Two developments would dramatically increase the efficacy of gene-based therapeutics: (1) efficient methods of introducing genes into specific cells, and (2) methods that effectively treat a disorder without aggravating the immune system or activating or inactivating other genes in the genome.

Two strategies for gene introduction are being used. In *ex vivo* gene therapy, cells are removed from the body, the gene of interest is inserted into them, the cells are cultured to multiply to a sufficient number, and they are then returned to the body by infusion or transplantation. If the patient's own cells (autologous cells) are used, rejection does not occur. The biggest technical hurdle for *ex vivo* gene therapy is the transplantation of transfected cells.

An example of *ex vivo* therapy is a nonviral approach called *transkaryotic therapy*. An individual's cells are obtained through a skin biopsy, and the gene of interest is inserted into them. Transformed cells are cultured *in vitro* to increase the number of cells expressing the desired protein and are then injected under the person's skin, where the therapeutic protein is synthesized. The protein is circulated throughout the body.

In *in vivo* gene therapy, the gene is introduced directly into specific cells within the body. Vectors usually are required to target the DNA to specific cells. *In vivo* therapy presents some technical difficulties. Sometimes, the transferred gene is unstable and the product is expressed only transiently, instead of for the life of the cell. In addition, because these cells remain in the body and are not isolated from other cells during transfection, these methods are not as controlled and specific as targeted *ex vivo* gene therapy.

Both *in vivo* and *ex vivo* approaches are being evaluated. *Ex vivo* procedures offer one benefit: because the vector is introduced only into the target cells and not into the body or circulatory system, the individual's immune system is not in direct contact with the vector. However, the method, in its current state of development, is time consuming and expensive. Injectable *in vivo* treatments are more practical and will probably become the method of choice in the future, although, because a specific population of cells must be targeted, bone marrow stem cells and other types of circulating progenitor cells are an exception.

Various modes of gene delivery are being investigated and, in some cases, tested in trials. The primary ways include the use of viruses such as adenovirus, adeno-associated virus, retrovirus and herpes simplex virus, liposomes, and naked DNA. Each offers both benefits and potential side effects.

Viral Vectors

Viral delivery seems to be a generally safe method used in most gene therapy experiments. Much research has focused on the development of safe and efficient viral vectors (Figure 10.1).

Retrovirus Retroviruses are RNA viruses that infect human cells. Retrovirus vectors infect only dividing cells and, therefore, cannot be used to target nondivid-

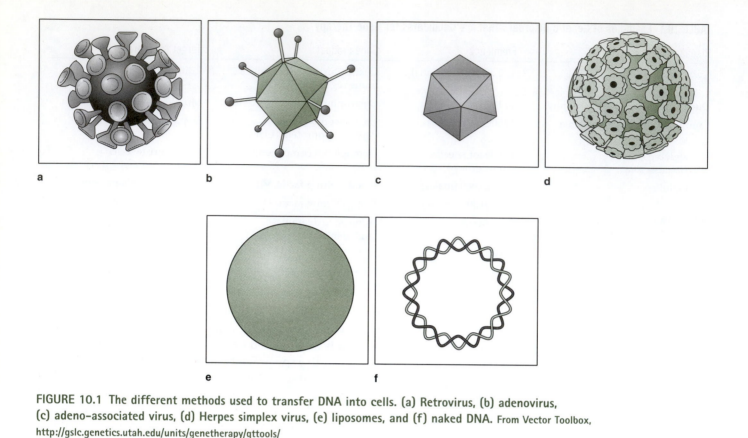

FIGURE 10.1 The different methods used to transfer DNA into cells. (a) Retrovirus, (b) adenovirus, (c) adeno–associated virus, (d) Herpes simplex virus, (e) liposomes, and (f) naked DNA. From Vector Toolbox, http://gslc.genetics.utah.edu/units/genetherapy/gttools/

ing cells. The maximum length of DNA that can be inserted is 8 kb. A retrovirus can insert into the host genome, although integration into the genome occurs randomly and the site of insertion cannot be predicted. Moreover, there is the possibility that viral insertion could inactivate indispensable genes (such as tumor suppressors) or activate oncogenes, thereby mutating the genome. A side effect is that the vector can elicit an immune response.

Adenovirus Adenovirus vectors show potential for *in vivo* gene therapy, because both dividing and nondividing cells are readily infected. As with any virus vector, it is possible to engineer proteins on the virus surface to target specific cells. The maximum length of DNA that can be inserted for transfer is 7.4 kb and can be highly expressed (that is, increased protein production). Adenovirus has become a desirable vector for the treatment of CF and other lung disorders, because nondividing cells of the respiratory tract seem to be preferentially infected. Adenovirus does not integrate into the genome, so there is little risk of mutation. Although these viruses lyse the cells they infect, they have been engineered not to kill their host cells.

There are several drawbacks. Genes may function transiently because they do not integrate into the host chromosome. Another concern is the potential for adenovirus to replicate in infected cells. Some gene prod-

ucts are involved in the malignant transformation of cells, are toxic to cells, or may lead to an adverse immune response. Newer adenoviral vectors have minimal viral DNA to decrease immunogenicity while retaining transfection efficiency.

Adeno-Associated Virus Adeno-associated viruses infect a wide variety of dividing and nondividing cell types with high efficiency. However, they require a helper virus for replication inside host cells. The maximum length of DNA that can be inserted is 5 kb. Although the adeno-associated virus integrates into the host cell genome, approximately 95% of the time it integrates into a specific region on chromosome 19. This reduces the possibility of a gene being activated or inactivated by virus insertion. Usually the vector does not cause an immune response.

Nonviral Delivery Methods

Nonviral delivery methods may be a safer approach to gene therapy than viral methods. Some nonviral methods are discussed in Chapter 3, Basic Principles of Recombinant DNA Technology, and Chapter 7, Animal Biotechnology: electroporation, microinjection, biolistics (microinjectile bombardment with a particle gun). Current research focuses on gene targeting to specific cell types, gene delivery methods that do not trigger an immune response, and vectors that can survive the circulatory system to reach target cells. Another method includes the use of liposomes (see Figure 10.1). DNA is packaged into membrane-bound spheres. When mixed with cells, liposomes fuse to cell membranes and the DNA is transferred. Although there is no maximum length of DNA that can be packaged and there is no cell-type specificity, the transfer efficiency is much lower than for viral vectors. Liposomes do not generally elicit an immune response, although liposome delivery is more appropriate for *ex vivo* therapy than *in vivo*.

Gene Therapy Examples

As gene therapy protocols are developed for specific disorders (see Table 10.1 for examples), clinical studies are performed to develop appropriate protocols for the targeting and expression of the gene, to assess the efficiency of the protocols, and to thoroughly evaluate the safety of the therapies. Many technical details are in the early stages of development; these include the construction of appropriate vectors and the design of efficient *ex vivo* and *in vivo* gene therapies. Gene therapies have been designed for 13 genetic diseases, as well as heart disease, AIDS, and cancer. Although more than $400 million is spent each year and more than 100 human clinical trials have been undertaken, to date there has not been one method that has resulted in a complete cure. Only some cases have had limited success—a temporary cure that requires treatment on a regular basis, and sometimes in conjunction with traditional drug therapy.

First Gene Therapy The first landmark gene therapy was performed in the United States at the National Institutes of Health (NIH) by Drs. W. French Anderson, Michael Blaese, and Kenneth Culver. In September 1990 they treated a 4-year-old girl, Ashanthi de Silva, for an inherited immunodeficiency called severe combined immunodeficiency (SCID, also referred to as *adenosine deaminase [ADA] deficiency*). SCID is caused by a defective adenosine deaminase gene that leads to an accumulation of adenosine, a toxic metabolic by-product. People without a normal ADA gene cannot produce a functional ADA enzyme. ADA is essential for normal immune system function. Children with this disorder have severe immune dysfunction and are prone to life-threatening infections. ADA deficiency can be treated with a synthetic form of the enzyme, polyethylene glycol (PEG)-ADA, although it costs more than $100,000 per year and must be taken by injection for the rest of the patient's life. ADA deficiency was an excellent first candidate for gene therapy because

1. Clinical research began before 1970, so much was known about the disorder

2. The disease is caused by a mutation in a single gene

3. The regulation of the gene was not complex—it is always turned on

4. The amount of ADA required in the body does not have to be precisely controlled. A small amount of ADA alleviates symptoms of the disease, but a larger amount is not toxic

Ashanthi was treated with modified T-lymphocytes containing the ADA gene. Four months later a second patient, 11-year-old Cynthia Cutshall, was treated. Both patients received gene-treated T cells to correct the genetic disorder. Each received a total of 11 to 12 gene infusions—the therapy worked for only a few months and then had to be repeated. Three years after treatment, more than 50% of Ashanthi's circulating T cells harbored the new gene and produced ADA. However, she still received injections of PEG-ADA. Only 0.1–1.0% of Cynthia's circulating T cells contained the ADA gene, demonstrating that effectiveness varies. In 2003 Ashanthi's cells still expressed ADA. Cynthia developed an immune response to the retrovirus gene delivery system and very few of her cells still produced ADA. Today Ashanti continues to lead a healthy, active life.

In 1993 clinicians used the umbilical cord blood stem cells from two infants born with ADA deficiency. The ADA gene was transferred to stem cells and used to treat these children. Both children continue to express the ADA gene in their T cells. Other children also have been treated successfully with this method.

Lung Disease—Cystic Fibrosis *In vivo* methods of gene therapy are being studied for the treatment of diseases that affect the airways. Genes have been efficiently transferred by adenovirus vectors to airway epithelial cells in the cotton rat. In these experiments, genes have been expressed only transiently. A method that may be useful is liposome-mediated gene transfer: A gene is encapsulated into an artificial phospholipid vesicle so that it diffuses into cell membranes and releases the gene into cells.

CF is a common genetic disorder that affects approximately 30,000 people in the United States. The symptoms of CF result from a defective ion transport molecule located in the plasma membrane of the cell. CF affects airways and other organs, such as the pancreas and intestines. In 1989, the CF gene was found on chromosome 7. Since then, researchers have elucidated the function of the encoded protein: the CF transmembrane conductance regulator (CFTR). Figure 10.2 shows a schematic of the CFTR protein. This protein is being used to develop effective treatments for both alleviating the symptoms of CF and permanently curing the genetic disorder. CFTR is a low-conductance chloride channel sensitive to cyclic AMP (adenosine 3′,5′-cyclic monophosphate, or cAMP) that is normally present in

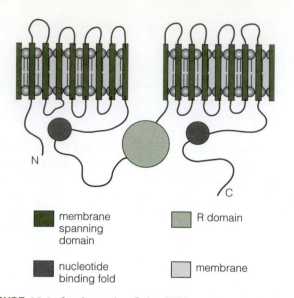

FIGURE 10.2 A schematic of the CFTR protein inserted into the cell membrane. A channel is formed that allows the movement of chloride ions. The R domain blocks the channel opening, possibly to regulate the flow of ions through the channel.

membrane spanning domain

R domain

nucleotide binding fold

membrane

the membranes of secretory epithelial cells in the pancreas, lung, intestine, and sweat glands. In patients with CF, the epithelial cells do not properly transport chloride ions in response to the presence of cAMP.

The relationship between gene mutation and expression of phenotypic characteristics has been intensely studied. Patients with CF express many physical characteristics such as increased mucus production, bacterial infections in the lungs, and altered epithelial cell transport that affects epithelial tissues such as the intestines. One important question is whether these problems are the direct effects of faulty CFTR protein and lack of cAMP-sensitive chloride conductance in membranes. More than 200 mutations that lead to CF have been identified—examples include improper protein processing, faulty regulation, overexpression of the protein, frame shift, and splice mutations; 70% of patients with CF harbor the mutation F508, which results in a three-nucleotide deletion and loss of the amino acid phenylalanine in the CFTR. This defective CFTR protein is not processed properly and so cannot be inserted into the cell membrane.

CF is an excellent candidate for gene therapy. In many genetic diseases, large amounts of the normal protein are required. However, in this case, only 10–100 molecules are required per cell—a very low level of expression. Gene therapy has been successful in both epithelial cell culture and transgenic CF mice. The CFTR gene was experimentally transferred by liposomes into lung epithelia, alveoli, and tracheas of various CF

model mouse strains with a disrupted CFTR gene. Although the CF **genotype** was still present, the CFTR gene was expressed, and the sodium-absorbing and chloride-secretion activities of epithelial cells were identical to those of normal cells. These studies also show only transient expression. An effective gene therapy will require longer-term expression of the CFTR; otherwise, repeated treatments will be required. Clinical trials have yielded mixed results to date.

One therapy that has mixed results is spraying genetically engineered adenoviruses (actually, a cold virus) harboring a normal CFTR gene into the noses of patients with CF. The CFTR gene is transferred to cells lining the nasal area. Treated cells express the gene and partly alleviate symptoms. Unfortunately, the transferred gene does not eliminate symptoms in distant organs (for example, the pancreas) and the cells with the new CFTR gene do not live more than a few weeks before they are shed. Thus, the patient must be continually re-treated.

Other therapies include treatment by inhalation of aerosols (such as DNA-liposome). Aerosols are less costly and time consuming and much simpler to use, especially when repeated treatments are required. Treatment of sites other than the lungs, such as the pancreas, will require a more complicated method, because the tissue is not directly accessible.

Liver Disease Both *ex vivo* and *in vivo* gene therapies for liver diseases are being studied. Transfected hepatocytes (liver cells) have not been reintroduced into the liver or spleen with success. Only approximately 10% of injected cells are incorporated into the liver. A more effective method of reintroducing transformed hepatocytes may be a hepatocyte graft.

The gene for the low-density lipoprotein **receptor** (LDLR) has been transferred into liver cells by retroviral and adenovirus vectors (by *ex vivo* and *in vivo* cell transduction, respectively). Although the expression of LDLR in cells has been transient, expression seems to be more prolonged if the gene therapy is preceded by a partial hepatectomy (excision of a piece of the liver).

In June 1992, at the University of Michigan Medical Center, a 29-year-old patient with a rare, life-threatening form of hereditary coronary artery disease made medical history. Familial hypercholesterolemia (FH) is caused by a deficiency in LDLR and leads to very high serum low-density-lipoprotein (LDL) cholesterol levels—almost 10 times normal—which in turn leads to heart disease and often sudden death. The patient was homozygous for a specific mutation in the LDLR gene that resulted in a defective LDLR. Consequently, cells could not take up cholesterol and metabolize it normally; the patient's LDL-to-HDL cholesterol ratio was much higher than normal. *Ex vivo* gene therapy was conducted in the patient after the procedure was tested in rabbits. Approximately 250 g of the patient's liver was removed and placed in culture. Hepatocytes were harvested and mixed with a retrovirus containing a normal copy of the human LDLR gene. Approximately 25% of the cells were infected with the virus. The transfected cells were then transferred to the patient by a catheter inserted in the portal vein. Tests later confirmed that cells were expressing the normal LDLR gene, and the patient's LDL-to-HDL cholesterol ratio, with the help of medication, was lowered. Two children were subsequently treated with this procedure. After preliminary data were submitted to the U.S. Recombinant DNA Advisory Committee (RAC) in December 1992, additional patients were approved for this therapy.

CLINICAL TRIALS

Before a gene therapy can be commercialized, it must be extensively tested on first animals and then humans to determine its safety and efficacy. A clinical trial is a research study using human volunteers to test the safety and efficacy of a particular treatment or therapy. There are several different types of clinical trials.

1. Diagnostic trial—identifies better tests for diagnosing diseases

2. Treatment trial—tests new therapies or drugs (for example, HIV treatment with antiviral cocktails)

3. Prevention trial—seeks new ways to prevent disease in people who have never had the particular disease (for example, meningitis vaccine, chicken pox vaccine)

4. Quality of life trial—seeks ways to improve the well-being and quality of life of chronically ill patients

5. Screening trial—finds ways to detect specific diseases or health conditions (for example, high blood pressure, high cholesterol, cancers)

Any therapeutic protocol or treatment must be tested at several levels before it can be established for widespread use. First, in preclinical trials, it is tested in *in vitro* experiments and on laboratory animal models. Clinical trials follow.

Phase I Trials—the procedure is tested on a small number (20–30) of human volunteers to determine dosage limits and route of delivery and to assess the procedure's toxicity and safety.

Phase II Trials—further trials determine efficacy and collect additional toxicity and safety information in a larger group of people (100–300). If the

therapy is shown to be effective and safe, phase III trials are conducted.

Phase III Trials—a larger group of people is tested (1000–5000). Information obtained from phase I and II clinical trials is incorporated, and a comprehensive investigation of the therapeutic role of the drug is conducted.

Seek FDA approval—after Phase III trials are completed, an application to the FDA is made for approval.

Phase IV Trials—when the therapy is approved, if any questions remain regarding safety and efficacy, as well as the use of the treatment, they are addressed in phase IV clinical tests.

This protocol for approval of new therapies is long and costly (Figure 10.3). It has been more than 30 years since the first studies of ADA deficiency were initiated and more than 15 years since gene therapy testing began. Interestingly, phase III clinical trials have not been conducted and, thus, no ADA gene therapy has become a generally accepted medical treatment for the disease.

ISSUES IN GENE THERAPY

Although gene therapy trials have demonstrated the potential of this type of treatment, more research is really required to determine the safety and efficacy of the protocols. In addition, the ethical, legal, and social implications of gene therapy must be addressed. Many questions come to mind and include

1. Who will be eligible to receive gene therapy (for example, children, the chronically ill, or only critically ill patients)?

2. Should gene therapy be used to prevent disease in addition to treating disease?

3. Who will have access to gene therapy? How will it be made available and affordable to those in need?

RECENT GENE THERAPY SUCCESS

Although no one has been completely cured by gene therapy despite the huge investment in time and money, recent successes have caused some individuals to believe gene therapy may one day provide an alternative therapy for diseases or disorders that are difficult to treat. It is now thought that if vector technology can be improved and more appropriate animal models (canines, primates) can be used, promising preclinical results can be obtained.

In October 2003 the Chinese equivalent to the FDA (the State Food and Drug Administration) approved the world's first commercially licensed gene therapy developed by Shenzhen SiBono Gene Technologies Co.

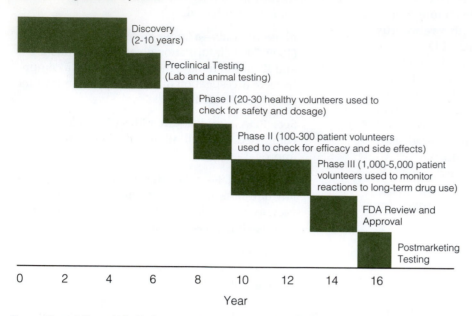

Biotech Drug Discovery Process

Discovery
(2-10 years)

Preclinical Testing
(Lab and animal testing)

Phase I (20-30 healthy volunteers used to
check for safety and dosage)

Phase II (100-300 patient volunteers
used to check for efficacy and side effects)

Phase III (1,000-5,000 patient
volunteers used to monitor
reactions to long-term drug use)

FDA Review and
Approval

Postmarketing
Testing

Year

Source: Ernst & Young LLP, *Biotechnology Industry Report:* Convergence, 2000

a

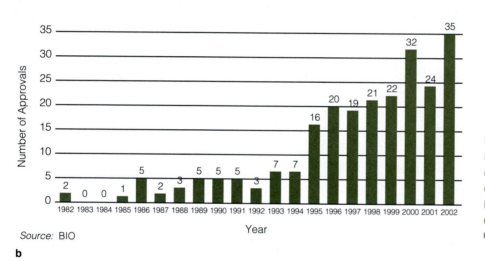

Source: BIO

b

FIGURE 10.3 **Drug discovery and
approvals. (a) The process of biotech
drug discovery. (b) New biotechnology
drug and vaccine approvals.** Used with
Permission of Biotechnology Industry
Organization, www.bio.org/er/statistics/asp

(Shenzhen, China). This therapy, marketed under the name Gendicine, treats head and neck squamous cell carcinoma. The drug was licensed after more than 5 years of clinical trials. The drug is expected to become a $100 million market within 1 year. Gendicine is an injectable therapy containing an adenovirus vector harboring the p53 tumor-suppressor gene. In clinical trials on late-stage patients, after weekly injections for 8 weeks, 64% of patients' tumors disappeared completely and 32% were reduced in size. Over a 3-year period, patients did not relapse. The drug's only reported side effect has been a low-grade fever. The drug can be used in conjunction with chemotherapy and radiation therapy, which increases the efficacy of the drug more than threefold.

NEW APPROACHES TO GENE THERAPY

Conventional gene therapy adds a correct copy of a defective gene. The normal gene then produces the functional protein and overrides the faulty one. Sometimes, however, adding a correct gene does not alleviate the disease. A disorder caused by a mutated gene whose protein prevents the normal protein from functioning properly cannot be fixed by adding a correct copy of the gene (the normal protein would not function properly). These mutated genes are referred to as *dominant negative* genes (Figure 10.4). A different approach to gene therapy must be used. Some examples of potential therapies are described here. Another therapy, RNA interference by microRNA, is described

Is Gene Therapy Safe?

The Case of Jesse Gelsinger. For 9 years, gene therapy seemed to promise cures to incurable genetic disorders. Then in 1999 a gene therapy experiment failed. Jesse Gelsinger, 18 years old, had ornithine transcarbamylase (OTC) deficiency, a rare metabolic disorder that causes a build-up of ammonia. He was able to control the disease with a low-protein diet and 32 pills per day. Although he was otherwise healthy at the time, he signed up for an experimental trial at the University of Pennsylvania that would benefit babies born with a fatal form of OTC

(severe OTC deficiency). The experiment was to test the safety of treatment and to determine the maximum tolerated dose of an adenovirus vector with the OTC gene. Four days after he received an infusion into his liver of recombinant virus, he died from complications resulting from a clotting disorder and organ failure due to the adenovirus. Interestingly, three monkeys had died of the same problem after they received a stronger version of the adenovirus at a dose 20× higher than any of the doses later used in the human experiment. Nevertheless, the experiment proceeded.

In 2003 nearly 30 gene therapy trials were canceled after two children treated for ADA deficiency developed leukemia (the first case of leukemia occurred in October 2002) in a French trial. It was discovered that the retrovirus had inserted next to an oncogene in a white blood cell and caused the cells to grow uncontrollably, causing leukemia. The children were treated with chemotherapy.

These cases raise issues about gene therapy safety, although some experts say the development of leukemia may have been specific to the SCID gene therapy trial.

---

in Chapter 3, Basic Principles of Recombinant DNA Technology.

Spliceosome-Mediated RNA Trans-splicing

Spliceosome-mediated RNA trans-splicing (SMaRT) targets and repairs mRNA transcribed from the mutated gene causing the disorder. The spliceosome is the RNA and protein machinery that removes the introns from a pre-mRNA after transcription and ligates the exons (see Chapter 2, From DNA to Proteins). Rather

than introducing a correct gene, the region of the mRNA that is affected by the mutation is repaired (Figure 10.5). In this method

1. An RNA strand that pairs (binds, therefore is complementary) with the intron next to the mutated region (exon) of the mRNA is introduced into cells. The RNA is genetically modified to contain the correct exon and a small region that binds to the neighboring intron.

2. When the RNA strand binds to the intron, the duplex causes the spliceosome to cut and remove the intron and the defective exon from the mRNA.

3. The exons are joined, with the corrected exon ligated into the mRNA, thereby generating a functional mature mRNA and protein.

Triplex-Helix–Forming Oligonucleotide Therapy

In triplex-helix–forming oligonucleotide therapy, synthetic triplex-forming oligonucleotides bind to target DNA regions, where they selectively block transcription into messenger RNA (Figure 10.6). A short single-strand DNA (oligonucleotide, approximately 15–21 nucleotides) that specifically binds in the groove between the double strands of the DNA of the mutated gene is used. This forms a triple helix where three DNA strands are bound together and prevents that region of DNA from being transcribed into mRNA. Therefore, a defective protein is not produced. The introduction of a correct gene in cells will allow a functional protein to be produced.

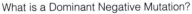

What is a Dominant Negative Mutation?

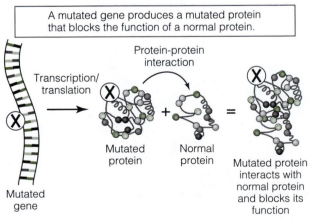

A mutated gene produces a mutated protein that blocks the function of a normal protein.

Protein-protein interaction

Transcription/ translation

X Mutated protein

Normal protein

X Mutated protein interacts with normal protein and blocks its function

Mutated gene

FIGURE 10.4 The effect of a dominant negative mutation on a normal protein. In this case, adding a normal gene would not alleviate the disorder.

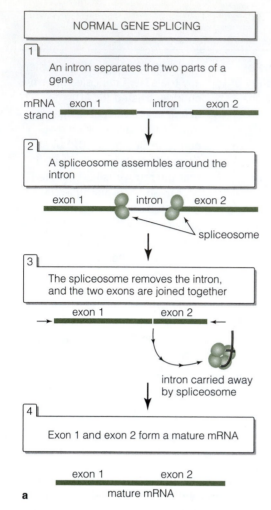

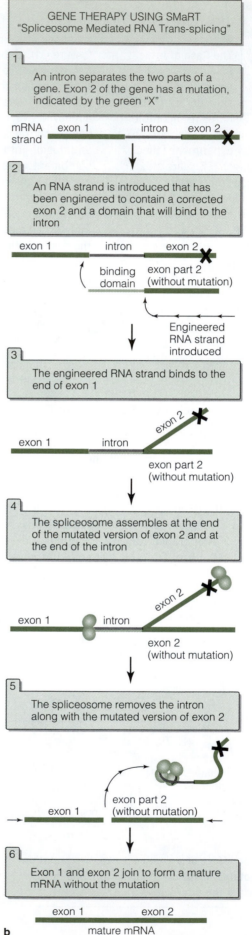

FIGURE 10.5 A new approach to gene therapy using spliceosome mediated RNA transplicing (SMaRT).
(a) An intron is excised from a normal transcript using a spliceosome (RNA–protein complex) and exons 1 and 2 are ligated. (b) In this example, a mutated exon 2 of the mRNA is repaired using SMART. When this mRNA is translated, a normal protein will be synthesized.
From http://gslc.genetics.utah.edu/units/genetherapy/gtapproaches

Antisense Therapy

Antisense therapy targets the mRNA of a mutated gene so that they cannot be translated into protein (Figure 10.7). During transcription, one strand of the DNA is used to synthesize a complementary mRNA. This "sense" mRNA (the gene region is the antisense DNA strand) encodes the information for the protein. An antisense RNA is complementary to the sense mRNA. In this method

1. RNA is introduced into cells that is just like the antisense sequence of the DNA (although uracil is used instead of thymine). This is called antisense RNA.

2. The antisense RNA binds to the sense mRNA strands synthesized by the cell during transcription.

3. The duplex RNA cannot be translated into a protein, thereby eliminating or dramatically reducing the production of a mutated protein.

In the early 1980s, scientific investigators learned that some microbes made antisense RNA to regulate gene expression; these microbes base paired with complementary sense RNA to prevent translation. It was later discovered that plants and animals also use antisense to regulate gene expression. In the course of studying this phenomenon, researchers realized that this cellular process could have important implications in medicine. They reasoned that antisense molecules could be designed as drugs by introducing them into the body, where they would decrease expression of a particular gene by selectively inhibiting translation.

Antisense therapy may be a treatment for diseases in which either there is a loss of control over gene regulation or a gene is overexpressed (for example, this happens in many types of cancerous cells). In the first case, mRNA (and therefore protein) is present all the time; in the second, there is too much mRNA. Cancer-specific antisense therapies are also being developed. In October 2003, Genta (Berkeley Heights, New Jersey), in collaboration with Aventis (Strasbourg, France), completed a phase III clinical trial of an antisense agent called Genasense. Genasense was used along with chemotherapy in patients with late-stage malignant melanoma. This antisense therapy is currently in 20 different clinical trials for a variety of cancers such as lung, prostate, liver, pancreas, colon, breast, kidney, and esophagus. Genasense targets the mRNA for a gene that is overexpressed in many different types of cancers. The protein (bcl-2) blocks the process of cell death (apoptosis). The protein is implicated in the resistance to chemotherapy.

Preventing Transcription of a Mutated Gene Using Triple-helix-forming Oligonucleotides

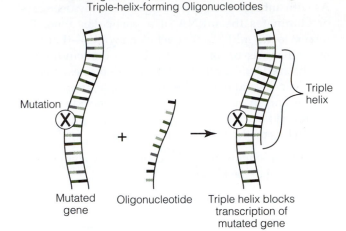

FIGURE 10.6 The prevention of transcription of mRNA with a mutation using a triplex–helix–forming oligonucleotide. A triple–helix DNA structure is not available for translation, thereby preventing a defecting protein from being synthesized. From http://gslc.genetics.utah.edu/units/genetherapy/gtapproaches

Preventing Translation Using Antisense Technology

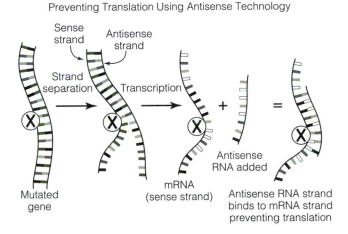

FIGURE 10.7 A mutated mRNA cannot be translated using antisense technology. An antisense RNA binds to the mRNA to prevent translation into protein. From http://gslc.genetics.utah.edu/units/genetherapy/gtapproaches

Preventing Translation Using Ribozyme Technology

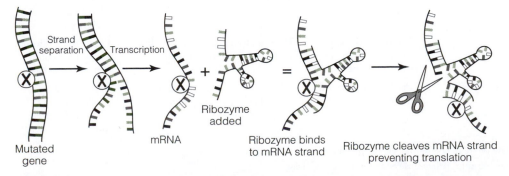

FIGURE 10.8 Ribozymes are catalytic RNA that act as enzymes to cut mRNA. In this way, a mutated mRNA cannot be translated into protein. From http://gslc.genetics.utah.edu/units/genetherapy/gtapproaches

Ribozyme Therapy

As with antisense therapy, ribozyme therapy decreases or eliminates the mRNA of a particular gene. Ribozymes are mRNAs that act as enzymes—that is, a molecular scissors or catalytic RNA—that cut mRNA synthesized by the cell (catalytic RNAs also exist naturally in cells and have roles, for example, in intron removal and the extension of the polypeptide by one amino acid at a time during translation).

In this method (Figure 10.8)

1. RNA is engineered to function as a ribozyme and to bind to the target mRNA.

2. The ribozyme is introduced into cells.

3. The ribozyme binds to the mRNA (target) encoded by the mutated gene.

4. The target mRNA is cut, thus preventing it from being translated into protein.

VIROTHERAPY

Viruses are highly insidious entities that have evolved efficient modes of entering and infecting cells (that is, they are good at what they do). They cause numerous diseases (examples include viral meningitis, SARS) and often cause great panic (most recently, the influenza surge of 2003). In general, the virus recognizes and binds to a specific receptor on the host cell surface and then, after infection, usually kills the host cell. Each type of virus recognizes a different type of receptor—thus, viruses exhibit cell specificity. Scientists are be-

ginning to exploit these characteristics and harness them to destroy cancer cells in a method called *virotherapy*. Viruses are engineered to selectively infect and kill tumor cells, while leaving normal cells intact. Several different methods are being studied and a number of these are in clinical trials (Table 10.2):

1. Transductional targeting to preferentially infect, reproduce, and then kill tumor cells. Released viruses from the doomed cells move on to infect other cancer cells.

2. Transcriptional targeting that uses engineered viruses that have a tumor-specific promoter linked to an essential virus gene. Although the virus can infect both normal and cancer cells, the gene turns on only in cancer cells and not normal cells. The virus kills the cancer cells.

3. Engineered viruses that render the tumor cells more susceptible to chemotherapy. In this method, viruses harbor genes that encode enzymes that become potent chemotherapies.

In addition, radioactively- or fluorescently-tagged viruses may be used diagnostically in imaging to identify very small tumors.

STEM CELLS

Many investigators are examining the potential of using genetically engineered undifferentiated stem cells (see Chapter 7, Animal Biotechnology) as therapeutic agents. These stem cells are the progenitors of

TABLE 10.2 Selected Companies Involved in Virotherapy

Company	Headquarters	Virus	Diseases	Viral Modifications	Clinical Trial Status
BioVex	Abingdon, Oxfordshire, U.K.	Herpes simplex virus (HSV)	Breast cancer and melanoma	Carries the gene for granulocyte-macrophage colony stimulating factor, an immune system stimulant	Phase I/II
Cell Genesys	South San Francisco, Calif.	Adenovirus	Prostate cancer	Targeted to prostate cancer cells using prostate-specific promoters	Phase I/II
Crusade Laboratories	Glasgow	HSV	Glioma (brain cancer), head and neck cancer, melanoma	Has a gene deletion that restricts it to actively dividing cells such as cancers	Phase II for glioma and head and neck cancer; Phase I for melanoma
MediGene	Martinsried, Germany	HSV	Glioma and colon cancer that has spread (metastasized)	Harbors two gene deletions that prevent it from reproducing in normal cells	Phase II for glioma; Phase I for colon cancer metastases
Oncolytics Biotech	Calgary, Alberta, Canada	Reovirus	Prostate cancer and glioma	Able to replicate only in cancer cells bearing the activated oncogene *ras*	Phase II for prostate cancer; Phase I/II for glioma

Note: Phase I tests are designed to evaluate safety in small numbers of patients.

Phases II and III are intended to determine the appropriate dose and efficacy, respectively.

Source: From *Sci. Amer.* Oct 2003. p. 75.

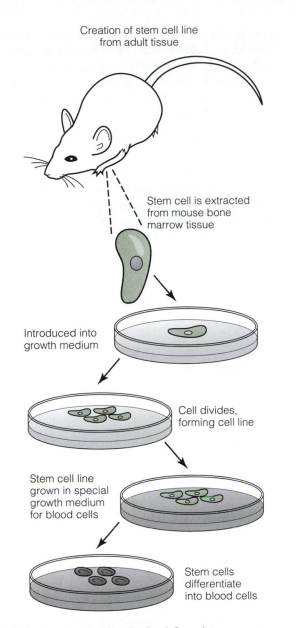

FIGURE 10.9 Stem cells obtained from bone marrow can be cultured and induced to differentiate into blood cells.

From http://gslc.genetics.utah.edu/units/stemcells/sccreate

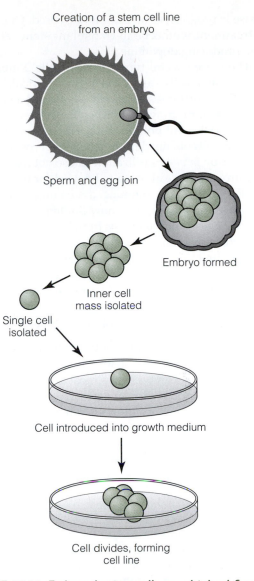

FIGURE 10.10 Embryonic stem cells are obtained from a blastocyst or embryo. Cells of the inner cell mass are placed in culture where they divide to form an embryonic stem cell line.

From http://gslc.genetics.utah.edu/units/stemcells/sccreate

many different cell types. For example, stem cells in bone marrow develop into red and white blood cells (Figure 10.9). Other stem cells become neural tissues of the brain. Methods to culture and transform stem cells have been developed. Experiments have demonstrated that stem cells can take on the form of different cell types—that is, they have plasticity. For example, bone marrow cells can become liver or kidney cells. If stem cells can be obtained from the patient to regenerate a tissue, this would prevent rejection. If stem cells harboring a correct gene are successfully introduced into a patient, these cells could divide and serve as a source of healthy cells for a variety of tissues.

Scientists are now able to isolate stem cells from the 5-day-old mass of cells called a *blastocyst* that develops into the embryo (Figure 10.10). These cells, called em-

bryonic stem cells, have the ability to develop into different cells of the body. Embryonic stem cells (obtained from aborted embryos or *in vitro* reproductive technologies) have been the source of much controversy. Much attention has focused on embryonic stem cells because only stem cells can develop into any cell type—unlike stem cells obtained from adult tissues.

The overall goal of stem cell therapy is the repair of damaged tissue. This might be achieved by transplanting stem cells into a region of the body so that they divide and differentiate into healthy tissue. An example is Parkinson's disease, in which brain cells that produce dopamine (called dopamine neurons) begin to die. Dopamine is a chemical messenger that helps control muscle movements. Although a patient can be treated with drugs that increase levels of dopamine, the disease is progressive and cannot be cured. One possible treatment would be to transplant stem cells that begin producing dopamine.

There are several types of stem cells. Embryonic and fetal stem cells are the most controversial. In 2000 the National Institutes of Health established a policy to allow federal funding of embryonic stem cell research using a strict set of policies with federal oversight. At that time, cells obtained from frozen embryos that were produced by *in vitro* fertilization (those not used, but to be destroyed) were allowed to be used for research. In 2001 President Bush announced a change to the NIH policy. Federal funding was now limited to nonembryonic stem cells. Only stem cells from blastocysts existing before that day (August 9, 2001, 9:00 PM) could be used. In the United States at this time, new research on embryonic and fetal cells cannot be funded using federal monies. Critics of this change express concern that current limitations on stem cell research will slow progress.

Stem cells comprise the following types:

1. Embryonic—These pluripotent cells are considered to be the most valuable type of stem cell because they can become almost any type of cell in the body. In 1998 human embryonic stem cells were first grown in culture. Human embryonic stem cells, cells from early embryos, have been obtained from *in vitro* fertilization (IVF) procedures where extra embryos that are not implanted are frozen. Couples can donate them to scientific research.

2. Fetal—These pluripotent cells are found in fetal brain tissue and are a natural source of dopamine neurons. They are an excellent source of cells for adults with Parkinson's disease. Fetal stem cells must be harvested from prematurely terminated human fetuses or late-stage embryos. A human embryo is considered a fetus 8 weeks after the egg is fertilized.

3. Umbilical cord blood—The cells from cord blood are called multipotent stem cells because, although they naturally become blood cells and immune system cells, they have the potential to become many different types of cells. Cord blood is rich in stem cells. Cord blood cells are highly desirable because they cause less rejection problems because they have not yet developed antigens that can be recognized by the immune system. Also, umbilical cord blood does not have mature immune system cells that can attack the recipient's body (this deadly reaction is called *graft versus host disease*). Today people can store cord blood (in a cord blood bank) in case they need stem cells; they can also be donated by women after giving birth.

4. Adult—Numerous multipotent adult stem cells exist in the body. Each type develops into cells of a specific type of tissue. It is thought that these cells are the least plastic in terms of having the potential to develop into any type of tissue. Much research today is aimed at developing treatment protocols and coaxing adult stem cells to become different cell types.

Most stem cell experiments use stem cells that are cultured in sterile dishes. These cells come from either human tissue samples or from animal models such as mice. Stem cells are cultured as stem cell lines that are continuously grown in the laboratory (that is, the cells keep dividing in culture). Stem cells naturally divide continuously and remain undifferentiated. For a transplantation to be successful, the cells must differentiate into the appropriate tissue and divided in a controlled manner.

Today stem cells are harvested from several sources for use in transplants.

1. Bone marrow

2. Peripheral blood

3. Umbilical cord blood

The most common and traditional type of transplant is the bone marrow transplant. This method is used to treat cancers such as leukemia and a variety of disorders of the blood. Because leukemia is a cancer of white blood cells, the source of these cells is bone marrow (see Chapter 4, Basic Principles of Immunology). Stem cells in bone marrow become white blood cells and are then released into the bloodstream. One way of treating abnormal white blood cells or leukemia is to use chemotherapy to kill the cells. If, however, this drug treatment cannot kill them all, a bone marrow transplant is used. After a healthy donor match is found, the patient's bone marrow and abnormal white blood cells are killed using powerful drugs and radiation. In a successful transplant, the donor stem cells will migrate to the patient's bone marrow and begin dividing and populating the bone marrow. These stem cells will begin producing healthy white blood cells.

Some stem cells can be obtained from the bloodstream. These stem cells can be used just like those from the bone marrow. One drawback, however, is that it is difficult to collect enough cells from blood because they are in low quantity.

Therapeutic Cloning and Embryonic Stem Cells

Manipulating a person's existing stem cells to be used therapeutically is one way to treat a disorder, as discussed earlier. Therapeutic cloning (rather than reproductive cloning) to produce embryonic stem cells is another method that has much potential. Embryonic stem cells can be obtained using methods for cloning whole animals (such as Dolly the sheep—see Chapter 7, Animal Biotechnology). This application of cloning has great medical potential and, therefore, is referred to as *therapeutic cloning* or *somatic cell nuclear transfer (SCNT).* In this process, the nucleus from an adult donor cell is transferred to a recipient egg cell whose nucleus has been removed (Figure 10.11). The cell is stimulated to begin dividing (therefore, not fertilized using sperm) and the transferred nucleus directs the development of the resulting embryo. These cells are never implanted in a female; rather they are kept in culture and continually divide, creating a cell line that is genetically identical to the adult donor cell. Stem cells can be collected 5–6 days after nuclear transfer. An embryonic cell line can be produced in culture. This method has much potential for a patient requiring stem cells and would be an exact genetic match.

Stem cell research has generated ethical debate from both opponents and proponents. Questions that have been asked include

1. Should we be using human embryonic stem cells in research?

2. What constitutes a human life? Do embryonic stem cells qualify?

3. Should the government regulate embryonic stem cell research?

4. Should frozen embryos from IVF be used for stem cell research?

5. Should therapeutic cloning be used medically to obtain embryonic stem cells?

VACCINES

Vaccines help the body recognize and fight infectious diseases and are essential to the ongoing struggle to eradicate infectious diseases. Conventional vaccines use weakened or killed forms of a virus or bacterium to stimulate the immune system to create antibodies that provide resistance to the disease. Through increased understanding of the process of infection, the immune response (see Chapter 4, Basic Principles of Immunology), and the factors that contribute to virulence, recombinant DNA technology has made it possible to develop a new generation of vaccines. New vaccine vectors and new delivery approaches, immunoenhancers, and even nucleic acids will be used. Vaccines are being developed against pneumococcal pneumonia, malaria, herpesvirus (Cytomegalovirus, varicella-zoster), diarrhea (rotavirus, prevalent in children), respiratory syncytial virus (respiratory infections), measles, and cholera. Recently, anthrax (because of the fear of terrorism and biological warfare) and severe acute respiratory syndrome (SARS), which became a worldwide concern in March 2003, have become areas of intense focus. Experiments also are investigating the possibility of using vaccines to prevent diabetes, Alzheimer's disease, and cancers.

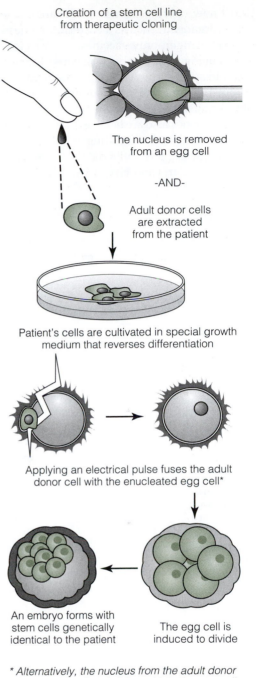

Creation of a stem cell line
from therapeutic cloning

The nucleus is removed
from an egg cell

-AND-

Adult donor cells
are extracted
from the patient

Patient's cells are cultivated in special growth
medium that reverses differentiation

Applying an electrical pulse fuses the adult
donor cell with the enucleated egg cell*

An embryo forms with
stem cells genetically
identical to the patient

The egg cell is
induced to divide

*Alternatively, the nucleus from the adult donor
cell could be directly injected into the egg cell.*

FIGURE 10.11 Therapeutic cloning enables stem cells to be produced that are genetically identical to the donor patient. Cells from a patient are placed in culture. The nucleus of an egg cell is removed and either the two cells are fused (enucleated egg and patient's cell) or the nucleus from the patient's cell is microinjected into an enucleated egg. The egg cell with the transferred nucleus begins to divide. Cells from the embryo are placed in culture to produce an embryonic stem cell line. These stem cells are genetically identical to the patient's cells. From http://gslc.genetics.utah.edu/units/stemcells/sccreate

Most new vaccines consist of the antigen and not the actual organism (such as a bacterium). These are referred to as antigen-only vaccines. The vaccine is generated by using genetically engineered yeast or other host cells to synthesize the protein or antigen. The protein is purified and marketed commercially. An example of an antigen-only vaccine is hepatitis B vaccine.

DNA vaccines also are being explored. Experiments have demonstrated that injecting small pieces of DNA from a microbe triggers antibody production. DNA vaccines against malaria and HIV are in clinical trials.

TABLE 10.3 The Incidence of Organ and Tissue Damage in the United States and the Number of Patients or Treatments

Disease/Injury	Number of Procedures or Patients/Year
Skin	
Burns*	2,150,000
Diabetic ulcers	600,000
Neuromuscular disorders	200,000
Spinal cord and nerves	40,000
Bone	
Joint replacement	558,200
Bone graft	275,000
Facial reconstruction	30,000
Cartilage	
Patella resurfacing	216,000
Meniscal repair	250,000
Arthritis (knee)	149,900
Arthritis (hip)	219,300
Fingers and small joints	179,000
Tendon repair	33,000
Ligament repair	90,000
Blood vessels	
Heart	754,000
Large and small vessels	606,000
Liver	
Liver cirrhosis	175,000
Liver cancer	25,000
Pancreas	
Diabetes	728,000
Intestine	100,000
Kidney	600,000
Bladder	57,200
Hernia	290,000
Breast	261,000
Blood transfusions	18,000,000
Dental	10,000,000

*Approximately 150,000 individuals are hospitalized and 10,000 die each year.

Reprinted with permission from R. Langer and J.P. Vacanti, *Science* 260:920–925. Copyright 1993 AAAS.

TISSUE ENGINEERING

The emerging field of tissue engineering focuses on the development of substitutes for damaged tissues and organs. Millions of people lose the function of their tissues or organs and maintain their health by resorting to mechanical devices (kidney dialysis machines, pacemakers, and artificial hearts are examples), organ transplants, or reconstructive surgeries (Table 10.3 presents a partial list). Tissue or organ damage often leads to death. The costs of treating patients are astronomical, with estimates exceeding $400 billion per year. More than eight million surgical procedures are performed each year in the United States to correct or alleviate problems associated with tissue or organ failure. The scarcity of suitable organ donors is a serious problem, and most transplant candidates die while waiting for an organ.

Tissue engineering provides a viable alternative to the problem; it combines the principles of engineering and biology to aid in the formation of new tissue or in the implantation of functional cells. Tissue-inducing compounds such as growth factors can serve as signals to stimulate the growth and development of different tissues. Isolated cells, engineered before transplantation if necessary, can replace nonfunctional cells. The cells can reside within an extracellular matrix, such as collagen or a synthetic **polymer**, that is sometimes incorporated into a patient's tissue. To reduce the chance of rejection, transplanted cells can remain isolated from host tissues. Isolated cells often re-form the correct tissue structure if the appropriate conditions are provided. Scaffolds or networks of polymers may serve as a substrate for cells (Figure 10.12) to help induce the formation of functional and morphologically correct tissues. Polymers may either be nondegradable and remain in the patient permanently or degrade over time and disappear from the tissue. A scaffold is seeded with living cells that divide and grow into the three-dimensional tissue. When implanted into the body, blood vessels grow and provide the new tissue with a supply of oxygen and nutrients.

Stem cells are a valuable source of cells for tissue engineering because they have the potential to develop into a variety of cell types. Ongoing studies are exploring tissue replacements for the nervous system, skin, cornea of the eye, liver, pancreas, bones and cartilage, muscles, and blood vessels.

One disease for which tissue engineering is showing potential is diabetes. Many Americans die from diabetes or its complications each year. When the islet cells of the pancreas are destroyed, a loss of blood glucose control occurs. The diabetic's glucose levels are normally controlled by diet or injections of **insulin**, depending on the severity of the disease. Unfortunately, the symptoms often worsen with age. Tissue engineering research has focused on the transplantation of encapsulated functional pancreatic islet cells. Islets of

a

b

FIGURE 10.12 Tissue engineering offers hope to hundreds of thousands of patients who need functional tissues and organs. Scaffolding allows cells to multiply in the proper location and/or form the appropriately shaped tissue. (a) Tubular biodegradable scaffolds to form tubular tissue such as veins, arteries, and intestines. Shown are different-sized tubing scaffolds. Tubes are tailored to different dimensions, mechanical properties, and degradation rates to meet a variety of needs. (b) Scanning electron micrograph of a biodegradable polymer scaffold made of poly(glycolic acid). Photos courtesy of P.X. Ma and Robert Langer, Massachusetts Institute of Technology, Cambridge, MA

Langerhans (insulin-producing pancreatic cells) that have been alginate-encapsulated are being used in experiments to develop a safe, efficient method of providing insulin to diabetics. The semiporous microspheres protect insulin-producing cells from attack by the immune system while allowing glucose to reach the cells and insulin to be secreted into the bloodstream. Usually, porcine islet cells (see xenotransplantation) are used as donor cells, but one day genetically engineered human islets that produce insulin may be used. In experiments using this method, rats have shown normal glucose and insulin levels for more than 2 years.

Various methods have been used to achieve functional transplants in animal models. Islet cells have been immobilized by a polysaccharide alginate within hollow semiporous fibers that are then implanted intraperitoneally in experimental animals. In other experiments, islet cells have been placed in a tubular membrane that enables glucose and insulin to pass freely through the pores; the membrane has been bound to a polymer that is grafted onto blood vessels. Antibodies and lymphocytes that would have generated an immune response were too large to pass into the membrane tube containing the islet cells.

Recently, a biohybrid kidney was developed that maintains patients with acute renal failure until the injured kidney recovers. Patients with only 10–20% probability of survival due to kidney injury completely recovered with the help of tissue engineering. The kidney was constructed of hollow tubes seeded with kidney stem cells that multiplied by cell division until they lined the inner walls of the tubes. These cells developed into functional kidney cells.

XENOTRANSPLANTATION

According to the United Network of Organ Sharing, more than 60,000 people are on organ-recipient lists, and approximately 100,000 additional people need new organs but have not been added to the lists. One way to alleviate the severe shortage of available organs for transplantations may be animal-to-human transplants, or **xenotransplants** (see Chapter 7, Animal Biotechnology). There may one day be organ farms, on which livestock are raised to provide organs for medical use. Although a major obstacle is the response by the immune system of the recipient—not to mention ethical and legal problems—researchers continue to look for ways to prevent rejection of animal organs, such as by identifying human shield proteins that protect organs from attack by the immune system. If genes encoding these proteins can be transferred to donor animals, their organs will express human proteins, thereby reducing the chance of rejection. Experiments with primates and mice seem promising. Shield proteins such as membrane cofactor protein (MCP) and decay accelerating factor (DAF) are bound to the surface of human cells. These proteins inactivate complement proteins to prevent them from marking cells for attack by the immune system. Pigs may be the most suitable organ donor because the major organs of pigs and humans are of similar size and shape, and because few diseases are transmitted from pigs to humans. In experiments, pig eggs have been transformed with MCP and DAF genes, as well as CD46 and CD59, and have produced pigs that manufacture the shield proteins. Astrid, born December 23, 1992, was the first transgenic pig to produce shield proteins (DAF and MCP). By 1994 two transgenic lines of pigs had been produced, each producing a variety of shield proteins. In 1996 a new generation of descendants from the original Astrid line were produced and may produce enough shield proteins to become a source of donor organs for clinical trials.

In 1999 a study was conducted with 160 people who received pig cells. None of the participants showed any adverse reactions to the transplant. Although progress in xenotransplantation has proceeded rapidly, ethical issues remain, such as whether it is appropriate to genetically manipulate animals for the use of their body parts. Although the question will continue to be debated, patients in need of organ transplants will benefit greatly from having an alternative source.

DRUG DELIVERY

An effective drug is useful only if a reliable method of delivery is available. How could a large protein or peptide be administered—other than by a painful injection—and protected from digestion or degradation? "Drugs" such as proteins, DNA, and even synthetic molecules could be inhaled as aerosols through the mouth or nose. Nebulized deoxyribonuclease (DNase), tested in clinical trials, has shown potential for patients with CF. DNase breaks down the mucus that accumulates in the lungs. A drug for osteoporosis, calcitonin, has been tested as a nasal spray. The lungs are a good route of entry for peptides and other molecules. They can readily reach the bloodstream through capillary-rich lung tissue, because only a thin epithelial layer separates the drug from the circulatory system. Oral delivery of peptides and larger proteins encapsulated by tiny microspheres ranging from 50 nanometers to 20 micrometers may protect these molecules so that they can be absorbed by the intestine.

The skin may also be an effective route for administering small molecules. A transdermal patch used in combination with an electric current can transfer peptides into the skin and circulatory system. Patches have been used effectively to administer, for example, medications for motion sickness, nitroglycerin for angina, and nicotine to help people quit smoking. However, the molecules of many drugs are too large and insoluble to be moved through skin by current methods. A process called iontophoresis, or electrotransport, can force molecules through the skin. Because like electric charges repel, a small electric current applied to the skin essentially pushes charged proteins through openings such as hair follicles and sweat glands.

Biosensors

Biosensors are biological components such as a cell, antibody, or protein and are linked to a very small transducer. The transducer receives a signal but then transforms the signal (or energy) into another form so that it can be detected. Biosensors identify and measure specific components or molecules at very low concentrations. When a specific substance binds to the biosensor, the transducer produces an electrical or optical signal (that can be detected) that is proportional to the concentration of the substance. Examples of components that can be measured rapidly and in very small concentration include

1. Blood components that can be identified and measured

2. Environmental pollutants that can be located, detected, and quantified

3. Toxins and biological warfare agents that can be detected and quantified

Nanotechnology (see Biotech Revolution) will eventually miniaturize biosensors so that the biological and electronic components will be integrated into one nanocomponent.

Medical biotechnology has come a long way since recombinant human insulin was commercialized in 1982. As scientists identify more genes and the mutations that lead to genetic diseases, unravel the complexities of their regulation, elucidate protein–protein interactions, and find the causes of acquired diseases, powerful biotechnological treatments will be developed.

Nanotechnology

In 2000 nanotechnology came into existence with the establishment of the National Nanotechnology Initiative. Nanotechnology is derived from nanometer (one-thousandth of a micrometer) and is the study, manipulation, and manufacture of extremely small tools (nanotools), structures (nanostructures), and machines (nanomachines) at the molecular and atomic levels. Nanotechnology requires the use of microscopic tools for imaging and manipulating single molecules. Engineers, physicists, chemists, and molecular biologists work in the area of nanotechnology. The huge potential of nanotechnology in industry, electronics, medicine, science, space technology, and many other areas has been recognized by many. Nanotechnology will allow many developments that were once impossible because of size limitations. President Bush requested $849 million to fund nanotechnology research in 2004.

Examples of potential applications in the area of nanotechnology include the following:

1. Nanowires use nanoscale molecules to conduct electricity (for example, tiny gold particles that self-assemble into wires). This could be adapted to function as a nanosized pacemaker for the heart or as an internal defibrillator.

2. Nanobots, or molecular machines, can assemble molecular building blocks to form products. Medical nanobots could move through the body to detect disease (for example, tumors), treat specific areas of the body (for example, destroy viruses, infected tissues, and cancers), and repair damaged tissue and organs. Nanoparticles with molecular sensors could move through blood vessels and into the ducts of organs to diagnose problems.

3. Nanomaterial production can create nanoscale materials such as single and multimetal powders.

4. Drug delivery by tiny nanocontainers made of nanopolymers can deliver substances to the appropriate cells, tissues, or organs.

5. DNA computers use DNA as both hardware and software. Proteins can form lattices and semiconductor quantum dots, giving promise for electronic devices.

Nanotechnology merging with molecular biology to exploit the nanostructures and nanomachines of the living organism can accomplish what would ordinarily be impossible in science. Rather than using silicon to build nanostructures, biological molecules can provide a natural framework for assembling nanostructures. DNA has been used to build nanostructures and nanomachines. Although difficult to comprehend, the genetic information stored in DNA (as deoxyribonucleotides) might one day serve as computers. As microprocessors and microcircuits are reduced to nanoprocessors and nanocircuits, DNA molecules fixed to a silicon chip may replace microchips. These biochips would be DNA-based processors that use DNA's immense storage capacity. These are difficult to conceptualize; however, biochips would utilize the properties of DNA to solve computational problems (that is, DNA could be used to solve math problems just like a computer). Scientists already have demonstrated that 1000 DNA molecules can solve in 4 months complex computational problems that would take a computer 100 years to solve.

General Readings

W. French Anderson. 1995. Gene Therapy. *Sci. Am.* 273:124–128.

A.R. Chapman and M.S. Frankel, eds. 2003. *Designing Our Descendants: The Promises and Perils of Genetic Modifications.* Johns Hopkins University Press, Baltimore, Maryland.

C. Dennis. 2003. Take a Cell, Any Cell. *Nature* 426:490–491.

Y.M. Elcin, ed. 2003. *Tissue Engineering, Stem Cells, and Gene Therapies.* Kluwer Academic/Plenum Publishers, New York.

D.S. Goodsell. 2004. *Bionanotechnology: Lessons from Nature.* Wiley-Liss, Hoboken, New Jersey.

D. M. Nettelbeck and D. T. Curiel. 2003. Tumor-busting viruses. *Sci. Amer.* 289:68–75.

M. Ratner and D. Ratner. 2003. *Nanotechnology: A Gentle Introduction to the Next Big Idea.* Prentice Hall, Upper Saddle River, New Jersey.

D.R. Salomon and C. Wilson, eds. 2003. *Xenotransplantation.* Springer, New York.

Additional Readings

S. Abramowitz. 1996. Towards inexpensive DNA diagnostics. *Trends Biotechnol.* 14:397–401.

S. Agrawal. 1996. Antisense oligonucleotides: Towards clinical trials. *Trends Biotechnol.* 14:376–387.

W.H. Allen. 1995. Farming for spare body parts. *Bioscience* 45:73–75.

E.W.F.W. Alton and D.M. Geddes. 1995. Gene therapy for cystic fibrosis: A clinical perspective. *Gene Ther.* 2:88–95.

M. Barinaga. 1993. Ribozymes: Killing the messenger. *Science* 262:1512–1514.

T.L. Beauchamp and L. Walters, eds. 2003. *Contemporary Issues in Bioethics*, 6th ed. Thompson/Wadsworth, Belmont, California.

S. Berry. 2002. Honey I've shrunk biomedical technology! *Trends Biotechnol.* 20:3–4.

R.M. Blaese. 1997. Gene therapy for cancer. *Sci. Am.* 276:111–115.

T.L. Blundell. 1994. Problems and solutions in protein engineering—towards rational design. *Trends Biotechnol.* 12:145–148.

G. Bock and J. Goode, eds. 2003. Tissue engineering of cartilage and bone. Wiley, Hoboken, New Jersey.

J. Bossart and B. Pearson. 1995. Commercialization of gene therapy: The evolution of licensing and rights issues. *Trends Biotechnol.* 13:290–294.

M.K. Brenner. 1995. Human somatic gene therapy: Progress and problems. *J. Intern. Med.* 237:229–239.

E. Chiellini et al., eds. 2002. *Biomedical polymers and polymer therapeutics.* Kluwer Academic/Plenum Publishers, New York.

J.S. Cohen and M.E. Hogan. 1994. The New Genetic Medicines. *Sci. Am.* 271:74–82.

P. Collas and A.-M. Hakelien. 2003. Teaching cells new tricks. *Trends Biotechnol.* 21:354–361.

Collection of articles. 1994. Frontiers in Medicine: Vaccines. *Science* 265:1371–1400.

S.T. Crooke, ed. 2001. *Antisense Drug Technology: Principles, Strategies, and Applications.* Marcel Dekker, New York.

A. Curtis and C. Wilkinson. 2001. Nanotechniques and approaches in biotechnology. *Trends Biotechnol.* 19:97–101.

W. Doerfler and P. Bohm, eds. 2004. *Adenoviruses: Model and Vectors in Virus-Host Interactions: Immune System, Oncogenesis, Gene Therapy.* Springer, New York.

S. Edington. 1994. A new force in biotech: Tissue engineering. *Bio/Technology* 12:361–365.

C. Eng. 1997. Genetic testing: The problems and the promise. *Nat. Biotechnol.* 15:422–6.

P.L. Felgner. 1997. Nonviral strategies for gene therapy. *Sci. Am.* 276(6):102–106.

J.A. Fishman. 2003. Xenotransplantation: A model for surveillance of unknown pathogens. *ASM News* 69:18–24.

C.K. Goldman et al. 1997. *In vitro* and *in vivo* gene delivery mediated by a synthetic polycationic amino polymer. *Nat. Biotechnol.* 15: 462–466.

F. Guilak et al., eds. 2003. *Functional Tissue Engineering.* Springer, New York.

C.A. Health. 2000. Cells for tissue engineering. *Trends Biotechnol.* 18:17–19.

J. Hubbell. 1995. Biomaterials in tissue engineering. *Bio/Technology* 13:565–575.

M. Kiehntopf, E.L. Esquivel, M.A. Branch, and F. Herrmann. 1995. Ribozymes: Biology, biochemistry, and implications for clinical medicine. *J. Mol. Med.* 73:65–71.

E.J. Kremer and M. Perricaudet. 1995. Adenovirus and adeno-associated virus mediated gene transfer. *Brit. Med. Bull.* 51:31–44.

P.E. Lacy. 1995. Treating diabetes with transplanted cells. *Sci. Am.* 273:50–58.

R. Langer and J.P. Vacanti. 1995. Artificial organs. *Sci. Am.* 273:130(3)–133.

R.P. Lanza, J.L. Hayes, and W.L. Chick. 1996. Encapsulated cell technology. *Nat. Biotechnol.* 14:1107–1111.

C.A. Machida. 2003. *Viral Vectors for Gene Therapy: Methods and Protocols.* Humana Press, Totowa, New Jersey.

C. Malvy, A. Harel-Bellan, and L.L. Pritchard, eds. 1999. *Triplix Helix Forming Oligonucleotides.* Kluwer Academic Publishers, Boston, Massachusetts.

E. Marshall. 1995. Gene therapy's growing pains. *Science* 269:1050–1055.

H. Maruta, ed. 2002. *Tumor-Suppressing Viruses, Genes, and Drugs: Innovative Cancer Therapy Approaches.* Academic Press, San Diego, California.

V. Mironov, T. Boland, T. Trusk, G. Forgacs, and R.R. Markwald. 2003. Organ printing: Computer-aided jet-based #D tissue engineering. *Trends Biotechnol.* 21:157–161.

R.C. Mulligan. 1993. The basic science of gene therapy. *Science* 260:926–931.

R. Nowak. 1994. Xenotransplants set to resume. *Nature* 266:1148–1151.

M. Ohtsu, ed. 2004. *Novel Devices and Atom Manipulation.* Springer, Berlin.

R.M. Ottenbrite and S.W. Kim. 2001. *Polymeric drugs and drug delivery systems.* Technomic Pub. Co., Lancaster, Pennsylvania.

L.S. Parker and E. Gettig. 1995. Ethical issues in genetic screening and testing, gene therapy, and scientific conduct. In J.L. Platt, ed. 2002. *Xenotransplantation: Basic Research and Clinical Applications.* Humana Press, Totowa, New Jersey.

N.R. Rabinovich, P. Mcinnes, D.L. Klein, and B.F. Hall. 1994. Vaccine technologies: View to the future. *Science* 265:1401–1404.

A. Rolland and S.M. Sullivan, eds. 2003. *Pharmaceutical gene delivery systems.* M. Decker, New York.

R.M. Sade. 1994. Issues of social policy and ethics in gene technology. *Methods Find. Clin. Pharmacol.* 16:477–489.

S.T. Sigurdsson and F. Eckstein. 1995. Structure-function relationships of hammerhead ribozymes: From understanding to applications. *Trends Biotechnol.* 13:286–289.

J.A. Smythe and G. Symonds. 1995. Gene therapeutic agents: The use of ribozymes, antisense, and RNA decoys for HIV-1 infection. *Inflamm. Res.* 44:11–15.

E.J. Sorscher and J.J. Logan. 1994. Gene therapy for cystic fibrosis using cationic liposome mediated gene transfer: A phase I trial of safety and efficacy in the nasal airway. *Hum. Gene Ther.* 5:1259–1277.

D.L. Vaux. 1995. Ways around rejection. *Nature* 377:576–577.

R.G. Vile and S.J. Russel. 1995. Retroviruses as vectors. *Brit. Med. Bull.* 51:12–30.

G.J. Waine and D.P. McManus. 1995. Nucleic acids: Vaccines of the future. *Parasitol. Today* 11:113–116.

B.M. Wallace and J.S. Lasker. 1993. Drug delivery: Stand and deliver—getting peptide drugs into the body. *Science* 260:912–913.

T. Wang et al. 1997. An encapsulation system for the immunoisolation of pancreatic islets. *Nat. Biotechnol.* 15:358–362.

A.B. Wedel. 1996. Fishing the best pool for novel ribozymes. *Trends Biotechnol.* 14:459–465.

D. White. 1996. Alteration of complement activity: A strategy for xenotransplantation. *Trends Biotechnol.* 14:3–5.

N.A. Wivel and R. Walters. 1993. Germ-line gene modification and disease prevention: Some medical and ethical perspectives. *Science* 262:533–538.

A.T. Woolley. 2001. Biomedical devices and nanotechnology. *Trends Biotechnol.* 19:38–39.

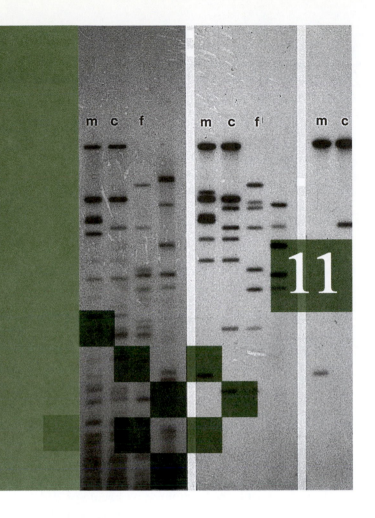

DNA PROFILING, FORENSICS, AND OTHER APPLICATIONS

11

Use of DNA as a means of identifying individuals is finding strong acceptance in criminal and civil proceedings, in searching for missing people, in determining close genetic relationships (such as establishing paternity), in medical diagnostics, in establishing evolutionary relationships, in determining the genetic diversity and relationships of plants, birds, and other animals, and in forensic archaeology. The very public criminal trial of O.J. Simpson greatly increased public awareness of forensic genetics as no previous event has to date. Although human DNA is 99% to 99.9% identical from one individual to the next, DNA identification methods use the 1% to 0.1% of DNA that is unique to generate a unique fingerprint, or identification tag, for an individual. DNA fingerprinting also is called *DNA profiling* or *DNA typing*.

Protein polymorphisms have been used to detect genetic differences among individuals since the late 1960s. Among the protein markers used have been the ABO blood groups, major histocompatibility complex (MHC) antigens or human leukocyte antigen (HLA), and red blood cell enzymes. The major problem with using protein polymorphisms is their limited variability. Thus, when they are used for forensic analysis, the results really exclude an individual rather than identify an individual, because the probability is high that a match between two samples could represent a chance event.

DNA profiling was first described in England in early 1985 by Dr. Alec Jeffreys at the University of Leicestershire. Jeffreys developed a method for identifying individuals by their DNA banding patterns and suggested that these DNAs would be useful in paternity disputes, immigration investigations, and forensics. At the time, fears were expressed that legal problems might preclude the use of this method. However, in April 1985, DNA fingerprinting was used to resolve an immigration dispute in England: A boy living outside the United Kingdom was granted entry because DNA profiling demonstrated that he was the son of a U.K. resident.

DNA was subsequently used as evidence in murder and rape cases. In October 1986 a prime suspect in a murder case was vindicated by DNA fingerprinting. When two schoolgirls were raped and murdered in Leicestershire, United Kingdom, Jeffreys' DNA probes were used in Southern hybridizations (described in Chapter 3, Basic Principles of Recombinant DNA Technology) of DNA from a collected semen sample. The tests demonstrated that the suspect was not the rapist. (A conventional criminal investigation eventually identified the murderer.)

Shortly after this crime, DNA testing was used in the United States, in the conviction of an individual for assault and rape. By 1987 DNA fingerprinting was admitted as evidence in both the United States and the United Kingdom. In 1988, the U.K. Home Office and Foreign and Commonwealth Office ratified the use of DNA fingerprinting to resolve family relationships in immigration disputes. However, by 1989, DNA fingerprinting was under attack in the United States. Most debate has centered on its use in criminal identification. Its use in paternity testing and determining familial relationships has generated less controversy.

During the late 1980s and early 1990s, U.S. courts of law questioned whether DNA profiles should be admissible in criminal trials. Questions of scientific validity and procedural consistency were raised. In response, the U.S. Office of Technology Assessment (OTA) issued a review finding that, used with the proper procedures and controls, DNA fingerprinting was a scientifically valid way of determining individuality. However, criticism and debate focused on the statistical validity of DNA fingerprints in the identification of criminals.

In May 1992 the U.S. National Research Council published a report that confirmed the scientific validity of DNA identity testing and issued appropriate guidelines. DNA evidence now is admissible, and reliability and accuracy have been continually improved. Many commercial laboratories have been created to meet the growing demand for forensic and parentage analysis (Orchid Cellmark is an example). More than 25 countries use DNA fingerprinting for forensics and paternity and maternity testing.

Today, DNA profiling has many applications in medical diagnostics, paleontology, archaeology, and many areas of the biological sciences such as population genetics, ecology, and taxonomy. Newly developed methods allow scientists to analyze ancient DNA samples (see the later discussion on forensic archaeology). For example, it has been used to match the goatskin fragments of the Dead Sea Scrolls. It also has been used to study extinct animals when only extremely small tissue samples remain.

SATELLITE DNA

Repetitive DNA

The human genome comprises 3 billion base pairs (bp) of DNA. Much of the human genome is composed of repetitive DNA that may not have any function or for which functions have not been completely established. Two major classes of repetitive DNA have been identified.

1. Tandemly repetitive sequences of satellite DNA make up approximately 10% of the genome

2. Interspersed repetitive DNA makes up 5–20% of the genome. Interspersed (so called because it is scattered throughout the genome) repetitive DNA is further subdivided into short and long sequences; sequences of fewer than 500 bp are called *SINES* (short interspersed elements), and

sequences of 500 bp or more are called *LINES* (long interspersed elements).

Satellite DNA was first defined as a fraction of DNA that banded out at a different place from the rest of the DNA during ultracentrifugation of genomic DNA in a CsCl (cesium chloride) gradient. The GC content of genomic DNA influences the position of the DNA in the tube during buoyant density gradient centrifugation. Eukaryotic DNA often forms more than one band during centrifugation, with the minor bands called "satellites."

Tandemly repeated DNA comprises long macrosatellite regions near the centromere and shorter microsatellite and minisatellite DNAs that are scattered within the genome. These repetitive DNA sequences are composed of a short sequence, or motif, sometimes called a *core* or unit sequence, that is repeated many times in tandem (that is, end to end). Tandem repeats form bands of a different buoyant density from the main genomic DNA because the GC content deviates significantly from other genomic DNA (although not all satellite DNA has a different GC content). The repeat units vary from very short ones, 2–6 bp long, to longer repeats of 300 bp or more. In this context, satellite DNA refers to the short head-to-tail tandem repeats of specific sequence motifs. The number of times the core sequences are repeated in tandem varies; thus these sequences are called **variable number of tandem repeats** (VNTRs) (Figure 11.1a and b). A VNTR refers to a single locus that has alternative alleles characterized by the differences in the number of times the core sequence is repeated in tandem. A VNTR locus is hypervariable when many alternative alleles exist in the population.

Many hypervariable loci have been identified in the human genome and are found on every chromosome—sometimes close to genes but more often in stretches of noncoding regions. In fact, much of our DNA comprises noncoding, short tandemly repeated sequences. For example, a 20-base sequence (a unit, or core, sequence) may be tandemly repeated 20–30 times or more. Often slight sequence variations may occur in some of the unit sequences. For example, the main unit may be

-AAGGGCACCAGAGACCCGGA-.

However, a single base change could yield

-AAGGGCACCAG**G**GACCCGGA-.

Other variations include changes not only in sequence but also in the number of times a unit is repeated in the locus. Thousands of VNTR loci are present throughout the human genome. A VNTR found at only one locus is a single-locus repeat, and sequences that are found at many loci along the chromosomes are called multilocus repeats.

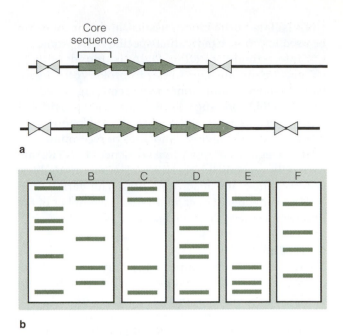

FIGURE 11.1 Variable number of tandem repeats (VNTR) are the most commonly occurring DNA sequence length polymorphisms. (a) Polymorphisms are generated by differences in the number of times the core sequence is repeated in a VNTR. (b) A person's VNTRs are inherited from the parents. The more VNTR probes that are used to analyze a person's DNA profile, the more distinctive and unique the DNA banding pattern.

Alec Jeffreys and his colleagues determined that when a region of the highly polymorphic sequences is homologous, probes can be used in hybridization. To detect polymorphisms, the DNA is digested with an appropriate restriction enzyme that cuts just outside the VNTR region and then the DNA fragments are separated by gel electrophoresis, blotted onto a nylon membrane, and hybridized with a complementary probe that yields the banding pattern, or fingerprint (Southern blot hybridizations are discussed in Chapter 3, Basic Principles of Recombinant DNA Technology). Digesting the DNA with a specific restriction endonuclease allows the difference in length of a VNTR locus between two individuals to be detected. PCR also is used to visualize the difference in length. The differences in the lengths of the fragments depend on the number of repeating sequence units and not on the length of the DNA flanking the repeats (which is invariable) where the enzyme cut. Hypervariable loci, reflected as length polymorphisms, are useful in distinguishing individuals and in determining parentage and other close relationships. When the variabilities from different loci of an individual are compiled, the composite profile becomes specific to an individual.

DNA isolated from blood, tissues, and hair roots can be used to produce individual-specific DNA profiles.

Interestingly, satellite DNA is found only in eukaryotic organisms and not in prokaryotic organisms. Biological functions have been suggested for satellite DNA; one suggestion is that they provide sequences necessary for the pairing of homologous chromosomes during meiosis and sites of recombination. Three categories of repetitive sequences (VNTRs) are generally recognized based on size: microsatellites, minisatellites, and macrosatellites. The variation in VNTRs is the result of changes in copy number of the unit composing the repeated sequences.

Microsatellites

Microsatellites are simple tandemly repeated sequences or short tandem repeats (STRs) with the highest rate of variation in copy number among individuals. Microsatellites are very short, with a repeat unit of 2–5 bp. The total length of the repeat is usually less than 1 kb. STR units are tandemly repeated so that the overall length usually varies from 70 to 200 bp. STRs are randomly distributed throughout the genome. The variation in length is due to replication slippage.

The most common motif observed in the human genome is $(CA)_n$–$(GT)_n$ (n equals the number of repeats). The motif is dispersed throughout the genome of all eukaryotes and is present in high copy number. Some 50,000 to 100,000 copies are found every 30 kilobases in human DNA.

One of the best described repeats is $(AC)_n$, where n can vary from 10 to 60. Approximately 50,000 $(AC)_n$ repeats exist in the human genome. Because they are widely distributed, these polymorphisms served as important markers in mapping the human genome. STRs are used in familial studies of genetic diseases and in paternity disputes. Because of their hypervariability (that is, their many allelic forms), they are more valuable than traditional restriction fragment length polymorphisms (RFLPs), which have only two allelic forms—present or absent (RFLPs are discussed in more detail later in the chapter).

Because the length of the repeats varies only slightly (especially when the repeat unit is based on only two nucleotides), only one microsatellite locus at a time is used in an analysis using PCR. Many microsatellites such as $(AGC)_n$ and $(AATG)_n$ have been identified, and yield somewhat greater differences in alleles (that is, fragment lengths), because the repeats are based on three or four nucleotides. Often several loci are examined simultaneously during polymerase chain reaction (PCR) amplification. This method is called multiplexing.

The Federal Bureau of Investigation (FBI) uses a set of 13 well-described STR loci for DNA profiling (see later discussion of CODIS). Today, STR sequences are used almost exclusively for forensic DNA profiling.

Minisatellites

Minisatellites were first characterized between 1986 and 1987; they were discovered in close association with the α-globin gene locus on chromosome 16 and the immunoglobin heavy-chain gene locus on chromosome 14. Minisatellites are usually located near the ends of chromosomes (called the *telomeres*) and vary due to intramolecular or interallelic (between alleles) recombination.

Many minisatellites share a core sequence (approximately 20 bp units) that is a part of the repeat unit of each minisatellite locus, although some minisatellites appear to lack a common core sequence. The rate of mutation at the loci can be very high, as can the number of alleles for each locus. DNA fragments are usually 1 to 30 kb; thus, the variation in repeat length is greater than in microsatellites. Because the repeats are hypervariable, the chance of finding polymorphic alleles is extremely high.

Loci may number in the thousands, although each locus has a distinct repeat unit. These sequences are used to detect length polymorphisms scattered throughout a genome. Thus, multilocus profiles are generated.

Macrosatellites

Macrosatellite DNA is located near the centromeres and telomeres. These DNA sequences are very large—megabases long—and usually require a special type of electrophoresis for separation (pulse-field gel electrophoresis). Their length makes these DNAs subject to breakage (thus resulting in short fragments), and they are therefore not used in forensic analysis, especially because most tissue used for forensic work is at least somewhat degraded.

POPULATION GENETICS AND ALLELES

A locus refers to the physical location of a segment of DNA encoding an RNA or a protein (that is, a gene) or to a specific sequence that is not a gene. The ability to detect alternative forms of a locus or sequence (that is, alleles) depends on the methods used. For forensic analysis, VNTRs are used. Because humans are diploid organisms, each individual has two alleles per locus. Alleles from both the mother and father contribute to an individual's genotype or genetic makeup. A homozygote has two copies of a VNTR of the same overall length, while a heterozygote has two copies of different lengths. In a homozygote, the VNTR alleles may be composed of different sequences and can be detected if the DNA is sequenced. Many alleles can exist in a population (although each individual would have only two alleles at each locus), with the maximum number of different alleles possible being two times the number of people in the population. Some DNA regions are hypervariable—that is, large numbers of alleles exist—whereas others are quite invariable and have only one or two alleles for a gene or locus. Typically, polymorphic genes encoding proteins are represented by two to six alleles. The VNTRs used in forensics and individual DNA typing are hypervariable and may have 100 or more alleles.

Allelic polymorphisms in a population are maintained by random genetic drift or natural selection. When the frequency with which a combination of banding patterns occurs is estimated to exceed the frequency found in the population, the evidence is overwhelming that the pattern is individualistic. Therefore, when conducted with appropriate controls, DNA profiling is an important and powerful identification tool.

Data on the prevalence of so-called individual-specific sequences in the population has demonstrated that the odds of two people sampled at random and having identical DNA fingerprints is extremely small—one in several billion. When a particular ethnic group is examined, although the odds increase, the chances of an exact match are still extremely small—one in several million.

MULTILOCUS MINISATELLITE VNTR

Alec Jeffreys developed the first minisatellite probes that were used in courts of law. These DNA probes, 33.6 and 33.15, comprised the core units (AGGGCTGGAGG), repeated 18 times, and (AGAGGTGGGCAGGTGG), repeated 29 times. These probes detected minisatellites at multiple loci in the genome. In Jeffreys's multilocus probes, approximately 17 variable DNA fragments for each individual are observed, ranging from 3.5 kilobases to greater than 20 kilobases (Figure 11.2). Many smaller DNA fragments are also present but are ignored in the analysis, because they are not well resolved by gel electrophoresis. These probes have also been useful in producing individualized DNA fingerprints in a wide variety of animal, plant, and bird species. Thus, fingerprinting also has applications in plant and animal breeding, conservation biology, and population genetics. Since the 33.6 and 33.15 tandemly repeated probes were developed, many other useful probes have been isolated that allow the detection of variable DNA fragments.

Multilocus human DNA fingerprinting is supported by genetic and population data. Jeffreys's multilocus probes have been statistically evaluated to determine the proportion of bands that are shared by unrelated individuals. The data are sufficiently reliable that DNA fingerprinting has been routinely used in paternity disputes (Figure 11.3) and immigration disputes (Figure 11.4). Multilocus fingerprints are more difficult to interpret than single-locus banding patterns for several reasons.

1. The large number of bonds usually generated

2. Incomplete cutting of the DNA

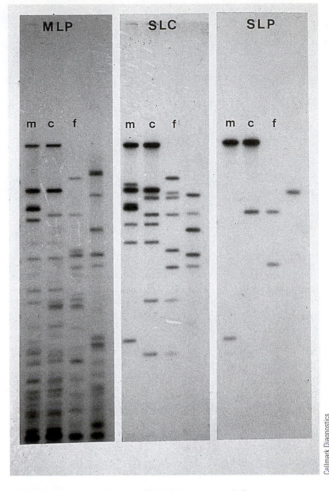

Cellmark Diagnostics

FIGURE 11.2 Two types of probes are used for identification testing: multilocus and single-locus. Shown here are three autoradiograms for a paternity test (m = mother's DNA, c = child's DNA, f = true father's DNA, unlabeled lane is excluded alleged father's DNA). Far left panel: Multilocus probes allow the detection of multiple repetitive DNA loci that are located on more than one chromosome. Since up to 20 to 30 bands often are obtained, the chances of two randomly selected people having a match at all band positions is extremely low; thus, the DNA profile constitutes a banding pattern that is unique for each person. Far right panel: Single-locus probes allow the detection of a single repetitive DNA locus on one chromosome. At most, two bands are observed, one for each of the two alternate alleles (that is, of different sizes) present at the locus on each member of the chromosome pair. Only one band is observed if the alleles are the same size. Middle panel: Many single-locus probes are available; to increase the sensitivity of discrimination, several are used to examine different loci. In this autoradiogram a single-locus cocktail (SLC) comprised of four different single-locus probes was used in the hybridization.

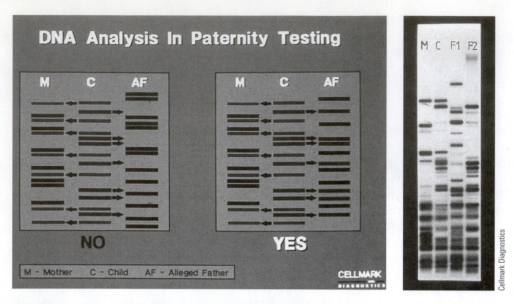

FIGURE 11.3 To establish paternity, DNA fingerprints of the mother, child, and alleged father are compared. The bands from the mother and child that co-migrate are identified, and the bands remaining in the child's fingerprint pattern are compared with those of the alleged father. The bands that do not match the mother's must come from the biological father. Left: schematic example: NO means alleged father is excluded and YES means alleged father is identified as the biological father. Right: autoradiogram (M = mother, C = child, F1 = alleged father #1, excluded, F2 = alleged father #2 identified as the biological father).

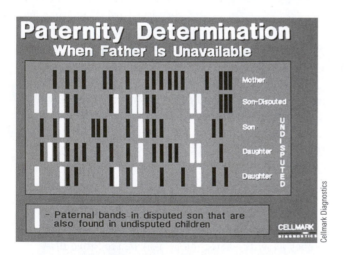

FIGURE 11.4 DNA fingerprinting can be used when a claim of relationship is in doubt, as, for example, in immigration disputes in which people seek entry into the United States on the grounds that they are blood relatives of U.S. citizens. Paternity can be established even when genetic information from only one of the parents is available. The DNA band pattern of the person in question can be reconstructed by analyzing the DNA of other blood relatives such as siblings. In this example, the mother's DNA fingerprint is compared to her children's. The remaining bands come from the biological father; thus, the paternal bands of the undisputed children can be compared to the child in question.

3. DNA degradation

4. Low DNA recovery

5. Identification problems with mixed DNA samples from more than one individual

However, despite the technical problems, DNA fingerprinting has been used successfully in a variety of ways—from forensic cases to the evaluation of the genetic diversity of plants and animals. Typically, the multilocus minisatellites are detected using Southern blot hybridization (see Chapter 3, Basic Principles of Recombinant DNA Technology).

SINGLE-LOCUS VNTR FOR STR

Single-locus VNTR patterns for microsatellites (STRs) generate only one DNA fragment (one allele for homozygous individuals) or two DNA fragments (two different alleles for heterozygous individuals), as Figure 11.2 shows. One band is generated for each allele. Thus, two bands are observed for heterozygous loci and one band for homozygous loci. A simple banding pattern (one or two different bands) is much easier to compare than is a pattern composed of many bands. The more loci used, however, the more informative the analysis (Figure 11.5). Thus, the banding pattern becomes a unique fingerprint that serves as an individual's signature.

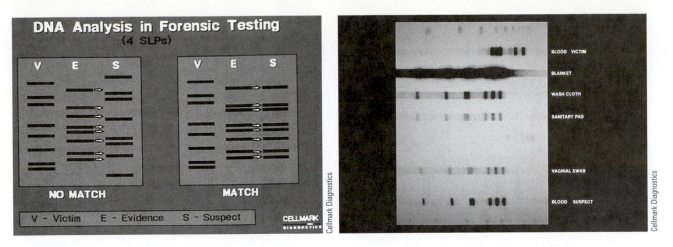

FIGURE 11.5 The use of four single-locus probes increases the discriminatory ability of DNA fingerprinting. Single-locus probes are approximately 10 times more sensitive than multilocus probes and can generate a fingerprint from as little as 20–50 nanograms of sample DNA. In this example from a criminal case, the DNA band patterns from a specimen retrieved at the crime scene and a sample from a suspect are compared. This information and additional circumstantial evidence provide compelling evidence that DNA band patterns from the two samples are not likely to be identical solely by chance. Left panel: schematic of DNA analysis showing exclusion of suspect as the source of evidence DNA (that is, no match) and a match indicating the suspect as the possible source of evidence DNA; Right panel: autoradiogram.

Multilocus probes using Southern blot hybridization have been used often in the past, because one probe reveals the pattern of VNTRs located throughout the genome. One probe yields much more information about the individual, providing that the bands can be resolved. The use of single-locus alleles (using PCR or Southern blot hybridization) is technically less difficult—for example, it eliminates problems with co-migration of alleles and provides better resolution of bands—and is less likely to produce artifacts (Figure 11.6). The use of PCR to amplify single-locus alleles is often the method of choice today. To reach the degree of variability that can be generated with one multilocus probe, at least four or five single-locus primer sets are used either in separate PCR reactions or together in one reaction (multiplexing). Co-amplification allows

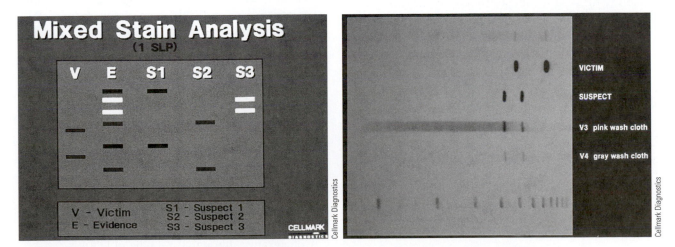

FIGURE 11.6 A major benefit of single-locus probes is their ability to resolve mixed stains, especially important in criminal evidence when blood samples may be mixed. The mixture can be analyzed with one single-locus probe to determine how many individuals were involved. For example, if six bands were observed, the mixture contained samples from at least three individuals. Several single-locus probes then could be used sequentially to increase the accuracy of identifications. (Left, schematic example; right, autoradiogram.)

the profiling of multiple STRs in one reaction (that is, targets multiple locations throughout the genome). Fluorescent labeling with multiple detection allows the use of STR alleles that are of the same size. The use of several loci generates a meaningful composite profile that is used to determine the statistical significance of a match. Although multiplexing is desirable when limited DNA samples are available and can save time and money, several problems must be overcome.

1. Primers from one locus can sometimes complex with those of other loci.

2. One locus may not amplify as well as another locus.

3. The optimization of PCR conditions can present a challenge—annealing temperature of primers and primer concentrations must be determined and may vary for the different STR alleles.

If problems are not resolved, the absence of a particular STR allele may not actually represent the individual's genotype. An alternative to multiplexing is to recover each PCR amplification and reuse it for the next STR locus amplification. This conserves the small amount of DNA that is available and allows the analysis of multiple loci on the genome. This method is referred to as *sequential multiplex amplification*. Less than 5 ng of template DNA is required to profile five different STR loci using PCR. Commercial sources for multiplexes are available (for example, Promega in Madison, Wisconsin, and Applied Biosystems in Foster City, California).

By determining the frequencies of various alleles at different loci in specific populations, one can calculate a frequency for combinations of alleles (Table 11.1). The probability for separate alleles to have come from the same individual then is calculated.

TABLE 11.1 Frequency Calculations of a Common Genotype, D2S44-D14S13-D17S79-D18S27, in Four Populations

Locus	DNA Fragment Size (kb)	
D2S44	11.5,	10.5
D14S13	6.0,	5.0
D17S79	3.5,	3.4
D18S27	6.2,	4.7
Group	Likelihood of Genotype	
African American	1:4,000,000	
Caucasion	1:700,000	
Hispanic	1:1,200,000	
Asian	1:400,000	

Adapted from K.C. McElfresh, D.V.-F., and I. Balazs. *Bioscience* 43:149–157 (1993).

The FBI has been a leader in developing DNA profiling technology for use in the identification of criminals. In 1997 the FBI selected 13 STR loci (CSF1PO, FGA, TH01, TPOX, vWA, D3S1358, D5S818, D7S820, D8S1179, D13S317, D16S539, D18S51, D21S11) to make up the core of the United States national database called the *Combined DNA Index System* (CODIS) (http://www.fbi.gov/hq/lab/codis/index1.htm). These 13 loci are tetrameric repeat sequences and have been typed in sample populations of U.S. Caucasians, African Americans, Hispanics, Bahamians, Jamaicans, and Trinidadians. The benefits of having a national STR typing database include

1. The adoption of this system by forensic DNA analysts, allowing standardization of methods

2. STR alleles are now readily available using commercially available kits

3. STR alleles have been well characterized and act according to known principles of population genetics

4. Laboratories worldwide are contributing data to the database so that the STR allele frequency in many different human populations can be determined

5. The data are digital (see later discussion on digital DNA typing) and therefore ideal for computer databases

6. The frequency for each of the 13 loci can be combined to calculate the frequency of an individual profile. For example, the frequency of an individual profile could be 1 in 10^{15} in a particular group (for example, Caucasian)! Thus, when any two samples (DNA profile of biological evidence from a crime scene and DNA from victim or suspect) have matching genotypes at all 13 CODIS loci, it is virtually certain that the two DNA samples come from the same individual (or from an identical twin).

STR-based DNA profiling has become the method of choice for several reasons.

1. STR analysis is much less labor intensive than RFLP analysis.

2. DNA profiles from badly degraded DNA can be obtained.

3. Very small amounts of DNA are required—less than 0.20 ng of target DNA (1 ng = one-billionth of a gram), although 0.5–1.0 ng of DNA is optimal.

4. Small amounts of contaminants will not yield anomalous PCR products.

5. Mixed DNA samples (for example, mixed blood stains) can be resolved.

6. Automated fluorescent detection of amplified STR fragments can be conducted using DNA sequencers. This greatly reduces the time between sample collection and generation of DNA profiles.

RESTRICTION FRAGMENT LENGTH POLYMORPHISMS

In a diploid organism—one with two sets of chromosomes—the alleles of a given gene may vary slightly. Recall that in an individual there are two alleles for each gene because there are two sets of chromosomes in each cell. When the alleles are identical on the pair of chromosomes, the organism is homozygous for the gene. If the alleles differ in sequence (even by one base pair), the organism is heterozygous for that gene. Sometimes a difference in the sequence alters a restriction endonuclease recognition site so that the enzyme can no longer cut. Alternatively, a new restriction site can be formed. Thus, different sizes of DNA fragments can be obtained based on differences in the sequences of alleles for any given gene. These differences are called *polymorphisms* and the difference in fragment pattern following restriction enzyme digestion is called *restriction fragment length polymorphisms,* or RFLPs. RFLPs are visualized in one of two ways: (1) by agarose gel electrophoresis and subsequent DNA hybridization with a probe (Figure 11.7a and b) or (2) by PCR amplification of a DNA fragment, subsequent digestion of the PCR product by a select restriction enzyme, and agarose gel electrophoresis. Thus, simple RFLP analysis (as opposed to the VNTR analysis just discussed) allows one to detect a single base change in a DNA sequence (that alters a restriction site on the DNA).

RFLP analysis is extremely powerful. DNA sequence polymorphisms are common in animal populations and, therefore, can be used as markers that are inherited (Figure 11.8a and b). This allows genetic mapping for paternity and maternity testing (it also has been used in forensics to exonerate a suspect in a crime before more modern STR methods were available) and in disease detection such as for the "sickle" allele (the cause of sickle cell anemia) of the β-globin genes or for cystic fibrosis. More than 100 genes have been mapped using information from RFLPs. RFLP analysis is beneficial when alleles (markers that do not have an observable effect on an individual) are linked to mutated genes that encode genetic diseases. In this way, genes can be localized and isolated (for example, the cystic fibrosis gene) (Figure 11.9). Also, individuals can be screened to determine whether they harbor a particular RFLP that is linked to a mutated gene. This can be used to determine whether a genetic defect might be passed on to progeny.

METHODS OF DNA PROFILING

The methods used to obtain a DNA fingerprint are discussed in Chapter 3, Basic Principles of Recombinant DNA Technology: (1) Southern blot hybridization—restriction enzyme digestion, gel electrophoresis of DNA fragments, hybridization with a VNTR probe, and autoradiography to visualize the bands on x-ray film and (2) PCR. To generate a DNA profile, DNA is isolated from body fluids such as urine, semen, or saliva and from tissue such as blood, bone, or skin.

For Southern blot hybridization, the DNA profile generated on x-ray film must be analyzed to determine the positions (that is, the sizes) of the DNA fragments or bands. DNA standards of known size are run on the same gel as the samples. The distance migrated is inversely correlated with the size or length of the fragment. A standard curve determines the relationship of the distance migrated and the length of the fragment, and is used to calculate the length of unknown band sizes. For PCR, the amplified fragments are electrophoresed on agarose gels and known DNA standards are used to determine size.

The goal of DNA profiling is to exclude particular suspects or to determine the probability of a match between samples collected from the crime scene and samples collected from a suspect. For the profile to be valid, the frequency of the selected sequences in the general population must be used for comparison.

TECHNICAL CONSIDERATIONS

The need to standardize methods to ensure reproducibility of DNA fingerprints has generated much discussion. When stringent methods are not followed, DNA profiles might wrongly exclude a suspect rather than falsely incriminate. Standardization of methods ensures reproducibility of results.

To prevent artifacts and maintain consistency, care should be taken to

1. Preserve the integrity of DNA

2. Completely digest the DNA with restriction enzymes

3. Standardize hybridizations

4. Select appropriate probes that have been demonstrated to be stably inherited

Errors can occur through

1. Contamination of the sample

2. DNA degradation

3. Difficulties in interpreting the bands on the x-ray film. Artifacts (for example, extra bands, missing bands, band shifting) could provide false information.

4. Statistical misinterpretation of a match

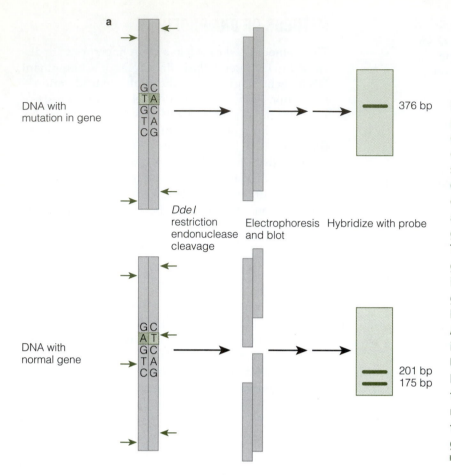

DNA with
mutation in gene

Ddel
restriction
endonuclease
cleavage

Electrophoresis
and blot

Hybridize with probe

376 bp

DNA with
normal gene

201 bp
175 bp

FIGURE 11.7 Restriction fragment length polymorphisms (RFLP) of a variable number of tandem repeats (VNTR). (a) A change in one of the base pairs in a given sequence sometimes eliminates (or adds) a restriction enzyme cleavage site. In this example, *Ddel* cannot cut within a mutated gene because of a one base change. The normal (wild type) gene contains a single *Ddel* restriction site. Two DNA fragments are formed when the gene is cut with *Ddel*. **(b)** RFLP patterns vary in different individuals and are detected by gel electrophoresis and Southern blot hybridization. In this example, in individual A, two fragments are generated, and in individual B, one large fragment is present because a restriction cleavage site is absent. Individual C has the same sized DNA fragment (1 kb) as A, but an additional AT repeating sequence has made the 2.1–kb fragment 200 bp longer. Mutations within genes can be detected in this manner.

Same segment of DNA from three different individuals:

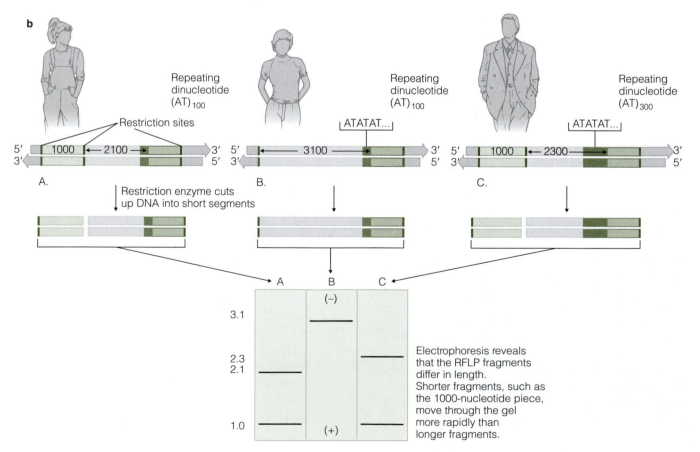

Electrophoresis reveals that the RFLP fragments differ in length. Shorter fragments, such as the 1000-nucleotide piece, move through the gel more rapidly than longer fragments.

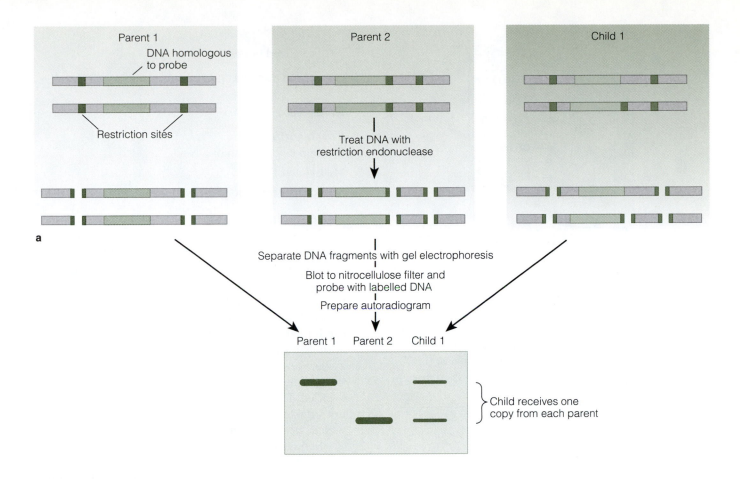

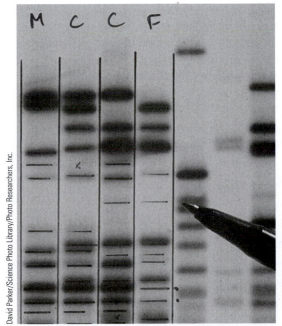

FIGURE 11.8 Restriction fragment length polymorphisms (RFLP) are inherited. DNA probes can be used to detect polymorphisms in the DNA. (a) RFLPs can be used to determine genetic relationships. Parent 1 harbors DNA with two restriction sites. Both alleles have the same sites. The DNA of Parent 2 has an additional restriction site on both alleles. The Child has inherited one allele from each parent. A probe is used to detect polymorphisms by Southern blot hybridization. Parent 1 has one large DNA fragment, Parent 2 has a smaller fragment because he/she has an extra restriction site within the DNA to be probed, and the Child has inherited one large DNA fragment from one parent and one small one from the other parent. (b) Gel electrophoresis of DNA showing an RFLP. Lanes are marked M (mother), C (child 1 and 2), and F (father). Every band that is present in one of the children is found in at least one of the parents.

Used with permission from University Science Books from Berg and Singer, "Dealing with Genes," © 1992.

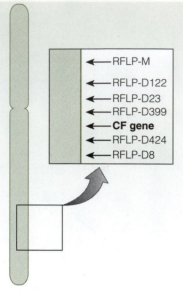

FIGURE 11.9 Localization of the gene associated with cystic fibrosis (CF) on chromosome 7. The CF gene is identified relative to restriction fragment length polymorphism (RFLP) markers. Used with permission from University Science Books from Berg and Singer, "Dealing with Genes," © 1992.

DNA

Considerations
Quality
Quantity
Contamination
Collection
Storage

Not all DNA from samples is of the same quality or quantity. The quantity of the DNA may be limited and the DNA may be partially or almost completely degraded. To further complicate the situation, samples from one person may be contaminated with samples from another, or even with bacterial DNA. To prevent degradation, tissue samples should be collected and packaged as soon as possible and placed on ice. DNA should be extracted upon reaching the laboratory.

If immediate DNA isolation is not possible, the tissue can be frozen. Experts recommend that samples be frozen either in an ultracold freezer at −70° C, on dry ice at −120° C, or in liquid nitrogen at −196° C. Although DNA degrades readily in warm, wet conditions (DNA from dried fluids such as blood and from cells is quite stable), it can survive for thousands of years in some environments. The small amount of DNA obtained is usually very fragmented. Examples of where ancient DNA has been isolated include

> Fossilized plants

> Insects and bacteria in amber

> Mummies

> Fossilized bones

> Frozen humans and animals

DNA Digestion

Considerations
Incomplete digestion of DNA
Methylated DNA
Suboptimal conditions for enzyme activity

If DNA is incompletely digested by restriction enzymes, the DNA profile will be inaccurate. The DNA fragments will be larger than they would have been if all sites had been cut. Reasons for incomplete digestion of DNAs include contaminants from tissues or reagents in the digestion solution, low enzyme activity, or suboptimal digestion conditions. Methylated DNA also can influence the digestions; if the DNA is methylated at the sites recognized by the selected restriction enzyme, the sites may not be cut. Enzymes that are sensitive to methylation (*Hpa*II, for example) should be avoided. The enzymes used in DNA fingerprinting are not affected by methylation (an exception is *Hinf*I, which cuts methylated DNA inefficiently).

Sometimes, if optimal conditions are not maintained (if, for example, the wrong salt concentration is in the buffer), enzymes cut at a site different from the normal recognition sequence, generating artifacts in the DNA profile. Under low salt conditions, the enzyme *Eco*RI recognizes another sequence, -AATT-, in addition to -GAATTC-. This recognition of the wrong sequence is typically called "star" activity. A profile generated with *Eco*RI "star" would show additional, smaller bands because of excessive digestion.

Gel Electrophoresis

Considerations
The choice of gel matrix
Concentration of agarose or polyacrylamide
The voltage used to separate DNA fragments
Choice and concentration of buffers
Thickness of gel
Position of the well of DNA in the gel
DNA overflow
DNA concentration in the well

The DNA banding pattern is strongly influenced by gel electrophoresis conditions. Numerous parameters influence the band patterns of DNA profiles. First, the choice of gel matrix is important. Typically, polyacrylamide is used to separate small DNA fragments and is usually the polymer of choice for the resolution of STR fragments. The concentration of agarose or polyacrylamide affects the migration of DNA fragments. Large DNA fragments require a lower concentration for better resolution, while a higher percentage is used to separate smaller fragments. Migration of DNA fragments in gels is influenced by the voltage gradient applied across the gel. A high voltage does not resolve large fragments well; a very low voltage decreases the separation of small fragments. The buffer type and concentration in the gel and buffer chamber also affect the banding pattern in the gel. A high salt concentration decreases band separation and raises temperature. The depth of the overlay buffer also affects the migration rate of DNA. If low voltage is used for a long time during electrophoresis, small fragments can diffuse from the gel into a deep overlay buffer. The temperature also affects the quality of the DNA profile, because migration rates increase as temperature increases. The thickness of the gels can also vary, but typically, thin gels allow better resolution. Even ethidium bromide staining significantly affects DNA fragment migration.

Finally, the position of the wells of the gel can affect DNA migration rate. Sometimes DNA loaded in the middle of the gel migrates more rapidly than DNA loaded closer to the edges. The FBI monitors migration differences by loading standard DNA molecular weight markers in every third lane (the first, fourth, seventh, tenth, and so on). During sample loading, care must be taken not to mix samples in adjacent lanes. Overflow from wells will mix samples and give erroneous DNA profiles (that is, similar band patterns). Overloading the well with too much DNA causes artifacts, such as broad bands, that decrease resolution and the accuracy of measurements.

Probe Selection

Considerations

Type, nucleotide sequence of probe

Quality of template DNA

When the exact sequence of the VNTRs is known, synthetic oligonucleotide probes usually are used. The probe must include all known sequence variants (between alleles) of the VNTR unit. The DNA also should be hybridized with a bacterial probe (usually a bacterial ribosomal RNA gene) to detect bacterial sequences. Often, forensic tissue is contaminated with bacteria, and VNTR probes are known to hybridize to bacterial DNA, thereby giving inaccurate DNA profiles.

POLYMERASE CHAIN REACTION

PCR amplifications of VNTR DNA regions also detect polymorphisms. PCR (described in Chapter 3, Basic Principles of Recombinant DNA Technology) is easy to conduct, and results are obtained relatively quickly.

Very small amounts of DNA—as little as the DNA from a single cell—from tissue or body fluids can be collected at a site and used as a template for amplification (that is, *in vitro* replication). DNA that has been degraded can also be amplified, because primers used in PCR amplification can anneal even to a small fragment of DNA (such as from skeletal remains or from the saliva on a postage stamp, as was obtained in the Unabomber investigation). DNA can be amplified in large enough quantity so that another laboratory can confirm results.

The extreme sensitivity of PCR makes cross-contamination a very serious problem. Trace amounts of contaminating DNA from tissues, fluids, bacteria, and laboratory equipment such as pipettes can be amplified along with the target DNA. Unfortunately, the conditions under which forensic samples are usually collected subject them to cross-contamination. Additional bands will be visible and confound results, possibly with devastating consequences. Careful selection and use of controls and adherence to appropriate protocol will ensure that cross-contamination is identified.

Usually PCR products (or sometimes the oligonucleotides) are spotted onto membrane and hybridized with the labeled probe (or the PCR product). To detect genetic variation, sequence-specific probes are used to detect the different sequences of alleles. Short sequence-specific oligonucleotides (15 to 30 bases) are hybridized against PCR products of the locus in question by dot blot hybridization. After exposure to x-ray film, or color development if a nonradioactive method of labeling is used, spots show where hybridization occurred. The sensitivity of this method allows PCR products differing by single bases to be detected.

Commercial forensic kits can distinguish the alleles that define different genotypes. The Cetus Corporation, a prominent biotechnology company, markets DQ typing kits. Its probe strip allows subtyping of samples and easy comparison of results. DQ locus variability is detected by dot blot hybridization with sequence-specific oligonucleotides. Degraded samples can be assayed by PCR. DQ is one of three families of genes on chromosome 6 that encode human leukocyte antigen (HLA) class II proteins. The DQ locus encodes the subunit of the DQ protein, a highly polymorphic protein of the MHC that is involved in the immune response. DNA analysis involves the detection of six alleles that define 21 genotypes. Sequence differences include at least 5 bp over a 30 bp length of the hypervariable region. Two other sequence differences in two of the alleles flank the hypervariable region. Much work has been conducted on polymorphisms and population frequencies. Population frequency data have been obtained for the Caucasian population (much less information is available for other ethnic populations) and demonstrate that there is approximately a 7% chance that two random individuals share the same DQ genotype.

The ideal marker allele for efficient amplification by PCR should have 100 to 500 bp—small enough that fragments can be amplified from degraded DNA. Also, all alleles should be resolved by electrophoresis and hybridization. Ideally, a large number of alleles should be available for use, and no single allele should be shared among individuals. National standards and controls on PCR testing as well as on methods of collecting samples have increased accuracy and reliability. DNA standards provide consistency. The method of choice for forensic laboratories is now STR typing using PCR (for example, CODIS loci as discussed previously).

DIGITAL DNA TYPING

Repeat coding is a digital method of DNA profiling that is appropriate for computer databases. DNA profiling typically involves the detection of the banding pattern. Digital typing, developed by Jeffreys in 1991 for minisatellites, does not have such problems as band shifting and the need to size specific DNA bands. Repeat coding or maps are not based on the number of repeat units in specific loci, but instead on the variation within one locus. One method of digital typing is using PCR amplification restriction digestion of the amplified DNA product. After PCR amplification of the locus, the DNA is treated with the restriction enzyme *Hae*III. Two classes of repeats are obtained, one that is cut with *Hae*III (5'GGCC3) and the other that is not cut because a *Hae*III recognition site is not present (for example, 5'GCCC3 instead). Numbers are assigned to each repeat. If *Hae*III cuts the PCR-amplified fragment, a 1 is assigned. An uncut fragment is assigned a 2. When one allele at a locus is cut by *Hae*III and the other allele remains uncut, a 3 is assigned (meaning both 1 and 2 are present). Several loci are combined to generate a complex numerical pattern, thus providing more information. In other words, if seven loci were used, the barcode would have seven numbers—for example, 2133212. Each locus is assigned one number depending on whether the PCR-amplified fragments are cut, uncut, or both cut and uncut by *Hae*III. A barcode for one sample (say from blood or tissue) or one individual can readily be compared with others. The primary benefit of this method is that by eliminating the need for sizing DNA bands it also eliminates the ambiguity due to band shifting and co-migrating bands.

POPULATION COMPARISONS

The biggest challenges in forensic analysis are (1) to calculate the probability of coincidental matches by comparison with a reference population and (2) to settle on the criteria to be used for determining what constitutes a relevant reference population. What constitutes appropriate reference populations is being debated.

Individual DNA polymorphisms must be used in the context of population data to determine the probability that a match occurs by chance alone. For comparisons to be valid, they must be made against an ethnic group relevant to the individual; the normal distribution of the different alleles in the population must be known. To demonstrate a match, identical patterns between two individuals or an individual and a sample must be established. The patterns then must be shown to be unique (that is, the probability of the bands being present in another person should be extremely low). What is the probability that a DNA match is random? The answer requires a thorough and careful study of different populations to determine the distribution of the particular RFLP. Unfortunately, data for only a limited number of samples have been collected.

Determining the probability of a DNA profile match requires that several assumptions be made. One assumption is that the alleles for loci of the VNTR probes used in the laboratory segregate independently and thus are not linked. Detecting linkage requires that a large population be sampled, which may not be possible in smaller ethnic groups. Population geneticists disagree about the significance of population comparisons. Many laboratories are establishing local population databases (for example, for African American, Caucasian, Hispanic, and Asian populations), as well as contributing to CODIS, the U.S. national database. These databases assume that random mating is occurring, which may or may not be the case in a particular group within a community.

ADMISSIBILITY OF SCIENTIFIC EVIDENCE IN A COURT OF LAW

The misuse of scientific evidence is one of great concern and attempts are made to prevent this in a trial. All trial courts make a preliminary determination of the admissibility of evidence. Pretrial hearings are conducted to assess the relevancy, validity, and admissibility of scientific evidence. At these hearings, the judge evaluates all evidence to determine whether the evidence is appropriate to the facts in the case before it is presented to the jury. Expert witnesses for both the defense and the prosecution evaluate the evidence. Potential problems with the methods and the statistical significance are presented.

Evidence is determined to be admissible in a court of law based on the Federal Rules of Evidence, which apply to all federal courts and have been adopted by most states. These rules were first drafted by the American Law Institute in 1942 and have been revised since then. They were signed into law in 1975. Rule 702 involves the admissibility of scientific evidence: "If scientific, technical, or other specialized knowledge will assist the trier of fact to understand the evidence or to determine a fact in issue, a witness qualified as an expert by knowledge, skill, experience, training, or education, may testify thereto in the form of an opinion or otherwise." Rule 702 requires that all scientific testimony be both relevant and reliable—and that the testimony be grounded in scientific methodology.

When evidence based on DNA and DNA profiling was first used in a court of law, certain standards had to be met before it was allowed. The type of standard or test used to determine admissibility depends on the state. For example, states accepting the Daubert standard include Connecticut, Indiana, Kentucky, New Mexico, and Texas. The Frye standard is used in Alaska, Arizona, Colorado, Kansas, and Pennsylvania. Other states have their own tests for admissibility, although the standards used are often consistent with federal standards (examples include Arkansas, Georgia, Montana, and Utah).

Although DNA profiling is now widely accepted in the courtroom, the parties in a case may have to submit to a Frye or Daubert hearing to demonstrate that a scientific principle is well established in the field. DNA evidence is admissible under either Frye or Daubert, although the testimony must include information about the protocols and probability estimates. Therefore, the defense often requests such a hearing to determine the validity of a company's DNA testing. Because the technology has been tested numerous times and the scientific basis is no longer challenged, rejection of DNA evidence is based primarily on such factors as methods used to collect DNA and possible contamination of samples. Perhaps the greatest area of controversy resides with the statistical probabilities of a match.

Frye Standard

The Frye standard, sometimes referred to as the *general-acceptance rule*, arose out of a judgment rejecting the admissibility of evidence based on polygraph testing because it was deemed insufficiently scientific to fall within the realm of "expert opinion evidence." The following is often cited from the 1923 *Frye v. United States* case:

While the courts will go a long way in admitting expert testimony deducted from a well-recognized scientific principle or discovery, the thing from which the deduction is made must be sufficiently established to have gained general acceptance in the particular field in which it belongs. We think the systolic blood pressure deception test [precursor to the modern polygraph test] has not yet gained such standing and scientific recognition among physiological and psychological authorities as would justify the courts in admitting expert testimony deducted from the discovery, development, and experiments thus far made.

Thus, the method or technique that is used to generate the evidence must be well established, reliable, accepted by the scientific community, and supported by scientific principles. The Frye test is intended to prevent the presentation to a jury of invalid opinions based on unreliable experimental procedures or evidence. To demonstrate that the method is foolproof is not required, however. Also, differing opinions as to the appropriate conclusions from the method or technique can be expressed.

Even though a court may accept the reliability of DNA profiling methods for forensic analysis, the reliability of the evidence generated by DNA profiling may be subject to doubt in a particular case. The court could rule the evidence is inadmissible on the grounds that the scientific consensus on the validity of the DNA analysis method did not extend to the material itself if the DNA was degraded or even contaminated.

Witnesses used by the proponents of a particular method must satisfy certain criteria: they must be experts in the field and possess the academic and professional credentials that allow them to understand the scientific principles involved and any controversy involving the reliability of the method. These expert witnesses are often scientific investigators or technicians who perform these methods as part of their work.

General acceptance of a method, however, does not indicate reliability; therefore, proof of reliability still must be demonstrated. Studies are required to determine the prevalence of sequences in populations. Sample collection, profiling methodology, and statistical interpretation is standardized and rigorous methods applied. Published guidelines, *Technical Working Group on DNA Analysis Methods*, for commercial laboratories have become the standard in the industry. Other guidelines have been published, for example, by the California Association of Criminologists. Standardization ensures that the methods are reliable and generate accurate results.

Daubert Standard

The Daubert standard evolved from a decision in 1993 by the U.S. Supreme Court involving a claim that the antinausea drug Benectin caused birth defects. In this case the Supreme Court rejected the Frye standard for the admissibility of evidence in a case involving *Jason Daubert v. Merrell Dow Pharmaceuticals, Inc.* The plaintiff, Daubert, claimed to have a birth defect (severely reduced limb length) because his mother had taken Benectin during pregnancy. Lower courts had rejected the scientific evidence presented by Daubert. However, the Supreme Court thought the Frye standard was too restrictive. In this case, Federal Rule 702 superseded the Frye standard. Six points were taken into account when considering the evidence.

1. Is the hypothesis presented by the scientific expert testable?

2. Has the theory or method been subjected to peer review and publication?

What About Privacy?

So much of our personal information is now available to the public—how much money we make, how much we paid for our homes, our credit history, the status of our health, our purchases in markets, as well as a number of other personal preferences. The computer age has made personal information readily accessible to banks, insurance companies, marketing companies, employers, and others. A DNA database is one more type of data catalog. Any type of personal information can be misused; however, the information obtained from DNA databases could have serious consequences. Genetic information from sample analysis and data about genetic disease, current medical problems, health risks (such as genetic susceptibilities and risky behaviors), and other personal information could be readily obtained. Stored materials are susceptible to unauthorized use if safeguards are not in place—and even when there are measures taken, databases may be susceptible to infringement (as we have sometimes seen when corporate computer databases have been sabotaged). Specific regulations must be implemented to prevent discrimination and breach of privacy. Donors must be told exactly what data are stored and must consent to the use of biological samples. In addition, current or potential employers or insurers must not base any decisions on genetic information and biological samples stored in banks. Issues of privacy are being addressed; however, it takes time for society and laws to catch up to technology.

3. What is the known or potential error rate of the method?

4. What are the expert's qualifications and stature in the scientific community?

5. Does the method rely on the special skills and equipment of one expert or can it be replicated by other experts elsewhere?

6. Can the method and the results be explained in such a way that the court and the jury can understand and evaluate the evidence? This is the Marx standard (established in 1975) and is incorporated in both Frye and Daubert.

The Daubert case (Daubert vs. Merrel Dow Pharmaceutical) established the Daubert standard. The role of the Frye standard is to prevent pseudoscience from being allowed as evidence. The expertise of scientific experts is used. Daubert does not rely on general acceptance; rather scientific knowledge must be produced using the scientific method and supported by peer review. Thus, the expert's testimony must be validated scientifically. This may help alleviate the so-called "battle of the experts" from the prosecution and defense teams.

DATABASES

A variety of information about individuals (including DNA and other means of identification) can be stored in databases for future retrieval. Genetic information about populations can be gathered from DNA databases, or data banks as they are often called (because tissue and blood also can be stored). Information databases store detailed personal information about individuals from medical records and from insurance and financial reports. These data are available to insurance companies (to prevent insurance fraud and often to set insurance premiums), hospitals, and employers for obtaining additional information about an individual.

Identification databases include descriptive information about a person such as physical characteristics (eye color, height, and weight), fingerprints, dental records, and genetic traits such as blood type and histocompatibility antigens. The government has used this information to identify missing individuals or even criminals.

The information provided to data banks comes from forensic collections, tissue samples taken during surgery or biopsies, blood samples, neonatal screenings, and samples taken from members of the armed forces. Databases and banks are being established across the country. The FBI, insurance companies, the military, and state facilities are collecting and storing a variety of personal information. DNA databases include CODIS, the Department of Defense's armed services pool of biological samples, and Virginia's saliva and blood bank of convicted felons. Although large-scale sample collection provides information on the population genetics of different ethnic groups, which is invaluable for calculating statistical probabilities of

DNA profile matches, it also has the potential to infringe on our privacy.

OTHER METHODS OF DNA PROFILING AND THEIR APPLICATIONS

Many other methods of profiling DNA have been developed. These methods are used for a variety of purposes such as determining the genetic diversity of plant cultivars, genetically characterizing animal populations, determining relationships of animals and birds in the wild, human disease diagnostics, and mapping disease-related genes.

Random Amplification of Polymorphic DNA

Methods described previously in this chapter refer to the detection of variability in specific targeted regions of DNA. An alternative method of detecting variation is to amplify random DNA sequences by PCR and look for the presence or absence of these randomly amplified polymorphic DNA sequences using gel electrophoresis. Random amplified polymorphic DNA (RAPD) sequences are used to detect variability or polymorphisms in the PCR priming sites. If there are base changes within the target DNA, the PCR product will be affected. This will change the DNA fragment profile produced by a specific primer. Typically, 8 to 10 nucleotide primers are used. These primers are randomly selected with arbitrary sequences. A computer is sometimes used to generate primers. Therefore, in a population of individuals, DNA sequences of various sizes will be amplified with some primers and with others there may not be a PCR product (Figure 11.10a and b). Thus, patterns of banding on gels are not the same for every individual in a population.

Some features of RAPD include the following:

1. RAPD methods are simple and rapid.

2. Because they are polymorphic they can be used as genetic markers.

3. They are dominant, meaning they do not distinguish between homozygous and heterozygous allele states.

4. RAPDs closely linked to a particular gene can be used by animal and plant breeders to ascertain whether they have transferred the desired trait.

5. RAPDs are useful for determining genetic variability within and between populations of bacteria, plants, or animals. This method is often used as a means to characterize different cultivars of crop plants or to identify potential mates for captive animals (for example, in zoos) to maintain or increase genetic diversity.

Amplified Fragment Length Polymorphism

Amplified fragment length polymorphism (AFLP) was designed by Keygene (based in The Netherlands) to be used in a variety of fields (medical diagnostics, plant and animal breeding, forensic analysis, bacterial typing) as a very sensitive method of DNA profiling. AFLP is really an improved variation of RAPD analysis (AFLP typically detects more polymorphisms per reaction than the RAPD method). AFLP is based on the selective PCR amplification of restriction enzyme-digested genomic DNA and is actually like a PCR-generated RFLP. This method assesses the entire genome for polymorphisms rather than targeting specific loci. AFLPs are very abundant, randomly distributed, and are inherited in a Mendelian fashion. In addition, the assay is very reproducible (>99%). This makes AFLP markers an excellent choice for population genetic studies. New methods also have been developed that help distinguish between homozygote and heterozygote genotypes in the population. AFLP analysis has been used to evaluate the genetic diversity of natural animal (for example, birds) and plant populations (population genetic studies), genotypes of captive animals for breeding programs, and the genetic diversity of plant cultivars. Commercial AFLP kits are available that offer different primers and standardized protocols (for example, Applied Biosystems, Foster City, California).

AFLP analysis is conducted as follows (Figure 11.11):

1. Genomic DNA is digested to completion (that is, all available restriction sites cut) with two restriction enzymes—one that cuts frequently, such as *MseI* (a 4-bp recognition sequence), and another that cuts infrequently, such as *Eco*RI (recognizes a 6-bp sequence). For example, *MseI* cuts approximately every 300–400 bp, while *Eco*RI cuts somewhere between every 2000 and 2500 bp in the experimental plant model *Arabidopsis thaliana*. Therefore, DNA fragments will be generated in the following descending order of abundance: *MseI*-*MseI* fragments, *MseI*-*Eco*RI fragments, *Eco*RI-*Eco*RI fragments. Only a subset of *MseI*-*MseI* and *Eco*RI-*Eco*RI fragments are used for AFLP analysis.

2. The ends of the resulting DNA fragments are ligated to specific 25–30-bp adaptor sequences (oligonucleotides) that act as priming sites for select primers.

3. A type of PCR called preselective PCR is conducted using primers that are complementary to the two end adaptor sequences. Each primer pair also has one extra nucleotide (sometimes up to three nucleotides) at the 3′ end of each primer. When PCR is conducted, only one-sixteenth of the *Eco*RI-*MseI* DNA fragments are amplified.

This DNA fragment contains 3 genes. A scientist is interested in amplifying only *gene B*:

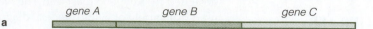

a

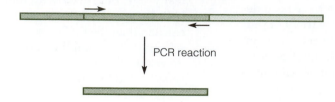

The scientist prepares 2 primers which will anneal to each end of *gene B*:

PCR reaction

Only *gene B* is amplified and can then be purified for further analysis.

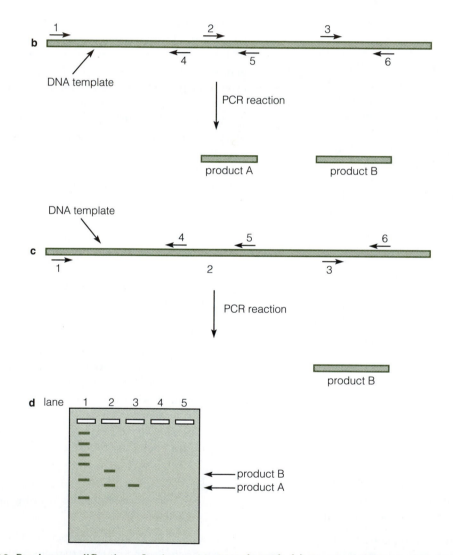

b

1 2 3

DNA template

4 5 6

PCR reaction

product A   product B

DNA template

c

4 5 6

1 2 3

PCR reaction

product B

d lane   1   2   3   4   5

product B
product A

FIGURE 11.10 Random amplification of polymorphic DNA (RAPD). (a) Standard PCR using a primer pair (arrows) to amplify DNA within the primer borders. Gene B is amplified. (b) Multiple copies of a single primer sequence are used in the reaction. Primers 2 and 5 anneal, and primers 3 and 6 anneal to the template DNA. Two RAPD PCR products are formed. (c) In this individual, only primers 3 and 6 anneal, forming one RAPD PCR product. (d) PCR products are run on an agarose gel to detect band differences. Lane 1 contains DNAs of known length (referred to as a DNA standard). Lane 2 is the RAPD pattern for the individual in (b) and lane 3 is the pattern for the individual in (c).

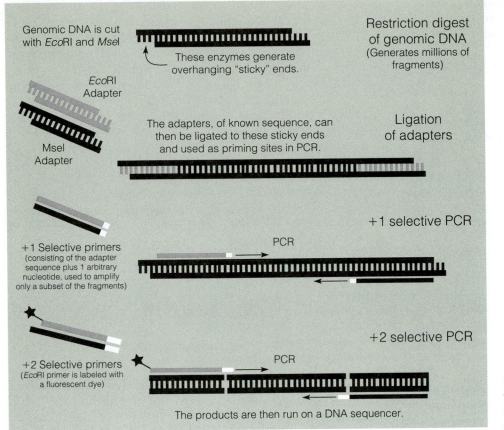

Genomic DNA is cut with *Eco*RI and *Mse*I

These enzymes generate overhanging "sticky" ends.

Restriction digest of genomic DNA (Generates millions of fragments)

*Eco*RI Adapter

*Mse*I Adapter

The adapters, of known sequence, can then be ligated to these sticky ends and used as priming sites in PCR.

Ligation of adapters

+1 Selective primers (consisting of the adapter sequence plus 1 arbitrary nucleotide, used to amplify only a subset of the fragments)

+1 selective PCR

PCR

+2 Selective primers (*Eco*RI primer is labeled with a fluorescent dye)

+2 selective PCR

PCR

The products are then run on a DNA sequencer.

FIGURE 11.11 The method used to generate amplified fragment length polymorphisms (AFLP).

Thus, the selective nature of AFLP-PCR is based on the extension at the 3′ end of the primers. Only those DNA fragments with the complementary sequence to the primer pair will be amplified. Taq polymerase, used in PCR, cannot extend DNA if mismatches between template and primer occur at the 3′ end of the DNA that is to be extended (the 3′ end of the primer). As a result, only a subset of the DNA fragments will be amplified. Because there will still be too many DNA fragments to be analyzed, the number is further reduced by a second PCR reaction (see step 4).

4. A second PCR is conducted (called selective PCR) using the PCR products from step 3 as a template. The primers used also contain two additional nucleotides (again, selected by the investigator). In addition, the primer that anneals to the *Eco*RI-adaptor sequence has a fluorescent or radioactive tag to allow only these PCR products to be seen on gels (step 5). This further reduces the number of DNA fragments to be analyzed.

5. If fluorescent tags are used, the PCR products can be quickly run in a DNA sequencer on a capillary gel. Bands are detected by the color of their fluo-rescence (for example, red or blue) and a DNA profile is generated. If radioactive-tagged primers are used, the PCR products are run on a gel and the gel is exposed to x-ray film.

Only approximately 1/4000 of the *Eco*RI-*Mse*I fragments are present from the original digested genome. In this way, the entire genome is randomly sampled, but the number of bands is greatly reduced to allow greater resolution of DNA fragments. Anywhere from 100–200 bands are often generated and, of these, only a subset of these will be polymorphic between individuals.

Single-Strand Conformation Polymorphism

Modifications in DNA sequence also can change DNA conformation or stability. This can be detected by the changes in mobility during gel electrophoresis. Because double-strand DNA is mostly independent of DNA sequence, single-strand DNA is used. The mobility of single strands varies dramatically as a result of small nucleotide changes in the sequence. This is the basis of single-strand conformation polymorphism (SSCP). Single-strand DNA is produced by asymmetric PCR where one primer is in higher concentration than the second primer. When the low-concentration primer is completely used, the PCR reaction continues, producing only one DNA strand (usually less than 600 bp long). The mobilities of different amplification products can be compared by native gel electrophoresis (using nondenaturing gels to maintain the native conformation of the sequence).

Features of SSCP include the following:

1. The mobility of single-strand DNA depends on intramolecular base pairing. This base pairing can form loops, stems, and other secondary structures. Thus, mobilities are influenced by the DNA sequence.

2. The exact sequence and location of the polymorphism can be unknown—only the mobilities are important and reflective of the DNA polymorphism.

3. Most SSCP methods analyze single loci of individuals.

4. SSCP is useful for detecting individual genetic variation in populations, detecting mutations in genomic DNA, and serving as molecular markers.

Single Nucleotide Polymorphism

At various sites within the genome of a population of individuals, single nucleotides are found to exist in two or more states (for example, G or C rather than A). Most single nucleotide polymorphisms (SNPs) are nucleotide substitutions, although insertions or deletions of a nucleotide also may occur. SNPs have a wide range of applications. SNPs are particularly important in diagnostics to detect mutations in genes and disease-related mutations mapped to chromosomes (through the analysis of families, populations, and case studies).

A variety of new innovative SNP genotyping methods have been developed and include the hybridization of allele-specific probes that fluoresce upon binding to their target sequence (dynamic allele specific hybridization), PCR amplification that utilizes chemiluminescence to detect extension during PCR, rapid minisequencing methods, and derivations of microarray analysis. Most are beyond the scope of this book. Applications of SNPs include the investigation of how genome variation contributes to phenotypic variation of disease in populations (to correlate SNPs with gene function), ecological research (for example, population structure of animals), evolutionary biology (for example, rate of divergence between species, human and primate evolution), and the expansion of SNP databases by data mining (using existing sequences from computer databases to identify SNPs—sequences are expressed sequence tags, or ESTs; see Chapter 9, Genomics and Beyond). Most recently, SNPs are being applied to pharmacogenomics (study of how the genome affects and responds to

drugs; drug discovery by studying the interactions of potential drugs on proteins in the body) by discovering functional implications of SNPs (that is, the development of new drugs, how SNPs in genomes affect the response to drugs). It is hoped that genetic profiling will enhance drug development, especially during clinical trials. Furthermore, proprietary rights to an SNP may be profitable if the SNP is licensed to other companies.

Mitochondrial DNA and the Y chromosome

Mitochondrial Eve The origin and migration of humans has been the topic of intensive research in recent years. The mitochondrial DNA of several human populations have been studied to determine that

1. Recent ancestors of modern humans originated in Africa, and therefore, had a recent African origin.

2. Modern humans appeared in one founding population.

3. Our closest ancestors evolved approximately 171,500 years ago.

4. Anatomically modern humans migrated to other parts of the world to replace other hominids (this term refers to bipedal, upright primates and includes extinct ancestral forms and early relatives of humans).

From the study of mitochondrial DNA sequences, many scientists believe that all modern humans arose from a few females approximately 171,500 years ago—the time when all the different mitochondrial sequences coalesced into one. This composite female is called mitochondrial Eve and, theoretically, is the mother of us all.

The organelle mitochondria (see Chapter 2, From DNA to Proteins) have their own genome of approximately 16,500 bp in humans and contain 13 protein encoding genes, 22 tRNAs, and 2 rRNAs. Mitochondria are present in large numbers in each cell so that there is much DNA obtained during isolations. There are several advantages to using mitochondrial DNA in constructing the evolutionary history of humans and animals.

1. Mitochondrial DNA undergoes mutations (the replacement of one nucleotide by another, for example, A to T) at a higher rate than nuclear DNA. Differences between closely related individuals can be resolved.

2. Mitochondria are inherited only from the maternal line so that a direct genetic line can be traced without the need to separate two different lineages (father *and* mother). During fertilization of an egg by a sperm, the sperm's mitochondria degrade.

3. Mitochondrial DNA does not undergo recombination. Nuclear DNA recombines during the formation of sex cells (eggs and sperm) by meiosis when sections of DNA from the mother's and father's chromosomes can exchange. This creates a jumbled genetic history by the mixing of two lineages (from the father and mother).

Y Chromosomal Adam The Y chromosome has been recently sequenced and is offering new opportunities for use in human evolutionary studies and forensic archaeology (such as sex typing; see the story of the Romanovs in Biotech Revolution). The X and Y chromosomes determine whether an individual is a male or female. Whereas a female (XX) has two X chromosomes (one each from the mother and father), a male (XY) has a Y inherited from the father and an X from the mother. Like the mitochondrial DNA, the Y chromosome is passed through one lineage only. The Y chromosome is passed on from father to son through the paternal lineage, with very few changes in the DNA. Small polymorphisms (mutations) do occur and can be used to determine the evolutionary history of humans. XX sex chromosomes of a female undergo DNA exchange during meiosis and the formation of gametes; however, the XY sex chromosomes in a male can exchange DNA only at the ends of their chromosomes because they are unmatched chromosomes (only matched, or homologous, chromosomes can exchange DNA).

The use of the Y chromosome for evolutionary studies is very recent because the chromosome had to be sequenced and DNA polymorphisms identified. Examples of markers for Y-chromosome profiling include the insertion YAP (Y chromosome *alu* polymorphism), SNPs, microsatellites, and minisatellites.

The male Y chromosome is a record of paternal inheritance. Thus, as with mitochondrial DNA, if we go far enough back in our ancestral lineage, all men are paternally related and come from potentially one common male, sometimes referred to as Y chromosomal Adam. The male counterpart to mitochondrial Eve, Y chromosomal Adam refers to a single male human ancestor, who lived 35,000–90,000 years ago, from whom all males are descended based on DNA analysis. As we have discussed, unlike genes of the nucleus, those of the Y chromosome are passed from father to son, just as mitochondrial DNA is passed from mother to daughter.

Y chromosomal Adam and mitochondrial Eve never met because they lived thousands of years apart. They are named after the Adam and Eve in Genesis but should not be identified with them. According to the DNA of people living in Africa today, both Y chromosomal Adam and mitochondrial Eve are believed to have lived in Africa.

Therefore, the Y chromosome is rooted in Africa and, like the results for mitochondrial DNA, supports the out-of-Africa theory. The modern Y chromosome had its recent origin in Africa and then migrated out and replaced the Y chromosomes elsewhere. Y-chromosome profiling shows that males moved south into Australia and north into Eurasia. Later migratory patterns, using Y-chromosome profiles, appeared to follow climate and lifestyle changes.

Other examples of Y-chromosome studies include

1. The evolutionary history of the Lemba, a South African Bantu-speaking population of possible Jewish ancestry. Preliminary analysis of Y chromosomes show a Semitic origin for more than half of the Y chromosomes profiled. At this time it is unknown whether the Y chromosomes were inherited from Jewish or Arab males because the Y profiles from these groups cannot be distinguished.

2. Using Y-chromosome polymorphisms, researchers determined that islands of Jewish priests (Cohanim) have been maintained in populations of Eastern European Jews (through distinction of priest and nonpriest Y-chromosome populations). Through oral tradition, Jewish priests (not the same as appointed rabbis) inherited their position through the male lineage. Thus, the priesthood should be passed on with the Y chromosome, creating distinct polymorphisms on the Y chromosome.

FORENSIC ARCHAEOLOGY

Forensic archaeology is the use of forensic science (often using DNA analysis) to examine and make conclusions about archaeological discoveries. Archaeological methods also are used to investigate crimes.

The development of DNA profiling methods on remains (many very ancient) has been a boon to the field of archaeology. DNA is used

1. For the identification of individuals (for example, Tomb of the Unknown Soldier in Washington, D.C.)

2. For the study of human and animal evolution (for example, in studying the evolutionary relationships between woolly mammoths and elephants)

3. To trace the migration of humans and animals (including extinct groups)

4. To trace the origins and relationships of different ethnic groups around the world

5. To determine family relationships of ancient remains

DNA Profiling and Human Remains

The use of DNA analysis has been used in a number of high profile cases such as identification of remains from the Tomb of the Unknown Soldier, people found in mass graves, war casualties, or casualties from large accidents (for example, mass graves in Iraq, human remains from the September 11, 2001, terrorist attack, and ancient mummified remains [dried or frozen]).

Romanov Family

Tsar Nicholas Romanov II of Russia was overthrown during the Bolshevik Revolution in 1917. He and his family (wife and five children) were taken prisoners and the fate of the family was unknown. In July 1918 the Bolshevik revolutionaries announced that they had executed the tsar and his family. However, many believed that one or more of the family members had survived the massacre. In fact, for many years one of the daughters, Anastasia, was thought to have lived in Europe or the United States. The mystery surrounding Anastasia has been portrayed in several books and even movies. Finally in 1991 nine skeletons were found in a shallow mass grave 20 miles from Ekaterinburg, Russia. They were thought to be the Romanov family. The British Forensic Science Service along with Russian authorities conducted extensive forensic examinations that included several DNA tests.

1. Sex testing using a gene that differs in the X and Y chromosomes to determine the sex of the remains

2. Comparison of mitochondrial DNA of living relatives of the Romanovs (including Prince Philip, the husband of Queen Elizabeth of England) with the remains found in the grave to determine maternal relationships

3. STR tests to determine the relationships among the bodies

Seven of the bodies were identified as the Tsar, his wife, and three of his children. Although two of the children—Tsarevitch and Alexei—were missing, historical reports suggest that they may have been buried separately. Thus, contrary to the many rumors, most likely none of the members of the family survived the execution in 1918.

Otzi the Ice Man

In 1991 a frozen body was found at 11,000 feet in the Alps, on the border of Austria and Italy. The man's body had been protected from predators and decay by freezing temperatures and snow that covered his body for approximately 5200 years, before a summer thaw exposed it. DNA testing from blood samples (and other physical evidence) indicated that he most likely died in a violent fight. The DNA of four other people was recovered—one on the knife blade, two different DNA sequences on one arrow, and a fourth DNA type on the iceman's goatskin coat.

DNA is obtained from teeth and bones of remains for a variety of research projects. PCR has allowed the isolation of very small amounts of DNA. Specific examples of research include the following:

1. Archaeologists in Egypt are comparing the DNA of dead pharaohs with other known human remains to determine how they were related.

2. The DNA of ancient humans in Africa and elsewhere is being studied to determine how and where humans evolved, as well as who our closest ancestors were.

3. The migration of groups into North America is being studied using DNA.

General Readings

P.R. Billings, ed. 1992. *DNA On Trial: Genetic Identification and Criminal Justice.* Cold Spring Harbor Laboratory Press, Cold Spring Harbor, New York.

B. Budowle, T. Moretti, J. Smith, and J. DiZinno. 2000. *DNA Typing Protocols: Molecular Biology and Forensic Analysis.* BioTechniques Books, Eaton Publishing, Natick, Massachusetts.

T. Burke, G. Dolf, A.J. Jeffreys, and R. Wolfe, eds. 1991. *DNA Fingerprinting: Approaches and Applications.* Birkhauser Verlag, Basel, Switzerland.

J.M. Butler. 2001. *Forensic DNA Typing.* Academic Press, San Diego, California.

M. Farley and J.J. Harrington, eds. 1991. *Forensic DNA Technology.* Lewis Publishers, Inc., Chelsea, Michigan.

R.M. Fourney. 1998. *Mitochondrial DNA and Forensic Analysis. A Primer for Law Enforcement. Can. Soc. Sci. J.* 31:45–53.

B. Herrmann and S. Hummel, eds. 1994. *Ancient DNA.* Springer-Verlag, New York.

C. Holden. 1997. DNA fingerprinting comes of age. *Science* 278:1407.

K. Inman and N. Rudin. 1997. *An Introduction to Forensic DNA Analysis.* CRC Press, Boca Raton, Florida.

H.C. Lee and R.E. Gaensslen, eds. 1990. *DNA and Other Polymorphisms in Forensic Science,* Year Book Medical Publishers, Inc. Chicago.

B. Sykes. 2001. *The Seven Daughters of Eve.* W.W. Norton and Company, New York.

B.S. Weir, ed. 1995. *Human Identification: The Use of DNA Markers.* Kluwer Academic Publishers, Boston.

Additional Readings

G.J. Annas. 1992. Setting standards for the use of DNA-typing results in the courtroom: The state of the art. *N. Engl. J. Med.* 326:1641–1644.

A. Badr, K. Muller, R. Schafer-Pregl, H. El Rabey, S. Effgen, H.H. Ibrahim, C. Pozzi, W. Rohde, and F. Salamini. 2000. On the origin and domestication history of barley (*Hordeum vulgare*). *Mol. Biol. Evol.* 17:499–510.

J. Baird. 1992. Forensic DNA in the trial court 1990–1992: A brief history. In P.R. Billings, ed., *DNA On Trial: Genetic Identification and Criminal Justice.* Cold Spring Harbor Laboratory Press, Cold Spring Harbor, New York, pp. 61–77.

P.L. Bereano. 1992. The impact of DNA-based identification systems on civil liberties. In P.R. Billings, ed. *DNA On Trial: Genetic Identification and Criminal Justice.* Cold Spring Harbor Laboratory Press, Cold Spring Harbor, New York, pp. 119–128.

D.A. Berry. 1992. Statistical issues in DNA identification. In P.R. Billings, ed. *DNA On Trial: Genetic Identification and Criminal Justice.* Cold Spring Harbor Laboratory Press, Cold Spring Harbor, New York, pp. 91–108.

M.J. Blears, S.A. De Grandis, H. Lee, and J.T. Trevors. 1998. Amplified fragment length polymorphism (AFLP): A review of the procedure and its applications. *J. Indust. Microbiol. Biotechnol.* 21:99–114.

B. Budowle, K.L. Monson, and J.R. Wooley. 1992. Reliability of statistical estimates in forensic DNA typing. In P.R. Billings, ed. *DNA On Trial: Genetic Identification and Criminal Justice.* Cold Spring Harbor Laboratory Press, Cold Spring Harbor, New York, pp. 79–90.

B. Budowle, T.R. Moretti, A.L. Baumstark, D.A. Defenbaugh, and K.M. Keys. 1999. Population data on the thirteen CODIS core short tandem repeat loci in African Americans, U.S. Caucasians, Bahamians, Jamaicans, and Trinidadians. *J. Forensic Sci.* 44:1277–1286.

K.H. Buetow, M.E. Edmonson, and A.B. Cassidy. 1999. Reliable identification of large number of candidate SNPs from public EST data. *Nat. Genet.* 21:323–325.

T. Burke and M.W. Bruford. 1987. DNA fingerprinting in birds. *Nature* 327:149–152.

J.M. Butler and B.C. Levin. 1998. Forensic applications of mitochondrial DNA. *Trends Biotechnol.* 16:158–162.

R. Chakraborty and K.K. Kidd. 1991. The utility of DNA typing in forensic work. *Science* 254:1735–1739.

L.C. Chen. 1994. O.J. Simpson case sparks a flurry of interest in DNA fingerprinting methods. *GEN (Genet. Eng. News)* 14:1 and 30.

B. Devlin, N. Risch, and K. Roeder. 1991. Estimation of allele frequencies for VNTR loci. *Am J. Hum. Genet.* 48:662–667.

J.T. Epplen. 1994. DNA fingerprinting: Simple repeat loci as tools for genetic identification. In: B. Herrmann and S. Hummel, eds. *Ancient DNA.* Springer-Verlag, New York, pp. 13–30.

A. Gibbons. 2001. Modern men trace ancestry to African migrants. *Science* 292:1051–1052.

P. Gill. 2002. Role of short tandem repeat DNA in forensic casework in the U.K.: Past, present, and future perspectives. *Biotechniques* 32:366–385.

D.B. Goldstein and L. Chikhi. 2002. Human migrations and population structures: What we know and why it matters. *Annu. Rev. Genom. Hum. Genet.* 3:129–152.

P. Helminen, C. Enholm, M.L. Lokki, A.J. Jeffreys, and L. Peltonen. 1988. Application of DNA "fingerprints" to paternity determinations. *Lancet,* March 12:574–576.

G. Herrin, Jr., L. Forman, and D.D. Garner. 1990. The use of Jeffreys' multilocus and single locus DNA probes in forensic analysis. In H.C. Lee and R.E. Gaensslen, eds. *DNA and Other Polymorphisms in Forensic Science,* Year Book Medical Publishers, Inc. Chicago, pp. 45–60.

W.G. Hill, A.J. Jeffreys, J.F.Y. Brookfield, and R. Semeonoff. 1986. DNA fingerprint analysis in immigration test-cases. *Nature* 322:290–291.

S. Hummel and B. Herrmann. 1994. General aspects of sample preparation. In B. Herrmann and S. Hummel, eds. *Ancient DNA.* Springer-Verlag, New York, pp. 159–168.

M. Ingman, H. Kaessmann, S. Paabo, and U. Gyllensten. 2000. Mitochondrial genome variation and the origin of modern humans. *Nature* 408:708–713.

A.J. Jeffreys et al. 1987. Highly variable minisatellites and DNA fingerprints. *Biochem Soc. Trans.* 53:165–180.

A.J. Jeffreys and S.D.J. Pena. 1993. Brief introduction to human DNA fingerprinting. In S.D.J. Pena, R. Chakraborty, J.T. Epplen, and A.J. Jeffreys, eds. *DNA Fingerprinting: State of the Science.* Birkhauser Verlag, Basel, Switzerland, pp. 1–19.

A.J. Jeffreys, V. Wilson, and S.L. Thein. 1985. Hypervariable "minisatellite" regions in human DNA. *Nature* 314:67–73.

M.A. Jobling and C. Tyler-Smith. 2003. The human Y chromosome: An evolutionary marker comes of age. *Nat. Rev. Genet.* 4:598–612.

Y. Ke. 2001. African origin of modern humans in East Asia: A tale of 12,000 Y chromosomes. *Science* 292:1151–1153.

V.N. Kristensen, D. Kelefiotis, T. Kristensen, and A.-L. Borresen-Dale. 2001. High-throughput methods for detection of genetic variation. *BioTechniques* 30:318–332.

M. Lynch. 2003. God's signature: DNA profiling, the new gold standard in forensic science. *Endeavor* 27:93–97.

L. Mueller. 1991. Population genetics of hypervariable human DNA. In M. Farley and J.J. Harrington, eds. *Forensic DNA Technology.* Lewis Publishers, Inc., Chelsea, Michigan, pp. 51–62.

J.K. Pritchard, M. Stephens, and P.J. Donnelly. 2000. Inference of population structure using multilocus genotype data. *Genetics* 155:945–959.

A.R. Purba, J.L. Nyer, L. Baudoun, X. Perrier, S. Hamon, and P.J.L. Lagoda. 2000. A new aspect of genetic diversity of Indonesian oil palm (*Elaeis guineensis* Jacq.) revealed by isozyme and AFLP markers and its consequences for breeding. *Theor. Appl. Genet.* 101:956–961.

N. Risch and B. Devlin. 1992. On the probability of matching DNA fingerprints. *Science* 255:717–720.

W. Thompson and S. Ford. 1991. The meaning of a match: Sources of ambiguity in the interpretation of DNA prints. In M. Farley and J.J. Harrington, eds. *Forensic DNA Technology.* Lewis Publishers, Inc., Chelsea, Michigan, pp. 83–152.

K. Weising, J. Ramser, D. Kaemmer, G. Kahl, and J.T. Epplen. 1991. Oligonucleotide fingerprinting in plants and fungi. In T. Burke, G. Dolf, A.J. Jeffreys and R. Wolfe, eds. *DNA Fingerprinting: Approaches and Applications.* Birkhauser Verlag, Basel, Switzerland, pp. 312–329.

E. Yee, K. Kidwell, G.R. Sills, and T.A. Lumpkin. 1999. Diversity among selected *Vigna angularis* (Azuki) accessions on the basis of RAPD and AFLP markers. *Crop Sci.* 39:268–275.

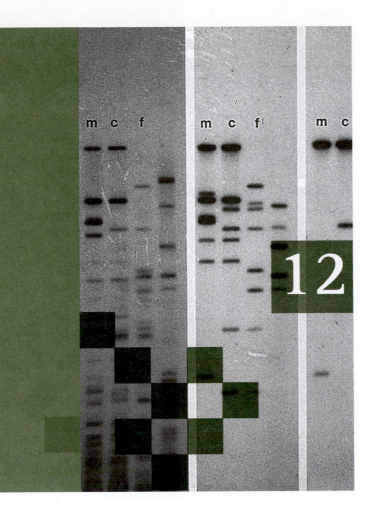

12

REGULATION, PATENTS, AND SOCIETY

Currently there are almost 1400 biotechnology companies in the United States, 339 of which are publicly held. The biotechnology industry has more than tripled since 1992. Revenues increased from $8 billion to $25 billion within 8 years. Biotechnology has become one of the most research-intensive sectors in the world. Consider that biotechnology in the United States spent almost $14 billion in 2000. Furthermore, the amount of funds invested in U.S. biotechnology increased 156% in 1 year alone—from almost $138 billion to $354 billion between 1999–2000! Many of the new products of biotechnology are either living organisms or the products of living organisms. Companies investing in research and development are seeking intellectual property rights to many of their living investments. The patenting of living organisms and the DNA of genomes has been steeped in controversy and has not been completely resolved, especially in Europe.

Rapid progress in biotechnology and the development of both new products and transgenics have raised legal, safety, and public policy issues. A coordinated effort among federal regulatory agencies and public education and dissemination of information will ensure that new products are evaluated, approved in a cost-effective and efficient manner, and accepted by the general public. A coordinated regulatory framework allows for a thorough evaluation of product safety and benefit to society and fosters a continuing open discussion of issues concerning safety and levels of tolerable public risk.

Although powerful new biotechnologies have the potential to alleviate medical, agricultural, and environmental problems, they can also pose problems if we do not reflect on the implications of these developments and evaluate their risks.

REGULATION OF BIOTECHNOLOGY

Modern biotechnology has been debated since the first recombinant DNA experiments in 1973 (described in Chapter 1, Biotechnology: Old and New). Such statements as "we are playing God" and "we are changing the course of evolution" have been made many times over the years. Although the focus of concern has changed through the years, from fear of recombinant molecules to release of genetically engineered organisms into the environment to genetically engineered foods and, most recently animal and human cloning, the public continues to express fears. Nevertheless, many concerns have diminished as the years have passed and predicted disasters did not occur.

The National Institutes of Health (NIH), a federal agency that provides funding for large medical research projects and oversees all federally funded research, also provides guidelines for laboratory research to ensure safety. In the late 1970s, as scientific knowledge increased and experience with recombinant DNA molecules and methods accumulated, the Recombinant DNA Advisory Committee (RAC) of the NIH began to relax laboratory guidelines pertaining to recombinant DNA research. Relaxed regulations and positive public perceptions have encouraged the development of products from recombinant DNA technology. Regulatory agencies now treat pharmaceuticals produced by this technology exactly the same as other pharmaceuticals. The process of production is no longer a regulatory consideration; only the product itself is evaluated for safety.

DNA REVOLUTION: PROMISE AND CONTROVERSY

Initially, recombinant DNA experiments generated much excitement among scientists because the method made it easier to study a cell's molecular processes. Recombinant DNA became a social issue when scientists began to discuss the potential risks that recombinant DNA experiments posed to human health and the environment.

Between 1971 and 1973, many molecular biologists began to use animal cells in culture and animal viruses in their research. They were especially interested in viruses that can cause tumors in animal hosts. As early as 1971, some scientists became concerned that such research might be a health hazard. Paul Berg also raised concerns by contemplating experiments to develop a vehicle for introducing foreign DNA into animal cells. Concerns further escalated during the 1971 Cold Spring Harbor Laboratory meetings when Janet Mertz proposed an experiment to introduce tumor-promoting DNA from SV40 into *Escherichia coli* cells to test whether SV40 could be used as a vector to transfer DNA into animal cells. However, because the bacterium naturally exists in human intestines and under experimental conditions could potentially induce tumors in laboratory rats, the idea that it would be engineered caused great concern. The SV40 experiments were postponed when a scientist at Cold Spring Harbor called Berg after the meetings to express his concern.

In response to concerns among a growing number of scientists, a conference was held in January 1973 at the Asilomar Conference Center in Pacific Grove, California, to specifically address the biohazards of working with recombinant DNA. The NIH responded in October 1974 by establishing the Recombinant DNA Molecule Program Advisory Committee. The mandate of this committee was to evaluate the risks in using recombinant DNA, to establish guidelines for recombinant DNA research, and to develop methods for preventing the accidental release of molecules to humans and the environment.

Asilomar Conference

The second international Asilomar Conference on Recombinant DNA Molecules was held February 24–27, 1975. The conference was sponsored by the National Academy of Sciences and funded by the NIH and the National Science Foundation to discuss the progress made in recombinant DNA technology and to evaluate the risks and potential dangers posed by recombinant DNA molecules. The recommendations that emerged from the conference were to be incorporated into the NIH guidelines. Guidelines were proposed that included laboratory safety and containment measures for a variety of recombinant DNAs, including host cells and vectors (Table 12.1). The committee's report was quickly drafted, and the guidelines appeared in *Science* on June 6, 1975.

Most scientists believed that they would be able to continue their recombinant DNA research without further delay, and were unprepared for the debate that ensued. Their research became a focal point for controversy. Research scientists, public interest groups, the U.S. Congress, corporations, and private citizens became engaged in a heated debate that would last almost 4 more years. Questions about risk could not be clearly answered because data were unavailable. Molecular biologists remained in disagreement among themselves about the safety of recombinant DNA technology and the restrictions to be placed on experimentation. The DNA debates rapidly became an international issue.

Drafting the NIH Guidelines

The RAC, established by the NIH just before the second Asilomar conference, had three tasks: (1) evaluate the potential biological and ecological hazards of DNA recombination, (2) develop procedures to minimize the spread of recombinant molecules in animal populations, and (3) devise guidelines for researchers using potentially dangerous recombinant molecules and organisms. In the spring and summer of 1975, RAC

TABLE 12.1 Classification of Experiments in the Provisional Statement of the Asilomar Conference

Containment Level	Techniques/Practices	Types of Experiments
Low containment	Good microbiological techniques, encompassing use of lab coats, use of mechanical pipettes, a prohibition on eating in the lab, and use of biological safety cabinets for procedures that generate large aerosols.	Experiments involving organisms that normally exchange genetic information. Most experiments involving DNA from prokaryotes, lower eukaryotes, plants, invertebrates, and cold-blooded vertebrates.
Moderate containment	Physical containment required for handling moderate-risk oncogenic viruses, encompassing use of gloves and lab coats, use of biological safety cabinets for transfer operations, vacuum lines protected by filters, negative pressure that is maintained in limited-access laboratories.	Experiments with bacterial genes that affect pathogenicity or antibiotic resistance. Experiments with viral DNA involving the linkage of viral genomes in whole or in part to prokaryotic vectos and their introduction into prokaryotic cells (if "disarmed" vectors are used). Experiments with segments of oncogenic viral DNA if these segments are nontransforming. Experiments with segments of nononcogenic viruses. Experiments with DNA from warm-blooded vertebrates. Experiments with any DNA and animal vectors.
High containment	Use of facilities that are isolated from other areas by air locks, clothing exchanges, and shower rooms and that have treatment systems to inactivate or remove biological agents that may be contaminant in exhaust air and liquid and solid wastes. All agents are handled in biological safety cabinets. Personnel wear protective clothing and shower on leaving. Facility is maintained under negative air pressure.	Experiments with high-risk viruses. Experiments linking eukaryotic or prokaryotic DNA to prokaryotic vectors when the resulting organism might express a toxic or pharmacologically active product.

Originally adapted from "Provisional Statement of the Conference and Proceedings," RDHC, Feb. 27, 1975 and used in S. Wright, *Molecular Politics*, University of Chicago Press, 1994. Used by permission of the publisher.

TABLE 12.2 Biological and Physical Containment Measures for Laboratory Safety

Biological Containment	Physical Containment
EK1 Standard laboratory *E. coli* K12 host strain and compatible plasmid and virus vectors	P1 Standard microbiological procedures
EK2 Genetically altered, disabled *E. coli* that cannot survive outside controlled laboratory conditions	P2 P1 standards plus KEEP OUT notice displayed at the laboratory entrance
EK3 An extremely weakened *E. coli* strain that is 10^6 times less likely to escape from the laboratory	P3 P2 standards plus negative-pressurized laboratory or the use of negative-pressure cabinets
	P4 Used for extremely dangerous host-vector systems; P3 plus investigators must change clothes and shower before exiting the facility; air locks must be used

Note: Both physical and biological containment would be matched to provide maximum laboratory safety. For example, an experiment might be classified as P2 + EK1 or P3 + EK2. Therefore, if mammalian DNA were used, especially if the investigator was working with primate DNA, a P3 + EK3 or P4 + EK2 level of containment might be required.

Adapted from an article by Nicholas Wade, *Science* vol. 190 (November 21, 1975).

---

drafted guidelines incorporating some of the recommendations made at Asilomar. RAC classified the many types of recombinant DNA experiments and the DNA of pathogenic organisms according to potential hazard. Both physical containment guidelines (P1–P4) and biological containment guidelines (EK1–EK3) were established on the basis of hypothetical hazards (Table 12.2). (Now we have B levels: B-1, B-2, and B-3.)

The NIH guidelines recognized four levels of risk, from minimal to high, and devised four corresponding containment levels, P1 to P4. P1 laboratories were to take simple precautions such as prohibiting workers from pipetting by mouth, whereas P4, the highest level of containment, was reserved for those experiments that used the most lethal and pathogenic microorganisms such as anthrax bacilli and smallpox viruses, as well as tumor viruses and toxin genes. Such work required a laboratory with negative-pressure air locks and extreme caution in handling biological material. All experiments were to be conducted in laminar-flow hoods with filtered or incinerated exhaust air.

Biological containment addressed the use of host cells and vectors that could not reproduce outside the laboratory if they escaped (disabled host–vector systems). Organisms would be disabled so that they would die outside the specialized laboratory environment. An example that occurs in nature is the bacterium *E. coli* strain EK2, which cannot survive outside the human gut because mutations prevent the cells from synthesizing thymine, necessary for DNA replication. The mutation is lethal because the cell could not replicate its DNA unless provided with thymine. Modified vectors also were to be used for recombinant DNA research so that the DNA could not be transferred accidentally from one host cell to other cells.

Levels of biological containment ranged from the lowest, EK1, to the highest, EK3. Initially, only EK1 host–vector systems were available for research.

An EK2 host–vector system was not developed and approved until April 1976.

These NIH guidelines were more stringent than those proposed at Asilomar. They were completed in December 1975 and released publicly on June 23, 1976 (see *Federal Register* vol. 41 no. 131, July 7, 1976 and Appendix A), one and a half years after Asilomar. At the time of their release, EK3 host–vector systems had not yet been developed and P4 levels of physical containment existed only in special military facilities or at medical laboratories where lethal pathogens were studied. Therefore no molecular biologists had access to these facilities. Because the NIH guidelines were relaxed in the years to come, an EK3 system never was developed to meet the most stringent requirements.

Because of speculation that the closer the phylogenetic relationship between humans and the species used in recombinant experiments, the greater the risk to humans, experiments that required the use of DNA from mammals or viruses were terminated and DNA samples were destroyed. Unfortunately, this measure precluded using the new technology for research on cancer or using bacteria as hosts for human genes to isolate important human proteins such as insulin and growth hormone. Only the DNA of fish, amphibians, and invertebrates could be used in experiments.

However, the NIH was enforcing the guidelines only at institutions and laboratories receiving federal funding. Compliance by facilities in the private sector, such as industrial laboratories, was voluntary. Individuals working in industry would have proprietary rights or patentable information to protect, and disclosure of experiments might preclude protection. Many research scientists complied voluntarily and destroyed recombinant DNA molecules when experiments did not meet NIH guidelines. In some cities additional guidelines were adopted, such as those specified by the Cambridge Experimentation Review Board. Anyone conducting recombinant DNA research in Cambridge, Massachusetts, had to follow both the NIH and board restrictions. Federal legislation was even proposed to regulate recombinant DNA research. Some scientists expressed fears that legislation would restrict intellectual freedom and impede important scientific progress.

Revision of the NIH Guidelines

In 1977 RAC began to reevaluate the NIH guidelines for revision. As part of the review process, experts assessed the biology and ecology of *E. coli*, the primary host organism for recombinant DNA technology. Reviewers concluded that the laboratory strain *E. coli* K12 was too defective genetically to survive outside the laboratory. In response to new information and additional discussion, RAC proposed revised NIH guidelines that reduced the stringency of some of the containment specifications for many types of experiments. The proposed guidelines were published in the *Federal Register* for public comment on September 27, 1977, but were further revised before the final version was published on December 22, 1978, and adopted on January 2, 1979.

In 1978, at the same time that research in laboratories across the nation was terminated or suspended, Genentech announced that *E. coli* could synthesize the mammalian hormone somatostatin. This was the first time a functional mammalian protein had been produced by recombinant DNA technology from a gene that had been synthesized chemically.

In the 1979 NIH guidelines, only a few types of experiments required P4-level containment, and *E. coli* K12 was subjected to only P1-level physical containment. Additional revised guidelines published in the *Federal Register* on January 29, 1980, further relaxed containment requirements. Other changes included granting experimental review to institutional biosafety committees established by organizations receiving federal funds and others voluntarily observing guidelines. These committees would be informed of *E. coli* K12 experiments either before or after initiation of experiments. This change represented a significant relaxation of regulations. In addition, although industrial compliance was voluntary, a code of practice that protected proprietary rights was included. In November 1980 the RAC recommended further that the review of recombinant DNA experiments covered by the NIH guidelines be left entirely to local institutional biosafety committees (IBC) (Appendix B).

REGULATORY AGENCIES IN THE UNITED STATES

Regulatory agencies of relevance to biotechnology regulate both processes and products, each responsible for specific and sometimes overlapping areas. The three primary agencies are discussed here. Each agency maintains biotechnology product databases of its activities that require applications and approvals. These databases are updated regularly and can be accessed by the public.

United States Department of Agriculture

The United States Department of Agriculture (USDA), created in 1862, is the primary agency for the regulation of food and agriculture. It oversees and regulates the meat and poultry industry and provides standards for the quality and purity of meat, poultry, dairy products, and eggs. The USDA also provides funding for research pertaining to agriculture. The USDA issues required field-release permits for genetically modified organisms (GMOs) and approval before commercial sale of GMO crops. The agency also monitors genetically engineered crops that produce pharmaceuticals, genetically engineered grasses, and genetically engineered insects. This agency examines

1. The origin and characteristics of the transgene (gene transferred into an organism)

2. The potential of the gene to outcross with other unrelated plants or wild relatives

3. Its impact on nontarget or beneficial organisms

The USDA is required to conduct environmental reviews of these genetically engineered products under the National Environmental Policy Act (NEPA). Interestingly, according to a number of watchdog groups (for example, Institute for Food and Development Policy), the USDA has not completed a complete environmental impact statement for any genetically engineered crop.

Environmental Protection Agency

The Environmental Protection Agency (EPA), created in 1970, is the federal agency responsible for providing guidelines, laws, and regulations to protect the environment. It monitors air and water quality and enforces limits of pollution (the Clean Air Act and Clean Water Act were passed by Congress). This agency also regulates the handling and disposal of industrial wastes. The EPA sets limits on the amount of toxic compounds that are allowed in food and potable water (for example, pesticides, lead). The EPA regulates pesticides, including chemical and biological (that is, GMOs), including those produced by plants such as Bt-producing plants (regulated as plant incorporated protectants). Like the USDA, the EPA assesses the impact on nontarget organisms. The EPA also examines the toxicity of the GMO and any risks to humans such as allergenicity.

Food and Drug Administration

The Food and Drug Administration (FDA) is a scientific regulatory agency that oversees and regulates marketable products and monitors their safety during consumer use. This includes the safety of the nation's domestically produced and imported foods, drugs, biologics, medical devices, cosmetics, and other consumables. The agency basically addresses consumer safety and protection. After the Federal Food and Drugs Act was passed in 1906, the FDA not only evaluated food safety but also assumed regulatory responsibilities. At one time the FDA was under the Department of Agriculture, but today it is part of the Public Health Service (PHS), under the jurisdiction of the Department of Health and Human Services. The FDA includes the Center for Biologics Evaluation and Research (CBER), the Center for Drug Evaluation and Research (CDER), the Center for Devices and Radiologic Health (CDRH), and the Center for Food Safety and Applied Nutrition (CFSAN). Some of the products regulated by the FDA and regulatory services it provides are listed in the following:

1. Food products (with the exception of meat and poultry)

2. Drugs for human and veterinary use

3. Drugs of biologic origin (such as human growth hormone and insulin)

4. Cosmetics

5. Animal feed such as packaged pet foods

6. Medical devices such as pacemakers, defibrillators, and diagnostic kits used in the medical field

7. Composition, quality, and safety of food and food additives (for example, coloring agents)

8. Labeling of drugs and medical devices to protect consumers from misinformation and wrongful manufacturing claims

The FDA regulates the introduction of all new foods, including GMOs, into the marketplace. This agency assesses any deviation of the GMO from typical foods, and evaluates safety issues. Tests for FDA clearance include

1. Gene source

2. History of use

3. Toxicity

4. Nutritional data

5. Allergenicity

6. Chemical composition

COORDINATED BIOTECHNOLOGY FRAMEWORK

As biotechnology became more commercialized and applications proliferated, most experiments fell outside the biomedical focus of the NIH and within the jurisdiction of other federal agencies. In the April 1982 NIH-RAC revisions, experiments in which GMOs would be deliberately released into the environment were subject to review by the RAC and approval by the Institutional Biosafety Committee (IBC) of the federally funded or voluntarily complying organization. Having established the first criteria for reviewing field-test proposals for GMOs, the NIH gradually reduced its role, inviting other regulatory agencies to review proposals and eventually transferring authority to the EPA and USDA. However, as the number of proposals for field tests increased, it was seen that a new coordinated regulatory structure was needed.

In November 1985 the Office of Science and Technology Policy (OSTP) established the Biotechnology Science Coordinating Committee (BSCC), a statutory interagency coordinating committee within the OSTP. Made up of representatives of the USDA, NIH, EPA, the National Science Foundation (NSF), and the FDA, the BSCC was to address scientific problems raised in regulatory and research applications, promote consistency in review and assessment, identify gaps in scientific knowledge, and facilitate cooperation among federal agencies.

The first tasks of the BSCC were to develop a federal policy on biotechnology and to define a multiagency regulatory structure that would assign jurisdictions to the various federal agencies. This policy was formulated by representatives from 18 agencies that made up the administration's Domestic Policy Council. The OSTP published the final version of the Federal Policy on Biotechnology in the *Federal Register* on June 26, 1986. This policy, known as the Coordinated Framework for the Regulation of Biotechnology, or simply the Coordinated Framework, was meant to regulate products in all stages, from research and development to marketing, shipment, and use. The framework incorporated policy statements from the three primary federal agencies (USDA, FDA, and EPA); established a systematic, coordinated review process for risk analysis and assessment; and defined the levels of review and regulation (therefore delineating how the regulatory responsibilities would be divided among the agencies). The framework was meant to encourage a common language in reviewing and assessing proposals, thereby ensuring consistency in regulation, with the ultimate goal of streamlining and simplifying the regulatory procedures. In addition, agency jurisdiction was defined for different products and consistency in regulatory review was encouraged to prevent unnecessary delays in the review process. (For example, if an organism was both a pest-control agent and a plant pathogen, the framework was supposed to establish which agency had primary jurisdiction.)

In setting policy for deciding whether a particular GMO should be granted a field trial, the BSCC determined how much scientific information was required and the level of risk. The policy goal was to minimize the risks while still encouraging innovation and product development. In this regulatory framework, agencies were directed to focus on the product and not on the mode of production. Products of recombinant DNA technology were treated exactly like organisms that were not genetically manipulated. Indeed, the primary feature of the Coordinated Framework was that it required no new legislation or agencies for regulating genetically engineered organisms; existing regulatory statutes and agencies were determined to be adequate. Among such statutes that now apply to many areas of biotechnology research, testing, and commercialization are the Toxic Substances Control Act, the Federal Insecticide Fungicide and Rodenticide Act, the Federal Plant Pest Act, the National Environmental Policy Act, the Plant Quarantine Act, and the Virus-Serum-Toxin Act. Although appropriate in some situations, existing statutes have shown increasingly apparent inadequacies.

Four federal agencies were charged with overseeing product regulation (Table 12.3). For example, the EPA oversaw the deliberate environmental release of genetically modified organisms for pest and pollution control; the USDA oversaw the use of genetically engineered agricultural plants and animals; the FDA evaluated the food safety of a new genetically engineered crop variety, and the NIH oversaw genetically engineered organisms that might affect public health. Interagency coordination was not entirely smooth, and conflicts often resulted in deadlocks and lack of progress.

The BSCC proved unable to resolve conflict among regulatory agencies, and, to facilitate cooperation, the Federal Coordinating Council on Science, Engineering, and Technology (FCCSET) replaced it in late 1990 with a new interagency committee: the Biotechnology Research Subcommittee (BRS) of the Committee on Health and Life Sciences. Although the BRS functions similarly to the BSCC, its membership is larger; in addition to the agencies listed for the BSCC, it includes representatives from the DOE (Department of Energy), AID (Agency for International Development), NASA (National Aeronautics and Space Administration), DOD (Department of Defense), OMB (Office of Management and Budget), DOC (Department of Commerce), and the Department of Interior.

Problems with the Coordinated Framework

The framework was created to regulate new, emerging biotechnologies using preexisting laws. Although this framework has served the nation well, numerous problems have become apparent over the years. A major problem has been inadequate legal authority for oversight of all recombinant products and genetic engineering research activities. Implementation of policies and regulations also presents problems. Statutes have not covered some areas of biotechnology, and there has at times seemed to be many areas of unclear jurisdiction. The original framework did not cover genetically engineered animals. Past critics cited the example of

TABLE 12.3 Regulatory Agencies and Related Statutes*

National Institutes of Health (NIH)
  Recombinant DNA Advisory Committee (RAC)**
Environmental Protection Agency (EPA)
  Toxic Substances Control Act (TSCA)
  Federal Insecticide, Fungicide, and Rodenticide Act (FIFRA)
  National Environmental Policy Act (NEPA)
Food and Drug Administration (FDA)
  Public Health Service Act (PHSA)
  Federal Food, Drug, and Cosmetic Act (FFDCA)
United States Department of Agriculture (USDA)
  Animal and Plant Health Inspection Service (APHIS)
  Agriculture Recombinant DNA Research Committee (ARRC)
  Food Safety Inspection Service (FSIS)
    Federal Plant Pest Act (FPPA)
    Plant Quarantine Act (PQA)
    Virus-Serum-Toxin Act

*For a discussion of federal agencies and their regulatory statutes see Appendix D.
**Now primarily oversees gene therapy.

laboratory-produced genetically engineered fish that could potentially be released without violating any regulations. The framework also has not been completely clear on the commercialization of transgenic crops. The Federal Plant Pest Act (FPPA) requires that the USDA determine the impact of crops on the environment, but does not clearly cover the commercial development of most crop plants.

No statutes slow the increased use of herbicides that has come about partially as a result of engineered herbicide-tolerant crops. With the use of pesticide-producing plants, insects can potentially become resistant to chemical pesticides. Neither the Federal Insecticide Fungicide and Rodenticide Act (FIFRA) nor the FPPA has clear application to these critical areas. How do we ensure that these issues are addressed appropriately? Another area of concern is the limited authority of the Toxic Substances Control Act (TSCA): this chemical control statute may not apply directly to living microorganisms.

Overall, the regulatory framework does not appear to meet current needs, especially in light of the new genetically engineered products headed for market within the next decade. According to critics, it does not provide companies with a clear idea of what they are allowed or not allowed to do in the course of marketing genetically engineered organisms and their products, especially those of agricultural and environmental significance. Confusion over which statute applies

More Effective Coordinated Framework?

Almost 20 years after the Coordinated Framework was established, the genetically engineered products that have been marketed have not caused problems—they appear to be safe to eat and environmental problems have not occurred. The framework used preexisting laws to regulate genetically engineered products. With the large number of genetically engineered crop plants in various stages of production, the advent of transgenic plants and animals producing pharmaceuticals, and the increased production of transgenic animals for a variety of purposes, some have claimed that the Coordinated Framework needs a tune-up.

The Coordinated Framework has been criticized for several reasons. One problem has been that there are sometimes different standards and procedures used by the agencies when regulating genetically engineered products. Furthermore, the public has not been involved in the regulatory process, and it is felt that this is a major roadblock to the acceptance of new genetically engineered products. It has been proposed that new legislation should be proposed to help close the gaps between agencies and to address the new generation of genetically engineered products. Some of the questions being asked are

1. Should some of the regulatory roles of the agencies be consolidated?

2. Does the current framework allow for regulation by consumer choice of products in the marketplace, formal government regulation, and holding companies accountable for product safety?

3. Can the framework provide a balance between the public's strong desire for both new products and complete product safety?

4. What about ensuring that all genetically engineered products receive a complete assessment before they can be released into the environment? Although the USDA regulates some genetically engineered crops under the Federal Plant Pest Act (appearing to imply that these plants are potential plant pests), these regulations do not cover all genetically engineered crop plants (for example, those produced with nonconventional methods such as a particle gun). In addition, the USDA review does not always include a comprehensive environmental assessment of the product.

Improvements to the regulatory framework are being implemented.

1. The FDA must better address the question of labeling genetically engineered foods. In January 2003 the FDA issued a draft on labeling guidelines to help manufacturers who wish to label foods that do or do not have bioengineered ingredients. They suggested that "bioengineered" and "derived through biotechnology" be used rather than "GM-free," "modified," and "GMO." The FDA believes that the latter terms can be misleading.

2. In January 2003 the FDA proposed mandatory premarket notification that increases the scrutiny of genetically engineered foods (in 1992 these were voluntary). Manufacturers of genetically engineered plant-derived food and animal feeds must notify the FDA at least 120 days before the products are marketed. Manufacturers must now submit safety and nutritional data demonstrating the foods are as safe and nutritious as their nongenetically engineered counterparts. This information is also required to be accessible to consumers (for example, through the Internet). This new ruling, however, does not change the nature of scientific review.

3. The FDA plans to regulate genetically engineered animals under its new animal drug jurisdiction—when the new, inserted gene and the protein it produces is a drug. However, there is no provision for the public to access any applications or provide input.

to particular activities has caused some to fear that the environment may eventually suffer.

Experience has revealed gaps, redundancies, and inconsistencies in the regulatory process. Coordination among the three main agencies (FDA, USDA, EPA) has sometimes been difficult. For example, if the same microorganism is both a plant pest and an animal pathogen, it is subject to both USDA and EPA review and regulation, as is a microorganism that is both a pesticide and a plant pathogen. A complex regulatory process can potentially create delays in approval, as could a regulatory dispute arising from a question of agency responsibility, for example, whether a product is considered a veterinary biologic (USDA) or a new animal drug (FDA).

Many of the problems described lie in the fact that the legislation used to regulate genetically engineered organisms was designed primarily for the chemical industry, such as the TSCA. No one could have predicted such a use—and organisms and chemicals present very different risks. Unfortunately, statutes within the Coordinated Framework have not always been able to efficiently regulate the products of biotechnology. Some regulatory procedures are extensive and costly.

Many new genetically engineered microorganisms, plants, and animals will be used to produce pharmaceuticals, new foods, and chemicals. How will these organisms and products be regulated? Which agencies will have oversight? Are we prepared for rapid advances in biotechnology? The framework currently is being reassessed, and steps are being taken to modify the regulations and regulatory roles of the agencies to ensure that the regulatory framework is consistent and comprehensive, reduces confusion, does not impede progress, and protects the environment and the public.

RELEASE OF GENETICALLY ENGINEERED ORGANISMS

Worldwide more than 25,000 field safety and production trials have been conducted on more than 60 genetically engineered plants and animals in 45 countries. In the United States, more than 6500 field tests of GMOs have been conducted at approximately 18,000 sites. In 2001 74% of soybeans, 71% of cotton, and 32% of corn planted in the United States were genetically engineered varieties. The Grocery Manufacturer's Association has estimated that approximately 70% of all foods on grocery shelves in the United States contain some component that was made from genetically engineered organisms. The list of FDA-approved genetically engineered foods includes corn, rice, soybean, potato, tomato, squash, canola, cotton, sugar beet, flax, and cantaloupe (although not all have been marketed). The first genetically engineered product approved for use in foods was the enzyme chymosin (rennin), approved for use in 1988. In the United States today, 80–90% of the cheese is made from genetically engineered chymosin. In the past, the enzyme was made from the stomach of calves.

Ice-Minus Bacteria

The first requests for approval to deliberately release genetically engineered organisms in field trials were submitted to the NIH-RAC in 1982. Despite critics' concerns about the environmental consequences, no specific guidelines and regulations had been drafted by federal agencies in anticipation of such ap-

plications. Among the first applications for field trials was one for testing the genetically modified soil microbe *Pseudomonas syringae* to determine its effect on frost damage to plants (see Chapter 5, Microbial Biotechnology). This landmark case stirred a great deal of controversy.

The *P. syringae* bacterium normally resides on plant surfaces and produces an ice-nucleation protein that facilitates the formation of ice crystals at low temperatures. Ice crystals cause tissue damage that costs U.S. citrus growers more than $1 billion a year in losses. Dr. Steven Lindow and his colleagues at the University of California at Berkeley reasoned that if an ice-minus strain were sprayed onto leaves before the natural bacteria could colonize the surfaces, plants could survive lower temperatures without significant damage. Ice-minus deletion strains of *P. syringae* were constructed by deleting the gene that encodes the ice-nucleation protein.

To obtain NIH-RAC permission, a proposal was submitted by the applicants, announced in the *Federal Register*, and reviewed by a panel of experts. In 1983, after carefully examining the proposal, the NIH-RAC granted permission to proceed with field tests. Concerned citizens immediately challenged the review process. The Foundation on Economic Trends, headed by Jeremy Rifkin, filed a suit to block the tests. The suit was successful, and a judge blocked the tests on the grounds that the NIH-RAC had not requested an environmental impact assessment or held a hearing as required by federal law.

In late 1984 the EPA claimed jurisdiction to regulate recombinant bacteria used as pesticides. Ice-minus *P. syringae* was considered a pesticide that controlled naturally occurring wild-type *P. syringae*. The EPA received two separate applications to apply ice-minus *Pseudomonas* species to potato plants as frost protectants. In November, Advanced Genetics Sciences, Inc. (now DNA Plant Technology Corp.) applied for clearance to spray ice-minus *P. syringae* and *P. fluorescens* on the flowers of the plants to determine whether the growth of ice-nucleation bacteria would be prevented or reduced, thus reducing frost injury between 0 and −5° C. In December Lindow applied for clearance to treat potato seed segments with ice-minus *P. syringae* just before planting them and then to spray the plants with the bacteria once foliage appeared, to assess the effect on frost damage.

The EPA conducted an extensive assessment of the ice-minus strain by evaluating its ecological impact, human health implications, and environmental fate. The EPA held public meetings, and commissioned evaluations by the NIH and FDA. The EPA's General Council Office, Office of Pesticide Program Review, and its committees on Toxic Substances, Research and Development Policy Planning, and Evaluation each assessed the proposed field trial plans. The California

Department of Agriculture also was involved in the evaluation, because the field tests would be conducted in California. In addition, a science advisory panel composed of a plant pathologist, a microbiologist, and an ecologist reviewed the proposals. In short, this review process was extensive, costly, and even redundant. (There was consensus, however, that in time the review process could be streamlined without sacrificing quality.)

Opponents of the tests expressed fears that deliberately released genetically engineered organisms might replace natural organisms in the environment, spreading into important niches and devastating the balance of nature, and might transfer introduced genes to other organisms (such as by conferring herbicide resistance to weed species). Both applications anticipated these objections. The vicinities of the test sites were to be monitored to detect dissemination by wind or insects, and after each experiment, the foliage was to be incinerated and plant debris was to be disked into the soil.

The Hazard Evaluation Division of the EPA concluded that neither of the proposed set of ice-minus field trials would pose a significant risk to the environment or to human health. Permission was granted (although fearful residents near the test site temporarily blocked the tests and further delayed experimentation through court orders). Field trials were finally conducted in 1987. Initially, those spraying the plants with bacteria took the precaution of wearing protective clothing (see Figure 1.20). Data indicated that the engineered bacteria were effective and neither persisted after application nor were dispersed into the environment. Nevertheless, engineered ice-minus microorganisms are not likely to be marketed. Natural deletions of the ice-nucleation gene have been isolated and will be marketed under the name Frostban. A nonengineered product will have an easier time making it through the regulatory maze.

The ice-minus experiments paved the way for the many field trials conducted since. The outdoor application of recombinant microorganisms seems to be safe. No transfer of genes to indigenous organisms has been detected, and the genetically engineered organisms seem to remain at the site of application. Genetically engineered crop plants have been introduced into the environment with little objection, although some fear that genes could be transferred to weedy species through cross-pollination. Some proponents argue that recombinant crop plants differ little from those generated by classical plant breeding methods.

First Genetically Engineered Food

The first genetically modified food product, the Flavr Savr tomato (see Chapter 6, Plant Biotechnology), generated much public debate. The controversy about Calgene's genetically engineered tomato arose over a marker gene that was used to identify plant cells that had been transformed with the antisense polygalacturonase (PG) gene. The marker gene conferred antibiotic resistance to kanamycin and was linked to the antisense gene. Opponents of genetically engineered foods feared that foods harboring antibiotic resistance genes could exacerbate the problem of increasing resistance to antibiotics in humans. Calgene scientists used the tomato in animal studies and demonstrated that most of the kanamycin resistance gene product was destroyed by the digestive system. Experimental results indicated that the expression level of the marker gene in the tomato was extremely low, and the likelihood of gene transfer to *E. coli* of the human gut was highly unlikely.

Under current FDA policy developed in 1992 (Statement of Policy: Foods Derived from New Plant Varieties, *Federal Register* May 29, 1992, 57 FR 22984), genetically engineered foods are not required to be reviewed and approved unless they contain additives. The characteristics of the food product and not the method of production (for example, genetic manipulation) are to be reviewed for safety. The FDA 1992 policy states that expression products are to be regulated as food additives if they are not generally recognized as safe. In July 1993 Calgene filed a food additive petition requesting the use of the kanamycin resistance gene product aminoglycoside 3'-phosphotransferase II (APH[3']II) in new plant varieties of tomato, oilseed rape, and cotton. Calgene submitted the findings from studies that provided evidence that APH(3')II posed no threat to either humans or the environment. The summaries of the data and deliberations were presented in the FDA document (Docket No. 93F-0232, May 23, 1994) "Secondary Direct Food Additives Permitted in Food for Human Consumption; Food Additives Permitted in Feed and Drinking Water of Animals; Aminoglycoside 3'-Phosphotransferase II." The FDA amended the food additive regulations to allow for the safe use of APH(3')II as a processing aid in the development of new plant varieties. The FDA also determined that the Calgene tomato did not differ significantly from a nonengineered variety and maintained the essential characteristics of a normal tomato, thus making it as safe as a normal tomato.

Although producers were not required to inform the FDA that they planned to sell their engineered products to the public, they were encouraged to do so. For example, Calgene voluntarily submitted Flavr Savr data for FDA review and assessment. Because the tomato was the first bioengineered food to be marketed, Calgene hoped that an FDA safety endorsement would help diminish the controversy. (Calgene had to obtain approval from the USDA before it could grow the tomatoes in commercial fields.) Opponents of genetically engineered foods expressed concerns that FDA approval of the tomato would open the door to many other genetically engineered foods in the near future.

The development of genetically engineered plants and animals has sometimes been controversial. Fears that these novel organisms might pose risks to the environment have led to federal policies requiring that thorough analyses of ecological risks be conducted. Before bioengineered plants and plant products can be commercialized, scientists must ensure that scientifically sound, experimental methods are developed to assess the ecological risks.

Thousands of U.S. field trials have been completed or are in progress on genetically engineered versions of cotton, maize, alfalfa, oilseed rape, cucumber, asparagus, apple, melon, cauliflower, and sugar beet (see Chapter 6, Plant Biotechnology). After many years of studies, the regulatory procedures for field testing transgenic plants have been relaxed, although some individuals continue to voice concern. In 1987 each genetically modified plant had to be field tested (required by the USDA's Animal and Plant Health Inspection Service, or APHIS) to demonstrate that the plant would not disperse into the environment—that is, would not become a weed—and disrupt the natural plant diversity. Approval for field testing required extensive review. In 1993 the USDA modified the restrictions for most genetically engineered plants. Researchers must inform the USDA that field tests will be conducted. An in-depth review is conducted only if a product has been altered in such a way as to be completely different from a naturally occurring plant, contains an unusual additive, or poses a potential health or environmental risk. The FDA has adopted a similar policy.

Approval for Release and Marketing

In the process of making and marketing a GMO today, there are many stages of review. The product must pass through all stages of review before it can be approved for the commercial market.

1. The NIH authorizes a review of any experiments that involve recombinant DNA by the IBC.

2. The USDA sometimes requires that the experiments involving recombinant DNA be reviewed.

3. The USDA requires that a field release permit be issued before a GMO is released from the laboratory into the field (environment).

4. The USDA requires that a shipping authorization be obtained before any GMO can be shipped between states.

5. The USDA requires a permit to commercialize the GMO or a formal decision to declare nonregulatory status of a GMO. These two processes require a specified period in which to consider public comment.

6. The EPA may require an experimental use permit if more than 10 acres will be used for the GMO.

7. The EPA may set limits on the amount of the product that can be expressed in the GMO. This also requires that an opportunity be provided for public comment.

8. The EPA requires product registration before commercialization. A time period is established for public comment.

9. The FDA requires a review and public comment before a GMO can be marketed as a food or included in a food or drug.

RISK ASSESSMENT

Risk assessment is the evaluation of a potential hazard such as flying in an airplane or riding a bicycle in a city. Science-based risk (for example, using radioisotopes in experiments in the laboratory or eating genetically engineered foods) assessment is the evaluation and interpretation of scientific data in which a particular hazard is identified and the risk is then quantified. The potential consequences associated with the hazard are examined.

The public and environmental scientists alike have voiced numerous fears—for example, that genetically modifying plants and animals will turn benign organisms into economically destructive pests or will enhance existing pests through hybridization (sexual outcrossing common in plants); that new biotechnology products (plant pesticides, for example) may cause harm to nontarget species such as insects; that a recombinant virus used in the production of recombinant organisms might detrimentally infect other organisms; and that recombinant organisms may eliminate indigenous species through competition, depletion of valuable resources and nutrients, or by incomplete degradation of toxic wastes to even more toxic by-products during bioremediation.

Those assessing risk must answer many questions. For example, when a transgenic organism is released into the environment, what is the probability of gene transfer, and how can the movement of genes be monitored? What information is needed to assess risk adequately? How much baseline data is required? Risk assessment must take into account numerous factors, both intrinsic and extrinsic. Intrinsic factors include the genetic modification, the molecular biology, natural history, and genetic background of the organism to be released. Characteristics that can be expressed include the conversion of a nonpathogen to a pathogen, the ability to resist host defense systems, and the ability to invade cells, increase host range, and survive adverse conditions. Extrinsic factors are the characteristics of the immediate environment such as selection pressure, habitat quality, nontarget organisms that might become recipients, and the density of organisms that directly or indirectly affect the organisms to be released. Introductions can be modified forms of resident organisms, forms that exist elsewhere, or resident organisms disabled in such a way as to require a supplement for survival.

The safety of our food and drugs, the environment, and agricultural production is evaluated by the FDA, EPA, and USDA. These regulatory agencies use scientific data to make regulatory decisions. New data is now required to assess the safety of genetically engineered organisms and their products. Since 1992 the USDA has funded research in areas that the FDA, EPA, and USDA consider very important to regulatory review and risk assessment. This research is administered through the Biotechnology Risk Assessment Research Grants Program (BRARGP). Through this program, scientific data is accumulating on the science of risk assessment, with a specific focus on GMOs.

HOW MUCH RISK?

Risk is defined as an estimate of the probability that an event will have an adverse effect and an estimate of the magnitude of that effect. In other words, what is the probability that something bad will happen and, if it does happen, how serious are the consequences?

Those charged with developing regulatory plans to ensure safety must balance the degree of oversight, or management, with the level of risk. Comprehensive risk analysis thus requires both risk assessment and risk management. Risk assessment uses scientific data to identify and estimate potential adverse effects; risk management is the process of weighing alternatives to select the most efficient and effective regulatory plan. Technical, economic, political, and social factors affect the manner in which assessment and management are carried out by regulatory agencies under legislative mandates. When weighing potential risks versus benefits versus regulatory costs, decision makers must make value judgments.

Risk can be broken down into several components. The first, *quantitative risk assessment*, has already been described; it is an empirical, or experimental, concept that provides an estimate of the probability that something negative will happen. *Acceptable level of risk* states a trade-off between risk and benefit. Do the costs, or risks, outweigh the benefits to society? The risk in question is weighed against both the benefit and the cost of eliminating the risk (for example, either by increasing safeguards or by forgoing the benefits—that is, not adopting the technology in question). The *cultural context of risk* encompasses the psychological and social aspects of risk. How do members of a community or society perceive the level of risk? How do they perceive the benefit? Where do they set the balance between the

two? Actual risk (that is, the experts' probabilistic risk assessment) does not usually parallel perceived risk. *Risk communication* is an essential component; the public perception of risk is influenced by the way that risk assessment is communicated by the media in a variety of contexts.

All of the components of risk just enumerated influence policy makers. The public is still grappling with the question of what constitutes an acceptable level of risk. How much risk will be tolerated, and more important, how much are we willing to pay to make a genetically engineered organism safe (zero risk can never be achieved)?

For an engineered organism to pose a risk to the environment (such as by displacing indigenous organisms) or to human and animal health (such as by the spread of a pathogen), it must be able to survive, reproduce, and disseminate beyond the area where it was released and intended to function or it must be able to transfer genetic material to other organisms. Humans, plants, animals, and ecological systems may be put at risk of injury and environmental displacement or disruption.

Risks to the environment also may arise from market forces. Suppose, for example, that a single company produced an herbicide and engineered herbicide-resistant crop plants that must be used with it. If farmers became completely dependent on the combination, the result might be the increased use of agricultural chemicals (pesticides and herbicides) or decreased genetic diversity among organisms.

PATENTS AND BIOTECHNOLOGY

The idea behind patents is to provide a reward (that is, money) for the development of new inventions. This is thought to promote the advancement of technology and at the same time protect the inventor. A patent is an exclusive government-granted right to the inventor that prevents others from using or making the invention without license to do so for a specific period of time. In exchange for the right to monopolize the invention, the inventor must disclose the details of the invention. The patent is basically a property (referred to as intellectual property).

The ethics of patenting living organisms and their genes have been hotly debated for several years. Proponents of such patents have argued that in the absence of patents, which demonstrate legal ownership of a product, there might not be incentive to invest large amounts of money for research and development. For a biotechnological development to make it into the marketplace, it takes 10–12 years and approximately $800 million. Some have also argued that the incentive of such patents may accelerate progress in biotechnology. The commercialization of biotechnology has benefited from the potential for patenting products and organisms resulting from recombinant DNA technology; companies are willing to take greater risks and invest more funds in research and development if they stand to benefit by receiving more profits on legally protected new products.

In addition to patents, legal protection of intellectual property rights can take the form of trade secrets (to protect technical details and formulas from unauthorized disclosure and use), trademarks (names for products) and service marks, trade dress (ornaments, shapes, esthetic objects), or copyrights. Patents are the most valuable type of intellectual property protection; by giving the owner exclusive rights to market a product or invention and thereby potentially earn substantial profits, they encourage technological innovation, investment, and the development of beneficial new products. In addition, any products that are derived from the original product also are protected. A patent is meant to exclude others from using, producing, or selling the legally protected product or invention without purchasing the rights to do so. In the United States, patent protection begins on the date the patent is granted and ends 20 years from the date the application was filed, after which others can use the invention freely.

Patent Process

In the United States, the patentability of inventions is governed by the U.S. Patent and Trademark Office (PTO) in the Department of Commerce. The role of the PTO is to promote the progress of science and technology by granting inventors an exclusive right to their inventions for a discrete period of time. They receive more than 300,000 patent applications each year.

Biotechnology innovations fall within all categories of patentable inventions:

1. Products or composition of matter (for example, new pharmaceutical, biopolymer)

2. Methods of use or process (utility patents) (for example, a new cancer treatment, a new method to express a recombinant protein)

3. A machine (for example, microarray detection equipment)

4. An article of manufacture (for example, DNA chip)

5. A useful improvement to an existing invention that falls within the first four categories

Several conditions must be met for patentability. The invention must be new (that is, not previously published or presented anywhere), useful (have a purpose), and nonobvious to one skilled in the field. In addition, the invention must be described in enough detail to allow someone skilled in the field to use it as stated. This is referred to as the *enablement criterion*.

A patent application must include several elements:

- The technical field to which the invention applies must be described.

- A background search must identify the problems to be solved and describe the prior art (including information that is publicly available); this search aids in determining the scope of the patent claim for the invention.

- How the invention improves upon the prior art must be described.

- A summary must enumerate the fundamental components of the invention.

- A description of the invention and the indispensable steps for constructing the invention must be sufficiently detailed for someone skilled in the field to reconstruct it.

- Finally, the application must include claims that outline the elements protected by law. (The claim must be clearly stated, must define the aspects of the invention to be protected, and must describe the scope, prior art, and infringement. A claim cannot be so broad that it infringes on prior art. On the other hand, if the claim is too narrowly defined, the applicant may risk losing property claims; an attorney familiar with both patent law and science can be invaluable in helping to define a claim.)

The PTO has 18 months after the filing date in which to issue a patent, similar to what occurs in Europe (the U.S. Congress in 1999 amended the pertinent statute to change the period from 3 years to 18 months). Patents cover an invention for 20 years from the date of filing, although Congress is now providing patent-term extensions beyond the 20 years from the filing date to compensate for delays caused by problems occurring during the patent process (delays, appeals, inefficiency). The actual patent term is actually less than 20 years because patent protection does not begin until the patent process has completed and the patent is granted.

In the United States, the "first to invent" rule is in effect. That is, whoever made the invention first, and can demonstrate this, is awarded the patent for 20 years. The inventor has a 1-year grace period to file after the invention is published. All other countries, with the exception of the Philippines, have a "first inventor to file" rule. Thus, in the United States information about the invention or patentable discovery cannot be disclosed in any manner—be it a scientific meeting or publication—more than one year before the application is submitted. In Europe, the application must be filed before any information can be disclosed by publication or other means.

Many biotechnology patents are granted as provisional patents. That is, a person or company that files the provisional patent application has up to 1 year to file the actual patent claim. In the provisional patent, there must be a written description of the invention and the names of the inventors. The provisional 1-year period is not included in the 20 years of the patent life.

In June 2002 the PTO released a 5-year strategic plan to increase patent quality. To meet this goal, the PTO plans to implement electronic filing and revise the filing fee system. Fees will be increased for longer and more complex patent applications. The PTO and biotechnology companies are working together to develop a fair fee structure.

HISTORY OF PATENTING LIVING ORGANISMS

In 1930 the U.S. Congress passed the Plant Patent Act, which allows asexually propagated plant varieties to be patented. Plant varieties produced by cuttings, budding, and grafting could be patented (but not seeds). More than 6500 plant patents have been granted—primarily for rose and fruit trees. With the Plant Variety Protection Act of 1970, Congress extended protection to sexually propagated plants (excluding first-generation hybrids). Plant breeders now could have exclusive rights over the propagation and sale of sexually propagated plant varieties. Important crops produced by the classical breeding methods of crossing, screening for desirable characteristics, and selection through seeds could be patented. Utility patents are granted for genetically engineered plants.

Patents for living organisms had been filed unsuccessfully until 1980, when after a landmark Supreme Court case, the PTO determined that life forms were patentable. In *Diamond v. Chakrabarty,* the Court ruled that there is no distinction between living and nonliving matter and that living matter is patentable. Ananda Chakrabarty had filed an application for a patent on a genetically modified bacterium from the genus *Pseudomonas* that could break down crude oil (see Chapter 5, Microbial Biotechnology). The bacterium harbored at least two plasmids that provided the genes for hydrocarbon degradation. In his application, Chakrabarty sought to patent the bacterium, the process of producing its characteristics, and the method of dispersal. The PTO rejected the organism claim, stating that the genetic manipulations that resulted in the oil-degrading product did not constitute a product of manufacture or a new composition of matter. The examiner operated on the supposition that living things are not patentable subject matter and that microorganisms are products of nature.

Because the inventor claimed the bacterium, the Supreme Court had to decide whether a living organism that had been modified by a person was patentable subject matter. The Court determined that "anything under the sun that is made by man" is patentable and ruled in favor of the patent, stating that the issue is not whether something is living or nonliving but whether it is a product of human modification or of nature.

In the past, microorganisms could be patented if they were a part of a process and the direct use was identified. Many such patents have been granted in the fermentation and pharmaceutical industries dating back to 1873, when Pasteur was issued a patent for purified yeast. In 1987, the PTO determined that all non-human multicellular living organisms, including animals, can be patented if certain criteria are met ("non-naturally occurring non-human multicellular living organisms including animals"). Shortly thereafter, scientists were granted a patent for a genetically altered triploid Pacific oyster that was produced by hydrostatic pressure, a process that did not use recombinant DNA methods. In 1988 the first genetically engineered animal was patented (both the process and product): the Harvard mouse (also known as the oncomouse), which had a breast cancer susceptibility gene that made the animal extremely sensitive to carcinogens. This mouse was to serve as a cancer model on which to test new anticancer drugs and to determine the role of environmental factors in cancer development. The patent covers all animals that have a variety of cancer genes inserted into the genome. The corresponding European patent was issued in 1992. In Europe, the road to patenting living organisms has not been as smooth; the European Patent Office has not readily granted authority to patent genetically modified plants or animals. Of the more than 300 patent requests, only a few have been granted patents.

The patenting of animals has generated more heated debate than perhaps any other area of patent law. Again, proponents argue that animal patents stimulate research and development by protecting investment during the development of new technologies. They argue that genetically engineered animals will be more resistant to disease and will go to market faster, thus reducing farmers' costs.

Opponents of animal patenting argue that private ownership of animals (that is, monopoly ownership) should not be allowed; that genetic manipulation may be cruel and cause undue harm to animals, may increase the use of animals in research and commercial endeavors, and may even decrease the genetic diversity of commercial animals. There have been claims that family farmers cannot afford to pay for the use of patented animals, and that the higher volume of animal production will drive prices lower and lead to the

DNA Patents

Because raw, unmodified products of nature cannot be patented, DNA usually is patentable after it has been isolated and modified in some way. The patentability of pieces of DNA with unknown function is highly controversial, as has been the patenting of human DNA without the demonstration of utility. At what point should a patent application be filed in the process of product development from the point of gene discovery?

The PTO has issued a few patents for gene fragments where functions are unknown. On pending applications of such DNA sequences, utility has been identified in vague terms such as its use as a probe to find a gene, or an expressed-sequence tag (EST) needed to map a chromosome. A broad patent may be granted for the nucleotide sequence of a protein-encoding gene, vectors, the transgenic organisms transformed with the gene, and possibly the protein product, provided it is new. Processes also are patentable. Some scientists have urged the PTO not to grant broad patents on DNA that has not been well characterized and that does not have an identified function or use.

In January 2001 the PTO issued stricter guidelines that require that product function (that is, usefulness) must be demonstrated before gene fragments can be patented. This new ruling requires "specific and substantial utility that is credible." Even with this change, some scientists feel

that patents should not be granted for discoveries such as DNA fragments (for example, ESTs). The isolation of short DNAs, only 300–500 base pairs long, takes much less work than a complete gene sequence for which both a function has been identified and a commercial application has been developed. If patents are granted for genome fragments, the cost of using patented portions of the genome will make research very costly. A scientist must spend time researching the various genome patents and determining which licenses (a legal contract between the owner of a patent and another individual or company who wants to use, make, or sell the patented invention) he or she will have to obtain. If most of the genome is patented, some fear that progress will be greatly impeded.

Proponents of DNA patenting have stated that patents reward scientists for their work and monetary gains will provide needed funds for their research; that all scientists will have access to the new discoveries and this will, ultimately, increase progress; and that duplication of effort will be prevented if sequence information is patented. Opponents of DNA patenting believe

1. That patents will reward those who make routine discoveries of uncharacterized cDNA sequences or partial sequences. Those who undertake complex functional studies will be penalized.

2. Patents may slow progress in diagnostics, therapeutics, and

other products because it might become too costly to use patented data.

3. Patent stacking allows one specific sequence to be patented in different ways, such as a gene, a single-nucleotide polymorphisms (SNP), a cDNA, and an EST, thereby making licensing very expensive. Consumers are likely to pay for these high development costs when purchasing a product developed from this sequence.

4. Because patent applications remain secret until they are granted, a company might invest a large amount of money in research and development only to discover that others have been granted patents on sequences and processes that were used. This will add unexpected costs to their work, as well as possible penalties for infringement.

5. If patents have been granted for small DNA fragments, a scientist who has sequenced a complete gene that encompasses the patented smaller fragments may have to obtain licenses to use the small fragments.

The PTO has issued thousands of patents for gene sequences encoding proteins. By the end of 2001, more than 6500 patents were issued, of which more than 1300 were for human genes. To date, more than 20,000 patent applications have been submitted for genes, DNA fragments, SNPs, and ESTs.

eventual demise of the small family farm. Farmers often pay a heavy price economically when surplus agricultural production drives prices down (although proponents have countered that the cost of production will be lower).

As a cautionary example, opponents of patents on living organisms cite the commercialization of hybrid crops that resulted from patent protection of seeds. Large, multinational companies invested heavily in plant breeding and the development of new varieties with desirable traits. Hybrid crop plants were patented and commercialized. Companies concentrated on a few high yield varieties and gained control of world markets. Farmers worldwide have become dependent on a small variety of high yield crop plants that require costly inputs of fertilizers and pesticides. One concern is that patents might encourage the development of too many pesticides and herbicides that are linked to the patented crops produced by the same companies (however, proponents point out that less toxic herbicides and perhaps reduced amounts of herbicides may be positive outcomes). Thus, chemical usage may be perpetuated by the patent system and the farmer might becomes even more dependent on the use of agricultural chemicals to maintain high crop productivity.

By embracing high-technology crops, farmers in the developing world often discard ancient crops that harbor resistance to disease and pests and have adapted to harsh (for example, nutrient-poor) environ-ments. The world increasingly is relying on a small number of food crops that have lost much genetic diversity. Whether patents are the impetus is difficult to determine, but some critics warn that we should proceed with caution. The PTO is not required to examine the environmental or social implications of a patent application, nor has Congress limited patents because a patented item might have adverse consequences. The public ultimately will be charged with determining which patented products are of social, medical, environmental, and agricultural benefit.

WHAT ABOUT THE FUTURE?

We are witnessing the beginning of a new era of patented products that will significantly affect our lives. We must determine how to proceed with new knowledge and technologies and how to use them within the confines of the law and after reflection on the moral and ethical consequences. If they are used wisely we can at the very least expect an increase in the quality of food, pharmaceuticals, and chemicals that can be used in a multitude of ways to enhance the quality of life; boost progress in human medicine and environmental cleanup; and increase agricultural productivity. The legal, ethical, economic, social, and global implications of modern biotechnology must be thoroughly examined if we are to ensure that biotechnology products contribute in positive ways.

General Reading

P. Barker, ed. 1995. *Genetics and Society.* The H.W. Wilson Company, New York.

L. Busch, W.B. Lacy, J. Burkhardt, and L.R. Lacy. 1992. *Plants, Power, and Profit: Social, Economic, and Ethical Consequences of the New Biotechnologies.* Blackwell Scientific, Cambridge, Massachusetts.

R.M. Cook-Degan. 1994. *The Gene Wars: Science, Politics, and the Human Genome.* W.W. Norton, New York.

R.S. Crespi. 1988. *Patents: A Basic Guide to Patenting in Biotechnology.* Cambridge University Press, Cambridge.

B.D. Davis, ed. 1991. *The Genetic Revolution: Scientific Prospects and Public Perceptions.* The Johns Hopkins University Press, Baltimore, Maryland.

K.A. Drlica. 1994. *Double-Edged Sword: The Promises and Risks of the Genetic Revolution.* Addison-Wesley, Reading, Massachusetts.

L.R. Ginzburg. 1991. *Assessing Ecological Risks of Biotechnology.* Butterworth-Heinemann, Boston, Massachusetts.

W.H. Lesser, ed. 1989. *Animal Patents: The Legal, Economic, and Social Issues.* Stockton Press, New York.

M.A. Levin and H.S. Strauss. 1991. Risk assessment in genetic engineering. McGraw-Hill, Inc., New York.

S.P. Ludwig. 2001. *Biotechnology Law: Biotechnology Patents and Business Strategies in the New Millennium.* Intellectual Property Course Handbook Series, No. 666. Practicing Law Institute, New York.

N. Miller, ed. 2002. *Environmental Politics Casebook: Genetically Modified Foods.* Lewis Publishers, Boca Raton, Florida.

National Research Council. 1989. *Field Testing Genetically Modified Organisms: Framework for Decisions.* National Academy Press, Washington, D.C.

G.T. Tzotzos, ed. 1995. *Genetically Modified Organisms: A Guide to Biosafety.* CAB International, UK.

S. Wright. 1994. Molecular Politics: Developing American and British Regulatory Policy for Genetic Engineering, 1972–1982. The University of Chicago Press, Chicago.

Additional Readings

■

R.G. Adler. 1984. Biotechnology as an intellectual property. *Science* 224:357–363.

J.H. Barton. 1991. Patenting life. *Sci. Am.* 264(3):40–46.

P. Berg, D. Baltimore, S. Brenner, R.O. Roblin III, and M.F. Singer. 1975. Asilomar conference on recombinant DNA molecules. *Science* 188:991–994.

P. Berg and M. Singer. 1995. The recombinant DNA controversy: Twenty years later. *Bio/Technology* 13:1132–1134.

D.B. Berkowitz. 1990. The food safety of transgenic animals. *Bio/Technology* 8:819–825.

A.H. Berks. 1994. Patent information in biotechnology. *Trends Biotechnol.* 12:352–364.

F.S. Betz, B.G. Hammond, and R.L. Fuchs. 2000. Safety and advantages of *Bacillus thuringiensis*–protected plants to control insect pests. *Regul. Toxicol. Pharmacol.* 32:156–173.

G. Bogosian and J.F. Kane. 1991. Fate of recombinant *Escherichia coli* K-12 strains in the environment. *Adv. Appl. Microbiol.* 36:87–131.

Canadian Biotechnology Advisory Committee. 2001. Biotechnological intellectual property and the patenting of higher life forms: consultation document 2001. Government Document No. C21-32/1-2001, Ottawa, Canada.

I. Carmen. 1992. Debates, divisions, and decisions: recombinant DNA Advisory Committee (RAC) authorization of the first human gene transfer experiments. *Am J. Hum. Genet.* 50:245–260.

K.G. Chahine. 1997. Patenting DNA: Just when you thought it was safe. *Nat. Biotechnol.* 15:586–587.

T. Cook. 2002. *A User's Guide to Patents.* Butterworths, London.

M.J. Crawley, S.L. Brown, R.S. Hails, D.D. Kohn, and M. Rees. 2001. Transgenic crops in natural habitats. *Nature* 409:682–683.

R.S. Crespi. 1997. Biotechnology patents and morality. *Trends Biotechnol.* 15:123–129.

R.S. Crespi. 1998. Patenting for the research scientist: Bridging the cultural divide. *Trends Biotechnol.* 16:450–455.

R.S. Crespi. 1999. The biotechnology patent directive is approved at last! *Trends Biotechnol.* 17:139–142.

P. Dale. 1999. Public reactions and scientific responses to transgenic crops. *Curr. Opin. Biotechnol.* 10:203–208.

W. Derek. 2001. European patents for biotechnological inventions: Past, present, and future. World Patent Information 23:339–348.

D.J. Drahos. 1991. Field testing genetically engineered microorganisms. *Biotechnol. Adv.* 9:157–171.

C. Eckenswiller and J. Morrow. 1996. Why patent life forms? *Policy Options* 17:11–15.

J. Enriquez. 2001. Green biotechnology and European competitiveness. *Trends Biotechnol.* 19:135–139.

P. Evans. 2001. Impact of U.S. PTO guidelines on gene patents. *Genet. Eng. News* 21:6 & 44.

G.J. Flattman and J.M. Kaplan. 2001. Patenting expressed sequence tags and single nucleotide polymorphisms. *Nat. Biotechnol.* 19:683–684.

J. Giles. 2003. Time to choose. *Nature* 425:655–659.

R.E. Gold. 2002. Biotechnology patents: Strategies for meeting economic and ethical concerns. *Nature Genetics* 30:359.

R.E. Gold, D. Castle, M.L. Cloutier, A.S. Daar, and P.J. Smith. 2002. Needed: Models of biotechnology intellectual property. *Trends Biotechnol.* 20:327–329.

D. Gurian-Sherman. 2003. Risks of genetically engineered crops. *Science* 301:1845–1846.

E. Johnson. 1996. A benchside guide to patents and patenting. *Nat. Biotechnol.* 14:288–291.

P.D. Kelly. 1991. Are isolated genes "useful"? *Bio/Technology* 10:52–55.

L. Kim. 1993. *Advanced Engineered Pesticides.* Section III, Regulatory, pp. 321–419. Marcel Dekker, New York.

T. Koppal. 2002. Inside the FDA. *Drug Discovery* 5:32–38.

E. Marshall. 1997. Companies rush to patent DNA. *Science* 275:780–781.

D.H. Mitten, R. MacDonald, and D. Klonus. 1999. Regulation of goods derived from genetically engineered crops. *Curr. Opin. Biotechnol.* 10:298–302.

National Academy of Sciences. 2000. *Genetically Modified Pest-Protected Plants: Science and Regulation.* National Research Council. National Academy Press, Washington, D.C.

G. Poste. 1995. The case for genomic patenting. *Nature* 378:534–536.

M. Ratner. 1990. Survey and opinions: Barriers to field-testing genetically modified organisms. *Bio/Technology* 8:196–198.

D. Resnik. 2003. Patents on human-animal chimeras and threats to dignity. *Amer. J. Bioethics* 3:35–36.

A.M. Shelton and M.K. Sears. 2001. The monarch butterfly controversy: Scientific interpretations of a phenomenon. *Plant J.* 27:483–488.

J.M. Tiedje, R.K. Colwell, Y.L. Grossman, R.E. Hodson, R.E. Lenski, R.N. Mack, and P.J. Regal. 1989. The planned introduction of genetically engineered organisms: Ecological considerations and recommendations. *Ecology* 70:298–315.

S.G. Uzogara. 2000. The impact of genetic modification of human foods in the 21st century. A review. *Biotechnol. Adv.* 18:179–206.

A.R. Williamson. 2001. Gene patents: socially acceptable monopolies or an unnecessary hindrance to research? *Trends Genet.* 17:670–673.

M. Wilson and S.E. Lindow. 1993. Release of recombinant organisms. *Annu. Rev. Microbiol.* 47:913–944.

L.L. Wolfenbarger and P.R. Phifer. 2000. The ecological risks and benefits of genetically engineered plants. *Science* 290:2088–2093.

M.D. Yolansky and W.J. Holmes. 1995. Patenting DNA sequences. *Bio/Technology* 13:656–657.

Readings for Thought and Discussion

The following section includes four readings on the subject of biotechnology and genetic engineering. A short introduction to each selection is provided to describe the topic and major themes of the reading, and to mention something about its author. Focus questions are provided to help stimulate thinking about the reading in an active sense.

INTRODUCTION

Technology and Science

IN THE MODERN WORLD, technology and science often go together, with science supporting technology and technology supporting science. Although they now share a great deal, their goals have been historically quite different. Traditionally, science was viewed as an elevated activity involving "pure contemplation" and the "value-free" pursuit of knowledge, whereas technology was associated with more practical concerns and with the arts. It was not until the beginning of the modern period in the seventeenth century that there was a decisive shift, when the view that scientific knowledge was valuable because it was useful in gaining mastery over nature became prominent.

This shift was largely due to the writings of several influential philosophers, such as René Descartes and

From *Society, Ethics, and Technology, Third Edition* by Winston and Edelbach. © 2006 Thomson Higher Education.

Francis Bacon. Bacon's works, particularly *Novum Organon* (1620) and *New Atlantis* (1624), are notable for their contempt for traditional speculative philosophy and their emphasis on the importance of empirical methods of investigation through which the secrets of nature could be revealed by means of judicious experiments. In 1637, Descartes wrote the *Discourse on Method,* in which he proclaimed that "it is possible to attain knowledge which is very useful in life; and that, instead of speculative philosophy which is taught in the Schools, we may find a practical philosophy by means of which, knowing the force and the action of fire, water, air, stars, the heavens, and all other bodies that environ us, as distinctly as we know the different crafts of our artisans, we can in the same way employ them in all those uses to which they are adapted, and thus render ourselves masters and possessors of nature."[1] This change in the dominant view of the nature and bases of human knowledge set the stage for the modern belief in progress, which was expressed by Bacon as the belief that "the improvement of man's mind and the improvement of his lot are one and the same thing."[2]

Despite the merger of science and technology in the modern period, there remain some significant differences between the two enterprises. Technologists primarily seek to answer the question, "How?" "How can we keep warm in the winter?" or "How can we see distant objects that are invisible to the naked eye?" Engineers seek to design and produce useful material objects and systems that will function under all expected circumstances for the planned lifetime of the product. Science, on the other hand, may be considered a form of systematic inquiry, seeking to understand the underlying laws governing the behavior of natural objects.

Scientists primarily try to answer the questions "What?" and "Why?" as in "What kind of thing is this?" and "Why does it behave the way it does?" In the early stages of a science, when little was known, the immediate goal of the science was to describe and classify the phenomena of the natural world. As more things became known, the sciences began asking, "How do these things change over time and interact with each other?" Scientists sought laws and principles that would enable them to predict and explain why things in nature behave as they do.

Science and technology have a symbiotic relationship, each one helping the other. Technology needs science to predict how its objects and systems will function so that it can tell if they will work. Science supplies predictive laws that apply to these objects and systems, a perfect match. However, although the laws of science are often simple, applying them to the complex objects of technology is often anything but simple. Sometimes the engineer must experiment with the complex objects that are the building blocks of a technology to find out what will happen. Perhaps the most important contribution that science makes to technology is in the education of engineers, supplying the conceptual framework upon which they build a body of more specific and practical knowledge.

Technology makes direct and obvious contributions to the progress of science. The laboratory equipment that the scientist uses is the product of technology. The biologist would discover little without a microscope, and the particle physicist even less without an accelerator. This direction of contribution is maintained throughout most of science. At the frontiers of science, however, the scientist may need equipment that has never been built before based on principles never used before in applied contexts. The design of the Tokamak (magnetic confinement torus) to study controlled nuclear fusion in hydrogen plasmas is without a doubt the product of technology, but it is technology that can only be done by the scientists who understand and can predict the behavior of high-temperature plasmas. Still, they do not do it alone; much of the design uses established technology, and this is the domain of the engineer.

But in recent years, the lines between the role of the scientist and that of the engineer or technologist have become increasingly blurred. Much of the current research agenda is dictated by the possible practical applications of new scientific knowledge. This merging of science and technology has led some writers, such as Bruno Latour, to speak of contemporary research as *technoscience*.[3] The complex relationship that presently exists between science and technology can be exemplified in the contemporary field of biotechnology.

Genetics is the science that studies inherited characteristics. Genetic engineering, by contrast, is the application of the knowledge obtained from genetic investigations to the solution of such problems as infertility, treatments for diseases, food production, waste disposal, or the improvement of crop species. Among genetic-engineering techniques are procedures that can alter the reproductive and hereditary processes of organisms. Depending on the problem, the procedures used in genetic engineering may involve artificial insemination, cloning, in vitro fertilization, species hybridization, or molecular genetics. Recent discoveries in molecular genetics have permitted the direct manipulation of the genetic material itself by the recombinant-DNA technique. Is this a scientific discovery or a technological one? Both know-that and know-how are essential elements of modern biotechnology, and both are a part of the training of every working geneticist and genetic-engineer.

The genome of an organism is the totality of genes making up its hereditary constitution. The Human Genome Project (HGP) was designed to map the human genome, and in 1990, the U.S. National Institutes of Health and the Department of Energy created the National Center for Human Genome Research. The goal is to determine the exact location of all the genes—about 35,000 of them, plus regulatory elements—on their respective chromosomes, as well as to establish the sequence of nucleotides—estimated to be about 6 million pairs—of all human genes. The HGP has been completed ahead of schedule at an estimated cost of $3 billion. It is highly doubtful that such a "big science" project would have been funded were it not for the expectation that it would result in new medical technologies for the treatment of diseases that currently afflict many people. Recently, patents have been granted for the production of erythropoietin, a hormone that stimulates blood-cell formation; tissue-plasminogen activator, an anticlotting agent used to treat heart attacks; and alpha-interferon, which has been found effective in treating hepatitis C.

Biotechnology also has nonmedical commercial applications. In 1980, a ruling by the U.S. Supreme Court permitted the U.S. Patent and Trademark Office to grant a patent on a genetically engineered oil-eating bacterium. The bacterium was categorized as a "nonnatural man-made microorganism." Over the following eight years, some 200 patents were granted for bacteria, viruses, and plants that had been genetically modified. In 1988, a patent was granted on a mouse strain in which the cells had been engineered to contain a cancer-predisposing gene sequence. Technically referred to as a "transgenic nonhuman eukaryotic animal," each mouse of this type can be used to test low doses of cancer-causing substances and to test the effectiveness of drugs considered as possibly offering protection against the development of cancer. In the field of agriculture, a number of plants with genetically engineered traits have been patented, including maize (corn) plants rich in the amino acid tryptophan, cotton plants resistant to weed-killing herbicides, tobacco plants resistant to various insects, and potato plants resistant to various viruses.

The trial of O. J. Simpson has brought to light another application of DNA technology—its use in identifying the assailants in violent crimes when the victims are no longer able to do so themselves. Each person, with the exception of identical twins, has a unique DNA fingerprint that can be detected by matching patterns of restriction enzyme DNA fragments. Using this technique, it is possible to extract samples of DNA from dried blood, hairs, or semen and produce a near certain match, with the DNA pattern obtained from cells of the accused. But the same technology also raises issues about personal privacy and genetic discrimination.

Each new application of biological knowledge to human reproduction has raised profound and troubling ethical questions. Already it is possible for children to be produced artificially by the use of in vitro fertilization (IVF) or gamete interfallopian transfer (GIFT). It is now possible for a child to have different genetic, gestational, and social parents, greatly complicating questions of custody. While it is still necessary to employ women's wombs in the gestational process, it may not be long before we have developed artificial wombs or incubators that will enable us to gestate mammals, such as ourselves, in an extra-uterine environment.

We already have the ability to create plants and animals in the laboratory by splicing together strands of DNA taken from different organisms. How long will it be before we start applying these same techniques to ourselves? Germ-line genetic therapies offer the promise of eliminating many dreaded genetic diseases from the human gene pool, but they also raise the specter of our playing God by redesigning the human species to suit our dominant social values. Do we really want to live in a world where prospective parents can choose whether they will have a boy or a girl and choose other characteristics such as height, body type, hair and eye color, or perhaps even intelligence and beauty? Do we really want to know what genetic diseases we are harboring? What will become of the notion of human dignity if we begin to clone ourselves, as if a person were just another artifact built according to our specifications?

Notes

1. René Descartes, "Discourse on the Method of Rightly Conducting the Reason and Seeking for Truth in the Sciences" (1637). In *The Philosophical Works of Descartes*, Vol. I, translated by Elizabeth S. Haldane and G. R. T. Ross (Cambridge, England: Cambridge University Press, 1970), p. 119.

2. Francis Bacon, "Thoughts and Conclusions." In Benjamin Farrington (ed.), *The Philosophy of Francis Bacon* (Chicago: University of Chicago Press, 1964), p. 93.

3. See Bruno Latour, *Science in Action* (Cambridge, MA: Harvard University Press, 1987).

BIOTECHNOLOGY AND GENETIC ENGINEERING

R-1 *A Glimpse of Things to Come*

Lee M. Silver

Princeton University molecular biologist Lee Silver opened his 1997 book *Remaking Eden: Cloning and Beyond in a Brave New World* with this provocative discussion of what human reproduction might look like in 2010, 2050, and 2350. He predicts that reproductive genetic engineering technology will make it possible for lesbian couples to have children that are genetically related to both "parents," that genetic resistance to diseases such as AIDS will be woven into an embryo's DNA, that human cloning will become widely accepted, and that, in the further future, the human race will divide into two classes: the Naturals and the Gene-enriched, or GenRich. The Naturals and the GenRich will grow up in separate worlds, and eventually, the genetic gulf between them will become so wide that they will become separate species. But Silver's vision differs from the dystopia imagined by Aldous Huxley in his 1932 novel *Brave New World*. In Silver's vision, it is not a totalitarian government that controls people's personal reproductive decisions; rather, it is the free market. Affluent parents will choose to give their children genetic advantages that science and biotechnology make available, just as they now choose to pay for private schools and tutors to prepare them to get into the elite colleges. Such a "free-market eugenics" is consistent with the American values of individual liberty and free markets, and if parentally selected genetic enhancement does not cause harm to the child being enhanced, what reason can there be not to embrace this libertarian version of a brave new world?

Focus Questions

1. What arguments does Silver give for thinking that human genetic enhancement should be regarded as morally permissible? What are the arguments used by opponents of genetic enhancement?

2. Compare Silver's position with that in the reading by Leon Kass (selection R-2). How does Kass attempt to argue against Silver's position regarding human reproductive cloning? Which of these arguments do you find more persuasive?

Keywords

cloning, disease resistance, genetic enhancement, individual freedom, reproductive technology, reprogenetics

DATELINE BOSTON: JUNE 1, 2010

SOMETIME IN THE NOT-SO-DISTANT FUTURE, you may visit the maternity ward at a major university hospital to see the newborn child or grandchild of a close friend. The new mother, let's call her Barbara, seems very much at peace with the world, sitting in a chair quietly nursing her baby, Max. Her labor was—in the parlance of her doctor—"uneventful," and she is looking forward to raising her first child. You decide to make pleasant conversation by asking Barbara whether she knew in advance that her baby was going to be a boy. In your mind, it seems like a perfectly reasonable question since doctors have long given prospective parents the option of learning the sex of their child-to-be many months before the predicted date of birth. But Barbara seems taken aback by the question. "Of course I knew that Max would be a boy," she tells you. "My husband Dan and I chose him from our embryo pool. And when I'm ready to go through this again, I'll choose a girl to be my second child. An older son and a younger daughter—a perfect family."

Now, it's your turn to be taken aback. "You made a conscious choice to have a boy rather than a girl?" you ask.

"Absolutely!" Barbara answers. "And while I was at it, I made sure that Max wouldn't turn out to be fat like my brother Tom or addicted to alcohol like Dan's sister Karen. It's not that I'm personally biased or anything," Barbara continues defensively. "I just wanted to make sure that Max would have the greatest chance for achieving success. Being overweight or alcoholic would clearly be a handicap."

You look down in wonderment at the little baby boy destined to be moderate in both size and drinking habits.

Max has fallen asleep in Barbara's arms, and she places him gently in his bassinet. He wears a contented smile, which evokes a similar smile from his mother. Barbara feels the urge to stretch her legs and asks whether you'd like to meet some of the new friends she's made during her brief stay at the hospital. You nod, and the two of you walk into the room next door where a thirty-five-year old woman named Cheryl is resting after giving birth to a nine-pound baby girl named Rebecca.

Barbara introduces you to Cheryl as well as a second woman named Madelaine, who stands by the bed holding Cheryl's hand. Little Rebecca is lying under the gaze of both Cheryl and Madelaine. "She really does look like both of her mothers, doesn't she?" Barbara asks you.

Now you're really confused. You glance at Barbara and whisper, "Both mothers?"

Barbara takes you aside to explain. "Yes. You see Cheryl and Madelaine have been living together for eight years. They got married in Hawaii soon after it

became legal there, and like most married couples, they wanted to bring a child into the world with a combination of both of their bloodlines. With the reproductive technologies available today, they were able to fulfill their dreams."

You look across the room at the happy little nuclear family—Cheryl, Madelaine, and baby Rebecca—and wonder how the hospital plans to fill out the birth certificate.

DATELINE SEATTLE: MARCH 15, 2050

You are now forty years older and much wiser to the ways of the modern world. Once again, you journey forth to the maternity ward. This time, it's your own granddaughter Melissa who is in labor. Melissa is determined to experience natural childbirth and has refused all offers of anesthetics or painkillers. But she needs something to lift her spirits so that she can continue on through the waves of pain. "Let me see her pictures again," she implores her husband Curtis as the latest contraction sweeps through her body. Curtis picks the photo album off the table and opens it to face his wife. She looks up at the computer-generated picture of a five-year-old girl with wavy brown hair, hazel eyes, and a round face. Curtis turns the page, and Melissa gazes at an older version of the same child: a smiling sixteen-year-old who is 5 feet, 5 inches tall with a pretty face. Melissa smiles back at the future picture of her yet-to-be-born child and braces for another contraction.

There is something unseen in the picture of their child-to-be that provides even greater comfort to Melissa and Curtis. It is the submicroscopic piece of DNA—an extra gene—that will be present in every cell of her body. This special gene will provide her with life-long resistance to infection by the virus that causes AIDS, a virus that has evolved to be ever more virulent since its explosion across the landscape of humanity seventy years earlier. After years of research by thousands of scientists, no cure for the awful disease has been found, and the only absolute protection comes from the insertion of a resistance gene into the single-cell embryo within twenty-four hours after conception. Ensconced in its chromosomal home, the AIDS resistance gene will be copied over and over again into every one of the trillions of cells that make up the human body, each of which will have its own personal barrier to infection by the AIDS-causing virus HIV. Melissa and Curtis feel lucky indeed to have the financial wherewithal needed to endow all of their children with this protective agent. Other, less well-off American families cannot afford this luxury.

Outside Melissa's room, Jennifer, another expectant mother, is anxiously pacing the hall. She has just arrived at the hospital and her contractions are still far apart. But, unlike Melissa, Jennifer has no need for a computer printout to show her what her child-to-be

From *Remaking Eden: Cloning and Beyond in a Brave New World* by Lee M. Silver, p. 1–11. © 1998 HarperCollins Publishers, Inc.

will look like as a young girl or teenager. She already has thousands of pictures that show her future daughter's likeness, and they're all real, not virtual. For the fetus inside Jennifer is her identical twin sister—her clone—who will be born thirty-six years after she and Jennifer were both conceived within the same single-cell embryo. As Jennifer's daughter grows up, she will constantly behold a glimpse of the future simply by looking at her mother's photo album and her mother.

DATELINE U.S.A.: MAY 15, 2350

It is now three hundred years later and although you are long since gone, a number of your great-great-great-great-great-great-great-great-great-great-great-grandchildren are now alive, mostly unbeknownst to one another. The United States of America still exists, but it is a different place from the one familiar to you. The most striking difference is that the extreme polarization of society that began during the 1980s has now reached its logical conclusion, with all people belonging to one of two classes. The people of one class are referred to as *Naturals*, while those in the second class are called the *Gene-enriched* or simply the *GenRich*.

These new classes of society cut across what used to be traditional racial and ethnic lines. In fact, so much mixing has occurred during the last three hundred years that sharp divisions according to race—black versus white versus Asian—no longer exist. Instead, the American populace has finally become the racial melting pot that earlier leaders had long hoped for. The skin color of Americans comes in all shades from African brown to Scandinavian pink, and traditional Asian facial features are present to a greater or lesser extent in a large percentage of Americans as well.

But while racial differences have mostly disappeared, another difference has emerged that is sharp and easily defined. It is the difference between those who are genetically enhanced and those who are not. The GenRich—who account for 10 percent of the American population—all carry synthetic genes. Genes that were created in the laboratory and did not exist within the human species until twenty-first century reproductive geneticists began to put them there. The GenRich are a modern-day hereditary class of genetic aristocrats.

Some of the synthetic genes carried by present-day members of the GenRich class were already carried by their parents. These genes were transmitted to today's GenRich the old-fashioned way, from parent to child through sperm or egg. But other synthetic genes are new to the present generation. These were placed into GenRich embryos through the application of genetic engineering techniques shortly after conception.

The GenRich class is anything but homogeneous. There are many types of GenRich families, and many subtypes within each type. For example, there are GenRich athletes who can trace their descent back to professional sports players from the twenty-first century. One subtype of GenRich athlete is the GenRich football player, and a sub-subtype is the GenRich running back. Embryo selection techniques have been used to make sure that a GenRich running back has received all of the natural genes that made his unenhanced foundation ancestor excel at the position. But in addition, at each generation beyond the foundation ancestor, sophisticated genetic enhancements have accumulated so that the modern-day GenRich running back can perform in a way not conceivable for any unenhanced Natural. Of course, all professional baseball, football, and basketball players are special GenRich subtypes. After three hundred years of selection and enhancement, these GenRich individuals all have athletic skills that are clearly "nonhuman" in the traditional sense. It would be impossible for any Natural to compete.

Another GenRich type is the GenRich scientist. Many of the synthetic genes carried by the GenRich scientist are the same as those carried by all other members of the GenRich class, including some that enhance a variety of physical and mental attributes, as well as others that provide resistance to all known forms of human disease. But in addition, the present-day GenRich scientist has accumulated a set of particular synthetic genes that work together with his "natural" heritage to produce an enhanced scientific mind. Although the GenRich scientist may appear to be different from the GenRich athlete, both GenRich types have evolved by a similar process. The foundation ancestor for the modern GenRich scientist was a bright twenty-first-century scientist whose children were the first to be selected and enhanced to increase their chances of becoming even brighter scientists who could produce even more brilliant children. There are numerous other GenRich types including GenRich businessmen, GenRich musicians, GenRich artists, and even GenRich intellectual generalists who all evolved in the same way.

Not all present-day GenRich individuals can trace their foundation ancestors back to the twenty-first century, when genetic enhancement was first perfected. During the twenty-second and even the twenty-third centuries, some Natural families garnered the financial wherewithal required to place their children into the GenRich class. But with the passage of time, the genetic distance between Naturals and the GenRich has become greater and greater, and now there is little movement up from the Natural to GenRich class. It seems fair to say that society is on the verge of reaching the final point of complete polarization.

All aspects of the economy, the media, the entertainment industry, and the knowledge industry are controlled by members of the GenRich class. GenRich parents can afford to send their children to private schools rich in the resources required for them to take advantage of their enhanced genetic potential. In contrast, Naturals work as low-paid service providers or as laborers, and their children go to public schools. But twenty-fourth-century public schools have little in common with their predecessors from the twentieth century. Funds for public education have declined steadily since the beginning of the twenty-first century, and now Natural children are only taught the basic skills they need to perform the kinds of tasks they'll encounter in the jobs available to members of their class.

There is still some intermarriage as well as sexual intermingling between a few GenRich individuals and Naturals. But, as one might imagine, GenRich parents put intense pressure on their children not to dilute their expensive genetic endowment in this way. And as time passes, the mixing of the classes will become less and less frequent for reasons of both environment and genetics.

The environmental reason is clear enough: GenRich and Natural children grow up and live in segregated social worlds where there is little chance for contact between them. The genetic reason, however, was unanticipated.

It is obvious to everyone that with each generation of enhancement, the genetic distance separating the GenRich and Naturals is growing larger and larger. But a startling consequence of the expanding genetic distance has just come to light. In a nationwide survey of the few interclass GenRich-Natural couples that could be identified, sociologists have discovered an astounding 90 percent level of infertility. Reproductive geneticists have examined these couples and come to the conclusion that the infertility is caused primarily by an incompatibility between the genetic makeup of each member.

Evolutionary biologists have long observed instances in which otherwise fertile individuals taken from two separate populations prove infertile when mated to each other. And they tell the sociologists and the reproductive geneticists what is going on: the process of species separation between the GenRich and Naturals has already begun. Together, the sociologists, the reproductive geneticists, and the evolutionary biologists are willing to make the following prediction: If the accumulation of genetic knowledge and advances in genetic enhancement technology continue at the present rate, then by the end of the third millennium, the GenRich class and the Natural class will become the GenRich humans and the Natural humans—entirely separate species with no ability to cross-breed, and with as much romantic interest in each other as a current human would have for a chimpanzee.

DATELINE PRINCETON, NEW JERSEY: THE PRESENT

Are these outrageous scenarios the stuff of science fiction? Did they spring from the minds of Hollywood screenwriters hoping to create blockbuster movies without regard to real world constraints? No. The scenarios described under the first two datelines emerge directly from scientific understanding and technologies that are already available today. The scientific framework for the last scenario is based on straightforward extrapolations from our current knowledge base. Furthermore, if biomedical advances continue to occur at the same rate as they do now, the practices described are likely to be feasible long before we reach my conservatively chosen datelines.

It's time to take stock of the current state of science and technology in the fields of reproduction and genetics and to ask, in the broadest terms possible, what the future may hold. Most people are aware of the impact that reproductive technology has already had in the area of fertility treatment. The first "test tube baby"—Louise Brown—is already eighteen years old, and the acronym for in vitro fertilization—IVF—is commonly used by laypeople. The cloning of human beings has become a real possibility as well, although many are still confused about what the technology can and cannot do. Advances in genetic research have also been in the limelight, with the almost weekly identification of new genes implicated in diseases like cystic fibrosis and breast cancer, or personality traits like novelty-seeking and anxiety.

What has yet to catch the attention of the public at large, however, is the incredible power that emerges when current technologies in reproductive biology and genetics are brought together in the form of *reprogenetics*. With reprogenetics, parents can gain complete control over their genetic destiny, with the ability to guide and enhance the characteristics of their children, and their children's children as well. But even as reprogenetics makes dreams come true, like all of the most powerful technologies invented by humankind, it may also generate nightmares of a kind not previously imagined.

Of course, just because a technology becomes feasible does not mean that it will be used. Or does it? Society, acting through government intervention, could outlaw any one or all of the reprogenetic practices that I have described. Isn't the *non*use of nuclear weapons for the purpose of mass destruction over the last half century an example of how governments can control technology?

There are two big differences between the use of nuclear technology and reprogenetic technology. These differences lie in the resources and money needed to practice each. The most crucial resources required to build a nuclear weapon—large reactors and enriched sources of uranium or plutonium—are tightly controlled by the government itself. The resources required to practice reprogenetics—precision medical tools, small laboratory equipment, and simple chemicals—are all available for sale, without restriction, to anyone with the money to pay for them. The cost of developing a nuclear weapon is billions of dollars. In contrast, a reprogenetics clinic could easily be run on the scale of a small business anywhere in the world. Thus, even if restrictions on the use of reprogenetics are imposed in one country or another, those intent on delivering and receiving these services will not be restrained. But on what grounds can we argue that they should be restrained?

In response to this question, many people point to the chilling novel *Brave New World* written by Aldous Huxley in 1931. It is the story of a future worldwide political state that exerts complete control over human reproduction and human nature as well. In this brave new world, the state uses fetal hatcheries to breed each child into a predetermined intellectual class that ranges from alpha at the top to epsilon at the bottom. Individual members of each class are predestined to fit into specific roles in a soulless utopia where marriage and parenthood are prevented and promiscuous sexual activity is strongly encouraged, where universal immunity to diseases has been achieved, and where an all-enveloping state propaganda machine and mood-altering drugs make all content with their positions in life.

While Huxley guessed right about the power we would gain over the process of reproduction, I think he was dead wrong when it came to predicting *who* would use the power and for what purposes. What Huxley failed to understand, or refused to accept, was the driving force behind babymaking. It is individuals and couples who want to reproduce themselves in their own images. It is individuals and couples who want their children to be happy and successful. And it is individuals and couples—like Barbara and Dan and Cheryl and Madelaine and Melissa and Curtis and Jennifer, *not governments*—who will seize control of these new technologies. They will use some to reach otherwise unattainable reproductive goals and others to help their children achieve health, happiness, and success. And it is in pursuit of this last goal that the combined actions of many individuals, operating over many generations, could perhaps give rise to a polarized humanity more horrific than Huxley's imagined Brave New World.

There are those who will argue that parents don't have the right to control the characteristics of their children-to-be in the way I describe. But American society, in particular, accepts the rights of parents to control every other aspect of their children's lives from the time they are born until they reach adulthood. If one accepts the parental prerogative after birth, it is hard to argue against it before birth, if no harm is caused to the children who emerge.

Many think that it is inherently unfair for some people to have access to technologies that can provide advantages while others, less well-off, are forced to depend on chance alone. I would agree. It is inherently

unfair. But once again, American society adheres to the principle that personal liberty and personal fortune are the primary determinants of what individuals are allowed and able to do. Anyone who accepts the right of affluent parents to provide their children with an expensive private school education cannot use "unfairness" as a reason for rejecting the use of reprogenetic technologies.

Indeed, in a society that values individual freedom above all else, it is hard to find any legitimate basis for restricting the use of reprogenetics. And therein lies the dilemma. For while each individual use of the technology can be viewed in the light of personal reproductive choice—with no ability to change society at large—together they could have dramatic, unintended, long-term consequences.

As the technologies of reproduction and genetics have become ever more powerful over the last decade, most practicing scientists and physicians have been loathe to speculate about where it may all lead. One reason for reluctance is the fear of getting it wrong. It really is impossible to predict with certainty which future technological advances will proceed on time and which will encounter unexpected roadblocks. This means that like Huxley's vision of a fetal hatchery, some of the ideas proposed here may ultimately be technically impossible or exceedingly difficult to implement. On the other hand, there are sure to be technological breakthroughs that no one can imagine now, just as Huxley was unable to imagine genetic engineering, or cloning from adult cells, in 1931.

There is a second reason why fertility specialists, in particular, are reluctant to speculate about the kinds of future scenarios that I describe here. It's called politics. In a climate where abortion clinics are on the alert for terrorist attacks, and where the religious right rails against any interference with the "natural process" of conception, IVF providers see no reason to call attention to themselves through descriptions of reproductive and genetic manipulations that are sure to provoke outrage.

The British journal *Nature* is one of the two most important science journals in the world (the other being the American journal *Science*). It is published weekly and is read by all types of scientists from biologists to physicists to medical researchers. No one would ever consider it to be radical or sensationalist in any way. On March 7, 1996, *Nature* published an article that described a method for cloning unlimited numbers of sheep from a single fertilized egg, with further implications for improving methods of genetic engineering. It took another week before the ramifications of this isolated breakthrough sank in for the editors. On March 14, 1996, they wrote an impassioned editorial saying in part: "That the growing power of molecular genetics confronts us with future prospects of being able to *change the nature of our species* [my emphasis] is a fact that seldom appears to be addressed in depth. Scientific knowledge may not yet permit detailed understanding, but the possibilities are clear enough. This gives rise to issues that in the end will have to be related to people within the social and ethical environments in which they live. . . . And the agenda is set by mankind as a whole, not by the subset involved in the science."

They are right that the agenda will not be set by scientists. But they are wrong to think that "mankind as a whole"—unable to reach consensus on so many other societal issues—will have any effect whatsoever. The agenda is sure to be set by individuals and couples who will act on behalf of themselves and their children. . . . The use of reprogenetic technologies is inevitable. It will not be controlled by governments or societies or even the scientists who create it.

There is no doubt about it. For better *and* worse, a new age is upon us. And whether we like it or not, the global marketplace will reign supreme.

R-2 *Preventing a Brave New World*

Leon Kass

In this article, Leon Kass, Addie Clark Harding Professor at the Committee on Social Thought at the University of Chicago and chairman of the President's Commission on Bioethics, argues that modern medical science is poised to cross an ethical boundary that will have momentous consequences for the future of humanity. Harking back to the dystopian vision of Aldous Huxley's classic *Brave New World* (1932), Kass argues that "the technological imperative, liberal democratic society, compassionate humanitarianism, moral pluralism, and free markets" are leading us down a path that places us at risk of losing our humanity. Kass argues that we should enact a worldwide ban on human cloning as a means of deterring "renegade scientists" from engaging in the practice. His proposed ban would apply to both reproductive and therapeutic cloning of human embryos because he believes that banning only reproductive cloning would prove impossible to enforce. Kass believes that we have the power to exercise control over the technological project but can only do so if we muster the political will to just say "no" to human cloning.

Focus Questions

1. What are the three main factors that Kass identifies as limiting our ability to control the onward march of the biomedical project? What other factors contribute to our inability to avoid the dangers Kass is concerned about?

2. What reasons does Kass give for "drawing the line" at human reproductive cloning by prohibiting the practice? Are these reasons convincing? Do the same arguments apply with equal force to therapeutic cloning?

3. Why does Kass believe that employing cloning or genetic engineering techniques to produce human children is "profoundly dehumanizing, no matter how good the product"?

4. Read the selections by Cummings (selection R-3) and Sagoff (selection R-4). Compare Kass's arguments to those offered by Cummings and Sagoff. Is there a moral difference between applying genetic engineering techniques to humans and to plants and nonhuman animals? Explain.

Keywords

dehumanization, genetic determinism, infertility, reproductive cloning, reproductive freedom, somatic cell nuclear transfer

I.

THE URGENCY OF THE GREAT political struggles of the twentieth century, successfully waged against totalitarianisms first right and then left, seems to have blinded many people to a deeper and ultimately darker truth about the present age: all contemporary societies are travelling briskly in the same utopian direction. All are wedded to the modern technological project; all march eagerly to the drums of progress and fly proudly the banner of modern science; all sing loudly the Baconian anthem, "Conquer nature, relieve man's estate." Leading the triumphal procession is modern medicine, which is daily becoming ever more powerful in its battle against disease, decay, and death, thanks especially to astonishing achievements in biomedical science and technology—achievements for which we must surely be grateful.

Yet contemplating present and projected advances in genetic and reproductive technologies, in neuroscience and psychopharmacology, and in the development of artificial organs and computer-chip implants for human brains, we now clearly recognize new uses for biotechnical power that soar beyond the traditional medical goals of healing disease and relieving suffering. Human nature itself lies on the operating table, ready for alteration, for eugenic and psychic "enhancement," for wholesale re-design. In leading laboratories, academic and industrial, new creators are confidently amassing their powers and quietly honing their skills, while on the street their evangelists are zealously prophesying a post-human future. For anyone who

From "Preventing a Brave New World," by Leon R. Kass, M.D., *The New Republic,* May 21, 2001. Copyright © 2001 by Leon R. Kass. Reprinted by permission of the author.

cares about preserving our humanity, the time has come to pay attention.

Some transforming powers are already here. The Pill. In vitro fertilization. Bottled embryos. Surrogate wombs. Cloning. Genetic screening. Genetic manipulation. Organ harvesting. Mechanical spare parts. Chimeras. Brain implants. Ritalin for the young, Viagra for the old, Prozac for everyone. And, to leave this vale of tears, a little extra morphine accompanied by Muzak.

Years ago Aldous Huxley saw it coming. In his charming but disturbing novel, *Brave New World* (it appeared in 1932 and is more powerful on each re-reading), he made its meaning strikingly visible for all to see. Unlike other frightening futuristic novels of the past century, such as Orwell's already dated *Nineteen Eighty-Four*, Huxley shows us a dystopia that goes with, rather than against, the human grain. Indeed, it is animated by our own most humane and progressive aspirations. Following those aspirations to their ultimate realization, Huxley enables us to recognize those less obvious but often more pernicious evils that are inextricably linked to the successful attainment of partial goods.

Huxley depicts human life seven centuries hence, living under the gentle hand of humanitarianism rendered fully competent by genetic manipulation, psychoactive drugs, hypnopaedia, and high-tech amusements. At long last, mankind has succeeded in eliminating disease, aggression, war, anxiety, suffering, guilt, envy, and grief. But this victory comes at the heavy price of homogenization, mediocrity, trivial pursuits, shallow attachments, debased tastes, spurious contentment, and souls without loves or longings. The Brave New World has achieved prosperity, community, stability, and nigh-universal contentment, only to be peopled by creatures of human shape but stunted humanity. They consume, fornicate, take "soma," enjoy "centrifugal bumble-puppy," and operate the machinery that makes it all possible. They do not read, write, think, love, or govern themselves. Art and science, virtue and religion, family and friendship are all passe. What matters most is bodily health and immediate gratification: "Never put off till tomorrow the fun you can have today." Brave New Man is so dehumanized that he does not even recognize what has been lost.

Huxley's novel, of course, is science fiction. Prozac is not yet Huxley's "soma"; cloning by nuclear transfer or splitting embryos is not exactly "Bokanovskification"; MTV and virtual-reality parlors are not quite the "feelies"; and our current safe and consequenceless sexual practices are not universally as loveless or as empty as those in the novel. But the kinships are disquieting, all the more so since our technologies of bio-psycho-engineering are still in their infancy, and in ways that make all too clear what they might look like in their full maturity. Moreover, the cultural changes that technology has already wrought among us should make us even more worried than Huxley would have us be.

In Huxley's novel, everything proceeds under the direction of an omnipotent—albeit benevolent—world state. Yet the dehumanization that he portrays does not really require despotism or external control. To the contrary, precisely because the society of the future will deliver exactly what we most want—health, safety, comfort, plenty, pleasure, peace of mind and length of days—we can reach the same humanly debased condition solely on the basis of free human choice. No need for World Controllers. Just give us the technological imperative, liberal democratic society, compassionate humanitarianism, moral pluralism, and free markets, and we can take ourselves to a Brave New World all by ourselves—and without even deliberately deciding to go. In case you had not noticed, the train has already left the station and is gathering speed, but no one seems to be in charge.

Some among us are delighted, of course, by this state of affairs: some scientists and biotechnologists, their entrepreneurial backers, and a cheering claque of sci-fi enthusiasts, futurologists, and libertarians. There are dreams to be realized, powers to be exercised, honors to be won, and money—big money—to be made. But many of us are worried, and not, as the proponents of the revolution self-servingly claim, because we are either ignorant of science or afraid of the unknown. To the contrary, we can see all too clearly where the train is headed, and we do not like the destination. We can distinguish cleverness about means from wisdom about ends, and we are loath to entrust the future of the race to those who cannot tell the difference. No friend of humanity cheers for a post-human future.

Yet for all our disquiet, we have until now done nothing to prevent it. We hide our heads in the sand because we enjoy the blessings that medicine keeps supplying, or we rationalize our inaction by declaring that human engineering is inevitable and we can do nothing about it. In either case, we are complicit in preparing for our own degradation, in some respects more to blame than the bio-zealots who, however misguided, are putting their money where their mouth is. Denial and despair, unattractive outlooks in any situation, become morally reprehensible when circumstances summon us to keep the world safe for human flourishing. Our immediate ancestors, taking up the challenge of their time, rose to the occasion and rescued the human future from the cruel dehumanizations of Nazi and Soviet tyranny. It is our more difficult task to find ways to preserve it from the soft dehumanizations of well-meaning but hubristic biotechnical "re-creationism"—and to do it without undermining biomedical science or rejecting its genuine contributions to human welfare.

Truth be told, it will not be easy for us to do so, and we know it. But rising to the challenge requires recognizing the difficulties. For there are indeed many features of modern life that will conspire to frustrate efforts aimed at the human control of the biomedical project. First, we Americans believe in technological automatism: where we do not foolishly believe that all innovation is progress, we fatalistically believe that it is inevitable ("If it can be done, it will be done, like it or not"). Second, we believe in freedom: the freedom of scientists to inquire, the freedom of technologists to develop, the freedom of entrepreneurs to invest and to profit, the freedom of private citizens to make use of existing technologies to satisfy any and all personal desires, including the desire to reproduce by whatever means. Third, the biomedical enterprise occupies the moral high ground of compassionate humanitarianism, upholding the supreme values of modern life—cure disease, prolong life, relieve suffering—in competition with which other moral goods rarely stand a chance. ("What the public wants is not to be sick," says James Watson, "and if we help them not to be sick, they'll be on our side.")

There are still other obstacles. Our cultural pluralism and easygoing relativism make it difficult to reach consensus on what we should embrace and what we should oppose; and moral objections to this or that biomedical practice are often facilely dismissed as religious or sectarian. Many people are unwilling to pronounce judgments about what is good or bad, right and wrong, even in matters of great importance, even for themselves—never mind for others or for society as a whole. It does not help that the biomedical project is now deeply entangled with commerce: there are increasingly powerful economic interests in favor of going full steam ahead, and no economic interests in favor of going slow. Since we live in a democracy, moreover, we face political difficulties in gaining a consensus to direct our future, and we have almost no political experience in trying to curtail the development of any new biomedical technology. Finally, and perhaps most troubling, our views of the meaning of our humanity have been so transformed by the scientific-technological approach to the world that we are in danger of forgetting what we have to lose, humanly speaking.

But though the difficulties are real, our situation is far from hopeless. Regarding each of the aforementioned impediments, there is another side to the story. Though we love our gadgets and believe in progress, we have lost our innocence regarding technology. The environmental movement especially has alerted us to the unintended damage caused by unregulated technological advance, and has taught us how certain dangerous practices can be curbed. Though we favor freedom of inquiry, we recognize that experiments are deeds and not speeches, and we prohibit experimentation on human subjects without their consent, even when cures from disease might be had by unfettered research; and we limit so-called reproductive freedom by proscribing incest, polygamy, and the buying and selling of babies.

Although we esteem medical progress, biomedical institutions have ethics committees that judge research proposals on moral grounds, and, when necessary, uphold the primacy of human freedom and human dignity even over scientific discovery. Our moral pluralism notwithstanding, national commissions and review bodies have sometimes reached moral consensus to recommend limits on permissible scientific research and technological application. On the economic front, the patenting of genes and life forms and the rapid rise of genomic commerce have elicited strong concerns and criticisms, leading even former enthusiasts of the new biology to recoil from the impending commodification of human life. Though we lack political institutions experienced in setting limits on biomedical innovation, federal agencies years ago rejected the development of the plutonium-powered artificial heart, and we have nationally prohibited commercial traffic in organs for transplantation, even though a market would increase the needed supply. In recent years, several American states and many foreign countries have successfully taken political action, making certain practices illegal and placing others under moratoriums (the creation of human embryos solely for research; human germ-line genetic alteration). Most importantly, the majority of Americans are not yet so degraded or so cynical as to fail to be revolted by the society depicted in Huxley's novel. Though the obstacles to effective action are significant, they offer no excuse for resignation. Besides, it would be disgraceful to concede defeat even before we enter the fray.

Not the least of our difficulties in trying to exercise control over where biology is taking us is the fact that we do not get to decide, once and for all, for or against the destination of a post-human world. The scientific discoveries and the technical powers that will take us there come to us piecemeal, one at a time and seemingly independent from one another, each often attractively introduced as a measure that will "help [us] not to be sick." But sometimes we come to a clear fork in the road where decision is possible, and where we know that our decision will make a world of difference—indeed, it will make a permanently different world. Fortunately, we stand now at the point of such a momentous decision. Events have conspired to provide us with a perfect opportunity to seize the initiative and to gain some control of the biotechnical project. I refer to the prospect of human cloning, a practice absolutely central to Huxley's fictional world. Indeed, creating and manipulating life in the laboratory is the gateway to a Brave New World, not only in fiction but also in fact.

"To clone or not to clone a human being" is no longer a fanciful question. Success in cloning sheep, and also cows, mice, pigs, and goats, makes it perfectly clear that a fateful decision is now at hand: whether we should welcome or even tolerate the cloning of human beings. If recent newspaper reports are to be believed, reputable scientists and physicians have announced their intention to produce the first human clone in the coming year. Their efforts may already be under way.

The media, gawking and titillating as is their wont, have been softening us up for this possibility by turning the bizarre into the familiar. In the four years since the birth of Dolly the cloned sheep, the tone of discussing the prospect of human cloning has gone from "Yuck" to "Oh?" to "Gee whiz" to "Why not?" The sentimentalizers, aided by leading bioethicists, have downplayed talk about eugenically cloning the beautiful and the brawny or the best and the brightest. They have taken instead to defending clonal reproduction for humanitarian or compassionate reasons: to treat infertility in people who are said to "have no other choice," to avoid the risk of severe genetic disease, to "replace" a child who has died. For the sake of these rare benefits, they would have us countenance the entire practice of human cloning, the consequences be damned.

But we dare not be complacent about what is at issue, for the stakes are very high. Human cloning, though partly continuous with previous reproductive technologies, is also something radically new in itself and in its easily foreseeable consequences—especially when coupled with powers for genetic "enhancement" and germline genetic modification that may soon become available, owing to the recently completed Human Genome Project. I exaggerate somewhat, but in the direction of the truth: we are compelled to decide nothing less than whether human procreation is going to remain human, whether children are going to be made to order rather than begotten, and whether we wish to say yes in principle to the road that leads to the dehumanized hell of *Brave New World.*

. . .

For we have here a golden opportunity to exercise some control over where biology is taking us. The technology of cloning is discrete and well defined, and it requires considerable technical know-how and dexterity; we can therefore know by name many of the likely practitioners. The public demand for cloning is extremely low, and most people are decidedly against it. Nothing scientifically or medically important would be lost by banning clonal reproduction; alternative and non-objectionable means are available to obtain some of the most important medical benefits claimed for (nonreproductive) human cloning. The commercial interests in human cloning are, for now, quite limited; and the nations of the world are actively seeking to prevent it. Now may be as good a chance as we will ever have to get our hands on the wheel of the runaway train now headed for a post-human world and to steer it toward a more dignified human future.

II.

What is cloning? Cloning, or asexual reproduction, is the production of individuals who are genetically identical to an already existing individual. The procedure's name is fancy—"somatic cell nuclear transfer"—but its concept is simple. Take a mature but unfertilized egg; remove or deactivate its nucleus; introduce a nucleus obtained from a specialized (somatic) cell of an adult organism. Once the egg begins to divide, transfer the little embryo to a woman's uterus to initiate a pregnancy. Since almost all the hereditary material of a cell is contained within its nucleus, the re-nucleated egg and the individual into which it develops are genetically identical to the organism that was the source of the transferred nucleus.

An unlimited number of genetically identical individuals—the group, as well as each of its members, is called "a clone"—could be produced by nuclear transfer. In principle, any person, male or female, newborn or adult, could be cloned, and in any quantity; and because stored cells can outlive their sources, one may even clone the dead. Since cloning requires no personal involvement on the part of the person whose genetic material is used, it could easily be used to reproduce living or deceased persons without their consent—a threat to reproductive freedom that has received relatively little attention.

Some possible misconceptions need to be avoided. Cloning is not Xeroxing: the clone of Bill Clinton, though his genetic double, would enter the world hairless, toothless, and peeing in his diapers, like any other human infant. But neither is cloning just like natural twinning: the cloned twin will be identical to an older, existing adult; and it will arise not by chance but by deliberate design; and its entire genetic makeup will be preselected by its parents and/or scientists. Moreover, the success rate of cloning, at least at first, will probably not be very high: the Scots transferred two hundred seventy-seven adult nuclei into sheep eggs, implanted twenty-nine clonal embryos, and achieved the birth of only one live lamb clone.

For this reason, among others, it is unlikely that, at least for now, the practice would be very popular; and there is little immediate worry of mass-scale production of multicopies. Still, for the tens of thousands of people who sustain more than three hundred assisted-reproduction clinics in the United States and already avail themselves of in vitro fertilization and other techniques, cloning would be an option with virtually no added fuss. Panos Zavos, the Kentucky reproduction specialist who has announced his plans to clone a child, claims that he has already received thousands of e-mailed requests from people eager to clone, despite the known risks of failure and damaged offspring. Should commercial interests develop in "nucleus-banking," as they have in sperm-banking and egg-harvesting; should famous athletes or other celebrities decide to market their DNA the way they now market their autographs and nearly everything else; should techniques of embryo and germline genetic testing and manipulation arrive as anticipated, increasing the use of laboratory assistance in order to obtain "better" babies—should all this come to pass, cloning, if it is permitted, could become more than a marginal practice simply on the basis of free reproductive choice.

What are we to think about this prospect? Nothing good. Indeed, most people are repelled by nearly all aspects of human cloning: the possibility of mass production of human beings, with large clones of look-alikes, compromised in their individuality; the idea of father-son or mother-daughter "twins"; the bizarre prospect of a woman bearing and rearing a genetic copy of herself, her spouse, or even her deceased father or mother; the grotesqueness of conceiving a child as an exact "replacement" for another who has died; the utilitarian creation of embryonic duplicates of oneself, to be frozen away or created when needed to provide homologous tissues or organs for transplantation; the narcissism of those who would clone themselves, and the arrogance of others who think they know who deserves to be cloned; the Frankensteinian hubris to create a human life and increasingly to control its destiny; men playing at being God. Almost no one finds any of the suggested reasons for human cloning compelling, and almost everyone anticipates its possible misuses and abuses. And the popular belief that human cloning cannot be prevented makes the prospect all the more revolting.

Revulsion is not an argument; and some of yesterday's repugnances are today calmly accepted—not always for the better. In some crucial cases, however, repugnance is the emotional expression of deep wisdom, beyond reason's power completely to articulate it. Can anyone really give an argument fully adequate to the horror that is father–daughter incest (even with consent), or bestiality, or the mutilation of a corpse, or the eating of human flesh, or the rape or murder of another human being? Would anybody's failure to give full rational justification for his revulsion at those practices make that revulsion ethically suspect?

I suggest that our repugnance at human cloning belongs in this category. We are repelled by the prospect of cloning human beings not because of the strangeness or the novelty of the undertaking, but because we intuit and we feel, immediately and without argument, the violation of things that we rightfully hold dear. We sense that cloning represents a profound defilement of our given nature as procreative beings, and of the social relations built on this natural ground. We also sense that cloning is a radical form of child abuse. In this age in which everything is held to be per-

missible so long as it is freely done, and in which our bodies are regarded as mere instruments of our autonomous rational will, repugnance may be the only voice left that speaks up to defend the central core of our humanity. Shallow are the souls that have forgotten how to shudder.

III.

Yet repugnance need not stand naked before the bar of reason. The wisdom of our horror at human cloning can be at least partially articulated, even if this is finally one of those instances about which the heart has its reasons that reason cannot entirely know. I offer four objections to human cloning: that it constitutes unethical experimentation; that it threatens identity and individuality; that it turns procreation into manufacture (especially when understood as the harbinger of manipulations to come); and that it means despotism over children and perversion of parenthood. Please note: I speak only about so-called reproductive cloning, not about the creation of cloned embryos for research. The objections that may be raised against creating (or using) embryos for research are entirely independent of whether the research embryos are produced by cloning. What is radically distinct and radically new is reproductive cloning.

Any attempt to clone a human being would constitute an unethical experiment upon the resulting child-to-be. In all the animal experiments, fewer than two to three percent of all cloning attempts succeeded. Not only are there fetal deaths and stillborn infants, but many of the so-called "successes" are in fact failures. As has only recently become clear, there is a very high incidence of major disabilities and deformities in cloned animals that attain live birth. Cloned cows often have heart and lung problems; cloned mice later develop pathological obesity; other live-born cloned animals fail to reach normal developmental milestones.

The problem, scientists suggest, may lie in the fact that an egg with a new somatic nucleus must reprogram itself in a matter of minutes or hours (whereas the nucleus of an unaltered egg has been prepared over months and years). There is thus a greatly increased likelihood of error in translating the genetic instructions, leading to developmental defects some of which will show themselves only much later. (Note also that these induced abnormalities may also affect the stem cells that scientists hope to harvest from cloned embryos. Lousy embryos, lousy stem cells.) Nearly all scientists now agree that attempts to clone human beings carry massive risks of producing unhealthy, abnormal, and malformed children. What are we to do with them? Shall we just discard the ones that fall short of expectations? Considered opinion is today nearly unanimous, even among scientists: attempts at human cloning are irresponsible and unethical. We cannot ethically even get to know whether or not human cloning is feasible.

If it were successful, cloning would create serious issues of identity and individuality. The clone may experience concerns about his distinctive identity not only because he will be, in genotype and in appearance, identical to another human being, but because he may also be twin to the person who is his "father" or his "mother"—if one can still call them that. Unaccountably, people treat as innocent the homey case of intrafamilial cloning—the cloning of husband or wife (or single mother). They forget about the unique dangers of mixing the twin relation with the parent-child relation. (For this situation, the relation of contemporaneous twins is no precedent; yet even this less problematic situation teaches us how difficult it is to wrest independence from the being for whom one has the most powerful affinity.) Virtually no parent is going to be able to treat a clone of himself or herself as one treats a child generated by the lottery of sex. What will happen when the adolescent clone of Mommy becomes the spitting image of the woman with whom Daddy once fell in love? In case of divorce, will Mommy still love the clone of Daddy, even though she can no longer stand the sight of Daddy himself?

Most people think about cloning from the point of view of adults choosing to clone. Almost nobody thinks about what it would be like to be the cloned child. Surely his or her new life would constantly be scrutinized in relation to that of the older version. Even in the absence of unusual parental expectations for the clone—say, to live the same life, only without its errors—the child is likely to be ever a curiosity, ever a potential source of déjà vu. Unlike "normal" identical twins, a cloned individual—copied from whomever—will be saddled with a genotype that has already lived. He will not be fully a surprise to the world: people are likely always to compare his doings in life with those of his alter ego, especially if he is a clone of someone gifted or famous. True, his nurture and his circumstance will be different; genotype is not exactly destiny. But one must also expect parental efforts to shape this new life after the original—or at least to view the child with the original version always firmly in mind. For why else did they clone from the star basketball player, the mathematician, or the beauty queen—or even dear old Dad—in the first place?

Human cloning would also represent a giant step toward the transformation of begetting into making, of procreation into manufacture (literally, "handmade"), a process that has already begun with in vitro fertilization and genetic testing of embryos. With cloning, not only is the process in hand, but the total genetic blueprint of the cloned individual is selected and deter-

mined by the human artisans. To be sure, subsequent development is still according to natural processes; and the resulting children will be recognizably human. But we would be taking a major step into making man himself simply another one of the man-made things.

How does begetting differ from making? In natural procreation, human beings come together to give existence to another being that is formed exactly as we were, by what we are—living, hence perishable, hence aspiringly erotic, hence procreative human beings. But in clonal reproduction, and in the more advanced forms of manufacture to which it will lead, we give existence to a being not by what we are but by what we intend and design.

Let me be clear. The problem is not the mere intervention of technique, and the point is not that "nature knows best." The problem is that any child whose being, character, and capacities exist owing to human design does not stand on the same plane as its makers. As with any product of our making, no matter how excellent, the artificer stands above it, not as an equal but as a superior, transcending it by his will and creative prowess. In human cloning, scientists and prospective "parents" adopt a technocratic attitude toward human children: human children become their artifacts. Such an arrangement is profoundly dehumanizing, no matter how good the product.

Procreation dehumanized into manufacture is further degraded by commodification, a virtually inescapable result of allowing baby-making to proceed under the banner of commerce. Genetic and reproductive biotechnology companies are already growth industries, but they will soon go into commercial orbit now that the Human Genome Project has been completed. "Human eggs for sale" is already a big business, masquerading under the pretense of "donation." Newspaper advertisements on elite college campuses offer up to $50,000 for an egg "donor" tall enough to play women's basketball and with SAT scores high enough for admission to Stanford; and to nobody's surprise, at such prices there are many young coeds eager to help shoppers obtain the finest babies money can buy. (The egg and womb-renting entrepreneurs shamelessly proceed on the ancient, disgusting, misogynist premise that most women will give you access to their bodies, if the price is right.) Even before the capacity for human cloning is perfected, established companies will have invested in the harvesting of eggs from ovaries obtained at autopsy or through ovarian surgery, practiced embryonic genetic alteration, and initiated the stockpiling of prospective donor tissues. Through the rental of surrogate-womb services, and through the buying and selling of tissues and embryos priced according to the merit of the donor, the commodification of nascent human life will be unstoppable.

Finally, the practice of human cloning by nuclear transfer—like other anticipated forms of genetically engineering the next generation—would enshrine and aggravate a profound misunderstanding of the meaning of having children and of the parent-child relationship. When a couple normally chooses to procreate, the partners are saying yes to the emergence of new life in its novelty—are saying yes not only to having a child, but also to having whatever child this child turns out to be. In accepting our finitude, in opening ourselves to our replacement, we tacitly confess the limits of our control.

Embracing the future by procreating means precisely that we are relinquishing our grip in the very activity of taking up our own share in what we hope will be the immortality of human life and the human species. This means that our children are not our children: they are not our property, they are not our possessions. Neither are they supposed to live our lives for us, or to live anyone's life but their own. Their genetic distinctiveness and independence are the natural foreshadowing of the deep truth that they have their own, never-before-enacted life to live. Though sprung from a past, they take an uncharted course into the future.

Much mischief is already done by parents who try to live vicariously through their children. Children are sometimes compelled to fulfill the broken dreams of unhappy parents. But whereas most parents normally have hopes for their children, cloning parents will have expectations. In cloning, such overbearing parents will have taken at the start a decisive step that contradicts the entire meaning of the open and forward-looking nature of parent-child relations. The child is given a genotype that has already lived, with full expectation that this blueprint of a past life ought to be controlling the life that is to come. A wanted child now means a child who exists precisely to fulfill parental wants. Like all the more precise eugenic manipulations that will follow in its wake, cloning is thus inherently despotic, for it seeks to make one's children after one's own image (or an image of one's choosing) and their future according to one's will.

Is this hyperbolic? Consider concretely the new realities of responsibility and guilt in the households of the cloned. No longer only the sins of the parents, but also the genetic choices of the parents, will be visited on the children—and beyond the third and fourth generation; and everyone will know who is responsible. No parent will be able to blame nature or the lottery of sex for an unhappy adolescent's big nose, dull wit, musical ineptitude, nervous disposition, or anything else that he hates about himself. Fairly or not, children will hold their cloners responsible for everything, for nature as well as for nurture. And parents, especially the better ones, will be limitlessly liable to guilt. Only the truly despotic souls will sleep the sleep of the innocent.

IV.

The defenders of cloning are not wittingly friends of despotism. Quite the contrary. Deaf to most other considerations, they regard themselves mainly as friends of freedom: the freedom of individuals to reproduce, the freedom of scientists and inventors to discover and to devise and to foster "progress" in genetic knowledge and technique, the freedom of entrepreneurs to profit in the market. They want large-scale cloning only for animals, but they wish to preserve cloning as a human option for exercising our "right to reproduce"—our right to have children, and children with "desirable genes." As some point out, under our "right to reproduce" we already practice early forms of unnatural, artificial, and extra-marital reproduction, and we already practice early forms of eugenic choice. For that reason, they argue, cloning is no big deal.

We have here a perfect example of the logic of the slippery slope. The principle of reproductive freedom currently enunciated by the proponents of cloning logically embraces the ethical acceptability of sliding all the way down: to producing children wholly in the laboratory from sperm to term (should it become feasible), and to producing children whose entire genetic makeup will be the product of parental eugenic planning and choice. If reproductive freedom means the right to have a child of one's own choosing by whatever means, then reproductive freedom knows and accepts no limits.

Proponents want us to believe that there are legitimate uses of cloning that can be distinguished from illegitimate uses, but by their own principles no such limits can be found. (Nor could any such limits be enforced in practice: once cloning is permitted, no one ever need discover whom one is cloning and why.) Reproductive freedom, as they understand it, is governed solely by the subjective wishes of the parents-to-be. The sentimentally appealing case of the childless married couple is, on these grounds, indistinguishable from the case of an individual (married or not) who would like to clone someone famous or talented, living or dead. And the principle here endorsed justifies not only cloning but also all future artificial attempts to create (manufacture) "better" or "perfect" babies.

The "perfect baby," of course, is the project not of the infertility doctors, but of the eugenic scientists and their supporters, who, for the time being, are content to hide behind the skirts of the partisans of reproductive freedom and compassion for the infertile. For them, the paramount right is not the so-called right to reproduce, it is what the biologist Bentley Glass called, a quarter of a century ago, "the right of every child to be born with a sound physical and mental constitution, based on a sound genotype . . . the inalienable right to a sound heritage." But to secure this right, and to achieve the requisite quality control over new human life, human conception and gestation will need to be brought fully into the bright light of the laboratory, beneath which the child-to-be can be fertilized, nourished, pruned, weeded, watched, inspected, prodded, pinched, cajoled, injected, tested, rated, graded, approved, stamped, wrapped, sealed, and delivered. There is no other way to produce the perfect baby.

If you think that such scenarios require outside coercion or governmental tyranny, you are mistaken. Once it becomes possible, with the aid of human genomics, to produce or to select for what some regard as "better babies"—smarter, prettier, healthier, more athletic—parents will leap at the opportunity to "improve" their offspring. Indeed, not to do so will be socially regarded as a form of child neglect. Those who would ordinarily be opposed to such tinkering will be under enormous pressure to compete on behalf of their as yet unborn children—just as some now plan almost from their children's birth how to get them into Harvard. Never mind that, lacking a standard of "good" or "better," no one can really know whether any such changes will truly be improvements.

Proponents of cloning urge us to forget about the science-fiction scenarios of laboratory manufacture or multiple-copy clones, and to focus only on the sympathetic cases of infertile couples exercising their reproductive rights. But why, if the single cases are so innocent, should multiplying their performance be so off-putting? (Similarly, why do others object to people's making money from that practice if the practice itself is perfectly acceptable?) The so-called science-fiction cases—say, Brave New World—make vivid the meaning of what looks to us, mistakenly, to be benign. They reveal that what looks like compassionate humanitarianism is, in the end, crushing dehumanization.

V.

Whether or not they share my reasons, most people, I think, share my conclusion: that human cloning is unethical in itself and dangerous in its likely consequences, which include the precedent that it will establish for designing our children. Some reach this conclusion for their own good reasons, different from my own: concerns about distributive justice in access to eugenic cloning; worries about the genetic effects of asexual "inbreeding"; aversion to the implicit premise of genetic determinism; objections to the embryonic and fetal wastage that must necessarily accompany the efforts; religious opposition to "man playing God." But never mind why: the overwhelming majority of our fellow Americans remain firmly opposed to cloning human beings.

For us, then, the real questions are: What should we do about it? How can we best succeed? These questions should concern everyone eager to secure deliberate human control over the powers that could re-

design our humanity, even if cloning is not the issue over which they would choose to make their stand. And the answer to the first question seems pretty plain. What we should do is work to prevent human cloning by making it illegal.

We should aim for a global legal ban, if possible, and for a unilateral national ban at a minimum—and soon, before the fact is upon us. To be sure, legal bans can be violated; but we certainly curtail much mischief by outlawing incest, voluntary servitude, and the buying and selling of organs and babies. To be sure, renegade scientists may secretly undertake to violate such a law, but we can deter them by both criminal sanctions and monetary penalties, as well as by removing any incentive they have to proudly claim credit for their technological bravado.

Such a ban on clonal baby-making will not harm the progress of basic genetic science and technology. On the contrary, it will reassure the public that scientists are happy to proceed without violating the deep ethical norms and intuitions of the human community. It will also protect honorable scientists from a public backlash against the brazen misconduct of the rogues. As many scientists have publicly confessed, free and worthy science probably has much more to fear from a strong public reaction to a cloning fiasco than it does from a cloning ban, provided that the ban is judiciously crafted and vigorously enforced against those who would violate it.

. . .

. . . I now believe that what we need is an all-out ban on human cloning, including the creation of embryonic clones. I am convinced that all halfway measures will prove to be morally, legally, and strategically flawed, and—most important—that they will not be effective in obtaining the desired result. Anyone truly serious about preventing human reproductive cloning must seek to stop the process from the beginning. Our changed circumstances, and the now evident defects of the less restrictive alternatives, make an all-out ban by far the most attractive and effective option.

Here's why. Creating cloned human children ("reproductive cloning") necessarily begins by producing cloned human embryos. Preventing the latter would prevent the former, and prudence alone might counsel building such a "fence around the law." Yet some scientists favor embryo cloning as a way of obtaining embryos for research or as sources of cells and tissues for the possible benefit of others. (This practice they misleadingly call "therapeutic cloning" rather than the more accurate "cloning for research" or "experimental cloning," so as to obscure the fact that the clone will be "treated" only to exploitation and destruction, and that any potential future beneficiaries and any future "therapies" are at this point purely hypothetical.)

The prospect of creating new human life solely to be exploited in this way has been condemned on moral grounds by many people—including *The Washington Post*, President Clinton, and many other supporters of a woman's right to abortion—as displaying a profound disrespect for life. Even those who are willing to scavenge so-called "spare embryos"—those products of in vitro fertilization made in excess of people's reproductive needs, and otherwise likely to be discarded—draw back from creating human embryos explicitly and solely for research purposes. They reject outright what they regard as the exploitation and the instrumentalization of nascent human life. In addition, others who are agnostic about the moral status of the embryo see the wisdom of not needlessly offending the sensibilities of their fellow citizens who are opposed to such practices.

But even setting aside these obvious moral first impressions, a few moments of reflection show why an anti-cloning law that permitted the cloning of embryos but criminalized their transfer to produce a child would be a moral blunder. This would be a law that was not merely permissively "pro-choice" but emphatically and prescriptively "anti-life." While permitting the creation of an embryonic life, it would make it a federal offense to try to keep it alive and bring it to birth. Whatever one thinks of the moral status or the ontological status of the human embryo, moral sense and practical wisdom recoil from having the government of the United States on record as requiring the destruction of nascent life and, what is worse, demanding the punishment of those who would act to preserve it by (feloniously!) giving it birth.

But the problem with the approach that targets only reproductive cloning (that is, the transfer of the embryo to a woman's uterus) is not only moral but also legal and strategic. A ban only on reproductive cloning would turn out to be unenforceable. Once cloned embryos were produced and available in laboratories and assisted-reproduction centers, it would be virtually impossible to control what was done with them. Biotechnical experiments take place in laboratories, hidden from public view, and, given the rise of high-stakes commerce in biotechnology, these experiments are concealed from the competition. Huge stockpiles of cloned human embryos could thus be produced and bought and sold without anyone knowing it. As we have seen with in vitro embryos created to treat infertility, embryos produced for one reason can be used for another reason: today "spare embryos" once created to begin a pregnancy are now used in research, and tomorrow clones created for research will be used to begin a pregnancy.

Assisted reproduction takes place within the privacy of the doctor-patient relationship, making outside scrutiny extremely difficult. Many infertility experts probably would obey the law, but others could and would defy it with impunity, their doings covered by the veil of secrecy that is the principle of medical confidentiality. Moreover, the transfer of embryos to begin a pregnancy is a simple procedure (especially compared with manufacturing the embryo in the first place), simple enough that its final steps could be self-administered by the woman, who would thus absolve the doctor of blame for having "caused" the illegal transfer. (I have in mind something analogous to Kevorkian's suicide machine, which was designed to enable the patient to push the plunger and the good "doctor" to evade criminal liability.)

Even should the deed become known, governmental attempts to enforce the reproductive ban would run into a swarm of moral and legal challenges, both to efforts aimed at preventing transfer to a woman and—even worse—to efforts seeking to prevent birth after transfer has occurred. A woman who wished to receive the embryo clone would no doubt seek a judicial restraining order, suing to have the law overturned in the name of a constitutionally protected interest in her own reproductive choice to clone. (The cloned child would be born before the legal proceedings were complete.) And should an "illicit clonal pregnancy" be discovered, no governmental agency would compel a woman to abort the clone, and there would be an understandable storm of protest should she be fined or jailed after she gives birth. Once the baby is born, there would even be sentimental opposition to punishing the doctor for violating the law—unless, of course, the clone turned out to be severely abnormal.

For all these reasons, the only practically effective and legally sound approach is to block human cloning at the start, at the production of the embryo clone. Such a ban can be rightly characterized not as interference with reproductive freedom, nor even as interference with scientific inquiry, but as an attempt to prevent the unhealthy, unsavory, and unwelcome manufacture of and traffic in human clones.

. . .

I appreciate that a federal legislative ban on human cloning is without American precedent, at least in matters technological. Perhaps such a ban will prove ineffective; perhaps it will eventually be shown to have been a mistake. (If so, it could later be reversed.) If enacted, however, it will have achieved one overwhelmingly important result, in addition to its contribution to thwarting cloning: it will place the burden of practical proof where it belongs. It will require the proponents to show very clearly what great social or medical good can be had only by the cloning of human beings. Surely it is only for such a compelling case, yet to be made or even imagined, that we should wish to risk this major departure—or any other major departure—in human procreation.

Americans have lived by and prospered under a rosy optimism about scientific and technological progress. The technological imperative has probably served us well, though we should admit that there is no accurate method for weighing benefits and harms.

And even when we recognize the unwelcome outcomes of technological advance, we remain confident in our ability to fix all the "bad" consequences—by regulation or by means of still newer and better technologies. Yet there is very good reason for shifting the American paradigm, at least regarding those technological interventions into the human body and mind that would surely effect fundamental (and likely irreversible) changes in human nature, basic human relationships, and what it means to be a human being. Here we should not be willing to risk everything in the naive hope that, should things go wrong, we can later set them right again.

Some have argued that cloning is almost certainly going to remain a marginal practice, and that we should therefore permit people to practice it. Such a view is shortsighted. Even if cloning is rarely undertaken, a society in which it is tolerated is no longer the same society—any more than is a society that permits (even small-scale) incest or cannibalism or slavery. A society that allows cloning, whether it knows it or not, has tacitly assented to the conversion of procreation into manufacture and to the treatment of children as purely the projects of our will. Willy-nilly, it has acquiesced in the eugenic re-design of future generations. The humanitarian superhighway to a Brave New World lies open before this society.

But the present danger posed by human cloning is, paradoxically, also a golden opportunity. In a truly unprecedented way, we can strike a blow for the human control of the technological project, for wisdom, for prudence, for human dignity. The prospect of human cloning, so repulsive to contemplate, is the occasion for deciding whether we shall be slaves of unregulated innovation, and ultimately its artifacts, or whether we shall remain free human beings who guide our powers toward the enhancement of human dignity. The humanity of the human future is now in our hands.

R-3 *Trespass*

Claire Hope Cummings

A great deal of attention has been given to the prospect of human genetic engineering, but comparatively little notice has been paid to the growing use of genetic engineering technologies on plants and animals. For more than a decade, food products made from genetically modified organisms (GMOs) have been on American supermarket shelves, and millions of people have eaten them with no apparent ill effects. So what is the fuss about when environmentalists and some others warn the public about the dangers of so-called "Frankenfoods"? Are these fears merely the concerns of overwrought Luddites, or might there be some good reasons to be concerned about the genetic manipulation of nonhuman organisms? Lawyer and environmental journalist Claire Hope Cummings describes the history of the development of genetically modified foods, paying particular attention to the lack of regulatory oversight by the U.S. government prior to their release into the general food supply. The rapid introduction into the market of GMOs was made possible by "regulatory capture" of the government by agribusiness interests coupled with the increasing trend toward privately funded research at our nation's leading universities. In the process, scant attention was paid to the risk of genetic contamination of non-GMO plant species by genetic materials from their bioengineered cousins. Cummings thinks that GMO foods have been literally forced down our throats by big business, with little or no way for the public to opt out of this uncontrolled experiment whose ultimate effects on human health and the environment remain unknown.

Focus Questions

1. What does Cummings describe as "the central dogma" of modern molecular biology? Is there evidence to suggest that this belief might turn out to be false?

2. What is the significance of the "revolving door" between government and agribusiness? How does this help explain why the federal government decided not to require labels on food products containing GMOs?

3. Does the fact that private corporations fund much of the academic research concerning the safety of GMOs concern you? Why or why not?

Keywords

academic freedom, agribusiness, genetic contamination, genetically modified organisms (GMOs), molecular biology, regulatory oversight, substantial equivalence, transgenic instability

HIDDEN INSIDE HILGARD HALL, one of the oldest buildings on the campus of the University of California at Berkeley, is a photograph that no one is supposed to see. It's a picture of a crippled and contorted corncob that was not created by nature, or even by agriculture, but by genetic engineering.[1] The cob is kept in a plastic bin called "the monster box," a collection of biological curiosities put together by someone who works in a secure biotechnology research facility.

What the photo shows is a cob that apparently started growing normally, then turned into another part of the corn plant, then returned to forming kernels, then went back to another form—twisting back and forth as if it could not make up its mind about what it was. It was produced by the same recombinant DNA technology that is used to create the genetically modified organisms (GMOs) that are in our everyday foods. When I saw this photo, I knew it was saying something very important about genetic engineering. I thought it should be published. But the person who owns it is frankly afraid of how the biotechnology industry might react, and would not agree. In order to get permission even to describe the photo for this article, I had to promise not to reveal its owner's identity.

What the distorted corncob represents is a mute challenge to the industry's claim that this technology is precise, predictable, and safe. But that this challenge should be kept hidden, and that a scientist who works at a public university should feel too intimidated to discuss it openly, told me that something more than just a scientific question was being raised. After all, if the new agricultural biotech were really safe and effective, why would the industry work so hard—as indeed it does—to keep its critics cowed and the public uninformed? Was there something about the way genetic engineering was developed, about how it works, that was inviting a closer look—a look that the industry would rather we not take? I had gone to Berkeley to see for myself what was going on behind biotechnology.

The University of California at Berkeley ("Cal") is the stage on which much of the story of genetic engineering has played out over the last 25 years. The biotechnology industry was born here in the San Francisco Bay area, and nurtured by scientists who worked at Berkeley and nearby universities. Critical controversies over the role genetic engineering and related research should have in society have erupted here. Even the architecture of the campus reflects the major scientific and policy divisions that plague this technology. Two buildings, in particular, mirror the two very different versions of biology that emerged in the last half of the twentieth century, and reflect two very different visions for agriculture in the future.

Hilgard Hall was built in 1918, at a time when mastering the classical form and celebrating beauty were important, perhaps even integral, to the accepted function of a building. Hilgard's facade is exquisitely decorated with friezes depicting sheaves of wheat, beehives, bunches of grapes, cornucopias, and bas relief sculptures of cow heads surrounded with wreaths of fruit. Above the entrance, carved in huge capital letters are the words, "TO RESCUE FOR HUMAN SOCIETY THE NATIVE VALUES OF RURAL LIFE." The massive front door opens to a grand two-story hall graced with granite, marble, and carved brass. But behind that elegant entrance is a building left in disrepair. Getting around inside Hilgard means navigating worn marble staircases and dark corridors laced with exposed pipes and heat-

ing ducts. The room where the monster box photograph is kept is small and dank. This building is home to the "old" biology—the careful observation of life, living systems, and their complex interactions. Being inside Hilgard is a visceral lesson in how Cal is neglecting the classic study of the intimate inter-relationships among agriculture, the environment, and human society.

Nearby, and standing in stark contrast to Hilgard's faded splendor, is a newer, modern office building, Koshland Hall. Koshland is not unattractive, with its pitched blue tile roof lines and bright white walls lined with blue steel windows, but it was built in the mid-1990s in a functional style that, like most new campus buildings, has all the charm and poetry of an ice cube. The interior is clean and well lit. Next to office doors hang plaques that name the corporations or foundations that fund the activities inside. This is the home of the "new biology"—the utilitarian view that life is centered in DNA and molecules can be manipulated at will. Molecular biology is clearly doing well at Cal.

Koshland Hall was named after a distinguished member of the faculty, Daniel Koshland, former editor of the journal *Science* and chair of Berkeley's Department of Biochemistry, now a professor emeritus. He has the unique distinction of having been present at the two most important scientific revolutions of our time: he participated both in the Manhattan Project, which developed nuclear weapons, and in the early development of molecular biology. He is credited with "transforming" the biological sciences at Berkeley.

THE NEW BIOLOGY

One hundred years ago, no one had heard of a "gene." The word was not recognized until 1909, and even after that it remained an abstraction for decades. At the time, scientists and others were making an effort to find a material basis for life, particularly heritability, the fundamental function of life. The story of genetic engineering in the United States begins with the decision to identify genes as the basis of life. But the ideological roots of this story go even deeper, into the nation's earlier history and attachment to the ideas of manifest destiny, eugenics, and social engineering.

Early in the twentieth century, the new "science" of sociology made its appearance—along with the highly appealing belief that social problems were amenable to scientific solutions. In time, sociology began to combine with genetic science, giving strong impetus to technocratic forms of social control, and particularly to eugenics—the belief that the human race could be improved by selective breeding. Until the 1930s, the science of genetics had not developed much beyond Mendelian principles of heredity, but eugenics was already being promoted as the solution to social problems. As the idea that genes determined traits in people took hold, eugenics twisted it to foster the concept that there were "good" genes and "bad" genes, good and bad traits. Eugenics eventually gained a powerful foothold both in the popular imagination and in the U.S. government, as well as in Nazi Germany. Even today, these notions underlie the decisions biotechnologists make about what genes and traits are beneficial, what organisms are engineered, and who gets to decide how this technology will be used.

According to Lily Kay, an assistant professor of the history of science at Massachusetts Institute of Technology, genetic engineering came about as the result of the concerted effort of a few scientists, who, along with their academic and philanthropic sponsors, had a shared vision about how they could use genetics to reshape science and society. In her book *The Molecular Vision of Life: Caltech, the Rockefeller Foundation, and the Rise of the New Biology*, Kay writes that this vision was not so much about underlying biological principles as it was about social values. The new biology that evolved from this thinking was founded on a strong belief in "industrial capitalism" and its perceived mandate for "science-based social intervention." The potential for this idea, and the intentional strategy to use it for social purposes was clearly understood from the outset, says Kay. The developers of "molecular biology" (a term coined by the Rockefeller Foundation) were confident that it would offer them a previously unimagined power and control over both nature and society.

Science was molded to this agenda in 1945, when Vannevar Bush, the head of President Franklin D. Roosevelt's wartime Office of Scientific Research and Development, wrote "Science, The Endless Frontier"— a landmark report that outlined how science could better serve the private sector. As Kay tells the story, at that point the search for a science-based social agenda began in earnest. It was funded and directed by business corporations and foundations acting together as "quasi-public entities" using both private and public funds to harness "the expertise of the human sciences to stem what was perceived as the nation's social and biological decay and help realize the vision of America's destiny." Eventually, the combined efforts of corporate, academic, and government interests began to bear fruit and "the boundary between individual and corporate self-interest, between private and public control, would be increasingly blurred."[2]

The story of how James Watson and Francis Crick described the structure of the DNA helix in 1953 is well known. Less known, but of considerable consequence, is what followed. With little hesitation, they announced that DNA is "the secret of life"—and began to promote what was to become known as "the central dogma"— the notion that genetic information flows in only one direction, from DNA to RNA to a protein, and that this process directly determines an organism's characteristics. This dogma was, as described by geneticist Mae-Wan Ho, author of *Living with the Fluid Genome*, "just

another way of saying that organisms are hardwired in their genetic makeup and the environment has little influence on the structure and function of the genes." In her book, Dr. Ho argues that the central dogma is too simplistic. She observes that not all DNA "codes for proteins" and that the genome is fluid and interactive. Similarly, in a 1992 *Harper's Magazine* article, "Unraveling the DNA Myth: The Spurious Foundation of Genetic Engineering," Queens College biologist Barry Commoner writes that "the central dogma is the premise that an organism's genome—its total complement of DNA genes—should fully account for its characteristic assemblage of inherited traits. The premise, unhappily, is false."

Still, the singular view of "life as DNA" dominated biology in the late twentieth century, in part because its very simplicity provided the biological rationale for engineering DNA. Technological advances in other fields—the study of enzymes that cut DNA, and bacteria that recombine it—were teamed up with high speed computers that provided the computational muscle needed. And yet, even as the old biology became the "new and improved" molecular biology, it was promoted with a social pedigree about how it would serve the public. Its mandate was the same one that was used to colonize the "new world" and to settle the Wild West—the promise that *this* progress would provide everyone a better life.

Judging by his comments, if James Watson had had his way, research would have proceeded undeterred by any concerns over the hazards that genetic engineering posed. He said he'd always felt that the "certain promise" of this revolutionary new technology far outweighed its "uncertain peril." But others, such as Paul Berg of Stanford University, were calling for a more measured approach. In 1975 Berg joined other scientists concerned about the risks of genetic engineering in a meeting held at the Asilomar conference center, near Monterey, California. It was a rare collective action, with participants coming from a spectrum of universities, government agencies, and research institutes.

In his introductory remarks, David Baltimore of MIT noted that the participants were there to discuss "a new technique of molecular biology," one that "appears to allow us to outdo the standard events of evolution by making combinations of genes which could be unique in natural history." He went on to say that they should design a strategy to go forward that would "maximize the benefits and minimize the hazards." They produced a 35-page report that detailed their concerns about creating new pathogens and toxins, the emergence of allergens and disease vectors that could cause cancer or immune disorders, as well as "unpredictable adverse consequences" and the specter of "wide ecological damages."

Then, in the last hours of the meeting, on the very last night, a couple of the participants pointed out that the public had the right to assess and limit this technology. What happened next was pivotal. These scientists believed they were entitled to benefit from the extraordinary potential of genetic engineering and they argued that they could find technological fixes for any problems that might emerge. Susan Wright, author of *Molecular Politics,* a history of biotech regulatory policy, recalls that there was virtual unanimity for the idea that scientists would create a central role for themselves in policymaking—to the exclusion of society in general. From then on, Wright says, this "reductionist discourse" became doctrine. Asilomar defined the boundaries of public discourse, and the questions about potential hazards that were raised there went unanswered.

PUBLIC POLICY: THE ENDLESS FRONTIER

The inoculation that Asilomar gave biotechnology against the ravages of government control was given a booster shot a few years later when executives from the Monsanto Corporation visited the Reagan White House. The industry sought and obtained assurance that they would not be blindsided by regulation. After all, these early developers of GMOs were agrochemical companies like Dow Chemical, DuPont, Novartis, and Monsanto, who were the sources of pervasive chemical pollution that resulted in the environmental laws that were passed in the 1960s. This time, they were intent on getting to the lawmakers before the public did.

The resulting "regulatory reform" was announced in 1992, by then Vice President Dan Quayle, at a press conference in the Indian Treaty Room near his office. It was custom-made for the industry. The new policy left just enough oversight in place to give the industry political cover, so that they could offer assurances to the public that the government was watching out for the public interest when in fact it was not. The regulatory system that was adopted, which is essentially what is still in place today, is basically voluntary and passive. It's a "don't look, don't tell" arrangement whereby the industry doesn't tell the government about problems with its products and the government doesn't look for them.

Quayle said that government "will ensure that biotech products will receive the same oversight as other products, instead of being hampered by unnecessary regulation." The rationale for this policy was a concept called "substantial equivalence," which means that GMOs are not substantially different than conventional crops and foods. The science journal *Nature* dubbed substantial equivalence a "pseudo-scientific concept . . . created primarily to provide an excuse for not requiring biochemical and toxicological tests." Nevertheless, it was adopted by all three agencies responsible for food and agriculture—the United States Department of Agriculture, the Environmental Protection Agency, and the Food and Drug Administration—and it is the reason

there have been no safety studies of GMO foods, no post-market monitoring, no labels, no new laws, no agency coordination, and no independent review.

Henry Miller, head of biotechnology at FDA from 1979 to 1994, told the *New York Times* in 2001 that government agencies did "exactly what big agribusiness had asked them to do and told them to do." During Miller's tenure at the FDA, staff scientists were writing memos that called for further testing and warning that there were concerns about food safety. But the man in charge of policy development at FDA was Michael Taylor, a former lawyer for Monsanto. And, according to Steven Druker, a public-interest lawyer who obtained three of these internal FDA memos, under Taylor "references to the unintended negative effects of bioengineering were progressively deleted from drafts of the policy statement."

Taylor went on to become an administrator at the USDA in charge of food safety and biotechnology, and then became a vice-president at Monsanto. All three agencies continue to employ people who are either associated with biotech companies or who formerly worked for them. At least 22 cases of this "revolving door" between government and industry have been documented. Biotech lawyers and lobbyists serve in policy-making positions, leave government for high paying jobs with industry, and in some cases return to government to defend industry interests again. Still, dismantling regulatory oversight was only part of industry's overall strategy to commercialize GMOs.

BREAKING THE BIOLOGICAL BARRIERS

All the big agrochemical seed companies—DuPont, Monsanto, Pioneer Hi-Bred, and Dekalb—were betting the farm on genetic technologies in the 1980s. But just one crop, corn, stood in their way. Corn was becoming the "Holy Grail" of agricultural biotechnology because these companies knew that if this idea was ever going to be commercially viable, it had to work with corn—which is of central importance to American agriculture. As they raced to find a way to genetically engineer corn, they perfected the complicated steps required to transform plants into transgenic crops. It all came together in June 1988, when Pioneer Hi-Bred patented the first viable and replicable transgenic corn plant.

In the end, the secret of recombining DNA was found not so much through a process of tedious, repetitive experimentation as of that traditional, Wild-West way of getting what you want—using stealth and brute force. The primary problem genetic engineers faced was how to get engineered DNA into target cells without destroying them. For some plants, like tobacco and soybeans, the problem was solved by the use of stealth. A soil microbe that produces cancer-like growths in plants was recruited to "infect" cells with new modified DNA. This *agrobacterium* formed a nonlethal hole in the wall of a plant cell that allowed the new DNA to sneak in. But that method did not work with corn. For corn, a more forceful cell invasion technique was called for, one that resulted in the invention of the gene gun.

One day in December 1983, during the Christmas break at Cornell University, three men put on booties, gowns, and hair coverings, picked up a gun, and entered the university's National Submicron Facility. John Sanford, a plant breeder at Cornell, and his colleagues, the head of the facility and a member of his staff, were about to shoot a bunch of onions to smithereens. For years, they had been looking for ways to speed up the conventional plant breeding process using genetic transformation techniques. Like other researchers, they had had difficulty forcing DNA fragments through the relatively thick walls of plant cells. They'd tried using lasers to drill mini-holes in cell walls and everything from ion beams to microscopic needles to electric shocks, but these methods either failed to deliver the payload or destroyed the cells in the process.

Then one day, while waging a backyard battle with some pesky squirrels, Sanford got the idea of using a gun. He figured out how to load the gun with specially coated microscopic beads, and then he and his friends tried the idea out on the onions. Soon, pieces of onion were splattered everywhere and the smell of onions and gun powder permeated the air. They kept up this odorous massacre until they figured out how to make it work. It seemed implausible, even laughable, at the time. But the gene gun, which uses .22-caliber ballistics to shoot DNA into cells, is now found in biotechnology laboratories all over the world.

Although it is clearly a "hit or miss" technique, transferring DNA is actually straightforward. The tricky part is getting the target plant to accept the new genes. That requires overcoming billions of years of evolutionary resistance that was specifically designed to keep foreign DNA out. You simply can't get a fish and a strawberry to mate, no matter how hard you try—or at least you couldn't until now. Genetic engineers are now able to take a gene that produces a natural anti-freeze from an arctic flounder and put it into a strawberry plant so that its fruit is frost resistant. But this feat can only be accomplished through the use of specially designed genes that facilitate the process. Along with the trait gene, every GMO also contains genetically engineered vectors and markers, antibiotic resistance genes, viral promoters made from the cauliflower mosaic virus, genetic switches and other constructs that enable the "transformation" process.

Once all these genes are inserted, where they end up and what they may do are unknown. The only precise part of this technique is the identification and extraction of the trait DNA from the donor organism. After that, it's a biological free-for-all. In genetic engineering, failure is the rule. The way you get GMO crops to look and act like normal crops is to do thousands and thousands of insertions, grow the ones that survive out, and then see what you get. What you finally select for further testing and release are those "happy accidents" that appear to work. The rest of the millions of plants, animals and other organisms that are subjected to this process are sacrificed or thrown out—or end up in some lab technician's monster box.

PROCESS, NOT PRODUCT

The public controversy over GMOs has focused largely on the products, on how they are marketed, and on what is planted where. But it now appears that the process used to make them, and the novel genetic constructs used in the process, may constitute greater threats to human and environmental health than the products themselves. There are documented reports of allergenic reactions to GMO foods. According to a report in *Nature Biotechnology*, for example, the commonly used cauliflower mosaic virus contains a "recombination hotspot" that makes it unstable and prone to causing mutations, cancer, and new pathogens. The British Medical Association and the U.S. Consumer's Union have both warned about new allergies and/or adverse impacts on the immune system from GMO foods. And public health officials in Europe are concerned that anti-bacterial resistance marker genes in GMOs could render antibiotics ineffective. There have been only about 10 studies done on human health and GMOs, and half of them indicate reasons for concern, including malformed organs, tumors, and early death in rats.

There are also increasing reports of a phenomenon previously thought to be rare, "horizontal gene transfer," which happens when genes travel not just "vertically" through the normal processes of digestion and reproduction, but laterally, between organs in the body or between organisms—sort of like Casper the Ghost floating through a wall. Geneticist Mae-Wan Ho, who has been documenting this phenomenon, says it's happening because the new technology "breaks all the rules of evolution; it short-circuits evolution altogether. It bypasses reproduction, creates new genes and gene combinations that have never existed, and is not restricted by the usual barriers between species."

In 2001, the world's most widely grown GMO, Monsanto's Round-up Ready soybean, was found to contain some mysterious DNA. Monsanto claimed it was native to the plant. When it was shown instead to be the result of the transformation process, Monsanto couldn't explain how it got there. And it has been shown that the nutritional profile of the transgenic soybean is different than that of the conventional variety.

A new report, based on peer-reviewed scientific literature and USDA documents,[3] has found that significant genetic damage to the integrity of a plant occurs when it is modified, including rearrangement of genes at the site of the insertion and thousands of mutations and random modifications throughout the transgenic plant. Another study, by David Schubert of the Salk Institute for Biological Studies in La Jolla, California, found that just one transgenic insertion can disrupt 5 percent of the genes in a single-cell bacterium. Translated into plant terms, that means 15,000 to 300,000 genes get scrambled. Industry was given a blank check by government allowing it to commercialize the technology prematurely, before science could validate the techniques being used to evaluate the safety of the products being developed.

STRATEGIC CONTAMINATION

Even before GMOs were released in the mid-1990s, they were thought by some scientists to be promiscuous. Now that GMO contamination is running rampant, it's hard to believe that the biotech industry wasn't aware of that risk. The industry would have had to ignore early warnings such as a study done at the University of Chicago which found one transgenic plant that was 20 times more likely to interbreed with related plants than its natural variety. But now, because herbicide-tolerant genes are getting into all sorts of plants, farmers have to contend with "super-weeds" that cannot be controlled with common chemicals, and American agriculture is riddled with fragments of transgenic material. The Union of Concerned Scientists recently reported that the seeds of conventional crops—traditional varieties of corn, soybeans, and canola—are now "pervasively contaminated with low levels of DNA originating from engineered varieties of these crops." One laboratory found transgenic DNA in 83 percent of the corn, soy, and canola varieties tested.

GMO contamination is causing mounting economic losses, as farmers lose their markets, organic producers lose their certification, and processors have to recall food products. The contamination is even beginning to affect property values. Consumers are eating GMOs, whether they know it or not, and even GMOs not approved for human consumption have shown up in our taco shells. New "biopharmaceutical" crops used to grow drugs have leaked into the human food supply. And across the nation, hundreds of open field plots are growing transgenic corn, rice, and soybeans that contain drugs, human genes, animal vaccines, and industrial chemicals, without sufficient safeguards to protect nearby food crops.

It's not only food and farming that are affected. Part of what makes GMOs such an environmental threat is that, unlike chemical contamination, GMOs are living organisms, capable of reproducing and recombining, and once they get out, they can't be re-

called. Now that there are genetically engineered fish, trees, insects, and other organisms, there's no limit to the kind of environmental surprises that can occur. The widespread ecological damage discussed at Asilomar is now a reality. In just one example of what can happen, a study found that when just 60 transgenic fish were released into a wild population of tens of thousands of fish, all the wild fish were wiped out in just 40 generations. And what will happen when there are plantations of transgenic trees, which can disperse GMO pollen for up to 40 miles and over several decades? Without physical or regulatory restraints, GMOs pose a very real threat to the biological integrity of the planet. As GMO activists say, it gives "pollution a life of its own."

The unasked question that lingers behind all the stories of GMO contamination is: what is the role of industry? How do the manufacturers of GMOs benefit from gene pollution? The fact is, the industry has never lifted a finger to prevent it and the biological and political system they have designed for it encourage its spread. The industry calls contamination an "adventitious presence," as if it were a benign but unavoidable consequence of modern life, like background radiation from nuclear testing.

In the United States, there are no legal safeguards in place to protect the public—not even labels. Labels would at least provide the consumer with a means for tracing the source of any problems that occur. Plus, without liability laws, the industry avoids accountability for any health or environmental damage it causes. It opposes independent testing and then takes advantage of the lack of data to make false assurances about its products' safety. The *Wall Street Journal* reported in 2003 that "makers of genetically modified crops have avoided answering questions and submitted erroneous data" on the safety of their products to the federal government. They have spent hundreds of millions of dollars on massive public relations campaigns that use sophisticated "perception management" techniques all aimed at falsely assuring the public, and government agencies, that their products are useful and safe.

Beyond their not having to label and segregate GMOs, biotech companies can manufacture, sell, and distribute them without having to take expensive precautions against contamination. They do not have to monitor field practices or do any post-market studies. When farms or factories are contaminated with GMOs, the industry is not held responsible for clean-up costs, as would be the case with chemical contamination. Instead, massive GMO food and crop recalls have been subsidized by taxpayers. Industry not only doesn't pay for a farmer's losses; it often sues the farmer for patent infringement and makes money on the deal. Monsanto, in particular, has profited richly by extorting patent infringement fines from farmers whose crops were inadvertently contaminated.

In September 2004, a study reported that herbicide-resistant genes from Monsanto's new bioengineered creeping bentgrass were found as far away as measurements were made—13 miles downwind. Monsanto's response was that there was nothing to worry about; it had proprietary herbicides that could take care of the problem, assuring more the sale of its products than a limit to the contamination. By assiduously avoiding any responsibility for the proliferation of GMOs, and by defeating attempts by the public to contain them, the agricultural biotechnology industry has thus virtually ensured that GMO contamination will continue unabated. A biotech industry consultant with Promar International, Don West-fall, put it this way: "the hope of industry is that over time the market is so flooded that there's nothing you can do about it. You just sort of surrender."

The most alarming case of GMO contamination is the discovery of transgenes in corn at the center of the origin of corn in Mexico. From the time GMO corn was first planted in the U.S. Midwest, it took only six years to make its way back home in the remote mountainous regions of Puebla and Oaxaca, Mexico. Ignacio Chapela, a Mexican-born microbial biologist, was the scientist who first reported this contamination in 2001. Early in 2002, I visited the area with Dr. Chapela to investigate the cultural and economic implications of his findings. While I was there I got a first-hand look at the complicity of government and industry in the spread of GMO contamination.

The genetic diversity of corn, the world's most important food crop after rice, has been fostered for thousands of years by Zapotec and hundreds of other indigenous farming communities who have lived in these mountainous areas since before the Spanish arrived. Now their traditional land-based ways of life, the sacred center of their culture, and the source of their economic livelihood, corn, has been imperiled by this new form of colonization. The farmers I talked to there were well informed, but worried about their cultural and economic survival. What they did not understand was how transgenic corn got into their fields.[4]

Early press reports blamed the farmers themselves, based on the observation that in order to help support their families and communities, some of them travel to the U.S. to work as migrant workers. But in fact, it turned out that the cause of the contamination was the Mexican government and "free trade" rules. Although Mexico had banned the commercial planting of transgenic corn, under pressure of NAFTA and the biotech industry it was importing corn from the U.S. that it knew was contaminated. It then distributed this whole-kernel corn to poor communities as food aid, without labels or warnings to rural farmers that it should not

be used for seed. This highly subsidized corn, which is being dumped on third world farmers at prices that are lower than the cost of production, undermines local corn markets. But instead of taking steps to stop the spread of this contamination, or to protect its farming communities, or even to guard its fragile biodiversity, the Mexican government, the international seed banks, and the biotech industry all deflected public and media attention to a convenient scapegoat—Dr. Chapela.

THE SUPPRESSION OF SCIENCE

Chapela and his graduate student, David Quist, had published their findings in the peer-reviewed journal *Nature*.[5] They had actually made two findings: first, that GMOs had contaminated Mexico's local varieties of corn—in technical terms, that "introgression" had occurred. And second, they found that once transgenes had introgressed into other plants, the genes did not behave as expected. This is evidence of transgenic instability, which scientists now regard with growing concern. But allegations of such instability can be dangerous to make because they undermine the central dogma's basic article of faith: that transgenes are stable and behave predictably. Not surprisingly, the industry attacked the first finding, but was foiled when the Mexican government's own studies found even higher levels and more widespread GMO contamination than the *Nature* article had reported. The industry then focused its attack to the finding of transgenic instability.

For over a year, the industry relentlessly assailed Quist and Chapela's work, both in the press and on the Internet. As the debate raged on, scientists argued both sides, fueled, Chapela says, by a well developed and generously funded industry public relations strategy that did not hesitate to make the attacks personal. Monsanto even retained a public relations firm to have employees pose as independent critics. The outcome was unprecedented. The editor of *Nature* published a letter saying that "in light of the criticisms . . . the evidence available is not sufficient to justify" the publication of the original paper. This "retraction" made reference to the work of two relatively unknown biologists, Matthew Metz and Nick Kaplinsky. At the time, Kaplinsky was still a graduate student in the Department of Plant and Microbial Biology at UC Berkeley. Metz had finished his work at Berkeley and was a postdoctoral fellow at the University of Washington. What few knew was that their role in the *Nature* controversy was linked to another dispute that they, Quist, and Chapela, had been involved in. That earlier dispute, too, was about the integrity of science. And in that case, Chapela had led the faculty opposition—and Quist had been a part of the student opposition—to private funding of biotechnology research at UC Berkeley.

THE PIE ON THE WALL

The University of California at Berkeley is a "land grant" institution, meaning that it was created to support California's rich agricultural productivity. But by the late 1990s, Cal had all but abandoned its original mission. Berkeley had become the national leader in collecting royalty payments on its patents, many of which related to the development of genetic engineering. This development was facilitated by the passage of the Bayh-Dole Act of 1980, which allowed universities to patent their research, even if it was publicly funded. By the fall of 1998, the private funding of research at Berkeley was in its full glory. That year, the dean of the College of Natural Resources, Gordon Rausser, announced that he had brokered an unprecedented research deal with the Novartis Corporation, then a multinational Swiss agrochemical and pharmaceutical giant.

Novartis was giving just one department of the College, the Department of Plant and Microbial Biology, $25 million over a 5-year period. The deal was fraught with conflicts of interest, not the least of which was that Novartis employees served on academic committees and got first license rights to the Department's research products. Novartis proudly announced that "the ultimate goal" of the agreement was "to achieve commercialization of products." This took private intrusion into the public sector to a new level, allowing private investors to profit directly from public investment in research, and arousing concerns about the increasing privatization of public research institutions across the country.

In true Berkeley fashion, the controversy erupted into protests. When the deal was announced in November 1998, I covered the press conference. It was held in a packed room upstairs in Koshland Hall, home of the Department of Plant and Microbial Biology. Novartis executives stood shoulder to shoulder with UC Berkeley administrators and leading faculty. They all looked on benevolently while the agreement was formally signed. Then the speeches started. Steven Briggs, president of Novartis Agricultural Discovery Institute, the foundation that funnels corporate money to the university and gets government research and tax credits for Novartis, signed the deal on behalf of Novartis. Briggs, who is an expert on the corn genome, called the agreement—without the least suggestion of irony—"the final statement in academic freedom."

The person most responsible for the Novartis deal, Dean Rausser, was proud of his considerable connections in the private sector. While he was dean, he built a consulting company worth millions. During the press conference, he stood at the front of the room with the other key participants. The press and other guests were seated in folding chairs facing them, and students sat on the floor along the walls. Hefty security men in blue blazers with wires dangling from their ears were lined up along the back wall. I was in the front row. Suddenly I felt a commotion erupting behind me. Something rushed past my head, missed its intended target, and splattered on the wall behind the front table. Then another object followed, grazed Dean Rausser, and landed on the floor at his feet. It all happened fast, but I soon realized that I was in the middle of a pie-throwing protest. In their hallmark style, which is humorous political theater, the "Biotic Baking Brigade" had tossed two vegan pumpkin pies (it was Thanksgiving week, after all) at the signers of the Novartis agreement.

As campus security guards wrestled the protesters to the floor and then pulled them out of the room, the AP reporter who was sitting next to me jumped up and ran out to call in her story. I stayed and watched Dean Rausser, who had been speaking at the time. He just looked down, brushed some pie off his suit, then smiled and shrugged. I got the distinct feeling he was enjoying the moment. He went on with his presentation, and for the rest of the time he was speaking, pie filling drooled down the wall behind him.

As a child of the '60s and a member of the UC Berkeley class of 1965, I was reminded of the winter of 1964, when Mario Savio gave his famous "rage against the machine" speech[6] on the steps of the campus administration building. When it began, the Free Speech Movement was about academic freedom but it enlarged into demonstrations against the war in Vietnam and support for the civil rights and women's movements. A lot was achieved, especially in terms of environmental protection. But it was always about who controls the levers of "the machine," as Savio called it. By 1998, however, the conservative backlash that was provoked by these protests was in full bloom. Private interests had successfully dismantled the regulatory system, invaded the ivy tower, and taken over the intellectual commons. The corporate executives and their academic beneficiaries who were there to celebrate the Novartis agreement clearly had nothing to fear—a fact that was neatly affirmed by Dean Rausser's shrug.

The Novartis funding ended in 2003. By then, faculty and graduate students who were on both sides of the debate had gone their separate ways. Dr. Chapela stayed, and continued to teach at Berkeley. As 2003 drew to a close, he was up for a tenure appointment. Even though he'd garnered extraordinary support from faculty, students, and the public, his role in opposing corporate funding on campus apparently cost him his teaching career. After an unusually protracted process, the University denied him tenure. In 2004, a 10-person team at Michigan State University that had spent two years evaluating the Novartis-Berkeley agreement concluded that the deal was indeed "outside the mainstream for research contracts with industry" and that Berkeley's relationship with Novartis created a conflict of interest in the administration that affected their tenure decision against Dr. Chapela.

Instead of applauding the bravery of scientists who question biotechnology, or at least encouraging further scientific inquiry, the industry and its cronies in the academic world denounce their critics. Dr. Chapela has now joined a growing number of scientists who have paid a high price for their integrity. Others have lost jobs, been discredited in the press, told to change research results or to repudiate their findings.[7] And for each victim whose story is told publicly, there are others who have been silenced and cannot come forward. The implications of the trend toward the privatization of research and the repression of academic freedom go far beyond the question of where the funds come from and who decides what gets studied. It's a trend that deeply undermines the public's faith in science, and the result is that society will lose the means to adequately evaluate new technologies. It may also mean that we adopt a view of the natural world so mechanistic that we will not even recognize the threats we face.

If science were free to operate in the public interest, it could provide the intellectual framework for innovations that work with nature, instead of against it. There already are technologists that use natural solutions to heal the wounds of the industrial age, formulate sustainable food production and energy solutions, create new economic opportunities through the imaginative use of ecological design, and build local self-reliant communities that foster both cultural and biological survival. So we do have a choice of technologies, and nature remains abundantly generous with us. What we do not have, given the perilous environmental state of the planet, is a lot of time left to sort this out. And as long as the critics are silenced, we can be lulled by the "certain promises" of genetic engineering, that it will provide magic answers to those age old problems of hunger and disease, and in doing so, be diverted from attending to its "uncertain perils."

THE NATURE OF TRESPASS

Trespass, in legal parlance, means "an unlawful act that causes injury to person or property." It connotes an act of intrusion, usually by means of stealth, force, or violence. It also implies the right to allow or to refuse an intrusion. A trespass occurs when that right has been violated. Genetic engineering technology is a trespass on the public commons. This is because of the way transgenics are designed and the way "the molecular vision" has been pursued. This vision required that science be compromised to the point where it would overlook the complex boundary conditions that form the very foundation of life. It had to have the hubris to break the species barriers and place itself directly in the

path of evolution, severing organisms from their hereditary lineage. And it requires the use of stealth and violence to invade the cell wall, and the implanting of transgenic life forms into an involuntary participant with organisms that are especially designed to overcome all resistance to this rude intrusion.

This trespass continues when ownership is forced on the newly created organisms in the form of a patent. The patenting of a life form was widely considered immoral, and until the U.S. Supreme Court approved the patenting of life in 1980, it was illegal. With that one decision, private interests were given the right to own every non-human life form on earth. We clearly are, as President Bush recently declared, "the ownership society." Now, when GMOs enter the borderless world of free trade and permeate every part of the web of life, they carry within them their owner's mark and effectively privatize every organism they infiltrate. This is made all the more unacceptable because this expensive technology is so unnecessary. Most of what agricultural biotechnology sells, such as insect-resistant plants and weed-control strategies, is already available by other means. Traditional plant breeding can produce all these advances and more—including increased yield, drought or salt resistance, and even nutritional enhancements. The whole point of the commercial use of the genetic engineering technology is the patents, and the social control they facilitate. The reason GMOs were inserted into crops is so that agbiochemical companies could own the seed supply and control the means and methods of food production, and profit at each link in the food chain.

Genetic engineering is a manifestation—perhaps the ultimate manifestation—of the term "full spectrum dominance." In this case, the dominance is achieved on multiple levels, first by exerting biological control over the organism itself, then by achieving economic control over the marketplace and then through "perceptual" control over public opinion. GMOs are disguised to look just like their natural counterparts, and then are released into the environment and the human food chain through a matrix of control that identifies and disables every political, legal, educational, and economic barrier that could thwart their owners' purpose. Arguably, this description suggests a more sinister level of intention than really exists. But the fact remains that denial of choice has been accomplished and it is crucial to this strategy's success. As a Canadian GMO seed industry spokesperson, Dale Adolphe, put it: "It's a hell of a thing to say that the way we win is don't give the consumer a choice, but that might be it."

Agricultural genetic engineering is dismantling our once deeply held common vision about how we feed ourselves, how we care for the land, water, and seeds that support us, and how we participate in decisions that affect us on the most intimate personal and most essential community level. The ultimate irony of our ecological crisis, says David Loy, a professor and author of works on modern Western thought, is that "our collective project to secure ourselves is what threatens to destroy us." But still, there are problems with making moral arguments like these. One is that we lack a practical system of public ethics—some set of common standards we can turn to for guidance. Another is that it does not address the most serious threat to our security, which is that no amount of science, fact, or even moral suasion is of any consequence when we are left with no options.

At the end of my inquiry I came to the conclusion that genetic engineering, at least as it is being used in agriculture is, by design, inherently invasive and unstable. It has been imposed on the American public in a way that has left us with no choice and no way to opt out, biologically or socially. Thus, the reality is that the evolutionary legacy of our lives, whether as human beings, bees, fish, or trees, has been disrupted. We are in danger of being severed from our own ancestral lines and diverted into another world altogether, the physical and social dimensions of which are still unknown and yet to be described.

Notes

1. Although I use the terms "biotechnology" and "genetic engineering" interchangeably, along with references to "transgenes" and "genetically modified organisms," I am, in all cases, referring to recombinant DNA technology used to cross species boundaries. I am not using the term "biotechnology" in its general sense, which can include natural processes. This analysis of genetic engineering will focus *only* on its agricultural applications. It does not address issues that might apply to medical or other uses.

2. Kay, Lily E. *The Molecular Vision of Life. Caltech, the Rockefeller Foundation, and the Rise of the New Biology,* Oxford University Press, 1993, p. 23.

3. "Genome Scrambling—Myth or Reality? Transformation-induced Mutations in Transgenic Crop Plants" by Drs. Wilson, Latham, and Steinbrecher is available at www.econexus.info.

4. The full story of how GMOs got into native corn in Mexico is told in "Risking Corn, Risking Culture," by this author, Claire Hope Cummings, *World Watch,* November/December 2002.

5. Quist, D. and Chapela, I., "Transgenic DNA Introgressed into Traditional Maize Landraces in Oaxaca, Mexico." *Nature,* 414:541–543, November 29, 2001.

6. In that speech Savio said that there comes a time when "the operation of the machine becomes so odious, makes you so sick at heart, that you can't take part. You can't even passively take part and you've got to put your bodies upon the gears and upon the wheels, upon the levers, upon all the apparatus, and you got to make it stop. . . ."

7. The stories of four such scientists and their reflections on their experiences can be heard on a recording of a remarkable conversation among them called "The Pulse of Scientific Freedom in the Age of the Biotech Industry" held on the UC Berkeley campus in December, 2003. A link to the web archive is available at http://nature.berkeley.edu/pulseofscience/plx/conv.txt1.html.

Genetic Engineering and the Concept of the Natural

Mark Sagoff

In the past several years, there has been a furor generated over the use of genetically modified organisms (GMOs) in the world food supply. Particularly in Europe, but also in several other countries, there are active consumer movements whose goal is to rid the supermarket shelves of any trace of GMO ingredients. The controversy has ignited a trade war between the United States, which has a rather laissez-faire attitude toward GMO crops, and its European trading partners, whose consumers are demanding GMO labeling, if not a complete ban, on GMO foods.

In this selection, philosopher Mark Sagoff takes a look at food labeling in general; he argues that the consumer's preference for "all-natural" foods has been stimulated in large part by the food industry itself, which has decided to sell its products along with the fantasy that our food is produced without technological intervention. Sagoff distinguishes four senses in which something may be said to be "natural" and argues that part of the problem with the food industry is that it equivocates on the meaning of this term and tries to have it both ways. Illustrating the different senses of "natural" with passages from Shakespeare's *The Winter's Tale,* Sagoff artfully reminds us that there are moral, aesthetic, and cultural value aspects to the debate over genetically engineered foods and that trade-offs between these values and those of convenience and consumerism can be papered over but not avoided by advertising claims made by corporate agribusiness.

Focus Questions

1. Why does Sagoff think that the current practice of advertising foods as "natural" is an example of consumer constructivism?

2. What are the four senses of "natural" that Sagoff distinguishes? Which of these senses correctly apply to genetically engineered foods? Which don't? Explain.

3. Is it possible to give consumers what they want with no trade-offs, as industry spokespersons suggest? Why or why not?

Keywords

agribusiness, consumerism, genetic engineering, GMOs, labeling, multinational corporations, nature

From "Genetic Engineering and the Concept of the Natural" by Mark Sagoff in *Genetic Prospect,* © 2003. Reprinted by permission of Roman & Littlefield.

WHY DO MANY CONSUMERS view genetically engineered foods with suspicion? I want to suggest that it is largely because the food industry has taught them to do so. Consumers learn from advertisements and labels that the foods they buy are all natural—even more natural than a baby's smile. "The emphasis in recent years," *Food Processing* magazine concludes, "has been on natural or nature-identical ingredients." According to *Food Product Design,* "the desire for an all natural label extends even to pet food."

The food industry, I shall argue, wishes to embrace the efficiencies offered by advances in genetic engineering. This technology, both in name and in concept, however, belies the image of nature or of the natural to which the food industry constantly and conspicuously appeals. It should be no surprise that consumers who believe genetically modified foods are not "natural" should for that reason regard them as risky or as undesirable. If they knew how much technology contributes to other foods they eat, they might be suspicious of them as well.

ALL-NATURAL TECHNOLOGY

Recently, I skimmed through issues of trade magazines, such as *Food Technology* and *Food Processing,* that serve the food industry. In full-page advertisements, manufacturers insist the ingredients they market come direct from primordial Creation or, at least, that their products are identical to nature's own. For example, Roche Food Colours runs in these trade magazines a full-page ad that displays a bright pink banana over the statement: "When nature changes her colours, so will we." The ad continues:

> Today more and more people are rejecting the idea of artificial colours being used in food and drink. . . .
>
> Our own food colours are, and always have been, strictly identical to those produced by nature.
>
> We make pure carotenoids which either singly or in combination achieve a whole host of different shades in the range of yellow though orange to red.
>
> And time and time again they produce appetising natural colours, reliably, economically, and safely.
>
> Just like nature herself.

Advertisement after advertisement presents the same message: food comes directly from nature or, at least, can be sold as if it did. Consider, for example, a full-page advertisement that McCormick and Wild, a flavor manufacturer, runs regularly in *Food Processing.* The words "BACK TO NATURE" appear under a kiwi fruit dripping with juice. "Today's consumer wants it all," the advertisement purrs, "great taste, natural ingredients, and new ideas. . . . Let us show you how we can put the world's most advanced technology in natural flavors at your disposal. . . ."

This advertisement clearly states the mantra of the food industry: "Today's consumer wants it all." Great taste. Natural ingredients. New ideas. The world's most advanced technology. One can prepare the chemical basis of a flavor, for example, benzaldehyde—almond—artificially with just a little chemical know-how, in this instance, by mixing oil of clove and amyl acetate. To get exactly the same compound as a "natural" flavor, one must employ far more sophisticated technology to extract and isolate benzaldehyde from peach and apricot pits. The "natural" flavor, an extract, contains traces of hydrogen cyanide, a deadly poison evolved by plants to protect their seeds from insects. Even so, consumers strongly prefer all-natural to artificial flavors, which sell therefore at a far lower price.

In its advertisements, the Haarmann & Reimer Corporation (H&R) describes its flavor enhancers as "HypR Clean Naturally." With "H&R as your partner, you'll discover the latest advances in flavor technology" that assure "the cleanest label possible." A "clean" label is one that includes only natural ingredients and no reference to technology. In a competing advertisement, Chr. Hansen's Laboratory announces itself as the pioneer in "culture and enzyme technologies. . . . And because our flavors are completely natural, you can enjoy the benefits of 'all-natural' labeling." Flavor manufacturers tout their stealth technology—i.e., technology so advanced it disappears from the consumer's radar screen. The consumer can be told he or she is directly in touch with nature itself.

The world's largest flavor company, International Flavors & Fragrances (IFF) operates manufacturing facilities in places like Dayton, New Jersey, an industrial corridor of refineries and chemical plants. Under a picture of plowed, fertile soil, the IFF Laboratory, in a full-page display, states, "Where Nature is at work, IFF is at work." The text describes "IFF's natural flavor systems." The slogan follows: "IFF technology. In partnership with Nature." Likewise, MEER Corporation of Bergen, New Jersey, pictures a rainforest under the caption, "It's A Jungle Out There!" The ad states that "true-to-nature" flavorings "do not just happen. It takes . . . manufacturing and technical expertise and a national distribution network . . . for the creation of natural, clean label flavors."

Food colors are similarly sold as both all natural and high tech. "VegetoneH colors your foods *naturally* for a healthy bottom line," declares Kalsec, Inc., of Kalamazoo, Michigan. Its ad shows a technician standing before a computer and measuring chemicals into a test tube. The ad extols the company's "patented natural color systems." The terms "natural" and "patented" fit seamlessly together in a conceptual scheme in which there are no trade-offs and no compromises. The natural is patentable. If you think any of this is contradictory, you will not get far in the food industry.

ORGANIC TV DINNERS

As a typical American suburbanite, I can buy not just groceries but "Whole Foods" at Fresh Fields and other upscale supermarkets. I am particularly impressed by the number of convenience foods that are advertised as "organic." Of course, one might think that any food may be whole and that all foods are organic. Terms like "whole" and "organic," however, appeal to and support my belief that the products that carry these labels are less processed and more natural—closer to the family farm—than are those that are produced by multinational megacorporations, such as Pillsbury or General Foods.

My perusal of advertisements in trade magazines helped disabuse me of my belief that all-natural, organic, and whole foods are closer to nature in a substantive sense than other manufactured products. If I had any residual credulity, it was removed by an excellent cover story, "Behind the Organic-Industrial Complex," that appeared in a recent issue of the *New York Times Magazine.* The author, Michael Pollan, is shocked, shocked to find that the prepackaged microwavable all-natural organic TV dinners at his local Whole Foods outlet are not gathered from the wild by red-cheeked peasants in native garb. They are highly-processed products manufactured by multinational corporations. Contrary to the impression created by advertisements, organic and other all-natural foods are often fabricated by the same companies—using comparable technologies—as those that produce Velveeta and Miracle Whip. And the ingredients come from as far away as megafarms in Chile—not from local farmers' markets.

Reformers who led the organic food movement in the 1960s wished to provide an alternative to agribusiness and to industrial food production, but some of these reformers bent to the inevitable. As Pollan points out, they became multimillionaire executives of Pillsbury and General Mills in charge of organic food production systems. This makes sense. A lot of advanced technology is needed to produce and market an all-natural or an organic ready-to-eat meal. Consumers inspect food labels to ward off artificial ingredients; yet they also want the convenience of a low-priced, pre-prepared, all-natural dinner.

At General Mills, as one senior vice president, Danny Strickland, told Pollan, "Our corporate philosophy is to give consumers what they want with no trade-offs." Pollan interprets the meaning of this statement as follows. "At General Mills," Pollan explains, "the whole notion of objective truth has been replaced by a value-neutral consumer constructivism, in which each sovereign shopper constructs his own reality."

Mass-marketed organic TV dinners do not compromise; they combine convenience with a commitment to the all-natural, eco-friendly, organic ideology. The most popular of these dinners are sold by General Mills through its subsidiary, Cascadian Farms. The ad-vertising slogan of Cascadian Farms, "Taste You Can Believe In," as Pollan observes, makes no factual claims of any sort. It "allows the consumer to bring his or her personal beliefs into it," as the Vice President for Marketing, R. Brooks Gekler, told Pollan. The absence of any factual claim is essential to selling a product, since each consumer buys an object that reflects his or her particular belief system.

What is true of marketing food is true of virtually every product. A product will sell if it is all-natural and eco-friendly and, at the same time, offers the consumer the utmost in style and convenience. A recent *New York Times* article, under the title, "Fashionistas, Ecofriendly and All-Natural," points out that the sales of organic food in the United States topped $6.4 billion in 1999 with a projected annual increase of 20 percent. Manufacturers of clothes and fashion accessories, such as solar-powered watches, are cashing in on the trend. Maria Rodale, who helps direct a publishing empire covering "natural" products, founded the women's lifestyle magazine *Organic Style.* Rodale told the *Times* that women want to do the right thing for "the environment but not at the cost of living well." Advances in technology give personal items and household wares an all-natural eco-friendly look that is also the last word in fashion. Consumers "don't want to sacrifice anything," Ms. Rodale told a reporter. Why should there be a trade-off between a commitment to nature and a commitment to the good life? "Increasingly there are options that don't compromise on either front."

The food industry does not sell food any more than the fashion industry sells clothes or the automobile industry sells automobiles. They sell imagery. The slogan, "Everything the consumer wants with no trade-offs," covers all aspects of our dreamworld. Sex without zippers, children without zits, lawns without weeds, wars without casualties, and food without technology. Reality involves trade-offs and rather substantial ones. For this reason, if you tried to sell reality, your competitor would drive you out of business by avoiding factual claims and selling fantasy—whatever consumers believe in—instead. Consumers should not be confused or disillusioned by facts. They are encouraged to assume that they buy products of Nature or Creation. In view of this fantasy, how could consumers view genetic engineering with anything but suspicion?

NATURE'S OWN METHODS

Genetic engineering, with its stupendous capacity for increasing the efficiencies of food production in all departments, including flavors and colorings, raises a problem. How can genetic recombination be presented to the consumer as completely natural—as part of nature's spontaneous course—as have other aspects of food technology? A clean label would tell consumers there is nothing unnatural or inauthentic about genetically engineered products. Industry has responded in two complementary ways to this problem.

First, the food industry has resisted calls to label bioengineered products. Gene Grabowski of the Grocery Manufacturers Association, for example, worries that labeling "would imply that there's something wrong with food, and there isn't." Michael J. Phillips, an economist with the Biotechnology Industry Organization, adds that labeling "would only confuse consumers by suggesting that the process of biotechnology might in and of itself have an impact on the safety of food. This is not the case."

Second, manufacturers point out that today's genetic technologies do not differ, except in being more precise, from industrial processes that result in the emulsifiers, stabilizers, enzymes, proteins, cultures, and other ingredients that do enjoy the benefits of a clean label. Virtually every plant consumed by human beings—canola, for example—is the product of so much breeding, hybridization, and modification that it hardly resembles its wild ancestors. This is a good thing, too, since these wild ancestors were barely edible if not downright poisonous. Manufacturers argue that genetic engineering differs from conventional breeding only because it is more accurate and therefore changes nature less.

For example, Monsanto Corporation, in a recent full-page ad, pictures a bucolic landscape reminiscent of a painting by Constable. The headline reads, "FARMING: A picture of the Future." The ad then represents genetic engineering as all natural—or at least as natural as are conventional biotechnologies that have enabled humanity to engage successfully in agriculture. "The products of biotechnology will be based on nature's own methods," the ad assures the industry. "Monsanto scientists are working with nature to develop innovative products for farmers of today, and of the future."

In this advertisement, Monsanto applies the tried-and-true formula to which the food industry has long been committed—presenting a technology as revolutionary, innovative, highly advanced, and as "based on nature's own methods." *Everything* is natural. Why not? As long as there are no distinctions, there are no trade-offs. Consumers can buy what they believe in. A thing is natural if the public believes it is. "There is something in this more than natural," as Hamlet once said, "if philosophy could find it out."

FOUR CONCEPTS OF THE NATURAL

If consumers reject bioengineered food as "unnatural," what does this mean? In what way are foods that result from conventional methods of genetic mutation and selection, which have vastly altered crops and livestock, more "natural" than those that depend in some way on gene splicing? Indeed, is anything in an organic TV

dinner "natural" other than, say, the rodent droppings that may be found in it? Since I am a philosopher, not a scientist, I am particularly interested in the moral, aesthetic, and cultural—as distinct from the chemical, biological, or physical—aspects of the natural world. I recognize that many of us depend in our moral, aesthetic, and spiritual lives on distinguishing those things for which humans are responsible from those that occur as part of nature's spontaneous course.

Philosophers have long pondered the question whether the concept of the natural can be used in a normative sense—that is, whether to say that a practice or a product is "natural" is somehow to imply that it is better to that extent than one that is not. Why should anyone assume that a product that is "natural" is safer, more healthful, or more aesthetically or ethically attractive than one that is not? And why is technology thought to be intrinsically risky when few of us would survive without quite a lot of it?

Among the philosophers who have questioned the "naturalistic fallacy"—the assumption that what is natural is for that reason good—the nineteenth-century British philosopher John Stuart Mill has been particularly influential. In his "Essay on Nature," Mill argues that the term "nature" can refer either to the totality of things ("the sum of all phenomena, together with the causes which produce them") or to those phenomena that take place "without the agency . . . of man." Plainly, everything in the world—including every technology—is natural and belongs equally to nature in the first sense of the term. Mill comments:

> To bid people to conform to the laws of nature when they have no power but what the laws of nature give them—when it is a physical impossibility for them to do the smallest thing otherwise than through some law of nature—is an absurdity. The thing they need to be told is, what particular law of nature they should make use of in a particular case.

Of nature in the second sense—that which takes place without the agency of man—Mill has a dour view. "Nearly all the things which men are hanged or imprisoned for doing to one another, are nature's every day performances," Mill wrote. Nature may have cared for us in the days of the Garden of Eden. In more recent years, however, humanity has had to alter Creation to survive. Mill concludes, "For while human action cannot help conforming to nature in one meaning of the term, the very aim and object of action is to alter and improve nature in the other meaning."

Following Mill, it is possible to distinguish four different conceptions of nature to understand the extent to which bioengineered food may or may not be natural. These four senses of the term include:

1. *Everything in the universe.* The significant opposite of the "natural" in this sense is the "supernatural." Everything technology produces has to be com-

pletely natural because it conforms to all of nature's laws and principles.

2. *Creation in the sense of what God has made.* The distinction here lies between what is sacred because of its pedigree (God's handiwork) and what is profane (what humans produce for pleasure or profit).

3. *That which is independent of human influence or contrivance.* The concept of "nature" or the "natural" in this sense, e.g., the "pristine," is understood as a privative notion defined in terms of the absence of the effects of human activity. The opposite of the "natural" in this sense is the "artificial."

4. *That which is authentic or true to itself.* The opposite of the "natural" in this sense is the specious, illusory, or superficial. The "natural" is trustworthy and honest, while the sophisticated, worldly, or contrived is deceptive and risky.

These four conceptions of nature are logically independent. To say that an item or a process—genetic engineering, for example—is "natural" because it obeys the laws of nature, is by no means to imply it is "natural" in any other sense. That genetically manipulated foods can be found within (1) the totality of phenomena does not show that they are "natural" in the sense that they are (2) part of primordial Creation; (3) free of human contrivance; or (4) authentic and expressive of the virtues of rustic or peasant life.

The problem of consumer acceptance of biotechnology arises in part because the food industry sells its products as natural in the last three senses. The industry wishes to be regulated, however, only in the context of the first conception of nature, which does not distinguish among phenomena on the basis of their histories, sources, or provenance. The industry argues that only the biochemical properties of its products should matter to regulation; the process (including genetic engineering) is irrelevant to food safety and should not be considered.

The food industry downplays the biochemical properties of its products, however, when it advertises them to consumers. The industry—at least if the approach taken by General Mills is typical—tries to give the consumer whatever he believes in. If the consumer believes in a process by which rugged farmers on the slopes of the Cascades raise organic TV dinners from the soil by sheer force of personality, so be it. You will see the farm pictured on the package to suggest the product is close to Creation, free of contrivance, and authentic or expressive of rural virtues. What you will not see on any label—if the industry has its way—is a reference to genetic engineering. The industry believes regulators should concern themselves only with the first concept of nature—the scientific concept—and thus with the properties of the product. Concepts re-lated to the process are used to evoke images that "give consumers what they want with no trade-offs."

SHAKESPEARE ON BIOTECHNOLOGY

I confess that, as a consumer, I find organic foods appealing and I insist on "all-natural" ingredients. Am I just foolish? You might think that I would see through labels like "all natural" and "organic"—not to mention "whole" foods—and that I would reject them as marketing ploys of a cynical industry. Yet like many consumers, I want to believe that the "natural" is somewhat better than the artificial. Is this just a fallacy?

Although I am a professional philosopher (or perhaps because of this), I would not look first to the literature of philosophy to understand what may be an irrational—or at least an unscientific—commitment to buying "all natural" products. My instinct would be to look in Shakespeare to understand what may be contradictory attitudes or inexplicable sentiments.

Shakespeare provides his most extensive discussion of biotechnology in *The Winter's Tale,* one of his comedies. In Act IV, Polixenes, King of Bohemia, disguises himself to spy upon his son, Florizel, who has fallen in love with Perdita, whom all believe to be a shepherd's daughter. In fact, though raised as a shepherdess, Perdita is the castaway daughter of the King of Sicily, a close but now estranged friend of Polixenes. Perdita welcomes the disguised Polixenes and an attendant lord to a sheep shearing feast in late autumn, offering them dried flowers "that keep/ Seeming and savour all winter long." Polixenes merrily chides her: "well you fit our ages/ With flowers of winter."

She replies that only man-made hybrids flourish so late in the fall:

> . . . carnations, and streak'd gillyvors,
> Which some call nature's bastards. Of that kind
> Our rustic garden's barren; and I care not
> To get slips of them.

Polixenes asks why she rejects cold-hardy flowers such as gillyvors, a dianthus. She answers that they come from human contrivance, not from "great creating nature." She complains there is "art" in their "piedness," or variegation. Polixenes replies: "Say there be;

> Yet nature is made better by no mean
> But nature makes that mean; so over that art
> Which you say adds to nature, is an art
> That nature makes. . . . This is an art
> Which does mend nature—change it rather; but
> The art itself is nature.

The statement, "The art itself is nature" anticipates the claim made by Monsanto that "The products of biotechnology will be based on nature's own methods." Polixenes, Mill, and Monsanto remind us that everything in the universe conforms to nature's own principles, and relies wholly on nature's powers. From a scientific perspective, in other words, all nature is one. The mechanism of a lever, for example, may occur in the physiology of a wild animal or in the structure of a machine. Either way, it is natural. One might be forced to agree, then, that genetic engineering applies nature's own methods and principles; in other words, "the art itself is nature."

The exchange between Perdita and Polixenes weaves together the four conceptions of nature I identified earlier in relation to John Stuart Mill. When Polixenes states, "The art itself is nature," he uses the term "nature" to comprise everything in the Universe, that is, everything that conforms to physical law. Second, Perdita refers to "great creating nature," that is, to Creation, i.e., the primordial origin and condition of life before the advent of human society. Third, she contrasts nature to art or artifice by complaining that hybrids do not arise spontaneously but show "art" in their "piedness." Finally, Perdita refers to her "rustic garden," which, albeit cultivated, is "natural" in the sense of simple or unadorned, in contrast to the ornate horticulture that would grace a royal garden. The comparison between the court and the country correlates, of course, with the division that exists in Perdita herself—royal in carriage and character by her birth, yet possessed of rural virtues by her upbringing.

Shakespeare elaborates this last conception of "nature" as the banter continues between Perdita and the disguised Polixenes. To his assertion, "The art itself is nature," Perdita concedes, "So it is." Polixenes then drives home his point: "Then make your garden rich in gillyvors,/ And do not call them bastards."

To which Perdita responds:

I'll not put
The dibble in earth to set one slip of them;
No more than were I painted I would wish
This youth should say 'twere well, and only therefore
Desire to breed by me.

Besides comparing herself to breeding stock—amusing in the context, since she speaks to her future father-in-law in the presence of his son—Perdita reiterates a fourth and crucial sense of the "natural." In this sense, what is "natural" is true to itself; it is honest, authentic, and genuine. This conception reflects Aristotle's theory of the "nature" of things, which refers to qualities that are spontaneous because they are inherent or innate.

Perdita stands by her insistence on natural products—from flowers she raises to cosmetics she uses—in spite of Polixenes' cynical but scientific reproofs. Does this suggest Perdita is merely a good candidate for Ms. Rodale's organic chic? Should she receive a free introductory copy of *Organic Style?* Certainly not.

There is something about Perdita's rejection of biotechnology that withstands this sort of criticism. Why have Perdita's actions a moral authority or authenticity that the choices consumers make today may lack?

HAVING IT BOTH WAYS

Perdita possesses moral authority because she is willing to live with the consequences of her convictions and of the distinctions on which they are based. By refusing to paint herself to appear more attractive, for example, Perdita contrasts her qualities, which are innate, to those of the "streak'd gillyvor," which owe themselves to technological meddling. This comparison effectively gives her the last word because she suits the action to it: she does not and would not paint herself to attract a lover. Similarly, Perdita does not raise hybrids, though she admits, "I would I had some flow'rs" that might become the "time of day" of the youthful guests at the feast, such as Florizel.

Perdita does not try to have it both ways—to reject hybrids but also to grow cold-hardy flowers. She ridicules those who match lofty ideals with ordinary actions—whose practice belies their professed principles. For example, Camillo, the Sicilian lord who attends Polixenes, compliments Perdita on her beauty. He says, "I should leave grazing, were I of your flock,/ And only live by gazing." She laughs at him and smartly replies, "You'd be so lean that blasts of January/ Would blow you through and through."

Many people today share Perdita's affection for nature and her distaste for technology. Indeed, it is commonplace to celebrate Nature's spontaneous course and to condemn the fabrications of biotechnology. Jeremy Rifkin speaks of "Playing Ecological Roulette with Mother Nature's Designs"; Ralph Nader has written the foreword to a book titled, *Genetically Engineered Food: Changing the Nature of Nature.* The Prince of Wales, in a tirade against biotechnology, said, "I have always believed that agriculture should proceed in harmony with nature, recognising that there are natural limits to our ambitions. We need to rediscover a reverence for the natural world to become more aware of the relationship between God, man, and creation."

While consumers today share Perdita's preference for the natural in the sense of the authentic and unadorned and spurn technological meddling, they do not share her willingness to live with the consequences of their commitment. They expect to enjoy year round fruits and vegetables of unblemished appearance, and consistent taste and nutritional quality. Gardeners wish to plant lawns and yards with species that are native and indigenous, and they support commissions and fund campaigns to throw back the "invasions" of exotic and alien species. Yet they also want lawns that resist drought, blight, and weeds, and—to quote Perdita again—to enjoy flowers that "come before the swallow dares, and take/The winds of March with beauty." In other words, the consumer wants it both ways. Today's consumers, as Ms. Rodale knows, "don't want to sacrifice anything." Today's consumers insist, as did Perdita, on the local, the native, the spontaneous. Yet they lack her moral authority because they are unwilling to live with the consequences of their principles or preferences. Consumers today refuse to compromise; they expect fruits and flowers that survive "the birth/Of trembling winter" and are plentiful and perfect all year round.

NAKED LUNCH

Those who defend genetic engineering in agriculture are likely to regard as irrational consumer concerns about the safety of genetically manipulated crops. The oil and other products of Roundup Ready soybeans, according to this position, pose no more risks to the consumer than do products from conventional soybeans. Indeed, soybean oil, *qua* oil, contains neither DNA nor protein and so will be the same whether or not the roots of the plant are herbicide resistant. Even when protein or DNA differs, no clear argument can be given to suppose that this difference—e.g., the order of a few nucleotides—involves any danger. Crops are the outcome of centuries or millennia of genetic crossing, selection, mutation, breeding, and so on. Genetic engineering adds but a wrinkle to the vast mountains of technology that separate the foods we eat from wild plants and animals.

The same kind of argument may undermine consumer beliefs that "natural" colors and flavors are safer or more edible than artificial ones. In fact, chemical compounds that provide "natural" and "artificial" flavors can be identical and may be manufactured at the same factories. The difference may lie only in the processes by which they are produced or derived. An almond flavor that is produced artificially, as I have mentioned, may be purer and therefore safer than one extracted from peach or apricot pits. Distinctions between the natural and the artificial, then, need not correspond with differences in safety, quality, or taste—at least from the perspective of science.

Distinctions consumers draw between the natural and the artificial—and preferences for the organic over the engineered—reflect differences that remain important nonetheless to our cultural, social, and aesthetic lives. We owe nature a respect that we do not owe technology. The rise of objective, neutral, physical and chemical science invites us, however, to disregard all such moral, aesthetic, and cultural distinctions and act only on facts that can be scientifically analyzed and proven. Indeed, the food industry, when it is speaking to regulators rather than advertising to consumers, insists on this rational, objective approach.

In an essay titled, "Environments at Risk," Mary Douglas characterizes the allure of objective, rational, value-neutral, science:

> This is the invitation to full self-consciousness that is offered in our time. We must accept it. But we should do so knowing that the price is William Burroughs' *Naked Lunch*. The day when everyone can see exactly what it is on the end of everyone's fork, on that day there is no pollution and no purity and nothing edible or inedible, credible or incredible, because the classifications of social life are gone. There is no more meaning.

Advances in genetic engineering invite us to the full self-consciousness that Douglas describes and aptly analogizes to the prison life depicted in *Naked Lunch*. It is the classifications of social life—not those of biological science—that clothe food and everything else with meaning. Genetic engineering poses a problem principally because it crosses moral, aesthetic, or cultural—not biological—boundaries. The fact that the technology exists and is successful shows, indeed, that the relevant biological boundaries (i.e., between species) that might have held in the past now no longer exist.

Given advances in science and technology, how can we maintain the classifications of social life—for example, distinctions between natural and artificial flavors and between organic and engineered ingredients? How may we, like Perdita, respect the difference between the products of "great creating nature" and those of human contrivance? Perdita honors this distinction by living with its consequences. Her severest test comes when Polixenes removes his disguise and threatens to condemn her to death if she ever sees Florizel again. Florizel asks her to elope, but she resigns herself to the accident of their origins—his high, hers (she believes) low—that separates them forever. Dressed up as a queen for the festivities, Perdita tells Florizel: "I will queen it no further. Leave me, sir; I will go milk my ewes and weep."

Perdita, of course, both renounced her cake and ate it, too. In Act IV, she gives up Florizel and his kingdom, but in Act V she gets them. Her true identity as a princess is eventually discovered, and so the marriage happily takes place. If you or I tried to live as fully by our beliefs and convictions—if we insisted on eating only those foods that come from great creating nature rather than from industry—we would not be so fortunate. "You'd be so lean that blasts of January/ Would blow you through and through."

Perdita is protected by a playwright who places her in a comedy. Shakespeare allows her to live up to her convictions without compromising her lifestyle. This is exactly what the food industry promises to do—"to give consumers what they want with no trade-offs." It is exactly what Ms. Rodale offers—to protect the environment "but not at the cost of living well." The food, fashion, and other industries work off stage to arrange matters so that consumers can renounce genetic engineering, artificial flavors, industrial agriculture, and multinational corporations. At the same time, consumers can enjoy an inexpensive, all-natural, organic, TV dinner from Creation via Cascadian Farms.

Perdita lives in the moral order of a comedy. In that moral order, no compromises and no trade-offs are necessary. You and I are not so fortunately situated. Indeed, we must acknowledge the tragic aspect of life — the truth that good things are often not compatible and that we have to trade off one for the sake of obtaining the other. The food industry, by suggesting that we can have everything we believe in, keeps us from recognizing that tragic truth. The industry makes all the compromises and hides them from the consumer.

This article is based on a presentation made at the National Agricultural Biotechnology Council's annual meeting, "High Anxiety and Biotechnology: Who's Buying, Who's Not, and Why?," held May 22–24, 2001. A version of this article is forthcoming in NABC Report 13 symposium proceedings.

The author acknowledges the support of the National Human Genome Research Institute program on Ethical, Legal, and Social Implications of Human Genetics, Grant R01HG02363; also the National Science Foundation, Grant 9729295.

Sources: *Food Processing,* February 1988. Lucy Saunders, "Selecting an Enzyme," *Food Product Design* (May 1995), online at: http://www.foodproductdesign.com/archive/1995/0595AP. html; Michael Pollan, "Behind the Organic-Industrial Complex," *New York Times Magazine* (May 13, 2001); Ruth La Ferla, "Fashionistas, Ecofriendly and All-Natural," *New York Times* (July 15, 2001); Bill Lambrecht, "Up To 50%+ of Crops Now Genetically Modified," *St. Louis Post-Dispatch,* Washington Bureau (August 22, 1999), and available on-line at http:// www.healthresearchbooks. com/articles/labels2.htm; Jim Wilson, "Scientific Food Fight," in *Popular Mechanics* on-line, which is available at http:// popularmechanics.com/popmech/ sci/0002STRSM.html; John Stuart Mill, "Nature" in *Three Essays on Religion* (New York, Greenwood Press, 1969), reprint of the 1874 ed.; Jeremy Rifkin, "The Biotech Century: Playing Ecological Roulette with Mother Nature's Designs," in *E Magazine* (May/June 1998); Martin Teitel and Kimberly A. Wilson, *Genetically Engineered Food: Changing the Nature of Nature: What You Need to Know to Protect Yourself, Your Family, and Our Planet* (Vermont: Inner Traditions, Int'l, Ltd., 1999); "Seeds of Disaster: An Article by The Prince of Wales," *Daily Telegraph* (June 8, 1998); Mary Douglas, "Environments at Risk," in *Implicit Meanings: Essays in Anthropology* (London: Routledge & Kegan Paul, 1975).

Summary of NIH Guidelines, July 1976

This summary of National Institutes of Health (NIH) guidelines of July 1976 defines four levels of physical containment, P1 through P4, and three levels of biological containment, EK1 through EK3, in order of increasing stringency. Experiments are assigned levels according to their potential risk.

a. Shotgun experiments using *E. coli* as the host	
Non-embryonic primate tissue	P3+EK3 or P4+EK2
Embryonic primate tissue or germ line cells	P3+EK2
Other mammals	P3+EK2
Birds	P3+EK2
Cold blooded vertebrates, non-embryonic	P2+EK2
embryonic or germ line	P2+EK1
If vertebrate produces a toxin	P3+EK2
Other cold blooded animals and lower eukaryotes	P2+EK1
If Class 2 pathogen*, produces a toxin, or carries a pathogen	P3+EK2
Plants	P2+EK1
Prokaryotes that exchange genes with *E. coli*	
Class 1 agents (non-pathogens)	P1+EK1
Low risk pathogens (for example, enterobacteria)	P2+EK1
Moderate risk pathogens (for example, *S. typhi*)	P2+EK2
Higher risk pathogens	banned
Prokaryotes that do not exchange genes with *E. coli*	
Class 1 agents	P2+EK2 or P3+EK1
Class 2 agents (moderate risk pathogens)	P3+EK2
Higher pathogens	banned

In all above cases, if DNA is at least 99% pure before cloning and contains no harmful genes, either physical or biological containment levels can be reduced one step.

b. Cloning plasmid, bacteriophage and other virus genes in *E. coli*

Animal viruses**	P4+EK2 or P3+EK3
If clones free from harmful regions	P3+EK2
Plant viruses	P3+EK1 or P2+EK2
99% pure organelle DNA, Primates	P3+EK1 or P2+EK2
other eukaryotes	P2+EK1

     Impure organelle DNA: shotgun conditions apply.

Plasmid or phage DNA from hosts that exchange
genes with *E. coli*

     If plasmid or phage genome does not contain harmful
     genes or if DNA segment 99% pure and characterised   P1+EK1

     Otherwise, shotgun conditions apply.

Plasmids and phage from hosts which do not exchange
genes with *E. coli*

     Shotgun conditions apply, unless minimal risk that
     recombinant will increase pathogenicity or ecological
     potential of the host, then   P2+EK2 or P3+EK1

c. Animal virus vectors

Defective polyoma virus+DNA from non-pathogen	P3
Defective polyoma virus+DNA from Class 2 agent	P4
If cloned recombinant contains no harmful genes and host range of polyoma unaltered, reduce to	P3
Defective SV40+DNA from non-pathogens	P4

If inserted DNA is 99% pure segment of prokaryotic
   DNA lacking toxigenic genes, or a segment of eukaryotic
   DNA whose function has been established and that
   has previously been cloned in a prokaryotic host-vector
   system, and if infectivity of SV40 in human cells unaltered
	P3

Defective SV40 lacking substantial section of the late
   region+DNA from non-pathogens, if no helper used and no
   virus particles produced
	P3

Defective SV40+DNA from non-pathogen can be used to
   transform established lines of non-permissive cells under
   P3 provided no infectious particles produced. Rescue of
   SV40 from such cells requires
	P4

d. Plant host-vector systems

    P2 conditions can be approximated by insect-free greenhouses, sterilization of
       plant, pots, soil and runoff water, and use of standard microbiological practice.

    P3 conditions require use of growth chambers under negative pressure and routine
       fumigation for insect control.

    Otherwise, similar conditions to those prescribed for animal systems apply.

*Classes for pathogenic agents as defined by the Center for Disease Control.

**cDNAs synthesised *in vitro* from cellular or viral RNAs are included in above categories.

Date	Containment Requirements (selected experiments)	Oversight for Government-Funded Work	Oversight for Work in the Private Sector	Prohibitions
December 1978	Containment levels work with *E. coli* K12 ranges from P1EK1 to P3EK2	Prior review and approval by IBCs; registration with NIH	Voluntary registration	Six classes; cloning of DNA from class 3, 4, or 5 from P1 to P3 pathogens or from oncogenic viruses classified as moderate risk
				Cloning of genes encoding toxins
				Creation of plant pathogens likely to have enhanced virulence or host range.
				Deliberate release into the environment
				Transfer of drug resistance trait not known to occur naturally
				Large-scale experiments (greater than 10 liters)
January 1980 (response to Rowe-Campbell proposal)	Experiments using *E. coli* K12 that are not prohibited and not exempt: P1	*E. coli* K12 experiments that are not prohibited and not exempt: registration with IBCs; no prior review, except for experiments involving expression of genes Other experiments: prior review and approval by IBCs; registration with NIH	Voluntary compliance scheme for private sector, with procedures for protection of trade secrets and commercial and financial information	Six classes retained
November 1980 (response to Singer proposal)	Experiments using *E. coli* K12 and *S. cerevisiae* systems that are not prohibited and not exempt: P1	Elimination of NIH oversight for all experiments assigned containment levels in the guidelines; IBCs responsible for reviewing research for compliance with the guidelines	RAC responsible only for setting containment requirements for L-S processes	Five classes of prohibition; prohibition of the creation of plant pathogens with increased host range and virulence deleted

Date	Containment Requirements (selected experiments)	Oversight for Government-Funded Work	Oversight for Work in the Private Sector	Prohibitions
July 1981 (response to IBC chairs' proposal)	Experiments using approved *E. coli*, *S. cerevisiae*, and *B. subtilis* systems that are not prohibited: exempt; P1 recommended	Nonexempt experiments assigned containment levels in guidelines; prior review and approval by IBCs		Prohibition on the formation of rDNA containing the genes for biosynthesis of toxins limited to a few lethal toxins; other four classes retained
October 1981 (response to Lilly proposal)				Prohibition lifted on large-scale processes using approved *E. coli*, *S. cerevisiae*, and *B. subtilis* systems; such processes require only IBC prior review and approval
April 1982 (response to Baltimore-Campbell proposal)	For nonpathogens: P1	Elimination of NIH oversight for all experiments and processes except for three classes of previously prohibited experiments; IBCs and/or principal investigators responsible for all other experiments and processes		All prohibitions removed; cloning of genes encoding highly lethal toxins, deliberate release into the environment, and transfer of drug resistance trait not known to occur naturally require RAC review and NIH and IBC approval before initiation

Sources: Department of Health, Education, and Welfare, National Institutes of Health, "Guidelines for Research Involving Recombinant DNA Molecules," *Federal Register* 43 (22 December 1978): 60108-31; Department of Health, Education and Welfare, National Institutes of Health, "Guidelines for Research Involving Recombinant DNA Molecules," *Federal Register* 45 (29 January 1980): 6724-49; Department of Health, Education, and Welfare, National Institutes of Health, "Recombinant DNA Research: Actions under Guidelines," *Federal Register* 45 (21 November 1980): 77372-81; Department of Health and Human Services, National Institutes of Health, "Guidelines for Research Involving Recombinant DNA Molecules," *Federal Register* 46 (1 July 1981): 34462-87; Department of Health and Human Services, National Institutes of Health, "Recombinant DNA Research: Actions under Guidelines," *Federal Register* 46 (30 October 1981): 53980-85; Department of Health and Human Services, National Institutes of Health, "Guidelines for Research Recombinant Involving Recombinant DNA Molecules," *Federal Register* 47(21 April 1982):17180–98.

Examples of Industrial Chemicals Produced by Fermentation

Organic Chemical	Microbial Sources	Selected Uses
Acetic acid	*Acetobacter*	Industrial solvent and intermediate for many organic chemicals, food acidulant
Acetone	*Clostridium*	Industrial solvent and intermediate for many organic chemicals
Acrylic acid	*Bacillus*	Industrial intermediate for plastics
Butanol	*Clostridium*	Industrial solvent and intermediate for many organic chemicals
2,3-Butanediol	*Aerobacter, Bacillus*	Intermediate for synthetic rubber manufacture, plastics and antifreeze
Ethanol	*Saccharomyces*	Industrial solvent, intermediate for vinegar, esters and ethers, beverages
Formic acid	*Aspergillus*	Textile dyeing, leather treatment, electroplating, rubber manufacture
Fumaric acid	*Rhizopus*	Intermediate for synthetic resins, dyeing, acidulant, antioxidant
Glycerol	*Saccharomyces*	Solvent, plasticizer, sweetener, explosives manufacture, printing, cosmetics, soaps, antifreeze
Glycolic acid	*Aspergillus*	Textile processing, pH control, adhesives, cleaners
Isopropanol	*Clostridium*	Industrial solvent, cosmetic preparations, antifreeze, inks
Lactic acid	*Lactobacillus, Streptococcus*	Food acidulant, dyeing, intermediate for lactates, leather treatment
Methylethyl ketone	*Chlamydomonas*	Industrial solvent, intermediate for explosives and synthetic resins
Oxalic acid	*Aspergillus*	Printing and dyeing, bleaching agent, cleaner, reducing agent
Propylene glycol	*Bacillus*	Antifreeze, solvent, synthetic resin manufacture, mold inhibitor
Succinic acid	*Rhizopus*	Manufacture of lacquers, dyes and esters for perfumes

Regulatory Agencies and Laws for Product Regulation

NATIONAL INSTITUTES OF HEALTH (NIH)

The Recombinant DNA Molecule Program Advisory Committee (RAC) of the National Institutes of Health (NIH) established voluntary guidelines for research involving recombinant DNA in 1976. These guidelines have been modified and relaxed over the years. The NIH lacks statutory authority to regulate research, and the guidelines have real significance only for recipients of federal grants; noncompliance resulted in loss of federal funding. Institutional Biosafety Committees are now responsible for reviewing, approving, and monitoring the research of principal investigators. The principal investigator is responsible for following appropriate safety standards when conducting experiments.

In the late 1970s the NIH, USDA, FDA, and NSF effected informal coordination of laboratory research; though implementing no new rules, they required research to comply with existing NIH guidelines. Coordination became more difficult when recombinant DNA technology reached the stage of commercial product development, presenting new questions regarding safety. With large-scale commercial production and the potential for deliberate release into the environment, additional regulatory agencies were required. Although existing federal statutes protected both human health and the environment, there was confusion about agency jurisdiction over various products. In fact, products were often regulated by more than one statute through more than one agency during development, testing, and marketing. Consequently, the NIH no longer had sole responsibility for the regulation of biotechnology products. Today, rather than overseeing the release of recombinant organisms, the NIH focuses almost exclusively on human gene therapy research.

ENVIRONMENTAL PROTECTION AGENCY (EPA)

The EPA was created in 1970 to administer executive authority over activities that have the potential to pollute the environment (air, water, and land) in the United States. At the time the USDA was already regulating biological pesticides. Before 1979 each microbial pesticide was registered on an *ad hoc* basis since a policy had not been developed. The EPA has jurisdiction over genetically engineered organisms (GEOs) as designated within the Toxic Substances Control Act (TSCA) and the Federal Insecticide, Fungicide, and Rodenticide Act (FIFRA). In 1986, the EPA stated that the FIFRA applies to all microbial pesticides, not just genetically engineered ones, and to small-scale releases of genetically modified organisms. Also in 1986 the TSCA was extended to cover genetically engineered microorganisms with a variety of uses (such as environmental cleanup and industrial uses). New regulations also applied to small-scale field testing. The EPA review and regulatory process is evolving to address current needs of modern biotechnology. The EPA proposed new regulations for plants that have been genetically modified to resist pests (plant pesticides) and disease (e.g., viral resistance). Recommended regulations were published on November 21, 1994, in the *Federal Register*.

Federal Insecticide, Fungicide, and Rodenticide Act (FIFRA)

The FIFRA regulates pesticides such as bacteria, viruses, fungi, algae, and protozoa and derivatives of any of these. The FIFRA prohibits the distribution, sale, and use of pesticides that have not been registered with the EPA. The EPA reviews all data submitted for microbial pesticides and determines whether registration is appropriate.

For registration, applicants must cite data on product composition, human health effects, environmental fate, and effects on nontarget organisms. The EPA evaluates this information in making a decision about safety. An experimental use permit (EUP) is required to conduct initial tests, even field studies. FIFRA has generally not required an EUP for small-scale testing in an area of 10 acres or less; however, in the case of genetically engineered microbial pesticides, the EPA does require an evaluation for potential risks to determine whether an EUP is required. In all other respects, microbial pesticides are afforded the same review as conventional pesticides.

Toxic Substances Control Act (TSCA)

The TSCA (1976) gives the EPA the authority to regulate chemicals in research and commerce that may pose a threat to human health and the environment, and requires that the EPA review test data within a specific period and determine whether the product is safe or poses a risk. TSCA was enacted to regulate organic and inorganic substances or a combination of substances, and applies to all microorganisms produced for industrial, consumer, and environmental uses with the exception of their manufacture, processing, or distribution as foods, food additives, cosmetics, medicines, or pesticides.

National Environmental Policy Act (NEPA)

The NEPA (1970) requires that all federal agencies prepare an environmental impact statement or an environmental assessment of any major federal action that might adversely affect the human environment. The impact statement must describe the predicted effect of the action, any adverse impacts, and alternatives for the proposed action. NEPA is not a regulatory statute; it simply ensures that agencies evaluate risks and environmental impact of genetically engineered organisms. Other federal statutes regulate the release of GEOs.

FOOD AND DRUG ADMINISTRATION (FDA)

The FDA, residing within the Department of Health and Human Services, regulates the manufacture of food, food additives, drugs, and cosmetics. This agency is charged with monitoring our food and drugs and ensuring that they are not contaminated. The Public Health Service Act (PHSA) ensures that the manufacturer is licensed to produce the product, and the Federal Food, Drug, and Cosmetic Act (FFDCA) ensures that products are manufactured within quality-control guidelines. The FFDCA gives the FDA jurisdiction over foods containing recombinant DNA. Products of biotechnology are evaluated like any other product; they must be approved, for example, under the New Drug Application (NDA), the New Animal Drug Application (NADA), or Product License Application (PLA). The FDA evaluates the product rather than the process, as demonstrated by the Flavr Savr tomato by Calgene, Inc.

UNITED STATES DEPARTMENT OF AGRICULTURE (USDA)

The USDA protects agriculture and forestry in the United States through research oversight and product regulation. This agency also encourages development of agriculture. The Animal and Plant Health Inspection Service (APHIS) regulates genetically engineered organisms within the USDA and administers several statutes to prevent the introduction of plant and animal diseases. APHIS reviews proposals for the release of genetically engineered organisms into the environment and prepares environmental impact statements as required by NEPA. In 1976, the Agriculture Recombinant DNA Research Committee (ARRC) was established to support the NIH-RAC on agricultural issues and to coordinate biotechnology regulation within the USDA. In 1985, the Committee on Biotechnology in Agriculture (CBA) replaced ARRC.

In 1986, research using genetically engineered meat became covered under the same regulations that apply to other experimental meat products by the Food Safety Inspection Service (FSIS). FSIS also regulates the commercial labeling and sale of meat products. The commercial development, sale, and labeling of transgenic meat must be addressed in the future.

Federal Plant Pest Act (FPPA) and Plant Quarantine Act (PQA)

The Federal Plant Pest Act (FPPA) and the Plant Quarantine Act (PQA) provide authority for regulating the movement into and within the United States of genetically engineered organisms that are potential plant pests (also of importance is the Federal Noxious Weed Act of 1974). Included in the 1986 biotechnology framework was a statement which applied the Federal Plant Pest Act (FPPA) to field testing of genetically engineered crops and plant pests (regulations were released in 1987 but relaxed by the Clinton Administration). Currently the role of the FPPA in the commercialization of transgenic crops is unclear.

Virus-Serum-Toxin Act

Within the Virus-Serum-Toxin Act, the USDA has authority over the exportation, importation, production, and distribution of veterinary biological products. Data that demonstrate the safety, purity, and efficacy of the product must be submitted to obtain a product license. If field testing is required to obtain these data, shipment of unlicensed products for experimental purpose is allowed if APHIS has determined that field trials carry little risk of spreading disease and the field trials are conducted under controlled conditions.

Summary of Typical Components of Prokaryotic and Eukaryotic Cells

		PROKARYOTIC	EUKARYOTIC			
	Archaebacteria,					
Cell Component	Function	Eubacteria	Protistans	Fungi	Plants	Animals
Cell wall	Protection, structural support	✔*	✔*	✔	✔	*None*
Plasma membrane	Control of substances moving into and out of cell	✔	✔	✔	✔	✔
Nucleus	Physical separation and organization of DNA	*None*	✔	✔	✔	✔
DNA	Encoding of hereditary information	✔	✔	✔	✔	✔
RNA	Transcription, translation of DNA messages into polypeptide chains of specific proteins	✔	✔	✔	✔	✔
Nucleolus	Assembly of subunits of ribosomes	*None*	✔	✔	✔	✔
Ribosome	Protein synthesis	✔	✔	✔	✔	✔
Endoplasmic reticulum (ER)	Initial modification of many of the newly forming polypeptide chains of proteins; lipid synthesis	*None*	✔	✔	✔	✔
Golgi body	Final modification of proteins, lipids; sorting and packaging them for use inside cell or for export	*None*	✔	✔	✔	✔
Lysosome	Intracellular digestion	*None*	✔	✔*	✔*	✔
Mitochondrion	ATP formation	**	✔	✔	✔	✔
Photosynthetic pigments	Light–energy conversion	✔*	✔*	*None*	✔	*None*
Chloroplast	Photosynthesis; some starch storage	*None*	✔*	*None*	✔	*None*
Central vacuole	Increasing cell surface area; storage	*None*	*None*	✔*	✔	*None*
Bacterial flagellum	Locomotion through fluid surroundings	✔*	*None*	*None*	*None*	*None*
Flagellum or cilium with 9 + 2 microtubular array	Locomotion through or motion within fluid surroundings	*None*	✔*	✔*	✔*	✔
Complex cytoskeleton	Cell shape; internal organization; basis of cell movement and, in many cells, locomotion	*Rudimentary***	✔*	✔*	✔*	✔

 * Known to be present in cells of at least some groups.

 ** Many groups use oxygen-requiring (aerobic) pathways of ATP formation, but mitochondria are not involved.

 *** Protein filaments form a simple scaffold that helps support cell wall in at least some species.

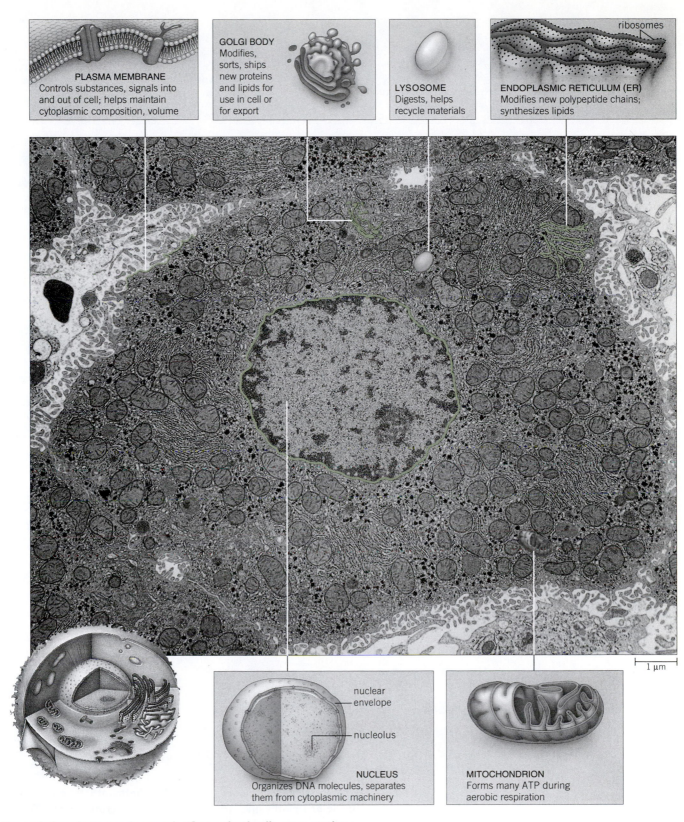

PLASMA MEMBRANE
Controls substances, signals into and out of cell; helps maintain cytoplasmic composition, volume

GOLGI BODY
Modifies, sorts, ships new proteins and lipids for use in cell or for export

LYSOSOME
Digests, helps recycle materials

ENDOPLASMIC RETICULUM (ER)
Modifies new polypeptide chains; synthesizes lipids

ribosomes

1 μm

nuclear envelope

nucleolus

NUCLEUS
Organizes DNA molecules, separates them from cytoplasmic machinery

MITOCHONDRION
Forms many ATP during aerobic respiration

Transmission electron micrograph of an animal cell, cross-section. The sketches highlight key organelles. The specimen is a cell from the liver of a rat.

CHLOROPLAST
Forms ATP and NADPH, then sugars, by photosynthesis

MITOCHONDRION
Forms many ATP during aerobic respiration

PLASMA MEMBRANE
Controls substances, signals into and out of cell; helps maintain cytoplasmic composition, volume

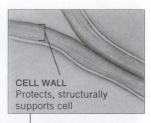

CELL WALL
Protects, structurally supports cell

primary cell wall

primary cell wall of a neighbor cell

middle lamella (layer between adjoining cell walls)

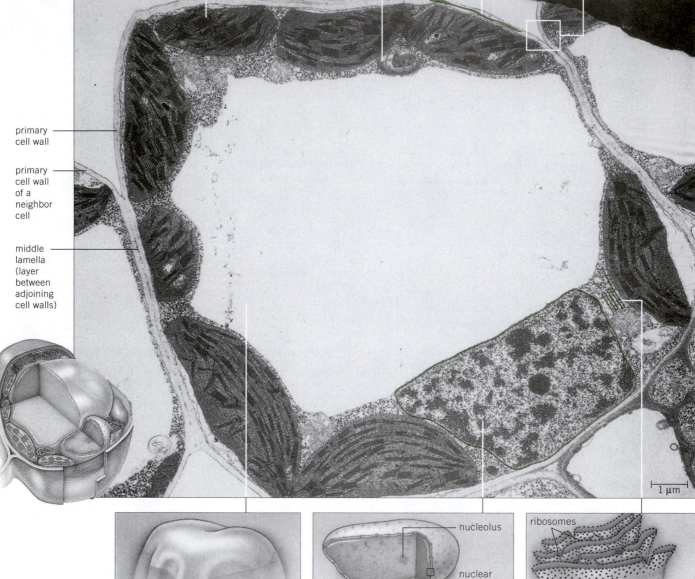

1 µm

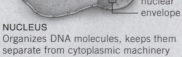

CENTRAL VACUOLE
Increases cell surface area, stores metabolic wastes

nucleolus

nuclear envelope

NUCLEUS
Organizes DNA molecules, keeps them separate from cytoplasmic machinery

ribosomes

ENDOPLASMIC RETICULUM (ER)
Modifies new polypeptide chains; synthesizes lipids

Transmission electron micrograph of a photosynthetic cell from a blade of Timothy grass *(Phleum pratense)*, in cross-section. The sketches highlight key organelles.

Glossary

airlift fermenter A fermenter in which cells are aerated and circulated by air mixing with the media. Air enters the column at the bottom of the vessel and rises as bubbles.

amino acid A small organic molecule that forms the unit of a polypeptide. A hydrogen atom, amino and carboxyl groups, and an R side group are covalently bonded to a central carbon atom.

androgen A male steroid hormone, such as testosterone, that is involved in the male reproductive system.

antibody-mediated allergy A hypersensitive response to an allergen, such as bee venom from a bee sting, that is due to an antibody response. This response usually takes only minutes to develop. This type of allergy is sometimes referred to as *immediate hypersensitivity*.

antibiotic A secondary metabolite, produced by some microorganisms, that kills or inhibits the growth of other microorganisms, often through transcriptional or translational regulation.

antibody An antigen-binding immunoglobulin produced by the B cells of the immune system.

anticodon The three-base-pair region of a tRNA molecule that base pairs with the complementary codon of the mRNA.

aquaculture The farming, often commercial, of a variety of shellfish, crustaceans, finfish, and algae. Aquaculture is a general term, but often refers to freshwater culture (*see* mariculture).

autoimmune disease An immunological disorder resulting from the attack on normal body cells and tissues by the body's lymphocytes.

autonomously replicating sequence (ARS) A sequence that can be cloned to allow extrachromosomal replication of a DNA molecule within an appropriate host cell (e.g., yeast cells); a component of yeast artificial chromosomes.

autoradiography A method by which a radiolabeled molecule such as DNA or RNA or a structure such as a chromosome produces an image on photographic film. The image is called an autoradiogram or autoradiograph.

auxin A class of plant growth-regulating hormones; an example is indoleacetic acid (IAA), which stimulates cell elongation, secondary growth, and fruit development.

bacteriophage A virus that infects bacteria; often called phage.

baculovirus A virus that specifically infects arthropods, most often insects.

baker's yeast A strain of *Saccharomyces cerevisiae* used in bread making.

batch culturing A method in which cells are cultured for a specific period of time after inoculation into the culture medium and then collected for processing.

bioconversion The conversion of a particular organic compound or molecule to an important product through the enzymatic reactions provided by a specific microorganism(s). This process minimizes the synthetic chemical steps required for the synthesis of a commercially important compound.

biodegradable A substance that can be broken down into organic molecules by organisms, usually microorganisms.

biodegradation The breakdown of compounds by living organisms, usually microorganisms and plants.

biolistics A method of delivering DNA into animal and plant cells by DNA-coated microprojectiles discharged under pressure at high velocity.

bioluminescence The ability of an organism to generate light, frequently through a symbiotic association with bioluminescent bacteria. Bioluminescent bacteria produce light when proteins called luciferins, in the presence of oxygen and the enzyme luciferase, are converted to oxyluciferins.

biomass The combined cell mass of a particular group of organisms, often at a particular trophic level in an ecosystem; can be used as an energy source, food supplement, or source of chemical compounds.

bioremediation A technology that uses microorganisms and sometimes plants (phytoremediation) to convert contaminates in soil and water to nontoxic by-products.

biotechnology The commercial use of living organisms or their components to improve animal and human health, agriculture, and the environment.

bivalve A member of a class of burrowing or sessile mollusks that includes oysters, clams, and mussels.

blastodisc A disklike area on the surface of the yolky eggs of birds that divides and gives rise to the embryo.

blunt end The DNA terminus produced, without overhanging 5' or 3' ends, after restriction endonuclease digestion.

CAP *See* catabolite activator protein.

catabolite activator protein (CAP) A protein involved in regulating the *lac* operon; in the absence of glucose, enhances transcription of the genes in the operon by interacting with cyclic AMP; the CAP-cAMP complex binds to the operon CAP site.

5'-cap A chemical modification of the 5' end of eukaryotic mRNA during post-transcriptional processing of the primary transcript (hnRNA).

cDNA Complementary double-stranded DNA synthesized *in vitro* by reverse transcriptase and DNA polymerase with mRNA as the template.

cDNA library Collection of expressed mRNAs, usually from specific cells in an organism, cloned as cDNAs in a vector.

cell-mediated allergy A hypersensitive response to an allergen that takes several hours to several days to develop. It is a cell-mediated reaction that involves T cells and macrophages. Examples include contact dermatitis from poison ivy and the rejection of an organ that has been transplanted from a donor. This type of allergy is sometimes referred to as *delayed hypersensitivity*.

centiMorgan A unit of genetic distance that represents how frequently two genes or loci on the same chromosome can be separated by crossing over. Two genes have a one percent chance of being separated (i.e. on different chromosomes) during recombination if they are located one centimorgan apart. One centimorgan of human DNA is approximately one million base pairs.

centromere A region of DNA on eukaryotic chromosomes to which the mitotic spindle fibers attach during mitosis and meiosis; the region where two chromosomes are attached after replication.

chimera A recombinant DNA molecule or organism that contains sequences from more than one organism.

chromatid One of the two daughter strands of a replicated chromosome joined together by the centromere.

clone Used for an individual or cell from a genetically identical population derived by asexual division. Also a cell harboring a recombinant DNA; a foreign DNA fragment contained within a recombinant clone.

codon engineering Modification of codons within a cloned gene to ones preferred by the host organism (codon bias); used to optimize translation within host cells.

cohesive ends The cohesive DNA termini produced, with overhanging 5' or 3' ends, after restriction endonuclease digestion; sometimes called sticky ends.

cohesive termini (cos) *cos* sites; the 12-base single-stranded complementary ends of bacteriophage λ DNA that are formed during packaging of concatemeric DNA into viral head particles.

colonies Clumps of bacterial cells on solid medium; each colony is composed of genetically identical cells that arose by cell division from a single cell; each colony is called a clone.

competent Describes cells, most often bacterial, that have the ability to take up DNA.

complementary Describes a pair of nucleotides that can base pair through hydrogen bonding; an example is cytosine (C) and guanine (G); strands of DNA also can base pair along their nucleotides.

complementation A method by which two DNA fragments (e.g., from two different regions of a gene) in a cell encode nonfunctional polypeptides that, when synthesized together in the cell, become a functional protein.

continuous fermentation A method in which culture medium is added to the fermenter continuously as medium with cells is removed to maintain exponential growth of cells or microorganisms.

cosmid A hybrid vector comprising plasmid sequences and the *cos* sites of lambda bacteriophage, enabling the vector to be packaged *in vitro* with large DNA inserts.

crossing over A process that generates new genetic combinations: Homologous chromosomes line up and exchange segments of corresponding DNA during meiosis.

crustacean A member of the class Crustacea that harbors gills and has an exoskeleton or shell; members include lobsters, shrimp, crabs, and barnacles.

cyclic AMP (cAMP) 3', 5'-cyclic adenosine monophosphate is a small ring-shaped molecule; a second messenger in cells that helps mediate a cell's response to an external signal; regulator of the *lac* operon.

cytokinin A member of a class of plant growth regulators that promote cell division and leaf expansion.

cytotoxic hypersensitivity This type of immune response occurs when the immune system attacks one's own cells (antibodies recognize the antigens as foreign). Examples include some autoimmune diseases and the reaction to a blood transfusion that involves incompatible blood types.

deoxyribonucleic acid (DNA) The nucleic acid encodes genetic information that is passed to progeny; most frequently a double-stranded helical molecule composed of four nucleotides. The nucleotide unit of DNA has one of four nitrogenous organic bases—adenine, guanine, cytosine, or thymine—attached to a deoxyribose sugar with a phosphate group (called a deoxyribonucleotide).

deoxyribonucleotide *See* deoxyribonucleic acid.

diagnostics Sensitive assays used to determine the presence of a disease, DNA mutation, or pathogen.

dideoxyribonucleotide A nucleotide that is missing a 3' hydroxyl group on the deoxyribose sugar; used in Sanger dideoxy sequencing for chain termination.

differential interference contrast microscopy A type of light microscopy that takes advantage of phase differences in light that has passed through a cell to produce a detailed image of the cell.

diploid Having two sets of chromosomes (2 n); chromosomes of each pair are referred to as homologous.

DNA *See* deoxyribonucleic acid.

DNA helicase Protein that hydrolyzes ATP while moving along the DNA molecule and melting the double strand.

DNA library A collection of cloned DNA fragments often from a specific genome.

DNA ligase An enzyme that re-forms the phosphodiester bond between two nucleotides of DNA (the 5' phosphate of one nucleotide and the 3'-OH group of the other).

DNA linker A synthetic oligonucleotide containing a restriction site ligated to the end of DNA fragments to create a restriction site for cloning.

DNA polymerase An enzyme that synthesizes a copy of a DNA template. DNA is synthesized in the 5' to 3' direction by adding nucleotides to the 3' hydroxyl group of the newly synthesized strand.

DNA repair Replacement of damaged or incorrect nucleotides with the original sequence; DNA polymerase, DNA ligase, and other enzymes are involved in repair.

DNA replication The process of duplicating DNA for distribution to daughter cells.

DNA topoisomerase An enzyme that during DNA replication removes twists to relieve tension in the double-stranded molecule by breaking phosphodiester bonds and then reseals the breaks.

echinoderm An invertebrate with spines, plates, or needles that exhibits radial symmetry with some bilateral characteristics; examples include sea urchins, sand dollars, sea stars, and sea cucumbers.

electroporation A method for introducing DNA into cells; exposure to a brief electric pulse is thought to open transient pores through which DNA enters.

embryoid A multicellular structure (such as a blastula and gastrula in animals, sporophyte in plants) formed by subsequent cell divisions after fertilization.

embryonic stem cells (ES) Cells from an early embryo that are not terminally differentiated but can divide to give rise to differentiated cells, including germ line cells.

endospore A dormant structure that forms within cells of certain bacteria during stressful conditions and can germinate and give rise to new cells when conditions improve.

enhancer A regulatory DNA sequence element that enhances transcription of a eukaryotic gene; functional from thousands of base pairs away and in either orientation.

enzyme A protein that catalyzes a specific chemical reaction.

ethidium bromide A molecule that binds to DNA by intercalating between the bases and fluoresces in the presence of ultraviolet light; used as a stain for DNA, especially in agarose gels.

eukaryotic cell A cell that is compartmentalized with a membrane-bound nucleus, other membrane-bound organelles, and a distinct cytoplasm; eukaryotic organisms are composed of one or more cells.

expressed Refers to a gene that is transcribed and translated to yield a product.

expression library A collection of DNA fragments inserted into a vector that has a promoter sequence adjacent to the insertion site; gene expression is controlled by the regulatory elements of the expression vector.

extrachromosomal Not part of the host cell chromosome; usually refers to DNA such as a plasmid.

fermentation The breakdown of organic compounds by cells or organisms in the absence of oxygen to generate ATP; in industrial applications, cells or microorganisms are cultured in fermenters (i.e., bioreactors) with or without oxygen to produce desirable products or to increase biomass.

fermenter A growth chamber used for cultivating cells and microorganisms; used for the production of important compounds; also called a bioreactor.

fingerprinting A method used to identify individual DNA banding patterns derived from hypervariable regions of DNA; used in forensics, to establish paternity, and in conservation biology.

flow cytometry A method for sorting cells and metaphase chromosomes.

gastropod A member of the phylum Mollusca; characteristics include a soft body that is supported by a flat, muscular foot and protected by a cap-shaped shell; examples are snails, abalones, and sea slugs.

gel electrophoresis A method used to separate DNA or RNA fragments by length; proteins can be separated by size and/or charge.

gene A region of DNA that encodes information for a discrete product that can be protein or RNA; includes coding sequences, introns, and noncoding regulatory sequences.

gene bank A facility where plant material is stored usually as seeds, tubers, or cultured tissue; see germplasm.

gene therapy The use of genes to correct a genetic or acquired disorder.

genetic code The 64 triplet codons that encode the 20 amino acids and three stop codons used in protein synthesis.

genetic linkage map A map that indicates the genetic distances between pairs of linked polymorphic DNA markers or genes that have alternative forms; determined by recombination frequencies.

genome The entire content of genetic material within an organism (including bacteria and viruses); sometimes, the amount of DNA in a haploid set of chromosomes in a eukaryotic organism.

genotype The genetic composition of an organism; can include any number of gene pairs from one up to all the genes.

germ cell An animal sperm or egg cell (i.e., gamete) or a precursor cell that gives rise to gametes.

germplasm The genetic material within an organism; usually meaning plant genetic material.

haploid Having only one set of chromosomes (n); found in gametes, some fungi and protists, and plant gameotphytes.

helper virus A virus that provides functional properties to another virus in the same cell that lacks specific functional proteins.

heterogeneous nuclear RNA (hnRNA) The primary product of transcription in eukaryotes that has not been processed to remove introns or to add a poly-A tail or a 5′ cap; after processing in the nucleus, the mRNA is transported to the cytoplasm for translation.

heterologous probe A DNA sequence from one organism that is used to hybridize to DNA sequences from a different organism; a DNA sequence used for hybridization that is not identical to the target sequence but has some degree of similarity.

heterozygous Having different forms or alleles of a gene at a specific locus on homologous chromosomes.

homologous recombination The reciprocal exchange of genetic information between homologous chromosomes during meiosis; see crossing over.

homozygous Having two identical forms of alleles of a gene at a specific locus on homologous chromosomes.

hybridization Hydrogen bonding between two complementary single-stranded DNA sequences or between DNA and RNA sequences.

hybridoma An immortal cell line produced from the fusion of antibody-secreting B lymphocytes to lymphocyte tumor cells and used in the production of monoclonal antibodies.

immune-complex hypersensitivity This type of immune response involves the production of antibodies that react to antigens from one's own body. The antigens are soluble molecules that are present in the circulation system.

immunoassay A method that uses the specificity of antibodies to detect the presence of a specific antigen (usually protein or glycoprotein) in a biological sample, such as blood.

insulin A hormone, secreted by β cells of the pancreas, that reduces the glucose levels in the blood by regulating glucose metabolism in animals.

intellectual property Once a traditional area of law that encompassed patents, trade secrets, trademarks, copy rights, and plant variety protection; intellectual property now can be living organisms.

intron A noncoding region of a eukaryotic gene that is removed during RNA processing after transcription.

karyotype A complete set of chromosomes from a somatic cell that is arrested in metaphase and arranged in pairs of homologous chromosomes and sex chromosomes; used to identify chromosomal abnormalities.

lagging strand The DNA strand that during replication is synthesized in short fragments that then are covalently joined by DNA ligase.

leading strand The DNA strand that during replication is synthesized continuously in the 5′ to 3′ direction.

linked Describes two or more genes or markers that are located on the same eukaryotic chromosome.

lysogenic Describes a bacteriophage DNA that integrates into a bacterial genome and does not become lytic; called a prophage.

lytic Describes a bacteriophage that replicates and lyses the host cell.

map *See* genetic linkage map and physical map.

mariculture The farming, often commercial, of a variety of marine shellfish, crustaceans, finfish, and algae.

meiosis Two successive nuclear cell divisions and one replication event leading to the formation of haploid gametes (eggs and sperm).

meristematic tissue A group of dividing cells, such as shoot and root apical meristems, that give rise to tissues and organs of flowering plants.

messenger RNA (mRNA) An RNA molecule that encodes the amino acid sequence of a polypeptide; translated into protein during the process of translation.

metabolite A small organic compound that is produced by an enzymatic reaction(s) or is required for metabolism.

metaphase chromosome A condensed chromosome that has replicated and is composed of two sister chromatids.

microinjection Direct injection of DNA or a therapeutic agent into a cell with a microcapillary.

microorganism A prokaryotic organism lacking membrane-bounded organelles and a true nucleus.

mismatch The lack of complementarity of two DNA sequences at one or more nucleotides leading to the inability of the two sequences to base pair or hybridize in that region.

molecular biology The study of molecular processes that encompasses DNA, RNA, and protein.

mollusk A member of the phylum Mollusca, which comprises seven classes and includes mussels, clams, snails, squid, and octopuses; members of the phylum have a tissue fold called a mantle around a soft, fleshy body.

monoclonal antibody (MAB) A single type of antibody specific for one antigenic determinant (portion of an antigen that binds to an antibody) that is secreted by a hybridoma clone derived from a single B cell.

mutation A change in the nucleotide sequence of a chromosome that is inherited.

nucleic acid A polymer of nucleotides joined by phosphodiester bonds; a general name given to DNA or RNA.

nucleotide The basic unit of DNA or RNA; a pentose sugar with a bound nitrogenous base and phosphate group.

oligonucleotide A short, single-stranded DNA that usually is synthesized *in vitro* and often used as a probe in hybridizations or as primers for the polymerase chain reaction.

operator A short region of DNA in bacteria that controls transcription of an operon (i.e., an adjacent gene or genes).

patent A legal document that grants exclusive rights to sell or use an invention for a defined period of time (17 years in the United States); *see* intellectual property.

pathogen An organism or virus that causes disease.

peptide bond A covalent bond between the carboxyl group of one amino acid and the amino group of an adjacent one; formed during protein synthesis.

phenotypic trait An observable characteristic of an organism or cell that is determined by the genotype or genetic composition, often in conjunction with the environment.

plaque A clear, circular area that forms after a bacterial colony cultured on solid medium has lysed by bacteriophage.

plasmid An extrachromosomal circular double-stranded DNA molecule that replicates independently from the chromosome; frequently modified for use as a cloning vector.

pluripotent Describes the ability of a cell to develop into a differentiated cell type; progenitor of other cells; often used in conjunction with stem cells (bone marrow, embryonic) in animals.

polymer A macromolecule composed of repeating units; examples include DNA, RNA, starch.

polymerase chain reaction (PCR) A method for selectively amplifying regions of DNA by *in vitro* replication involving repeated denaturation and renaturation of the DNA template.

polymorphic Describes variation within a population; frequently refers to DNA sequences. *See* polymorphic DNA marker.

polymorphic DNA marker A DNA sequence along a chromosome that varies in a population of the same species. Markers are inherited in a Mendelian fashion.

polypeptide A long linear sequence of amino acids; sometimes called a protein.

primer A short oligonucleotide sequence that is hybridized to a complementary DNA template and to which a nucleotide can be added at the 3' end to initiate replication; used in the polymerase chain reaction (PCR).

probe A labeled molecule, frequently DNA or RNA, that is used for hybridization.

promoter A region of DNA where RNA polymerase and other proteins involved in the regulation of transcription bind.

pronuclear microinjection A method by which one of the nuclei in a fertilized egg is injected with DNA prior to fusion of the gamete nuclei. *See* microinjection.

protease An enzyme that degrades proteins by hydrolyzing the peptide bonds; an example is trypsin.

protein A linear macromolecule that is composed of amino acids that are connected by peptide bonds; sometimes used interchangeably with polypeptide, although a protein can consist of more than one polypeptide (e.g., dimeric protein is composed of two polypeptides).

protoplasts Plant, bacterial, or yeast cells that have no cell wall; cell walls usually are removed enzymatically.

protozoan A single-cell, free-living or parasitic, nonphotosynthetic, and motile eukaryotic organism; examples are amoebas and paramecia; a type of protist.

purine One of two classes of nitrogen-containing ring structures in DNA and RNA; adenine and guanine have double-ring structures.

pyrimidine One of two classes of nitrogen-containing ring structures in DNA and RNA; cytosine, thymine (DNA), and uracil (RNA) have single-ring structures.

reading frame Triplet bases in a nucleotide sequence; a messenger RNA can be read in any of three reading frames, although only one will generate the correct amino acid sequence of the polypeptide.

receptor A protein, located on the cell surface or intracellularly, that binds to a specific type of molecule (ligand) and elicits a cellular reaction or response.

recombinant DNA A sequence comprising DNA from different sources that have been joined together.

replica plating The transfer of cells in bacterial colonies (or plaques) from one petri dish, or plate, to another; used for screening libraries so that a master plate is retained for the isolation of positive clones.

replication fork The Y-shaped area of a replicating DNA molecule where the DNA strands are separated and the two daughter strands are synthesized.

replicative form (RF) A viral nucleic acid that serves as the template for replication in a host cell.

reporter A gene that encodes a product that can be assayed to determine whether a specific DNA construct is present and functional in a host organism or cell.

repressor protein A protein that binds to a gene promoter or operator site and blocks the binding of RNA polymerase, thereby preventing transcription.

restriction endonuclease A member of several classes of enzymes that cut DNA internally; each recognizes a specific short double-stranded DNA sequence and hydrolyzes phosphodiester bonds on both strands (type II enzymes are used in recombinant DNA technology).

restriction fragment length polymorphism (RFLP) Nucleotide sequence variations in a region of DNA that generates fragment length differences according to the presence or absence of type II restriction endonuclease recognition site(s).

retrovirus A virus with an RNA genome that replicates in a eukaryotic host cell by synthesizing a double-stranded DNA intermediate; the double-stranded form can integrate into the host chromosome.

reverse transcriptase An RNA-dependent DNA polymerase that uses an RNA template for the synthesis of a complementary DNA (cDNA); present in viruses with RNA genomes.

ribonuclease An enzyme that hydrolyzes RNA by hydrolyzing the phosphodiester bonds between the nucleotides (RNase).

ribonucleic acid (RNA) A heteropolymer of ribonucleotides. The nucleotide unit of RNA has one of four nitrogenous organic bases—adenine, guanine, cytosine, or uracil—attached to a deoxyribose sugar with a phosphate group.

ribosome A two-subunit structure that associates with mRNA and catalyzes the synthesis of protein; composed of rRNA and ribosomal proteins.

ribosomal RNA (rRNA) RNA molecules that associate with proteins to form the ribosome; types are distinguished by their sedimentation coefficient (e.g. 28S rRNA).

RNA *See* ribonucleic acid.

RNA polymerase The enzyme that catalyzes the synthesis of RNA on a DNA template.

RNA primer A short RNA sequence that is involved in the initiation of DNA synthesis during replication. The RNA is synthesized by the enzyme complex DNA primase and is removed by DNA polymerase.

sequence-tagged sites Two hundred to 300-base pair regions of known DNA sequences determined to be unique and that can be amplified by the polymerase chain reaction; chromosomal DNA sequences used to integrate genetic linkage maps with physical maps.

shellfish Marine and freshwater invertebrates; usually meaning crustaceans and mollusks.

single-cell protein Dried cells of microorganisms that can be ingested by animals and humans as a source of protein.

Southern blotting A method of transferring denatured DNA from an agarose gel after gel electrophoresis to a membrane for hybridization to detect specific sequences.

sponges Primarily marine invertebrates that exist in many forms, such as fingerlike branches, vases, and hollow columns; surfaces have tiny pores leading to interior chambers lined with ciliated cells that help flush water through the chambers where bacteria and food particles are trapped; from the phylum Porifera.

sporangium A hollow unicellular or multicellular structure in which spores form.

spore *See* endospore. (Spores are present in fungi and plants, but have a different function from that of endospores.)

sticky ends *See* cohesive ends.

stirred tank reactor A fermentor that uses internal mechanical agitation of the culture medium by a series of rotor blades.

substrate An energy source for cells and microorganisms that can be transformed into valuable products; also a compound that is modified to a product by an enzyme in a reaction.

T-DNA A region of DNA within the Ti plasmid of *Agrobacterium tumefaciens* that is transferred to and integrated into the chromosome of plant cells.

telomere The end of a eukaryotic chromosome having a special sequence that enables it to replicate to prevent shortening of the chromosomal end with each round of replication.

therapeutic Any agent (DNA, protein, synthetic compound, etc.) that is used in the treatment of disease.

Ti plasmid A large plasmid in *Agrobacterium tumefaciens*; harbors tumor inducing genes that lead to the formation of crown galls.

transcription The synthesis of RNA that is catalyzed by RNA polymerase and requires a DNA template.

transfection The transfer of DNA into a eukaryotic cell.

transformation The uptake of DNA into a cell, usually a bacterium or yeast, and its expression; also refers to the conversion of a normal cell to a cancerous one.

transgene A gene that has been introduced into the genome of a eukaryotic cell(s) or organism that originated from another source.

transgenic organism A living organism that harbors an inserted functional gene in its germ line.

translation Protein (polypeptide) synthesis by the formation of peptide bonds between amino acids encoded by a messenger RNA and translated with the aid of ribosomes and transfer RNA molecules.

triploid Three copies of homologous chromosomes (3n).

tunicates Marine animals from the phylum Chordata, mostly sessile, that are anchored on rocks or other substrates (some are planktonic); also called sea squirts. Although adults do not have a notochord, nerve cord, or tail, the larvae of some tunicates exhibit these characteristics.

vaccine An agent that is used to promote an antibody response that protects the organism against infection; confers immunity to disease. Examples include protein, DNA, living or nonliving virus or microorganism.

variable number of tandem repeats (VNTR) Short DNA sequences that are repeated in tandem (adjacent to one another). The variation is determined by the number of times the short sequence is repeated at a particular locus on the chromosome. VNTRs are used in DNA fingerprinting.

virulent Describes a microorganism or virus that causes disease and even death of an organism; also a bacteriophage that causes host cell lysis.

xenobiotics Chemicals or compounds that are not produced by living organisms.

xenotransplant A tissue or organ removed from one species and transplanted into another species.

x-ray diffraction pattern The three-dimensional arrangement of atoms in a molecule (such as protein and DNA) that is generated by the diffraction pattern of x-rays passing through a crystal molecule. The method of generating the patterns is called x-ray crystallography.

yeast artificial chromosome A linear recombinant DNA molecule that replicates in a yeast host cell; important components include a large fragment of foreign DNA (up to 1 million base pairs), a centromere, a yeast origin of replication, and yeast telomeres at the ends of the chromosome.

zygote A diploid cell that is produced by the fusion of nuclei from a male and female gamete (i.e., sperm and egg); also called a fertilized egg.

Credits

Index

f indicates terms located in figures.
t indicates terms located in tables.